THE CHEMISTRY OF ESSENTIAL OILS AND ARTIFICIAL PERFUMES

Volume 1

FOURTH EDITION, REVISED AND ENLARGED

BY

Ernest J. Parry

Wexford College Press

2007

PREFACE TO THE THIRD EDITION.

THE mass of accumulated information in regard to essential oils and their constituents has necessitated the division of this work into two volumes. The present volume deals entirely with the essential oils themselves in a series of monographs.

The second volume covers the necessary ground relating to the analyses of essential oils and the characters of their isolated constituents. It also deals with artificial aromatic bodies which are not present in essential oils, but which form the groundwork of synthetic perfumery. It is believed that this division of the subject will be found more convenient than the arrangement of the subject-matter in previous editions.

I have to acknowledge my indebtedness to Messrs. Baker and Smith of the Technological Museum, Sydney, who have been exceedingly kind in placing much information and several illustrations at my disposal.

I am also indebted to M. Gattefossé, Messrs. Roure-Bertrand Fils, the Monopoly Bureau of the Government of Formosa, the Indian Forestry Department, the Editor of the *Chemist and Druggist*, and the Editor of the *Perfumery and Essential Oil Record* for permission to reproduce numerous illustrations and much useful information.

ERNEST J. PARRY.

56A GREAT DOVER STREET,
LONDON, S.E., *April*, 1918.

PREFACE TO THE FOURTH EDITION.

THE rapid sale of the Third Edition of this work has necessitated the publication of this, the Fourth Edition. No alteration has been made in the arrangement of the subject-matter, which has been brought up to date, so that all work of importance published up to the end of 1920 has been included.

ERNEST J. PARRY.

56A GREAT DOVER STREET,
LONDON, S.E., *March*, 1921.

CONTENTS OF VOLUME I.

PARMELIACEÆ.

Oak Moss Oil.

Under the name "Mousse de Chêne" an extract of various lichens growing on oak and other trees has been placed upon the market, and is of the greatest value in perfumery. The principal plants yielding this extract are *Evernia furfuracea*, *Evernia prunastri* and *Ramalina calicaris* and numerous varieties of the latter. Various types of *Usnea*, *Alectoria*, and *Parmelia* also yield a similar extract in less quantity and of poorer odour. If the plants are extracted by means of a volatile solvent, and the extract is steam distilled, a colourless essential oil is obtained, which has an intense odour and, according to Gattefossé,[1] consists almost entirely of a phenol which he has termed lichenol. On standing the lichenol crystallises out, and when pure melts at 72° to 73°. Gattefossé originally considered this phenol to be identical with carvacrol, but now (*in a private communication to the author*) considers that it is of lower molecular weight, and closely related to cetrarine, a phenol-ketone occurring in the oil from *Cetraria islandica*.

JUNGERMANNIACEÆ.

Oils of Liverworts.

When the dry plant *Mastigobryum trilobatum*, one of the well-known liverworts, is distilled, a small quantity of an orange-yellow essential oil is obtained, having a powerful odour recalling those of sandal and cedar-wood oils. The oil has, according to Müller, the following characters:—[2]

Specific gravity	0·978 at 12°
Optical rotation	+ 12° 88′
Saponification value	5·4

The oil probably consists of sesquiterpenes or sesquiterpene alcohols.

Leioscyphus Taylori yields a thick green essential oil which has the following characters:—

Specific gravity	0·982 at 20°
Optical rotation	− 3° 44′

It contains two sesquiterpene alcohols, $C_{15}H_{26}O$, one boiling at 265° and the other at 275°.

Madotheca levigata yields an orange-yellow essential oil having the following characters:—

[1] *Parfumerie Moderne* **4**, (**1911**), 4. [2] *Jour. Pharm. Chim.* (6), **22**, 555.

Specific gravity	0·856 at 16°
Optical rotation	− 72° 14′
Saponification value	5·56

Alicularia scalaris yields a lemon-yellow essential oil having the following characters:—

Specific gravity	0·965 at 15°
Optical rotation	− 33° 49′

It appears to contain a sesquiterpene alcohol.

CONIFERÆ.

Cedar-Wood Oil.

The oil usually known as cedar oil or cedar-wood oil is obtained by distillation of the wood of *Juniperus virginiana*, the Virginia cedar, one of the family of the *Cupressineæ*. Other cedar-wood oils are occasionally met with, and will be mentioned later, but the present remarks apply to the above-mentioned oil.

The oil is generally distilled from the waste shavings from lead-pencil manufacture, which are usually plentiful when the pencils are cut from logs. When thin sawn boards are imported and used, the refuse is much smaller in amount, and the price naturally rises. Owing to the methods of distillation, the oil manufactured from pencil waste is of less value to perfumers than normally prepared oil. The yield of oil obtained varies from 2·5 to 5 per cent. *Juniperus virginiana* occurs over a very wide area in America, and some of the finest oil is obtained from the Florida cedar. This oil is one of those which are indispensable in certain classes of perfumery, especially in fine soap-making, where intensely "sweet" odours are not required. The wood in fine powder finds its way into much of the incense and similar preparations used either for ceremonial or fumigating purposes.

The oil is usually of a brownish colour, but it can be obtained almost water-white, and although we know nothing definitely of what change occurs, it is certainly amongst those oils which improve in quality by keeping. Occasionally samples will be found to deposit a small amount of crystalline "cedar camphor".

Pure cedar-wood oil has a specific gravity of from 0·940 to 0·962, and is always laevo-rotatory. The limits, however, are wide, the oil varying from − 25° to − 47°. According to Schimmel, 80 per cent. distils at between 125° and 155° at a pressure of 14 mm. The refractive index is very high, usually exceeding 1·5, and often rising to 1·51. It is very insoluble in alcohol, 1 part requiring 10 to 20 parts of 90 per cent. alcohol for solution. The acid value is from 0 to 2, and the ester value from 2 to 7. A small amount of free alcohols is present, the ester value of the acetylated product being from 15 to 44. The known constituents of the oil are the sesquiterpene cedrene, $C_{15}H_{24}$, and the oxygenated bodies cedar camphor, or cedrol, $C_{15}H_{26}O$, cedrenol $C_{15}H_{24}O$, and *pseudo*-cedrol, $C_{15}H_{26}O$. Cedrol, when pure, is a silky crystalline mass, with pleasant aromatic odour, melting at 84°. It is suggested by Schimmel that cedrol is not a normal constituent of fresh wood, but is formed in it by keeping it under favourable conditions. This suppositition has not, however, been experimentally justified except in so far as very old wood yields a semi-solid oil, which deposits much cedrol on cooling. At all events, only a small quantity occurs in normal oils, as is shown by the low acetylation figure. If the oil be acted upon by acetic anhydride, and the resulting product saponified, no more than from 6 to 8 per cent. of cedrol will be indicated. It must be remembered, however, that dehydration takes place to a small extent with acetic anhydride, so that the result may be somewhat under-estimated. Ten per cent. may be regarded as the maximum usually occurring in good oils. The figures found by the author for four pure samples (percentage of KOH to saponify the acetylated oils) are as follows: 2·32, 2·21, 2·13, 1·72.

Cedrenol, which was isolated from cedar-wood oil by Semmler [1] and his pupils is a primary sesquiterpene alcohol, having the following characters :—

Boiling-point at 9·5 mm.	166° to 169°
Refractive index	1·5212
Specific gravity at 20°	2·0083
Optical rotation	±0°

It forms an acetate of specific gravity 1·0168 at 20° and refractive index 1·5021.

According to Semmler cedrenol stands in the same relation towards cedrene as the two primary alcohols of the santalol series towards the sesquiterpenes, $C_{15}H_{24}$, the santalanes; and as myrtenol and the ginger-grass alcohol stand towards pinene and limonene. The primary CH_2OH group in the cedrenol molecule occupies the same position which is occupied by the CH_3 group in cedrene and in solid cedrol. Hence the relation of cedrene towards cedrol and cedrenol is as follows :—

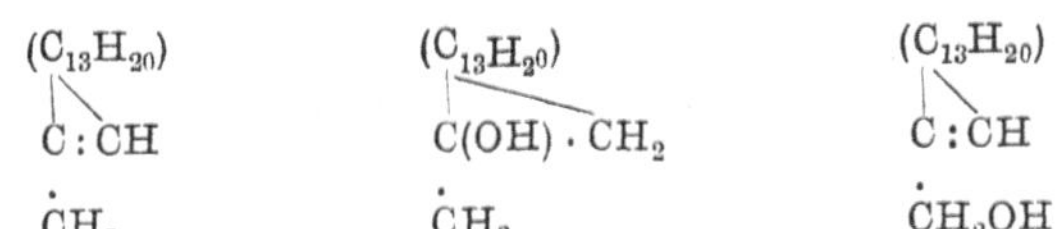

Cedrene, $C_{15}H_{24}$. Cedrol, $C_{15}H_{26}O$. Cedrenol, $C_{15}H_{24}O$.

Pseudo-cedrol is a saturated alcohol, boiling at 147° to 152° at 10 mm., and having a specific gravity 0·9964 at 20°, optical rotation + 21·5°, and refractive index 1·5131. It appears to be a physical isomeride of cedrol.

Rousset [2] has studied the oxidation products of cedrene, from which he obtained the ketone cedrone, $C_{15}H_{24}O$, by means of chromic acid. On reduction this yields an alcohol, isocedrol, isomeric with cedrol.

Schimmel & Co. [3] have examined a cedar-wood oil sent over from Hayti, of whose botanical origin they were unable to ascertain anything definite, but which according to microscopical results was derived from a conifer, which gave on distillation a yield of 4·33 per cent., of oil of lemon-yellow colour and the odour of the common cedar-wood obtained from *Juniperus virginiana* L. But it differs from the latter by its higher specific gravity ($d_{15°}$ 0·9612), the lower rotation (α_D − 14°58′),

[1] *Berichte*, 45 (**1912**), 355, 786, 1553. [2] *Bull. Soc. Chim.*, iii., 17 (**1897**), 485.
[3] *Bericht*, April, **1906**, 10.

and the higher content of alcoholic constituents (ester number after acetylation 64·0). The acid number was 2·7, the ester number 5·0. The oil was not completely soluble in 10 volumes of 90 per cent. alcohol, but dissolved in every proportion in 95 per cent. alcohol.

A cedar-wood oil has been distilled during the past few years from the wood of *Cedrus Atlantica*, the so-called Atlas cedar, which is probably a variety of *Cedrus Libani*, the cedar of Lebanon. This oil has the following characters:—

Specific gravity at 15°	0·950 to 0·970
Optical rotation	+ 45° ,, + 62°
Refractive index	1·5119 ,, 1·5175
Acid value	0 to 2
Ester ,,	3 ,, 11
,, ,, (after acetylation)	30 ,, 48

It is soluble in 1 to 10 volumes of 90 per cent. alcohol. According to Grimal,[1] the oil contains traces of acetone, and about 5 per cent. of a ketone, $C_9H_{14}O$, which he terms libanone. This substance yields a semi-carbazone, melting at 159° to 160°, and a liquid oxime, which on bromination yields a dibromide melting at 132° to 133°. The principal constituent of the oil is dextro-cadinene, which was obtained from it in a sufficiently pure condition to have an optical rotation of + 48°.

The wood of the ordinary Lebanon cedar, *Cedrus Libani*, yields about 3 to 4 per cent. of oil, having an odour recalling those of methyl heptenone and thujone. Its characters are as follows:—

Specific gravity at 15°	0·940 to 0·950
Optical rotation	+ 66° ,, + 86°
Refractive index at 20°	1·5120 ,, 1·5140
Acid value	0 to 1·8
Ester ,,	2 ,, 4
,, ,, (after acetylation)	18 ,, 25

A sample examined by Messrs. Schimmel & Co. gave the following results on fractionation:—

270° to 275°	30 per cent.
275° ,, 280°	40 ,,
280° ,, 285°	14 ,,
285° ,, 290°	6 ,,
Residue	10 ,,

A cedar wood from East Africa, the product of *Juniperus procera*, yields an essential oil, which has been examined by Schimmel & Co.[2] and found to have the following characters:—

	Oil from Sawdust.	Oil from Boards.
Yield of essential oil	3·2 per cent.	3·24 per cent.
Density at 15° C.	0·9876	1·0289
Refractive index at 20° C.	1·50893	1·51011
Optical rotation	− 37° 10′	− 32° 30′
Acid value	14·9	27·06
Ester ,,	8·4	7·93
,, ,, (after acetylation)	70·0	89·6
Solubility in 80 per cent. alcohol	1·6 vol.	2 vols.
,, 90 ,, ,,	0·5 vol.	0·5 vol.

[1] *Comptes rend.*, 135 (**1902**), 582, 1057.

[2] *Bericht*, October, **1911**, 110.

The essential oil derived from the boards is filled with crystals which have been identified as *cedar camphor ;* after purification they melt at 86° to 87° C., and have a specific rotation of + 10·12°

Roberts[1] has examined the cedar oil distilled from the wood of the Deodar tree, *Cedrus deodara*, a native of the Western Himalayas, which is widely distributed in Northern India. The samples examined were of a reddish colour and characteristic balsamic odour. They had the following characters :—

Specific gravity	0·9549	0·9756
Optical rotation	+ 52° 16′	+ 34° 6′
Refractive index at 21°	1·5195	1·5225
Acid value	5·6	4·5
Ester ,,	19·3	4·9
,, ,, (after acetylation)	30·8	34·4

One was miscible in all proportions with 90 per cent. alcohol; the other was not soluble in twenty volumes. A ketone was isolated which did not combine with acid sulphites, but had an odour resembling amyl acetate, and formed a semi-carbazone melting at 163° to 164°, and a dibromoxime melting at about 130°. Schimmel[2] considers that it is possibly *p*-methyl-tetrahydroacetophenone. Decomposition of the semi-carbazone by dilute sulphuric acid gave a ketone differing from the original, boiling at 220° to 224°, and forming a semi-carbazone melting at 204° to 205°, and forming *p*-toluic acid on oxidation. A phenol was isolated which gave a blood-red colour with ferric chloride and possessed a sweet phenolic odour. It yielded a benzoyl derivative which crystallised from dilute alcohol in small needles melting at 70°. The quantity found was less than 10 per cent. of the oil. On hydrolysis of the oil, hexoic, heptoic, and stearic acids were obtained. The principal constituents of this oil are sesquiterpenes, associated with a variable amount of sesquiterpene alcohols. The sesquiterpene fraction boiled at 151° to 153° at 19 mm., or 262° to 265° at ordinary pressure. It was a colourless, mobile liquid, with a specific gravity of 0·9319, optical rotation (1) + 16° 51′, (2) + 58° 34′, and refractive index 1·5150 (23°). It gave a colour reaction with sulphuric acid similar to that of cadinene. Attempts to prepare solid derivatives were unsuccessful. The composition of the oil was as follows: Ketone 2 per cent., phenol 0·07 to 0·4 per cent., esters of hexoic, heptoic, and stearic acids 3 to 12 per cent., and sesquiterpenes 50 to 70 per cent., the remainder of the oil consisting chiefly of sesquiterpene alcohols, with high boiling viscous decomposition products.

Oil of Cryptomeria.

The wood of the Japanese cedar, *Cryptomeria japonica*, yields about 1·5 per cent. of an essential oil having the following characters :—

Specific gravity	0·945
Optical rotation	− 23°
Ester number	3·9

The oil contains about 60 per cent. of sesquiterpenes and 40 per cent. of sesquiterpene alcohols. The principal sesquiterpene is probably cadinene, as indicated by the following figures in comparison with those for cadinene obtained from oil of cade :—

[1] *Jour. Chem. Soc.* (**1916**), 791. [2] *Bericht*, **1917**, 23.

	Sesquiterpene from Cryptomeria Oil.	Cadinene from Oil of Cade.
Specific gravity	0·925	0·927
Optical rotation	− 98° 45′	− 97° 30′
Melting-point of dihydrochloride	117° to 118°	117° to 118°

A second sesquiterpene is present, which only yields liquid halogen acid derivatives. This body, to which the discoverer has given the name suginene, has a specific gravity 0·918 and specific rotation − 10° 34′. It yields a liquid hydrobromide of specific gravity 0·988 and specific rotation − 11° 15′. One of the fractions obtained on distillation was analysed and found to correspond with the formula $C_{15}H_{25}OH$, and is apparently one or more sesquiterpene alcohols.

By conversion into the xanthogenic esters the alcoholic portion of the oil was separated into two distinct portions, which on saponification yielded two isomeric sesquiterpene alcohols. One of these, isocryptomeriol, $C_{15}H_{25}OH$, is a liquid of specific rotation − 3° 25′. The isomeric alcohol has been termed cryptomeriol, and is an oil of specific rotation − 37° 5′, and has a specific gravity 0·964.

By dehydration by means of formic acid, cryptomeriol yields a sesquiterpene having a specific gravity 0·918, and a specific rotation − 1° 5′. It boils at 143° to 144° at 12 mm. pressure. By dehydration by means of phosphorus pentoxide an entirely different sesquiterpene results, having a specific gravity 0·917 and a specific rotation + 56° 26′.

The essential oil from the leaves has been investigated by So Uchida.[1] On distillation with steam the green leaves yielded 0·70 per cent. of volatile oil, which was brownish-yellow in colour and had a fresh aromatic odour. It has the following characters:—

Specific gravity	0·9217
,, rotation	+ 19·29°
Refractive index	1·4895
Acid value	1·0
Ester ,,	6·56
,, ,, (after acetylation)	14·35

On fractional distillation the following constituents were separated: d-α-pinene, together with a small quantity of dipentene formed about 34 per cent. of the oil. An alcohol, $C_{10}H_{18}O$, having a fresh odour resembling that of camphor and peppermint, was found to the extent of 4·5 per cent., partly combined as the caprylic ester. It had a specific gravity of 0·9414, refractive index 1·4832, and specific rotation + 56·07° in chloroform solution. This alcohol, though small in amount, is regarded as the chief odorous constituent of the oil. Sesquiterpenes were present to the extent of about 30 per cent., consisting partly of cadinene and partly of a sesquiterpene with two double linkings, boiling-point 266° to 268°, specific gravity 0·9335, refractive index 1·5041, and specific rotation + 15·19° in chloroform solution. A sesquiterpene alcohol, occurring to the extent of about 12 per cent., had boiling-point 284° to 286°, specific gravity 0·9623, refractive index 1·5048, and specific rotation + 16·76° in chloroform solution. A new crystalline diterpene, representing about 18 per cent. of the oil, has been isolated, to which the Uchida has given the name α-cryptomerene. On passing dry hydro-

[1] *Jour. Amer. Chem. Soc.* (**1916**), 687.

chloric acid into a well-cooled solution of this dipentene in ether a crystalline isomer was obtained, which is termed β-cryptomerene. Traces of a lactone, $C_{20}H_{32}O_2$, free fatty acids, and a blue oil, azulene, were also separated.

OIL OF TAXODIUM DISTICHUM.

The oil extracted from the wood of *Taxodium distichum* ("Southern Cypress") by means of alcohol, has been examined by Odell.[1] He fractionated the resinous residue *in vacuo* and has thus isolated a hydrocarbon $C_{15}H_{24}$ which he calls *cypressene;* it is an inodorous oily liquid, boiling-point (35 mm.) 218° to 220° C., boiling-point (778 mm.) 295° to 300° C.; $[\alpha]_D^{20}$ + 6·53°. He has also extracted a compound $C_{12}H_{20}O$, a bright yellow oil with a strong odour, possessing the properties of an aldehyde; it has received the name of *cypral* (boiling-point, 35 mm., 182° to 185° C.); it is dextro-rotatory.

Cypressene yields, on oxidation with nitric acid, an amorphous yellow product and an acid possessing the odour of isovaleric acid; the amorphous product dissolves in alkalies with a red coloration.

Odell has also examined the oil from the cones of the same tree. When harvested in September the cones yield 1 per cent. of a greenish-yellow essential oil, with a strong odour of pinene, whilst the cones harvested at a later period yield on distillation 1·5 to 2 per cent. of a darker oil possessing an odour of lemon. These two oils possess the following characters:—

Specific gravity	0·860	0·850
Optical rotation	+ 18°	+ 35° 30′

They contain about 85 per cent. of *d*-α-pinene, 5 per cent. of *d*-limonene, and 2 per cent. of an alcohol. Small amounts of carvone, and of a sesquiterpene, which is probably cypressene, are also present.

CHAMÆCYPARIS OIL.

Chamæcyparis obtusata, the hinoki tree which is extensively grown in Japan, furnishes a valuable timber wood, and the tree thrives well in the mountainous districts of Formosa. Uchida[2] has examined a sample of the crude oil obtained by the dry distillation of the wood in Formosa. It is a reddish-brown mobile liquid possessing a woody and empyreumatic smell and containing tarry matter. The yield was 2·4 per cent. of the wood. The rectified oil obtained by distillation with steam was lemon yellow in colour, and after treatment with sodium carbonate solution to remove pyroligneous acids it had a specific gravity of 0·8821, refractive index, 1·4990, and specific rotation + 50·37° in chloroform solution. The constituents identified were *d*-α-pinene and cadinene, with a small amount of oxygenated compounds, the amount of terpenes being about 70 per cent., and that of the sesquiterpenes about 24 per cent.

The leaves of *Chamæcyparis Lawsoniana* yield about 1 per cent. of oil having the following characters:—

[1] *Jour. Amer. Chem. Soc.*, **33**, 755. [2] *Jour. Amer. Chem. Soc.* (**1916**), 699.

Specific gravity	0·9308
Optical rotation	+ 23° 48′
Refractive index	1·4884
Acid value	3·7
Ester ,,	61·6
,, ,, (after acetylation)	78·8

The oil contains laurinic aldehyde.

The oil distilled from the wood of *Chamæcyparis Lawsoniana* and rectified by steam has been examined by Schorger.[1] The oil has the following characters:—

Specific gravity	0·8905
Refractive index	1·4758 at 15°
Optical rotation	+ 39·6°
Acid value	0·3
Ester ,,	32·8
,, ,, (after acetylation)	71·6

The oil behaved as follows on fractional distillation:—

(760 mm.)	155° to 157°	60·5 per cent.
,,	157° ,, 170°	3·0 ,,
,,	170° ,, 180°	4·0 ,,
(15 mm.)	100° ,, 130°	20·5 ,,
,,	130° ,, 160°	7·0 ,,
,,	160° ,, 190°	1·0 ,,

The oil contains *d*-*a*-pinene to the extent of 60 per cent., dipentene (6 to 7 per cent.), *l*-borneol (11 per cent.), cadinene (6 to 7 per cent.), formic and acetic acids in the free state, and formic and capric acids as esters.

Oil of Juniperus Oxycedrus Wood.

Juniperus oxycedrus is the juniper which furnishes the wood from which the tarry empyreumatic oil, known as cade oil, is obtained by destructive distillation. Huerre[2] has submitted the same material, suitably comminuted, to ordinary steam distillation. The essential oil thus obtained comes over very slowly; the yield varies from 1·6 to 3·4 per cent. The higher yield was obtained from the autumn cut wood. The oil is a dark yellow viscous liquid; having the following characters:—

Specific gravity	0·915 to 0·927
Optical rotation	− 31° 42′

At normal atmospheric pressure it boils between 260° and 300° C. Soluble 1 : 1 in 95 per cent. alcohol, and 1 : 12 in 90 per cent. alcohol. On rectifying the crude oil in a current of steam, under reduced pressure, 70 per cent. of the original oil distilled as a bright yellow liquid with a penetrating odour; it was much less viscous than the original oil. The residue (30 per cent.) was very viscous, dark in colour, and had a feeble sweet odour.

Oils of Librocedrus Decurrens.

Librocedrus decurrens, a tree found in California and known as the Incense cedar, yields a leaf and twig oil and a bark oil. The leaf and twig oil has the following characters:—

[1] *Jour. Ind. Eng. Chem.*, 6 (**1914**), 631.
[2] *Jour. Pharm. Chim.* (1915), **12**, 273.

Specific gravity	0·865 to 0·874
Optical rotation	− 3° 20′ „ + 36·68°
Ester value	18·5 „ 27·8
„ „ (after acetylation)	28·6 „ 48·2

It contains furfural, a-pinene, sylvestrene, limonene, dipentene, borneol, esters, and a sesquiterpene, which has been named librocedrene, having the following characters:—

Specific gravity at 20°	0·929
Boiling-point	270° (approximate)
Refractive index	1·4994
Optical rotation	+ 6·4°

The bark oil has the following characters:—

Specific gravity	0·862
Refractive index	1·4716 at 15°
Optical rotation	+ 1·1°
Ester value	3·2
„ „ (after acetylation)	9·5

The oil contains furfural, a-pinene, dipentene, and borneol.

OIL OF TURPENTINE.

Numerous members of the family of the *Coniferæ* contain large quantities of resin, and an oil consisting almost entirely of terpenes, in well-characterised resin ducts in the wood. This is especially the case with the various species of pine. The exudation from the wood, an oleo-resin, is often known as crude turpentine. Oil of turpentine, or, as it is generally called, turpentine or "turps," is the oil obtained by distilling this, leaving the crude resin or "rosin" behind.

At the present time American turpentine is the chief commercial variety, although a considerable amount is produced in France, Russia, and many other countries, and India produces a fair quantity which, although at present consumed in the East, may before long find its way to Western markets.

In the United States the turpentine (*i.e.* the oleo-resin) is obtained either by what is known as the "box system" or by the "cup and gutter" system. The box system, briefly, is carried out as follows: As spring approaches, a groove is cut diagonally across the trunk of the tree, and inclined inwards, ending in a hollow excavation about 12 to 18 ins. above the ground level. This wound in the sapwood induces the flow of the oleo-resin, and as the warmer weather approaches, the bark above and on both sides of the excavation or "box" is peeled off and shortly afterwards the flow of oleo-resin commences. Every two or three weeks fresh cuts are made so as to open up fresh passages for the exudation. The oleo-resin collects in the "box" and is baled out into barrels and transferred to the distillery. Trees are not considered worth tapping after four to six years. The cup and gutter system is a combination of the older box system and of the usual method employed in France. The cutting of a large hollow in the trunk is obviated, and its place is taken by a receptacle which is suspended from a nail in the trunk. As in the "box" system, here also the first work is cutting strips in the bark, and as the work proceeds, the space above the receptacle is gradually cleared of bark. In order to let the exuding resin collect in the earthenware pots or vessels placed below, two strips or gutters of galvanised sheet-iron about 2 ins. wide by 6 ins. to 12 ins. long are employed. These are placed in grooves cut into the trunk with a broad axe in such manner, that each of the two cuts forms an angle of 60° with the longitudinal axis of the tree, and the two together therefore

FIG. 1.—Still of a "Turpentine Farm" in the "Landes".
[*Roure Bertrand Fils.*

form an angle of 120° open at the top. The two gutters, however, are not placed at the same height, but one about 1 in. or 2 ins. above the other, so that the resin flowing along the upper gutter only passes the other one when flowing down, and then passing through the lower gutter reaches the vessel placed below the latter. When the cuts in the trunk and the peeling of the bark are continued in an upward direction, the gutters and the vessel are also moved upwards. In older trees contain-

FIG. 2.—Collection of the crude turpentine in the "Landes".
[*Roure Bertrand Fils.*

ing more oleo-resin, several such vessels can of course be placed around the tree.

The French turpentine industry is practically confined to the Landes district, where the following method of collection is followed :—[1]

[1] Vèzes, *Roure-Bertrand's Bulletin* (1909), 2, 9, 3.

This crude material—the "*gemme*," to give it the name under which it is known in the Landes—exudes during the warm season (from March to October) from an incision made by the axe of the resin collector in the stem of the maritime pine. Kept open by the removal of a thin slice every week, this incision—the "*carre*"—rises gradually up the trunk of the tree until it reaches, after five years' working, a height of nearly 4 metres. The tree is then left alone for two or three years, after which time a fresh "carre" is made at another point of the base. The same tree is tapped in this manner for about forty years, after which it is "bled to death" (that is, it is worked by means of several "carres" simultaneously), before being felled and delivered to the timber merchants. The forest is, moreover, rapidly reconstituted, by means of sowings made at the proper time, so that the resin production of the Landes district is not diminished by the work of the wood-cutters and the operations of the mechanical saw-mills. So far, this principle has not been followed in the United States, where every tree was "tapped to death" from the beginning, and every plot which was subjected to tapping was thereby practically destroyed after four years' working, without any steps being taken for its reconstitution. For some years, however, a serious movement has been set on foot in America to save what remains of the pine forests by a more rational and less destructive system of exploitation; and this appears to be producing good results.

From the top of the "carre," the drops of "gemme" as they exude from the resin ducts of the tree, run down into the receiver placed to collect them; this was in former times in the Landes (and was until three or four years ago in America) a hole—the "*crot*"—hollowed out of the stem of the tree itself, at the base of the "carre"; at the present day it is, both in America and in the Landes, an earthenware or metal *pot*, suspended at the base of the "carre" and raised every year so as to bring it near the upper part of the incision and reduce the distance over which the drops of the exudation have to run. The importance of this change and why it is advantageous that the path of the "gemme" down the "carre" should be as short as possible, is obvious when one remembers how easily the turpentine is oxidised.

From the pots the "gemme" is transferred, about once a month, into casks and carried from the forest to the factory; this harvest, "*l'amasse*," takes place in the Landes from five to eight times a year.

In India, the *chir*, as the *Pinus longifolia* is termed, is treated by a method based on the French method.[1]

An initial blaze about 6 ins. long, 4 ins. broad and not more than 1 in. deep is cut near the base of the tree, and the blaze is freshened every six or seven days throughout the summer, until it is about 18 ins. long by the end of the year: if the freshening is carefully done the use of a ladder is thus as a rule unnecessary till the fifth year. The resin collects in a cup fixed at the base of the current years' blaze, and the contents are emptied periodically. A curved channel, in place of the former rectangular channel, is now universally adopted in the United Provinces, as it works easily, does not injure the tree so much, gives a cleaner cut, and so far shows promise of giving a higher out-turn.

There are two classes of tapping: (1) *light tapping* and (2) *heavy tapping*, sometimes termed *tapping to death*. The latter is carried out in

[1] *Indian Forest Memoirs* (Sylviculture series), Vol. I, Part I, p. 100.

FIG. 3.—"Turpentine Farm" in the "Landes"—Interior. [Roure Bertrand Fils.

the case of trees due to be felled within five years, and consists in making as many blazes as the tree will hold: the former consists in making a moderate number of blazes, and is carried out in the case of trees not due to be felled within the next few years. In the Naini Tal Division light tapping under the present system is commenced when the trees reach a girth of 3½ ft., and the following number of blazes are cut :—

(1)	Trees over 6 ft. in girth	3 blazes
(2)	,, ,, 4½ to 6 ft. in girth	2 ,,
(3)	,, ,, 3½ to 4½ ft. in girth	1 blaze

In the maritime pine forests of Bordeaux no tree is tapped lightly till it reaches a girth of 1·10 metres (3 ft. 7 ins.), when one blaze is made: the tree is then about fifty years old. A second blaze is added when the tree reaches a girth of 1·30 metres (4 ft. 3 ins.). In the Chakrata Division tapping is confined to fire-protected forests of the first and second periodic blocks in Jaunsar-Bawar; no tapping is at present done in the Tehri-Garhwal forests owing to the cost of carriage. At present no heavy tapping is done in this Division: light tapping commences when the trees reach 4½ ft. in girth. In periodic block I. trees 4½ to 6 ft. in girth have two blazes, and those over 6 ft. in girth have four blazes. In periodic block II. trees 4½ to 6 ft. in girth have one blaze, and those over 6 ft. in girth have two blazes.

The object of heavy tapping is to obtain the largest possible out-turn of resin before felling the tree, and for this purpose as many blazes are made as there is room for. Experiments carried out so far show that only a comparatively small proportion of trees tapped in this way die: recent experiments in the Kangra Division showed a mortality of 6 per cent. in two years. In the maritime pine forests of Bordeaux trees to be removed in thinnings are heavily tapped as soon as they are large enough to hold a cup: this ordinarily commences when they are fifteen years old.

In the United Provinces the season's tapping commences in March and continues till November, after which the flow of resin practically ceases. In order to promote the free outflow of resin the blazes require to be freshened with an adze at intervals. Recent experiments carried out by Mr. E. A. Smythies in the Naini Tal Division showed that the out-turn increased steadily up to at least ten freshenings per month: this is borne out by similar experiments made in the Punjab. In Naini Tal, however, it was found that the yield did not increase in proportion to the extra labour required beyond five freshenings per month, and this number (*i.e.* once in about six days) is now universally adopted in that Division in place of the four freshenings per month formerly in vogue. As regards labour, in the Naini Tal Division the most economical working is found to be attained by an allotment of about 600 trees per man, that is, 100 trees a day per man.

In the case of light tapping, it is customary in India to tap for a certain number of years and then to allow a period of rest before tapping is recommenced. In the Naini Tal Division, and in Kumaon generally, the rule hitherto has been to tap for five years and then to give a rest for ten years before recommencing tapping: this complete period of fifteen years may be termed the tapping-cycle. In the Chakrata Division a similar procedure was in force till 1914, when a nine years' tapping-cycle was introduced, with tapping for three years followed by a rest for six years. The cycle was shortened in order to avoid the use of a ladder, which was found necessary in the fifth year. The highest yield of resin is obtained in the third and fourth years of tapping, and thus by reducing the tapping period to three years considerable loss is involved, which is avoidable considering that a ladder is not required in the fourth season if reasonable care is taken in freshening the blazes.

If a tapping-cycle at all is to be decided on, twelve years (*i.e.* four years' tapping and eight years' rest) appears to be preferable to either nine or fifteen years. The objection to the former has been mentioned, while in the case of the latter it cannot be made to fit in with a suitable scheme of thinnings.

In the maritime pine forests of Bordeaux light tapping, once begun, continues without interruption, no rest being given unless a tapped tree shows signs of becoming sickly, when a rest of five years is given: such cases are rare. Now in the case of the *chir* pine, experiments have shown that the tree possesses remarkable vitality even under the strain of heavy tapping. There seems to be no reason, therefore, why the continuous light tapping system of France should not be employed in the *chir* forests. Not only would it increase largely the out-turn of resin, but it would avoid the complication in the general scheme of management which is inevitable if a tapping-cycle is adopted.

Much has been written in reference to the adulteration of oil of turpentine, most of which is of little more than historical interest to-day. The chief difficulty the expert has to contend with at the present time is the detection of the so-called wood or stump turpentine oil. This oil is clearly entitled to the description "turpentine oil," on account of its origin and character, which are exceedingly close to those of normal or "gum" turpentine; but it is generally conceded that it properly requires the qualification "wood" or "stump," so that users may know exactly what they are dealing with.

Wood turpentine, "stump turpentine," or "wood spirits of turpentine," as the product is variously known, is turpentine oil made from cut pine, fir, or spruce, by distilling the wood in closed retorts. When properly refined it closely resembles gum spirits of turpentine, obtained by distilling the oleo-resin exuding from the cut surface of the living tree. The turpentine made by distilling the wood with steam below a temperature of 150° C. more closely resembles gum spirits than that obtained by destructively distilling the wood. The latter is always contaminated with other oils derived from the breaking down of the wood and the resins which it contains.

As a result of the reckless devastation of the pine forests which has been practised for many years, in the United States, the tree-stumps which have been left standing, and which are known as "light wood" are now often distilled. The stumps which are being used up in America at present have remained in the ground many years after the trunk had been removed, and yield a pine wood which is very rich in resin. It would appear that a chemical process is taking place in these stumps by which oil is formed subsequent to the removal of the trunk, for the oil-yield from new stumps is very low. This fact would also explain why the oil thus obtained is somewhat different from ordinary oil of turpentine, and why the percentage is so high.

The stumps which are mostly employed for distilling purposes are those of *Pinus palustris* (long leaf pine), whence the oil is called "long leaf pine oil". Similar products are brought to market under the names of "light wood oil" and "oil of fir"; while the designations "white" and "yellow pine oil" appear to apply to separate fractions of the oil.

According to Teeple[1] ordinary commercial long leaf pine oil in its crude state has a specific gravity 0·935 to 0·944, optical rotation − 11°, and refractive index 1·4830. These values, however, are often exceeded, and genuine samples may have a specific gravity 0·954 and may be dextro-rotatory up to + 8°. A sample was examined by Schimmel & Co.[2] which had the following characters:—

Specific gravity	0·9536
Optical rotation	− 3° 26′
Refractive index	1·48537
Ester number	14·2
Total alcohols (by acetylation)	58 per cent.

The bulk of the oil boiled between 190° and 220°. The following terpenes were identified in the oil: α-pinene, β-pinene, *l*-limonene, dipentene, and λ-terpinene. A considerable amount of α-terpinene was found (this had previously been identified by Teeple), as well as borneol and methyl-chavicol. Fenchyl alcohol was also found as a constituent of the oil, and traces of camphor and cineol. The presence of fenchyl alcohol is of particular interest as it occurs in the inactive variety. It is well known that this body results from the reduction of fenchone, but it has never before been discovered as a natural constituent of an essential oil. The table below shows the character of the fenchyl alcohol occurring in this oil and of those prepared by the reduction of active and inactive fenchone.

FENCHYL ALCOHOL AND ITS DERIVATIVES.

From Yellow Pine Oil.					From Active Fenchone.				From Inactive Fenchone.			
m.p.	b.p.	d.	n_D.		m.p.	b.p.	d.	n_D.	m.p.	b.p.	d.	n_D.
33 to 35°	202 to 203°	—	—	Fenchyl Alcohol	45°	201 to 202°	—	—	33 to 35°	201 to 202°	—	—
—	192 to 193°	0·9501 (15°)	1·47023 (20°)	Fenchone	+ 6°	192 to 193°	0·952 (15°)	1·46337 (17°)	—	—	—	—
159 to 160°	—	—	—	Fenchone Oxime	164 to 165°	—	—	—	153 to 160°	—	—	—
142 to 143°	—	—	—	Acid Phthalic Ester	145 to 145·5°	—	—	—	—	—	—	—
—	154 to 156°	0·8687 (15°)	1·47240 (20°)	Fenchene	—	154 to 156°	0·8660 to 0·8665 (15°)	1·46733 to 1·46832 (17°)	—	154 to 156°	0·8665 (15°)	1·46733 to 1·46832
93 to 94°	—	—	—	*Iso*fenchyl Phenyl-urethane	106 to 107°	—	—	—	94°	—	—	—

[1] *Jour. Amer. Chem. Soc.*, 30 (**1908**), 412.

[2] *Report*, April, **1910**, 104.

The crude oil obtained by any distillation process is redistilled or refined before it is marketed. Crude oils from the destructive process are agitated with alkali whereby the greater part of the phenols, cresols, and related bodies are dissolved, and separated by settling from the turpentine; rosin spirits and rosin oils are run off and separated by redistillation with steam as steam-distilled wood turpentine is. This is usually done in a copper-pot still of suitable size, live steam being conducted directly into the crude oil. Ordinarily this distillation is not conducted with care, and but little attempt is made to ensure that the wood turpentine obtained does not contain considerable quantities of the heavier "pine oils," as the heavier oils present in the crude product are called.

The principal source of American turpentine is the long-leafed pine, *Pinus palustris* (*Pinus australis* Mchx.); *Pinus heterophylla* and *Pinus echinata* also furnish turpentine.

American oil of turpentine consists practically entirely of terpenes, of which pinene is the chief. Dextro-α-pinene, laevo-α-pinene and β-pinene are all present, as well as small quantities of camphene. The question as to whether firpene,[1] which has been described as a constituent of the oil, is a distinct individual or not, is a matter of some doubt.

Crude wood turpentine differs from normal turpentine primarily in that it contains additional terpenes, and terpene derivatives, together with other non-terpene derivatives. In that obtained by the destructive distillation of long-leaf yellow pine, dipentene, pentane, pentene, toluene, heptine, etc., have been identified, in addition to pinene.

Steam-distilled wood turpentine consists essentially of pinene, together with camphene, limonene, dipentene, cineol, and (depending on care of fractionating), terpineol, borneol, terpinene, etc.

The pine oils contain chiefly terpineol, borneol, and fenchyl alcohol, with small quantities of limonene, dipentene, terpinene, cineol, and even less pinene and camphene.

American oil of turpentine varies in character to a considerable extent, the optical rotation, especially, having steadily decreased in the past few years. Such variations are to be expected when one remembers that the gathering grounds have been extended in various directions, the methods of preparation alter from time to time, and the use of different species of trees is not uncommon.

Any standards for these oils must therefore be understood to be fair working figures, to which by care, oils can be made to attain, although pure distillates may, obviously, give values widely outside the accepted figures.

The United States Department of Agriculture in 1911 issued suggested standards for pure turpentine, of which that for No. 1 or "standard" turpentine is the most important. The suggested standards are as follows: Specific gravity at 20° = 0·862 to 0·870; refractive index at 20° = 1·4680 to 1·4760; 95 per cent. should distil below 170°. On polymerisation with sulphuric acid (thirty-eight times normal) the residue should not exceed 1 per cent. and should have a refractive index of 1·500 to 1·520.

The author,[2] as the result of the examination of a very large number of pure commercial samples, finds the following limits to embrace practically all genuine normal samples as met with in commerce :—

[1] Frankforter and Frary, *Jour. Amer. Chem. Soc.*, 28 (1906), 1461.

[2] E. J. Parry, *Chemist and Druggist*, 24 August, 1912.

Specific gravity at 15°	0·862 to 0·870
Refractive index	1·4680 „ 1·4730
Initial boiling-point	154° „ 155·5°
Distillate under 160°	72 „ 74·5 per cent.
„ „ 170°	95 „ 97·5 „
Bromine value	1·96 „ 2·31
Iodine value (Hübl)	360 „ 375
„ „ (Wijs)	335 „ 350
„ „ of last 10 per cent. on distillation (Hübl)	349 „ 369
„ „ „ „ „ „ (Wijs)	350 „ 365
Refractive index of 1st 20 per cent.	1·4700 „ 1·4722
„ „ 2nd „	1·4700 „ 1·4724
„ „ 3rd „	1·4710 „ 1·4735
„ „ 4th „	1·4710 „ 1·4740
„ „ 5th „	1·4780 „ 1·4821

The characters of wood turpentine, and the effect of this and of other adulterants on these figures is shown in the following table by the author (*loc. cit.*):—

	(1) Pure American Turpentine.	(2) Wood Turpentine.	(3) 50 per cent. each (1) and (2).	(4) "Petroleum" Turpentine.	(5) 50 per cent. each (1) and (4).
Specific Gravity	0·866	0·873	0·869	0·808	0·838
Refractive Index	1·4720	1·4745	1·4737	1·4490	1·4610
Initial Boiling-point	155°	159°	157°	98°	99°
Distillate under 160°	74%	61%	68%	48%	62%
„ „ 170°	94%	78%	84%	75%	82%
Bromine Value	2·2	1·46	1·82	0·05	1·1
Iodine Value (Hübl)	372	264	321	9·0	190·5
„ „ (Wijs)	350	240	298	8·4	179
„ „ of last 10 per cent. (Hübl)	360	251	304	9·0	184
„ „ „ „ (Wijs)	355	242	298	8·5	177
Refractive Index of 1st 20 per cent.	1·4719	1·4731	1·4728	1·448	1·4660
„ „ 2nd „	1·4700	1·4730	1·4720	1·419	1·4675
„ „ 3rd „	1·4712	1·4734	1·4722	1·4481	1·4721
„ „ 4th „	1·4712	1·4732	1·4721	1·4470	1·4721
„ „ 5th „	1·4781	1·4842	1·4821	1·4495	1·4735

As Hertz[1] has shown, the optical rotation of American oil of turpentine is very variable.

The experiments on the opposite page by the United States Government Chemists with the oil from named trees show how widely the rotation may vary even with the product of the same tree:—

[1] *Jour. Amer. Chem. Soc.*, **30**, 863.

Tree Designation.	Species.	Diameter (inches).	Character of Chipping.	Optical Rotation 100 mm. Tube. 20° ″C.
A1	*P. heterophylla* . .	7·0	1st year, normal depth .	– 20° 50′
A2	,, . .	14·5	,, ,, .	+ 0° 15′
A3	,, . .	24·5	,, ,, .	– 15° 0′
A4	*P. palustris* . . .	7·3	,, ,, .	+ 15° 40′
A5	,, . . .	15·0	,, ,, .	+ 8° 9′
A6	,, . . .	21·0	,, ,, .	+ 18° 18′
C1	*P. heterophylla* . .	12·3	2nd year, shallow . .	– 27° 11′
C2	,, . .	8·2	,, ,, . .	– 26° 28′
C3	*P. palustris* . . .	13·0	,, ,, . .	– 7° 26′
C4	,, . . .	8·7	,, ,, . .	+ 7° 31′
D1	,, . . .	9·0	2nd year, normal depth .	+ 10° 50′
D2	,, . . .	13·5	,, ,, .	+ 1° 23′
D3	*P. heterophylla* . .	13·0	,, ,, .	– 18° 35′
D4	,, . .	9·0	,, ,, .	– 29° 26′

In a general way the figures support the view that the oils of *Pinus palustris* are dextro-rotatory and those from *Pinus heterophylla* laevo-rotatory. That this is not strictly true, however, is evidenced by the dextro-rotation of A2 (*P. heterophylla*), and more especially by the laevo-rotation of C3 (*P. palustris*).

With these variations in the first collection from the several trees, the question naturally arose, would the variations change as the season advanced or would the figures prove constant for the individual trees? The rotations for the successive collections are as follows:—

Optical Rotation in 100 mm. Tube, 20° C.

Collection.	A1.	A2.	A3.	A4.	A5.	A6.	C1.
1	– 20° 50′	+ 0° 15′	– 15° 0′	+ 15° 40′	+ 8° 9′	+ 18° 18′	– 27° 11′
2	– 22° 5′	– 0° 30′	– 14° 26′	+ 15° 22′	+ 8° 50′	+ 17° 43′	– 26° 48′
3	– 21° 45′	+ 0° 15′	– 15° 55′	+ 14° 15′	+ 8° 27′	+ 19° 30′	– 26° 25′
4	– 21° 7′	– 1° 15′	– 15° 50′	+ 14° 20′	+ 8° 34′	+ 18° 46′	– 23° 32′
5	– 20° 30′	– 2° 5′	– 15° 15′	+ 14° 21′	+ 8° 32′	+ 19° 24′	– 21° 12′
6	– 20° 15′	– 3° 30′	– 15° 27′	+ 14° 35′	+ 8° 4′	+ 15° 16′	– 21° 46′
7	– 22° 15′	– 5° 45′	– 17° 52′	+ 12° 49′	+ 7° 6′	+ 14° 47′	– 21° 35′

Collection.	C2.	C3.	C4.	D1.	D2.	D3.	D4.
1	– 26° 28′	– 7° 26′	+ 7° 31′	+ 10° 50′	+ 1° 23′	– 18° 35′	– 29° 26′
2	– 25° 37′	– 6° 42′	+ 7° 20′	+ 11° 23′	+ 2° 40′	– 17° 0′	– 27° 45′
3	– 26° 20′	– 4° 45′	...	+ 13° 7′	+ 2° 25′	– 15° 20′	– 28° 19′
4	– 26° 30′	– 4° 29′	...	+ 12° 46′	+ 2° 22′	– 15° 0′	– 27° 38′
5	– 26° 7′	– 3° 55′	...	+ 13° 0′	+ 1° 13′	– 14° 38′	– 27° 48′
6	– 26° 0′	– 4° 5′	...	+ 13° 0′	+ 1° 15′	– 14° 7′	– 26° 11′
7	– 26° 28′	– 6° 6′	...	+ 10° 48′	– 0° 55	– 14° 19′	– 26° 12′

Wood turpentine can, however, be so rectified as to distil at temperatures much closer to those of normal turpentine than is shown in the table on p. 18, so that whilst positive results indicate the presence of wood turpentine, negative results do not necessarily exclude its presence.

The lower iodine value, especially of the last 10 per cent. remaining on distillation is very characteristic of wood turpentine, and is, in the author's opinion, the most valuable evidence that can be obtained of its presence.

The fact that wood turpentine usually contains dipentene is also of considerable assistance to the analyst in some cases. In normal turpentine, if the successive fractions of 5 per cent. be examined, they will be found to increase regularly in specific gravity and refractive index, with each rise in the distillation temperature. Dipentene, however, has a higher boiling-point than pinene, but a lower specific gravity. So that in a wood turpentine containing dipentene, when about 70 per cent. has distilled over, some of the fractions will be found to have increased in boiling-point, increased in refractive index, but decreased in specific gravity. In such a case, the presence of wood turpentine may safely be assumed.

When adulterated with most types of petroleum products, the temperature of distillation rises gradually, and no large fractions are obtained at any definite temperature when the adulteration is at all excessive. The presence of ordinary petroleum spirit lowers the flash-point of turpentine. When pure it flashes at 92° to 95° F. when tested in Abel's flash-point apparatus. With only 1 per cent. of ordinary petroleum spirit this temperature is reduced by 10°.

According to Armstrong, a good indication of the presence of the usual adulterants is obtained by distillation with steam. A current of steam is allowed to pass into a definite volume of the turpentine contained in a flask attached to a condenser. Unless it has been allowed free access to the air for some time, the genuine oil leaves only traces of non-volatile matter, but old samples may leave up to 2 per cent. Usually, however, the presence of more than ·5 per cent. after steam distillation indicates the presence of unvolatilised petroleum oil. This is easily recognised by its low specific gravity and its fluorescence when dissolved in ether. If the residue consists of resin oil, it will form a bulky soap when rubbed with slaked lime. The specific gravity of the fractions coming over with the steam will largely assist in determining the presence of volatile adulterants.

For the approximate estimation of the amount of petroleum naphtha in adulterated turpentine, Armstrong[1] recommends the following process: 500 c.c. of the sample is placed in a separator and treated with about 150 c.c of sulphuric acid (2 volumes of acid to 1 of water). The mixture is cautiously agitated, and, if much rise of temperature is observed, the separator must be placed in cold water for a short time. The turpentine is gradually converted into a viscid oil, and when this has taken place, and no more heat is developed on repeated agitation, the acid is tapped off. The oily layer is then transferred to a flask and subjected to steam distillation. When all that is volatile with steam has passed over, the oily portion of the distillate is separated from the aqueous layer, and heated with half its volume of sulphuric acid previously diluted with one-fourth of its measure of water. The mixture is well agitated, the acid liquid separated, and the oily layer again distilled with steam. When genuine turpentine is operated upon, the volatile portion of this second treatment consists merely of cymene

[1] *Jour. Soc. Chem. Ind.*, 1882, 1, 480.

and a small quantity of paraffinoid hydrocarbons. It never exceeds 4 to 5 per cent. of the volume of the original sample, and with care is as low as 3 per cent. If the volume notably exceeds 5 per cent., it is advisable as a precaution to repeat the treatment with the (4 to 1) acid. When treated in this manner, petroleum naphtha is not appreciably affected; hence the proportion may be fairly estimated by making an allowanee of 4 to 5 per cent. from the volume of volatile oil which has survived the repeated treatment with sulphuric acid. A further purification may be effected by violently agitating the surviving oil with several times its volume of concentrated sulphuric acid heated to 50° or 60°. This treatment can be repeated if necessary, after which the residual hydrocarbon is separated, steam distilled and again measured, when the surviving oil from pure turpentine oil will not exceed from ½ to 1 per cent. of the original sample. Any excess over this will be the minimum quantity of petroleum naphtha present. Shale naphtha cannot be at all estimated in this way. The behaviour of the oil on distillation is the best indication of the presence of rosin spirit, as the temperature rises gradually, and no considerable fraction is obtained at 158° to 160° if much rosin spirit be present.

Allen (*Commercial Organic Analysis*) gives the table on page 22 as indicating some of the properties of some substances which have been used as adulterants for turpentine oil.

More empirical methods for detecting adulteration in turpentine have been published than is the case for any other essential oil. These are principally based on colour reactions, and are, in very many cases, quite useless, especially when the adulterant is carefully purified. Many modifications of processes described above have been published, but do not appear to possess any advantages over those which have been described. It is, therefore, considered unnecessary to reproduce so large a bulk of published work, which has so little bearing on the practical examination of this oil; but readers who wish for information on these or other little-used processes or tests, or for details of methods which for any other reason are not dealt with here, are referred to the following papers :—

J. M. McCandless (*J. Amer. Chem. Soc.*, 1904 (**26**), 981).
Marcusson and Winterfeld (*Chem. Zeit.*, **1909**, 987).
W. Burton (*J. Amer. Chem. Soc.*, 1890 (**12**), 102).
P. H. Conradson (*J. Soc. Chem. Ind.*, 1897 (**16**), 519).
H. Wolff (*Farbenzeit.*, 1912 (**17**), 1553).
Grimaldi (*Chem. Zeit.*, 1908 (**32**), 8, and 1910 (**34**), 721).
Dunwody (*Amer. Jour. Pharm.*, 1892 (**33**) 131).
H. Herzfeld (*Chem. Zentralblatt*, 1903 (**1**), 258).
P. van der Wielen (*Pharm. Weekblad.*, 1911 (**8**), 35).
Nicolardot and Clément (*Bull. Soc. Chim.*, iv. (1910), **7**, 173).
Vèzes (*Bull. Soc. Chim.*, iii. (1903), **29**, 896).
C. Piest (*Chem. Zeit.*, 1912 (**36**), 198).
Utz (*Chem. Zentralblatt*, 1905 (**1**), 1673).

A. W. Schorger[1] has examined the oils from a number of pine-trees

[1] *U.S. Department of Agriculture, Bulletin* 119.

	Turpentine Oil.	Rosin Spirit.	Petroleum Naphtha.	Shale Naphtha.	Coal-tar Solvent Naphtha.
1. Optical Activity	active	usually none	none	none	none
2. Specific Gravity	·860 to ·872	·856 to ·880	·700 to ·740	·700 to ·750	·860 to ·875
3. Temperature of Distillation .	156° to 180°	gradual rise	gradual rise	gradual rise	gradual rise
4. Action in the Cold on Coal-tar Pitch	readily dissolves	readily dissolves	very slight action	very slight action	readily dissolves
5. Behaviour with Absolute Phenol at 20°	homogeneous mixture	homogeneous mixture	no apparent solution	homogeneous mixture, crystallises on cooling	homogeneous mixture
6. Behaviour on Agitating 3 vols. with 1 vol. Castor Oil . .	homogeneous mixture	homogeneous mixture	two layers of nearly equal volume	like petroleum naphtha	—
7. Bromine Absorption (Dry) . .	203 to 236	184 to 203	10 to 20	60 to 80	—
8. Behaviour with H_2SO_4 . .	almost entirely polymerised	polymerised	very little action	considerable action	moderate action

found in the Western parts of the United States, and Adams[1] has reported upon similar "wood turpentines". These results are as follows:—

TURPENTINE FROM PINUS PONDEROSA (LAWS).

The specific gravity of the oil was 0·8625; the index of refraction 1·4772; and the specific rotation $[\alpha]_D = -14{\cdot}39°$. On fractionation of 500 grams of oil the following results were obtained:—

Temperature (corrected).	Distillate.		Specific Gravity at 15°.	Index of Refraction at 15°.	Specific Rotation at 20°.
	Observed.	Cumulative.			
°C.	Per Cent.	Per Cent.			
164·0-168·2	5·01	5·01	0·8578	1·4740	-13·81
168·6	9·94	14·95	·8595	1·4750	-13·70
169·0	9·36	24·31	·8601	1·4752	-13·17
169·2	9·20	33·51	·8606	1·4753	-12·92
170·5	10·45	43·96	·8608	1·4754	-12·73
171·0	10·71	54·67	8612	1·4757	-12·98
171·6	10·28	64·95	·8615	1·4760	-13·33
172·4	9·69	74·64	·8619	1·4762	-14·17
175·0	10·36	85·20	·8621	1·4763	-17·15
179·5	7·04	92·24	·8631	1·4775	-24·13
179·5+	7·74	99·98	·9186	1·5063	+0·86

Laevo-α-pinene, β-pinene, and laevo-limonene were detected in the oil.

TURPENTINE FROM PINUS PONDEROSA SCOPULORUM (ENGELM).

The oil had the following characters: specific gravity 0·8639 at 15°; index of refraction 1·4727 at 15°.

The physical constants of the fractions obtained on distillation are as follows:—

Temperature (corrected).	Distillate.		d15°. Specific Gravity at 15°.	N_D15° Index of Refraction at 15°.
	Observed.	Cumulative.		
°C.	Per Cent.	Per Cent.		
157·6-158·5	2·03	2·03	0·8586	1·4686
159·5	9·69	11·72	·8621	1·4701
159·9	9·97	21·69	·8620	1·4702
160·1	10·21	31·90	·8620	1·4705
160·9	10·06	41·96	·8620	1·4706
161·7	9·39	51·35	·8620	1·4711
162·6	9·97	61·32	·8623	1·4714
165·2	9·12	70·44	·8626	1·4722
170·2	10·50	80·94	·8626	1·4735
173·0	5·94	86·88	·8615	1·4752
177·1	6·25	93·13	·8593	1·4758
183·3	2·19	95·32	·8576	1·4769
183·3+	4·60	99·92	—	—

α-Pinene, β-pinene, and limonene were found in the oil.

[1] *Jour. Ind. Eng. Chem.*, 1915 (**7**), 957.

Turpentine from Pinus Sabiniana.

The oil had a pleasant orange-like odour; specific gravity, 0·6971; and index of refraction, 1·3903. On fractional distillation of 500 grams of the oil the following results were obtained:—

Temperature (corrected).	Distillate.		$d15°$ Specific Gravity at 15° C.	$N_D15°$ Index of Refraction at 15° C.
	Observed.	Cumulative.		
°C.	Per Cent.	Per Cent.		
96·1-97·3	4·24	4·24	0·6889	1·3890
97·9	7·93	12·17	·6885	1·3898
98·3	9·04	21·21	·6882	1·3898
98·3	8·44	29·65	·6882	1·3898
98·4	8·89	38·54	·6880	1·3898
98·4	8·97	47·51	·6881	1·3898
98·5	8·06	55·57	·6881	1·3898
98·5	7·13	62·70	·6882	1·3898
98·5	6·92	69·62	·6881	1·3898
98·5	6·94	76·56	·6880	1·3899
98·6	6·83	83·39	·6879	1·3899
98·7	6·19	89·58	·6879	1·3899
98·8	5·66	95·24	·6881	1·3900
98·8+	4·66	99·90	·7393	1·4140

This oil is very noteworthy, as it consists almost entirely of the hydrocarbon heptane, a low boiling paraffin hydrocarbon, quite distinct from the terpenes, which are the normal constituents of practically all other turpentine oils.

Turpentine from Pinus Contorta.

Two samples of this oil had the following characters:—

	1.	2.
Specific gravity	0·8518	0·8549
Rotation	− 20·12°	—
Refractive index	1·4862	1·4860

About 400 grams were submitted to fractional distillation with the following results:—

Temperature (corrected).	Distillate.		$d15°$ Specific Gravity at 15° C.	$N_D5°$ Index of Refraction at 15° C.
	Observed.	Cumulative.		
°C.	Per Cent.	Per Cent.		
164·4-170 6	2·89	2·89	0·8547	1·4793
171·6	8·55	11·44	·8546	1·4814
172·6	9·08	20·52	·8542	1·4823
173·6	10·77	31·29	·8534	1·4835
175·3	10·22	41·51	·8522	1·4852
177·2	9·79	51·30	·8514	1·4861
177·6	10·03	61·33	·8507	1·4871
177·9	10·23	71·56	·8502	1·4880
178·4	9·26	80·82	·8498	1·4891
178·9	4·15	84·97	·8498	1·4894
178·9+	14·83	99·80	—	—

The oil had a very characteristic odour; it was pungent when freshly distilled, and, after standing some time, was orange-like. The oil polymerised very considerably at its boiling-point. The 15 per cent. residue remaining in the distilling flask solidified to a solid, amber-coloured mass resembling colophony, insoluble in alcohol, but soluble in ether and chloroform. To avoid loss of oil by polymerisation it must be distilled under reduced pressure. The oil carefully purified and finally distilled over metallic sodium, had the following properties: Boiling-point, 60° at 11 mm.; specific gravity, $\frac{21^\circ}{15^\circ}$ 0·8460; refractive index, 1·4861; specific rotation $[\alpha]_D = -12{\cdot}36^\circ$.

This oil is also noteworthy, as it consists mainly of β-phellandrene which has not previously been detected in any turpentine oil.

Turpentine from Pinus Lambertiana.

An oil obtained from oleo-resin collected during 1911 had the following properties: specific gravity, 0·8663; index of refraction, 1·4728; specific rotation $[\alpha]_D = +10{\cdot}42^\circ$. Another sample of the oil had a specific gravity of 0·8658 and an index of refraction of 1·4727, and gave the following results on fractionation:—

Temperature (corrected).	Distillate.		d15° Specific Gravity at 15° C.	ND15° Index of Refraction at 15° C.
	Observed.	Cumulative.		
°C		Per Cent.		
155·8-157·3	2·98	2·98	0·8531	1·4657
158·6	9·33	12·31	·8620	1·4693
158·9	10·78	23·09	·8629	1·4698
159·3	10·34	33·43	·8632	1·4701
159·7	10·70	44·13	·8633	1·4703
160·3	9·79	53·92	·8633	1·4708
161·3	9·94	63·86	·8634	1·4716
162·5	9·54	73·40	·8633	1·4726
168·4	7·67	81·07	·8624	1·4736
178·8	5·11	86·18	·8548	1·4721
178·8+	13·62	99·80	·9011	1·4912

The oil contains α-pinene, β-pinene, a very small amount of phellandrene (?), a hydrocarbon, probably of the aliphatic series, and a sesquiterpene which is probably aromadendrene.

Turpentine from Pinus Edulis.

The oil had the characteristic odour of thuja leaf oil, due to the presence of a sesquiterpene; the specific gravity was 0·8680; the index of refraction, 1·4707; and the specific rotation, $[\alpha]_D = +19{\cdot}26^\circ$.

The oil contains α-pinene, β-pinene, and cadinene, a compound which has not previously been recorded as occurring in turpentine oils.

On fractional distillation the following results were obtained:—

Temperature (corrected).	Distillate.		d15° Specific Gravity at 15° C.	ND 15° Index of Refraction at 15° C.
	Observed.	Cumulative.		
°C.	Per Cent.	Per Cent.		
155·9-157·4	5·10	5·10	0·8617	1·4677
157·6	10·66	15·76	·8617	1·4683
157·9	10·28	26·04	·8621	1·4687
158·4	8·99	35·03	·8619	1·4689
158·9	10·54	45·57	·8616	1·4689
160·0	11·04	56·61	·8611	1·4693
162·4	8·76	65·37	·8609	1·4696
176·5	9·72	75·09	·8605	1·4713
194·9	3·87	78·96	·8616	1·4723
194·9 +	20·74	99·70	·9129	1·4837

Turpentine from Pinus Jeffreyi.

Schorger[1] states that the normal turpentine from this tree consists of about 95 per cent. of heptane and 5 per cent. of an aldehyde, probably citronellal. Schimmel & Co.[2] have examined this oil and found it to have the following characters:—

Specific gravity	0·7051
Optical rotation	1·39653
Ester value	5·2
,, ,, (after acetylation)	8·7

They agree with Schorger that the principal constituent of the oil is heptane. They do not agree, however, that citronellal is present, but have identified decylic acid; possibly traces also of octylic and nonylic aldehydes may be present. Linalol and methyl-chavicol are also probably present.

The "wood turpentine" of *Pinus Jeffreyi* was found to consist of 90 to 95 per cent. of heptane, with traces of limonene and citronellal. That from *Pinus monophylla* had a specific gravity 0·9702, rotation + 21·15°, and refractive index 1·4771. In it were identified α-pinene, β-pinene, and cadinene. The corresponding oil from *Pinus ponderosa* had a specific gravity 0·8626, rotation − 13·15°, and refractive index 1·4727. It contains α-pinene, β-pinene, and limonene.

Russian oil of turpentine is a product, the importance of which has increased during the past few years, and appreciable quantities are now imported into this country. The oil is obtained principally from *Pinus sylvestris*, but *Pinus Ledebourii* also contributes to its production. The crude oleo-resin is, whilst still mixed with a large percentage of impurities, distilled over a naked fire without the use of water (essence de térébenthine de résine) or the wood is crudely distilled directly (essence de térébenthine de four).

The resulting oil is brownish in colour with a most unpleasant odour, due to the presence of decomposition products, which include phenols, furfural, and benzenoid hydrocarbons.

The principal constituents of Russian turpentine are, according to Schindelmeiser,[3] β-pinene, sylvestrene, and dipentene. A little *dextro*-α-

[1] *Jour. Ind. Eng. Chem.* **6** (1914), 541. [2] *Report*, October, **1914**, 45.
[3] *Chemiker Zeitung*, 1908 (**32**), 8.

pinene is also present. Zelinski and Alexandroff[1] isolated a highly laevo-rotatory hydrocarbon $[\alpha]_D = -70° 45'$, which they believed to be laevo-pinene, but Schindelmeiser[2] has shown it to be *laevo*-camphene contaminated with some impurity. Acetone and α-terpineol have been found in some samples of Russian turpentine oil.

According to Schkatelow[3] the following trees yield "Russian" turpentine oils, having the following characters :—

	Field of Oil.	[α]D.	Specific Gravity.
Pinus Sylvestris . .	15 to 16 per cent.	+ 22° to + 24°	0·867 at 15°
,, *Abies* . . .	13·4 ,,	− 13·2°	0·873 ,, 15°
(*Abies excelsa*) . .			
Larix Sibirica . . .	14·1 ,,	− 14·3°	0·870 ,, 19°
Pinus Cembra . . .	6 ,,	+ 14·04°	0·865 ,, 15°
,, *Taurica* . . .	20 ,,	− 75·9°	0·861 ,, 19°
Abies Sibirica . . .	28 ,,	− 35·6°	0·875 ,, 19°

The oil of turpentine arriving on this market from Russia does not correspond in character with any of the above oils. This is due to two facts; firstly, that the oil is a mixture derived from more than one species, and, secondly, that the bulk of the exported oil has been deprived of its "middle runnings" in Russia, the so purified middle fractions being used in the country, so that much of the Russian turpentine as we know it is a fractionated oil. The author[4] has examined samples of pure unfractionated Russian turpentine supplied to him by Prof. Schindelmeiser, and found them to contain over 60 per cent. of oil distilling between 155° to 160°, which fractions are missing in nearly all the commercial oil exported. Two such typical samples had the following characters :—

	1.	2.
Specific gravity	0·867	0·865
Optical rotation	+ 7° 50′	+ 10°
Refractive index	1·4718	1·4736
Absorbed by 5 per cent. KOH . .	5 per cent.	6 per cent.
Distilled below 155°	traces only	traces only
,, 155° to 160°	65 per cent.	63 per cent.
,, 160° ,, 165°	11 ,,	9 ,,
,, 165° ,, 170°	13 ,,	15 ,,
,, 170° ,, 180°	7·5 ,,	7 ,,
,, above 189°	3·5 ,,	6 ,,

Sample No. 1, after removal of the acid bodies and tarry matter, gave a rectified turpentine oil having the following characters :—

Specific gravity	0·8646
Optical rotation	+ 8°
Refractive index	1·4890
Absorbed by KOH	none
Distils below 155°	none
,, 155° to 160°	68 per cent.
,, 160° ,, 165°	13 ,,
,, 165° ,, 170°	10 ,,
,, 170° ,, 180°	7 ,,
,, above 180°	2 ,,

[1] *Chemiker Zeitung*, 1902 (**26**), 1224. [2] *Chem. Zentralblatt*, 1908 (**1**), 2097.
[3] *Moniteur Scientifique*, 1908, iv., **22**, i., 217.
[4] *Chemist and Druggist*, 26 October, **1912**.

These facts are recognised in the turpentine market, and users are content to accept the fractionated oil as Russian turpentine. But it is considered necessary that it should be in a reasonably good condition and as what is known as of good merchantable quality. The difficulty of fixing empirical standards under these conditions is obvious, but as the result of considerable experience in disputes on shipments of this oil the author is able to say that the points to which objection is taken are (1) the presence of phenoloid bodies (absorbed by caustic potash); (2) the presence of much oil boiling over 180° which renders the oil unremunerative to refine, as these fractions are useless for most purposes; (3) the presence of impurities distilling below 155°. As a result of this experience, the following values may be laid down as representing what is accepted on the London markets as Russian turpentine of fair merchantable quality. The figures relate to four typical samples:—

	1.	2.	3.	4.
Initial boiling-point	157°	156°	157°	158°
Distils below 155°	none	none	none	none
,, 155° to 160°	1 per cent.	1 per cent.	5 per cent.	11 per cent.
,, 160° ,, 165°	44 ,,	45 ,,	40 ,,	18 ,,
,, 165° ,, 170°	37 ,,	35 ,,	42 ,,	48 ,,
,, 170° ,, 180°	15 ,,	16 ,,	10 ,,	19 ,,
,, ,, above 180°	3 ,,	3 ,,	3 ,,	4 ,,
Specific gravity at 15° . .	0·863	0·8635	0·863	0·868
Refractive index at 20° . .	1·4730	1·4726	1·4725	1·4748
Optical rotation . . .	+ 4° 28′	+ 4° 30′	+ 9°	+ 8°
Absorbed by 5 per cent. KOH	nil	nil	nil	nil

A very large number of samples, however, have been even more largely deprived of their middle runnings, and contain a considerable amount of hydrocarbons boiling over 180°, and also a considerable amount of acid bodies, which are absorbed by caustic potash. Such samples are quite useless to the rectifier, as their redistillation must ensure the removal of the acid bodies and also of the bodies boiling over 180°, with a resulting loss which, as mentioned above, causes the rectification to be unremunerative.

The following are typical samples of this kind which represent numerous deliveries on the London market:—

	1.	2.	3.	4.
Initial boiling-point . .	148°	146°	146°	153°
Distils below 155° . .	2 per cent.	3 per cent.	5·5 per cent.	1 per cent.
,, 155° to 160° . .	3 ,,	3 ,,	5 ,,	3 ,,
,, 160° ,, 165° } ,, 165° ,, 170° } . .	35 ,,	34 ,,	22 ,,	36 ,,
,, 170° ,, 180° . .	48 ,,	50 ,,	46 ,,	50 ,,
,, above 180° . .	12 ,,	10 ,,	21·5 ,,	10 ,,
Specific gravity at 15° .	0·868	0·8665	0·878	0·869
Refractive index at 20° .	1·4762	1·4756	1·4780	1·4792
Optical rotation . . .	+ 8°	+9°	+ 11°	+ 12° 45′
Absorbed by KOH . .	6 per cent.	8·5 per cent.	7 per cent.	3 per cent.

Finnish turpentine oil closely resembles Russian oil. Aschan[1] has identified it in the following compounds; diacetyl, furane, methyl-furane, methyl-isobutyrate, benzene and its homologues, pinene, sylvestrene, *l*-limonene, and dipentene. He has also recently isolated a new terpene from this oil, boiling at 163° to 165°, specific gravity 0·8628 at 20°, and specific rotation + 7·7°. It is a saturated bicyclic terpene, closely related to

[1] *Farm. Notisblad.*, **1907**, 93, *and Chem. Zeit.* (1919), **1**, 284.

pinene. A terpene alcohol, and a sesquiterpene (cadinene ?) are also present. A somewhat similar product is manufactured in Norway, as a by-product in cellulose manufacture. It has a specific gravity 0·874 and optical rotation + 13°. It contains pinene, cadinene, and sylvestrene.[1]

French oil of turpentine is characterised by a high laevo-rotation. It is obtained principally from *Pinus Pinoster* (*Pinus maritima* Poir) (*vide supra*), and has the following characters :—

Specific gravity at 15°	0·863 to 0·875
Optical rotation	− 20° ,, − 38°
Initial boiling-point	152° ,, 155°
Distillate below 165°	80 to 90 per cent.

It resembles American oil in other respects, the pinene present merely differing by consisting more largely of the laevo-rotatory variety.

Normal Italian turpentine is the product of *Pinus picea* and *Pinus pinaster*. Palazzo[2] gives the following values for a typical pure commercial sample :—

Specific gravity	0·863
Optical rotation	− 48° 15′
Refractive index	1·4678 at 25°

Eighty per cent. distilled between 155° and 165·5°. The first fraction, distilling up to 159° had a specific gravity 0·8576 at 20°, optical rotation − 46° 54′, and refractive index 1·4653 at 25°. The pure samples from *Pinus pinaster* only had the following characters :—

Specific gravity at 15°	0·867	0·871
Refractive index at 20°	1·4700	1·4707
Optical rotation	− 35°	− 35°

The high optical rotation is stated to be due to the presence of *l*-limonene.

Greek turpentine oil is a commercial article in the South of Europe and is obtained from the Aleppo pine (*Pinus halepensis*). It is produced in nearly all the provinces of Greece. It is usual either to make deep incisions in the trunks through which the resin quickly flows away, or in the French manner to cut rather shallow notches and place a wooden receptacle below these, in which the oleo-resin is collected. *Pinus halepensis* is the only tree in Greece which is used for the production of turpentine.

The crude oleo-resin contains from 20 to 25 per cent. of oil and 70 per cent. of resin. The essential oil consists principally of dextro-α pinene. Gildemeister and Kohler[3] isolated this body in a fairly pure condition, but Tsakalotos and Papaconstantinou[4] have more recently separated it in an apparently purer condition. The characters of the hydrocarbon so separated are as follows :—

	Gildemeister.	Tsakalotos.
Specific gravity	0·8642 at 15°	0·8548 at 25°
Refractive index	1·46565 ,, 20°	1·4633 ,, 25°
Optical rotation	+ 40·23°	+ 47·7°
Boiling-point	156° at 760 mm.	155° to 156° at 760 mm.

Greek turpentine has the following characters :—

[1] Fosse, *Ber. Deutsch. pharm. Ges.*, 25 (**1915**), 303.
[2] *Ann. Chim. Applic.* (**1917**), 7, 88.
[3] *Wallach-Festschrift*, Göttingen, **1909**, 429.
[4] *Jour. de Pharm. et de Chim.*, 1916, **4**, 97.

Specific gravity	0·860 to 0·866
Optical rotation	+ 34° ,, + 47°
Refractive index	1·4670 ,, 1·4740
Distillate from 152° to 156°	5 to 10 per cent.
,, ,, 156° ,, 160°	80 ,, 85 ,,

The turpentine oil from *Pinus longifolia* has recently been examined by Simonsen.[1] He found present the following bodies: laevo-α-pinene, β-pinene, a new terpene for which he proposes the name *d*-carene, and a new sesquiterpene which he terms longifolene. In the previous edition of this work (p. 28) it was stated that *d*-sylvestrene was present in comparatively large amount.

It is probable that this is incorrect, but that another terpene has been mistaken for sylvestrene on account of the fact that by the action of hydrochloric acid, a molecular rearrangement takes place, with the formation of sylvestrene hydrochloride. This hydrocarbon is that isolated by Simonsen and named by him *d*-carene. It is an oil having the following characters:—

Specific gravity @ 30°	0·8586
Refractive Index @ 30°	1·4690
Optical Rotation	+ 7·69°
Boiling-point at 705 mm.	168° to 169°

It yields sylvestrene hydrochloride when treated with hydrogen chloride in ethereal solution. It yields a crystalline nitrosate, decomposing at 141·5°.

Its constitution is not yet settled. The sesquiterpene *d*-longifolene boils at 150° to 151° at 36 mm., or at 254° to 256° at 706 mm. Its specific gravity is 0·9284 at 30°, refractive index, 1·4950 at 30°, and optical rotation + 42·73°.

It yields a hydrochloride melting at 59° to 60°, and having a specific rotation + 7·1°. Its hydrobromide melts at 69° to 70°, and its hydriodide at 71°.

The characters of this oil fall within the following limits:—

Specific gravity	0·865 to 0·875
Optical rotation	0° to + 4°
Initial boiling-point	165°

Two samples on fractionation by Rabak[2] and by Schimmel & Co.[3] gave the following results:—

	Rabak.	Schimmel.
165° to 170°	56 per cent., α_D = − 2°	33 per cent., α_D = − 3° 30′
170° ,, 175°	20 ,, ,, = + 2° 48′	31 ,, ,, = + 0° 40
175° ,, 180°	9 ,, ,, = + 6° 50′	14 ,, ,, = + 4° 32′
above 180°	15 ,, ,, = + 18° 12′	22 ,, ,, = + 16° 22′

Austrian turpentine oil is derived principally from *Pinus laricio*. Its composition has not been exhaustively studied, but it consists of terpenes and is similar in composition to French turpentine oil. Its characters are as follows:—

[1] *Jour. Chem. Soc.* (**1920**), 570.
[2] *Pharm. Rev.*, 23 (**1905**), 229.
[3] *Report*, April, **1906**, 66.

Specific gravity	0·863 to 0·870
Optical rotation	− 35° ,, − 40°
Refractive index	1·4680 ,, 1·4725
Distillate from 155° to 158°	45 to 50 per cent.
,, ,, 158° ,, 160°	about 30 ,,
,, ,, 160° ,, 165°	,, 15 ,,
,, ,, 165° ,, 175°	,, 8 ,,

Turpentine from Agathis Robusta.

The oleo-resinous exudation from this Australian tree, known as the "Queensland Kauri," yields an essential oil, which has been examined

Fig. 4.—Tree of *Agathis robusta*, Queensland.
[P. 373, *Pines of Australia.*

by Baker and Smith.[1] The yield of oil was 11·64 per cent. having the following characters :—

Specific gravity at $\frac{16}{15}$° C.	0·8629
Rotation α_D in 100 mm. tube	+ 20·2°
Refractive index at 16° C.	1·4766

Thirty c.c. of the oil were distilled under atmospheric pressure, when nothing came over below 155° C.; between 155° and 156° C., 53·3 per cent. distilled; and between 156° and 159° C., 33·3 per cent. more came over. The residue in the flask, 13·3 per cent., was also determined.

The first fraction had the following characters :—

Specific gravity at $\frac{17}{15}$° C.	0·8625
Rotation α_D	+ 14·4°
Refractive index at 17° C.	1·4755

The second fraction had the following characters :—

Specific gravity at $\frac{17}{15}$° C.	0·8603
Rotation α_D	+ 20·4°
Refractive index at 17°C.	1·4763

The portion remaining in the flask had the following characters :—

Specific gravity at $\frac{17}{15}$° C.	0·8610
Rotation α_D	+ 38·6°
Refractive index at 17° C.	1·4791

Pinene was the only constituent identified.

In addition to the above described turpentine oils, the following, although not of considerable importance, are of some interest :—

Origin.	Specific Gravity.	Rotation.	Refractive Index.	Constituents.
Spanish—				
Pinus halepensis }	0·860	− 9°	1·4654	—
,, *pinaster* }				
[2] Algerian—				
Pinus halepensis	0·860-0·863	+ 45° to + 48°	1·4638-1·4660	*d*-pinene
Mexican	—	+ 34°	—	limonene (?)
[3] Philippine—				
Pinus insularis	0·869	+ 13° to + 28°	1·4700	*d*-α-pinene; β-pinene
Burmese—				
Pinus Khasya	0·863	+ 36°	— }	*d*-α-pinene
,, *Merkusii*	0·861	+ 32°	— }	
[4] Japanese—				
Pinus Thunbergii	—	—	—	—
Various—				
Picea excelsa	0·863-0·870	+ 3° to − 12°	—	α-pinene, dipentene, limonene
[5] "Strasburg Turpentine Oil"—				
Abies Alba	0·860-0·861	—	—	α-pinene
[6] Venetian Turpentine Oil—				
Larix decidua	0·865-0·878	− 8° to − 12°	1·4700	α-pinene
[7] Canada Balsam Oil—				
Abies balsamea (and other species)	0·862-0·865	− 25° to − 38°	1·4730-1·4765	*l*-α-pinene
Oregon Balsam Oil—				
Pseudotsuga Douglasii	0·858-0·882	− 30° to − 48°	1·4730	*l*-α-pinene

[1] *The Pines of Australia*, 386.

[2] 80 per cent. distils at 155° to 156°.

[3] 96 per cent. distils between 154° and 165·5°.

[4] Distils chiefly at 165°.

[5] Distils chiefly at 162° to 163°.

[6] 60 per cent. distils from 157° to 161°; 20 per cent. from 161° to 164°, and 6 per cent. from 164° to 168°.

[7] Distils chiefly between 160° and 167°.

Oil of Juniper.

The plant yielding this oil is *Juniperus communis*, a native of Greece, and widely distributed over Europe. It is a small shrub or tree varying from 2 to 6 ft. in height, but in parts of Norway it forms a forest tree some 30 to 40 ft. in height. The Italian berries, however, are most valued for oil. According to Pereira, the juniper of the Bible was a leguminous plant, probably the broom or furze tree. Fraas considers the κέδρος μικρα of Dioscorides was our juniper. The plant is usually diœcious, and the scales of the cones, when ripe, become succulent, and cohere to form a kind of berry (galbulus). These berries are rich in sugar, and by their fermentation and distillation the well-known beverage, gin, is obtained, which owes its characteristic flavour to the oil of juniper. It has been stated that the oil is distilled from the full-grown, but unripe, fruit. But with reference to this point, Schimmel and Co.[1] states that "this is an error, at least no such oil is distilled on a large scale. It is an antiquated idea, originating with Zeller, that unripe fruits render a larger yield in oil than ripe ones. In any case, the oil distilled from unripe berries in all essential qualities is inferior to normal oil of ripe fruits." The following are given by Schimmel as the average yields of oil from plants grown in certain districts:—

Bavarian	1·2 per cent.
East Prussian	·6 ,,
Thuringian	·7 ,,
Swedish	·5 ,,
Italian	1 to 1·5 ,,
Polish	·9 ,,
Hungarian	·8 to 1 ,,

The oil is distilled on a very small scale in England, but, according to the *Perfumery and Essential Oil Record*,[2] hitherto it has not been possible to produce the oil competitively with southern Europe, because of the relative cheapness of labour and the vast tracts of land over which the trees grow wild. It also must be remembered that the foreign oil is produced under somewhat different conditions, and may be almost considered a by-product. There is a considerable demand on the Continent for an aqueous extract of the berries, called "Roob," or "Rob of Juniper," and the distilled oil is in this case a by-product, the berries being first crushed and macerated with water and then distilled with water, and the residue in the still evaporated to a soft consistence.

Much of the oil met with in commerce also is probably not normal in composition, but is obtained as a by-product in the manufacture of gin and similar spirits.

There can be no reason why it should not be possible for our northern dependencies to produce all that is required.

The juniper plant is a small shrub 4 to 6 ft. high, or in sheltered woods growing higher, widely distributed throughout the northern hemisphere, in Europe from the Mediterranean to Scandinavia, in Asia northwards from the Himalaya Mountains, in North America from the southern states to Greenland. It requires a certain amount of moisture, with some drainage, and apparently prefers some lime in the soil. It occurs freely on the slopes of the chalk downs near London, and on healthy, and therefore silicious, soil where a little lime occurs. On mountains in the Arctic regions a small form of the plant occurs, *viz.*,

[1] *Report*, October, **1898**, 30. [2] *P. and E.O.R.*, March, **1915**, 63.

Juniperus nana Willd., with rather longer berries and a prostrate habit. Especially where bands of limestone occur, as on some of the Scottish mountains and on the limestone hills in the Lake district, it is a common plant.

The berries take two years to come to maturity, so that there are always mature and immature berries on the same plant, and the collection by hand, unless the branches are beaten over a tray, is, owing to the sharp points of the leaves, rather a prickly matter. The blackish berries that are imported are apparently all mature, indicating that the green immature berries are separated. The immature berries, however, darken on keeping.

Whether imported oil of juniper is always distilled from *Juniperus communis* is open to question, since the oil of some red juniper berries imported from Smyrna, and attributed to *Juniperus phœnicea*, has been found to correspond in all its properties with those of *Juniperus communis*. But on examination of an oil distilled in Cyprus from the berries of *Juniperus phœnicea* at the Imperial Institute, it was shown to differ from that of *Juniperus communis* in optical rotation, which was + 5°, whilst that of *Juniperus communis* distilled from the berries is usually − 3° to − 15°.[1]

The yield of oil from imported berries varies from 0·5 to 1·5 per cent, the Italian averaging 1 to 1·5 per cent., the Bavarian 1 to 1·2 per cent., the Hungarian 0·8 to 1 per cent. The berries imported from East Prussia, Poland, and Thuringia average 0·6 to 0·9 per cent., and Swedish berries only 0·5 per cent. of oil, the berries of the more southern plants yielding the most oil. This indicates that sunny slopes are likely to be the best places to cultivate the plant for the berries. The yield of oil, however, varies considerably in different years.

The known constituents of juniper oil are the terpenes, α-pinene, and camphene, the sesquiterpene cadinene, and a small amount of oxygenated constituents. The principal of these bodies is terpinenol. A second alcohol (or possibly a mixture of alcohols) has also been separated, having the following characters: boiling-point, 218° to 226°; specific gravity, 0·9476; optical rotation, − 4° 30′; and refractive index, 1·4825. In old oils a crystalline substance is sometimes to be found, which has been described as juniper-camphor. Its constitution is unknown, but it is possibly a sesquiterpene alcohol. Traces of esters are also present.

The physical characters of juniper oil vary so enormously according to the age of the oil, that it is very difficult to suggest limit values. For reasonably fresh oil that has been kept under proper conditions the following figures will usually apply :—

Specific gravity	0·865 to 0·890
Optical rotation	− 3° ,, − 20° (rarely slightly dextro-rotatory)
Refractive index . . .	1·4750 to 1·4880
Acid value	1 to 4
Ester ,, ...	2 ,, 8
,, ,, (after acetylation)	15 ,, 25

The oil is usually soluble, when freshly distilled, in four volumes of 95 per cent. alcohol. But too much reliance must not be placed on this test, as insolubility rapidly results from age. The approximate proportions of pinene and cadinene should be determined by a fractional distillation. The results vary according to the fractionating apparatus

[1] *P. and E.O.R.*, **1913**, 291.

used, but with a series of bulbs, from 25 to 35 per cent. is obtained between 155° and 160°, and from 10 to 20 per cent. between 270° and 280°. With our present knowledge of this oil, no further analytical methods are yet available. A good juniper oil should contain sufficient cadinene for the residue after distilling off 75 per cent. to have a refractive index of 1·4950.

Umney[1] has reported on the effect of age on an English distilled juniper oil, which is no doubt identical with what takes place with other oils. The oil when distilled in 1898 had a specific gravity 0·871. In 1914, this figure had risen to 0·977, and the refractive index was 1·4960.

Juniper wood oil, as met with in commerce, is not a natural product at all. It is usually a mixture of turpentine with sufficient juniper berry oil to give it an odour of juniper oil.

The berries of *Juniperus phœnicea*, which are found in Cyprus, yield an oil having the following characters :—

Specific gravity	0·867 to 0·870
Optical rotation	+ 3° ,, + 5°
Refractive index	about 1·4710

The leaves of *Juniperus phœnicea* yield an oil which has been incorrectly offered in the market as oil of savin.

The berries of *Juniperus oxycedrus* yield about 1·5 per cent. of essential oil having the following characters :—

Specific gravity	0·850 to 0856
Optical rotation	– 4° ,, – 10°
Refractive index	1·4710 ,, 1·4735

The wood of this plant yields up to 3 per cent. of oil of specific gravity 0·925, and optical rotation – 30°.

Oil of Cypress Leaf.

Oil of cypress is distilled from the leaves of *Cupressus sempervirens* (*fastigata*), the yield being from 0·2 to 1·2 per cent. *Cupressus lusitanica* also yields a similar essential oil.

This oil is popularly supposed to be a remedy for whooping-cough, but its efficacy in this respect can scarcely be said to be soundly established. The oils have the following characters :—

French Oils.

	Cupressus sempervirens.	*Cupressus lusitanica.*
Specific gravity	0·868 to 0·884	0·872
Optical rotation	+ 12° ,, + 31°	+ 9° 10′
Refractive index	1·4710 ,, 1·4760	—
Acid value	0·7 ,, 2	1
Saponification value	3 ,, 16	9·8
,, ,, (after acetylation)	9 ,, 32	26·6

German Oils.

	Cupressus sempervirens.
Specific gravity	0·880 ,, 0·900
Optical rotation	+ 4° ,, + 18°
Refractive index	1·4740 ,, 1·4800
Acid value	1·5 ,, 3
Ester ,,	13 ,, 22
,, ,, (after acetylation)	36 ,, 51

[1] *P. and E.O.R.*, **1914**, 5.

The oil contains *d*-*α*-pinene, *d*-camphene, *d*-sylvestrene, a ketone resembling thujone, terpinyl esters, a sesquiterpene, cedrol, and alcohols which have been investigated by Schimmel & Co.[1] with the following results:—

The principal constituent of the mixture is an alcohol, terpinenol-4 of the formula $C_{10}H_{18}O$. Sabinol is possibly present in the alcohol mixture, but has not been characterised with certainty. A third alcohol is present which has the following features:—

It is esterifiable with benzoyl chloride (using pyridine) and was recovered from the steam-distillation residue by saponification. For purposes of purification it was again treated with phthalic anhydride at an increased temperature. Unfortunately, in this manipulation part of it was decomposed. The pure alcohol $C_{10}H_{18}O$ has a pleasant, rose-like odour. Its constants were as follows: boiling-point 76° to 77° (4 to 5 mm.), 210° to 212° (ordinary pressure), $d_{15°}$ 0·9422, α_D + 43° 38′ $n_{D_{20°}}$ 1·46678.

With phenyl and naphthyl *iso*cyanates only liquid derivatives were obtained. No diphenylurethane was formed. Judging by the odour, oxidation with Beckmann's mixture appears to give rise to an aldehyde.

With regard to the cypress oil fractions with the highest boiling-point, which possess in a more pronounced degree the peculiar balsamic odour, it would appear that they contain, besides cadinene and cypress-camphor, a sesquiterpene alcohol, boiling at 136° to 138° at 5 mm. pressure.

THUJA-LEAF OIL.

This oil is obtained by distillation with steam of the leaves and possibly also the small twigs of *Thuja occidentalis*, the well-known *Arborvitæ*, sometimes also called the white cedar. The yield of oil is from 0·4 to 1 per cent. It is almost colourless, but by oxidation becomes slightly greenish or yellow. Its odour is not particularly pleasant, somewhat recalling that of tansy oil. It is frequently adulterated with cedar oil, pine-leaf oils, and turpentine. The earliest scientific investigations of this oil are those of Schweizer and Jahns,[2] but that of Wallach,[3] undertaken when our knowledge of the group of compounds contained in the oil was a little more advanced, is now the standard work. The chief constituents of the oil are *α*-pinene, laevo-fenchone $C_{10}H_{16}O$, and the isomeric ketone, dextro-thujone, which also occurs in the oils of tansy, wormwood, and sage. Traces of esters, probably of acetic and formic acid, are also present. A small quantity of laevo-borneol or its esters is also present. The highest boiling constituents of the oil have not yet been thoroughly examined, but Wallach believes them to contain the last-named body in the inactive form.

Carvotanacetone was isolated from the oil by Wallach during a fractional distillation, but it is probable that this is the result of the action of heat on the thujone present, and that it is not a natural constituent of the oil.

Pure thuja-leaf oil has the following characters:—

Specific gravity	0·910 to 0·935
Optical rotation	− 5° „ − 15°
Acid value	0 „ 1
Ester „	18 „ 25
„ „ (after acetylation)	about 40

[1] *Report*, April, **1913**, 50.
[2] *Jour. prakt. Chem.*, 30 (**1843**), 376; *Arch. Pharm.*, 221 (**1883**), 749.
[3] *Annalen*, 272 (**1892**), 99.

It is soluble in 3 to 4 volumes of 70 per cent. alcohol. The greater part of the oil distils between 180° and 208°.

Rose and Livingstone[1] obtained about 1 per cent. of essential oil from the leaves and young branches of *Thuja plicata*, the so-called Washington cedar. This oil has the following characters:—

Specific gravity	0·913
Optical rotation	– 4·77°
Refractive index	1·4552
Acid value	0·5
Ester ,,	2·28
,, ,, (after acetylation)	8·8

About 85 per cent. of the oil distilled between 100° and 110° at 40 mm., and consisted of α-thujone. From 3 to 5 per cent. of *d*-α-pinene is also present, as well as about 3 per cent. of tanacetyl alcohol, probably in the form of its acetate. An oil distilled by Schimmel & Co. from the dried leaves had an optical rotation + 5·4°, but as the sample examined by Rose and Livingstone was prepared from fresh material in the country of its origin, it may be taken to possess the normal characters of the oil.

Savin Oil.

Oil of Savin is obtained from the fresh twigs of *Juniperus sabina*. It is an oil which finds but a limited employment in legitimate pharmacy as a uterine stimulant and emmenagogue. It is of no value for perfumery purposes, as its odour is somewhat unpleasant. It is occasionally used for criminal purposes, and one case is recorded of a medical man being sentenced to transportation for having used it with the intention of procuring abortion.[2] It was official in the 1885 *British Pharmacopœia*, but it has been wisely omitted from the later editions. The yield of oil from the twigs is fairly high, varying from 3 to 5 per cent. An oil is distilled in the south of France from various other species of juniperus (*Juniperus phœnicea* and *Juniperus thurifera*, for example) and is incorrectly called savin oil.

The oil contains a number of constituents, of which Wallach first characterised the sesquiterpene, cadinene. A considerable proportion of terpenes is present, which appear to consist of sabinene, terpinene, and traces of pinene. Fromm[3] separated the oil into three main portions by fractional distillation. The earliest distillate, below 195°, contains furfurol, di-acetyl, and methyl alcohol, but consists mainly of terpenes; the middle portion, 195° to 235°, consists chiefly of ethereal salts, and the last portion, 235° to 310°, consists of cadinene and some resinous bodies, which are possibly the effects of heat on the terpenes. The middle fraction yields an oil boiling at 222° to 224°, which was shown to be the acetate of an alcohol $C_{10}H_{15}OH$, which the author terms sabinol and which is present to the extent of 10 per cent. (Schimmel had previously identified an alcoholic acetate, but the formula of the alcohol was given as $C_{10}H_{17}OH$). The alcohol, obtained by hydrolysing the acetate, is a colourless oil, boiling at 210° to 213° (but see below), and with only a faint odour. In the higher boiling fraction of the esters, Fromm also detected a liquid dibasic acid boiling at 255°, of the formula $C_{20}H_{36}O_5$ and a solid acid, $C_{14}H_{16}O_8$, melting at 181°. According to

[1] *Jour. Amer. Chem. Soc.*, 34 **(1912)**, 201.
[2] *Medical Times and Gazette*, **1852**, 404.
[3] *Bericht*, 31 **(1898)**, 2025; 33 **(1900)**, 1210.

Schimmel,[1] the amount of this ester present is about 40 per cent., and there is in addition about 10 per cent. of the free alcohol in the oil. There is also present a small amount of either an aldehyde or a ketone boiling between 220° and 250°, yielding a hydrazone melting at 41° and an oxime melting at 85°.

Schimmel & Co.[2] have found citronellol in this oil, and Elze[3] has detected geraniol and dihydrocumic alcohol, and also traces of normal decyl-aldehyde.

Agnew and Croad[4] have examined this oil and have isolated the terpene sabinene from it in a pure condition. Contrary to Semmler[5] and Wallach[6] these chemists found it to be present as the laevo-rotatory variety, having a specific rotation − 42·5°; the hydrocarbon isolated by Semmler had an optical rotation + 63° and that of Wallach + 67·5°.

According to Agnew and Croad the characters of sabinol are as follows:—

Boiling-point	208° to 209°
Specific gravity	0·9391
Specific rotation	+ 17·04 (= + 16° observed for 100 mm.)

But according to Pastini and Retora,[7] sabinol has the following characters:—

Boiling-point	208°
Specific gravity	0·9518
Optical rotation	+ 7° 56′
Refractive index	1·4895′

Savin oil has the following characters:—

Specific gravity	0·907 to 0·930
Optical rotation	+ 38° to + 63° (rarely to + 68°)
Refractive index	1·4720 to 1·4800
Acid value	0 ,, 3
Ester ,,	105 ,, 140
,, ,, (after acetylation)	125 ,, 155

From 25 to 30 per cent. distils under 175°. The oil is soluble in 5 to 15 volumes of 80 per cent. alcohol, but not always to a quite clear solution.

According to Umney and Bennett,[8] the French so-called savin oil is probably distilled from the leaves and twigs of *Juniperus phœnicea* which accounts for its difference from the German and English distillates. This French oil has a specific gravity about 0·890; rotation + 4° 30′; esters 9·3 per cent.; total alcohols as sabinol 17·1.

About 64 per cent. of this oil distils below 165°, due to the presence of a large amount of pinene. Cadinene and traces of an aldehyde were also found in this oil, which has also been investigated by Rodie[9] who found that it contained over 90 per cent. of terpenes, which consisted of pinene, *l*-camphene and phellandrene. Acetic and capronic acids were identified as constituents of the esters. The oil from the leaves and twigs of *Juniperus phœnicea* has the following characters, which will indicate the effect of its addition to the true oil of savin:—

[1] *Report*, October, **1895**, 44.
[2] *Bericht*, October, **1907**, 80.
[3] *Chem. Zeit.*, 34 (**1910**), 767.
[4] *Analyst*, **1912**, 295.
[5] *Bericht*, 33 (**1900**), 1464.
[6] *Ibid.*, 40 (**1907**), 587.
[7] *Att. R. Acad. Lincei*, 1916, **25**, ii., 377.
[8] *Pharm. Jour.*, 75 (**1905**), 827.
[9] *Bull. Soc. Chim.*, iii., 35 (**1906**), 922.

Specific gravity	0·863 to 0·892
Optical rotation	+ 2° „ + 8°
Ester value	0 „ 26
„ „ (after acetylation)	5 „ 60

These figures have reference to samples of apparently authentic origin, but it must be remembered that some admixture of the leaves of allied species may have taken place before distillation. Most specimens of absolutely authentic origin have a specific gravity below 0·875, an ester value not exceeding 5, and ester value after acetylation below 20.

Cedar-leaf Oil.

The true cedar-leaf oil may be regarded as the distillate of the leaves of *Juniperus virginiana*, the red cedar, but the confusion implied in the name "cedar" is reflected in the enormous variations met with in commercial samples of so-called cedar-leaf oils. No doubt the leaves of *Thuja occidentalis* and *Chamæcyparis sphæroidea* contribute largely to the commercial oil. Brant, indeed, describes the oil as having a "penetrating, disagreeable odour," and as "therefore not suitable for perfumery purposes". Cedar leaves yield about 0·2 per cent. of essential oil, of pleasant aromatic odour, having the following characters:—

Specific gravity	0·883 to 0·900
Optical rotation	+ 55° „ + 62°
Saponification value	8 „ 14
„ „ (after acetylation)	35 „ 45

It is not soluble in even 10 volumes of 80 per cent. alcohol.

The examination of a pure sample by Schimmel[1] showed that it contains α-pinene, limonene, cadinene, borneol, and bornyl esters, probably including bornyl valerianate; other terpenes are also present. Commercial oils have specific gravities of 0·860 to 0·920 and optical rotations of − 3° to − 24°.

The true *Juniperus virginiana* leaf oil is more expensive than the corresponding oil from cedar wood—being worth from three to five times as much, in fact, and as the oils from thuja and other leaves are also less costly, it is a matter of importance that the oil should be carefully examined. Six commercial samples and one pure sample, examined by Schimmel & Co., gave the following results:—

	Specific Gravity.	Optical Rotation.	Solubility in 70 per Cent. Alcohol.
Pure oil	·887	+ 59° 25′	very insoluble.
1	·897	− 12° 25′	insoluble.
2	·886	− 3° 40′	„
3	·887	− 24° 10′	„
4	·920	− 10° 25′.	soluble in 4 parts.
5	·918	− 10° 55′	„ „ 5 parts.
6	·905	− 10°	insoluble.

Sequoia Oil.

Sequoia oil is obtained from the leaves of *Sequoia gigantea* (*Wellingtonia gigantea*, Lindl.), the Californian mammoth tree. Lunge and Steinkauler[2] obtained the oil from smaller ornamental trees grown in Zurich. The first portion of the distillate contained a crystalline hydrocarbon $C_{13}H_{10}$, melting at 105°, which these chemists termed sesquoiene. The later distillate contained a dextro-rotatory terpene boiling at 155°, and

[1] *Bericht*, April, **1898**, 14. [2] *Ibid.*, 13 (**1880**), 1656; 14 (**1881**), 2202.

having a specific gravity 0·852 and optical rotation + 24°; and an oxygenated body, $C_{10}H_{20}O_3$. This body boils at 227° to 230° and has a specific gravity 1·045 and optical rotation + 6°. It has an odour recalling that of peppermint. These bodies require further investigation, but as the oil has no commercial importance, it has not attracted much attention so far.

ATHROTAXIS-LEAF OIL.

The leaves of *Athrotaxis selaginoides*, a member of the dense scrub which abounds in the region of Williamsford, Tasmania, yield about 0·076 per cent. of essential oil, which has been examined by Baker and Smith,[1] who give the following description of it:—

The crude oil is of a light amber colour, is somewhat mobile, and has a secondary lemon-like odour not well defined. The oil is a terpene one, and consisted very largely of dextro-rotatory limonene, which had the very high specfic rotation $[\alpha]_D = + 112·2°$. The oil is somewhat insoluble in 90°/₀ alcohol, but it formed a clear solution with absolute alcohol in all proportions. Pinene was probably present, but in traces only. A small amount of an ester was found, but sufficient material was not available to enable either the alcohol or the acid to be identified.

The oil contained a fair amount of a constituent boiling above 270° C. —evidently a sesquiterpene or similar body. The reactions for cadinene were not satisfactorily obtained, although some of the results would seem to indicate the presence of that sesquiterpene.

The specific gravity of the crude oil at $\frac{16°}{15}$ C. = 0·8755; rotation $\alpha_D = + 74·8°$; the refractive index at 16° C. = 1·4905. The saponification number for the esters was 8·6, equal to 3 per cent. of ester as bornyl-acetate.

Only a small quantity of the oil could be spared for analysis, but this on redistillation gave a very small amount boiling below 174° C. Between 174° and 177° C., 47 per cent. distilled; between 177° and 200°, 23 per cent. distilled; the temperature then quickly rose to 275°, and between that and 295° C. 12 per cent. distilled.

The specific gravity of the first fraction at $\frac{18°}{15}$ C. = 0·8446; of the second, = 0·8494; of the third, 0·9373. The rotation of the first fraction, $\alpha_D = + 90·2°$; of the second, + 91·8°; of the third, + 29·6°.

On again distilling the first two fractions, 28 per cent. of the total oil came over between 174° to 175° C., and 18 per cent. between 175° to 176° C. (cor.). The specific gravity of the first fraction at 19° C. = 0·8427; and of the second, 0·8425.

The rotation of the first fraction $\alpha_D = + 91·4°$, or specific rotation $[\alpha] = + 108·5°$; of the second, + 94·5°, or specific rotation $[\alpha]_D = + 112·2°$. The refractive index of the first fraction at 20° C. = 1·4783; of the second, 1·4785.

The tetrabromide was readily prepared from both fractions, and this melted at 104° C.

From the above it is evident that the larger portion of the lower boiling constituents of the oil of this tree is dextro-rotatory limonene. Dipentene does not occur. From the slightly lower rotation and boiling-point of the first fraction, it is probable that a small amount of pinene was present, but it can only occur in traces. A trace of the sesquiterpene

[1] *The Pines of Australia*, 303.

evidently still remained with the fraction, as indicated by the refractive index, although the results, taken as a whole, are very close to those required for pure limonene.

Fig. 5.—Athrotaxis selaginoides. King William Pine. *Athrotaxis selaginoides.* Don. Tasmania.
The Pines of Australia.] [*Baker and Smith.*

A small amount of a phenolic body was isolated from the higher boiling portion of the oil (carvacrol ?).

OILS OF ARAUCARIA CUNNINGHAMII.

This Australian tree, known as the Moreton Bay Pine, is found in New South Wales and Queensland. Its leaves yield a very small quantity of essential oil (0·005 per cent.) having the following characters :—

Specific gravity at 21°	0·8974
Refractive index at 21°	1·4977
Saponification value	4·4

FIG. 6.—Araucaria Cunninghamii. Fasciation at top of a tree under cultivation at Beecroft, New South Wales. (A rare instance of teratology.)
The Pines of Australia]. [*Baker and Smith.*

A latex exudes from the tree, which has been examined by Baker and Smith,[1] and is of interest in that it yields an essential oil containing a

FIG. 7—Araucaria Cunninghamii. Cultivated tree at Beecroft, New South Wales. *The Pines of Australia.*] Fasciation at top. [*Baker and Smith.*

hydrocarbon $C_{10}H_{20}$, which is a colourless liquid having a specific gravity 0·7927 at 19°, refractive index 1·4437 and boiling-point 155°. It has an

[1] *The Pines of Australia*, 334.

odour recalling that of menthene. It is probable that a second hydrocarbon of the formula $C_{10}H_{18}$ is also present. The oil itself has the following characters :—

Specific gravity at 22°	0·8058
Refractive index	1·4670
Optical rotation	+ 3·2°

FIG. 8.—Araucaria Cunninghamii, growing on the Ranges, New South Wales. [*Baker and Smith. The Pines of Australia.*]

Oils of the Genus Callitris.

A large number of essential oils derived from various parts of this important family of Australian pines, have been examined by Baker and Smith.[1] These authorities summarise a number of interesting points which their researches on these oils have established as follows:—

The rotation of the terpenes of the oil from the leaves of some species of *Callitris* is in the opposite direction to that obtained from the fruits, even if collected from the same tree.

The acetic ester of geraniol is more pronounced in the leaf oils than is that of borneol, and it continues to increase in the several members of one section, until a maximum of over 60 per cent. is reached in the oil of *Callitris Tasmanica*.

An ester of terpineol was found in the leaf oil of *Callitris gracilis*.

The limonenes and dipentene occur in the leaf oils, the dextro-rotatory form reaching a maximum in *Callitris arenosa*, and the laevo-form in *Callitris intratropica*. In these oils is seen a well-defined illustration of the formation in nature of the two active forms of limonene in the same plant, as well as the racemic modification.

The leaf oil of *Callitris Macleayana* contains a constituent which has a marked resemblance to menthene, and is apparently a member of that group of hydrocarbons.

The leaf oil distilled from some species of *Callitris* is comparable with the best "Pine-needle oils" of commerce.

The oil obtained by steam distillation from the timber of the *Callitris* generally, contains the sesquiterpene alcohol Guaiol in some quantity; the corresponding sesquiterpene is also present.

The characteristic odour of *Callitris* timber is due to a phenol. This has distinctive colour reactions and is evidently new. It appears to be the constituent which renders *Callitris* timber objectionable to white-ants. The name Callitrol is proposed for it.

The following species have been examined:—

Callitris Robusta.

The leaves of this tree yielded 0·261 per cent. of oil having a specific gravity 0·8825, optical rotation + 10·3°, and refractive index 1·4752 at 19°. The saponification value was 49·59, equal to 17·35 per cent. ester as bornyl and geranyl acetates. In the cold, with three hours' contact, the saponification number was 22·78, equal to 7·97 per cent. ester. On redistilling, practically nothing came over below 155° C.; between 155° and 160°, 35 per cent. distilled; between 160° and 165°, 17 per cent.; between 165° and 200°, 20 per cent.; between 200° and 250°, 12 per cent. The somewhat large percentage of the oil boiling above 250° indicated the presence of a sesquiterpene or allied body, but it was not isolated.

The specific gravity of the first fraction at 15° C. = 0·8613; of the second, 0·8616; of the third, 0·8651; of the fourth, 0·907. The rotation of the first fraction a_D = + 12·2°; of the second, + 12·7°; of the third, + 14·15°. With the fourth fraction the light did not pass well, but it was more highly dextro-rotatory than the third fraction, thus indicating the presence of the dextro-rotatory bornyl acetate, common to these oils. The saponification number for the esters of the fourth fraction

[1] *A Research on the Pines of Australia.* Published by the authority of the Government of the State of New South Wales.

was 206·33, equal to 72·2 per cent. of ester. In the separated alcohols both borneol and geraniol were identified. The high percentage of ester in this fraction did not leave much room for the sesquiterpene or similar bodies.

The fruits yielded 0·363 per cent. of oil having a specific gravity 0·877 at 16°, optical rotation − 17·9° and refractive index 1·4774 at 18°. The esters present were 5·88 per cent. calculated as bornyl acetate.

Callitris Verrucosa.

The leaf and fruit oils have the characters tabulated below.

In the leaf oil both forms of limonene are present, dextro-limonene predominating. About 3 per cent. of free alcohols (borneol ?) is present in the oil.

Oil from the Leaves.

No.	Specific Gravity, ° C.	Rotation α_D.	Refractive Index, ° C.	Ester, per Cent.	Yield, per Cent.
1	0·8596 @ 23	+ 44·2	1·4809 @ 20	3·13	0·331
2	0·8591 @ 23	+ 47·5	1·4809 @ 19	3·8	0·266

Oil from the Fruits.

Specific Gravity, ° C.	Rotation α_D.	Refractive Index, ° C.	Ester, per Cent.	Yield, per Cent.
0·8608 @ 22	+ 0·3	1·4738 @ 19	1·78	0·44

Callitris Propinqua.

The essential oil from the leaves alone (1) and the leaves and fruits together (2) have the following characters:—

No.	Specific Gravity, ° C.	Rotation α_D.	Refractive Index, ° C.	Ester, per Cent. by Boiling.	Ester, per Cent., in the Cold.	Yield, per Cent.
1, Without Fruits.	0·8662 @ 19	+ 32·4	1·4752 @ 19	12·2	8·84	0·41
2, With Fruits.	0·8709 @ 20	+ 20·5	1·4749 @ 19	11·29	—	0·326

Callitris Glauca.

Dextro-pinene, dextro-limonene, and dipentene were identified in the leaf oil of this species. Free borneol is present, and probably both borneol and geraniol in the form of esters (acetate and butyrate ?).

Seven samples gave the results set out in Table I, and five of these samples yielded fractionation results as set out in Table II.

TABLE I.

No.	Specific Gravity, ° C.	Rotation α_D.	Refractive Index, ° C.	Ester, per Cent.	Yield, per Cent.
1	0·8729 @ 18	+ 27·9°	1·4747 @ 18	16·46	0·562
2	0·8665 ,, 18	+ 31·3°	1·4779 ,, 16	11·96	0·616
3	0·8631 ,, 24	+ 30·8°	1·4755 ,, 20	12·76	0·610
4	0·8659 ,, 17	+ 28·4°	1·4774 ,, 19	12·10	0·635
5	0·8657 ,, 19	+ 31·25°	1·4749 ,, 19	13·27	0·569
6	0·8665 ,, 24	+ 25·2°	1·4720 ,, 24	14·07	0·563
7	0·8782 ,, 24	+ 22·7°	1·4774 ,, 19	14·21	0·532

TABLE II.

No.	Fractions.				Specific Gravity and Rotation of Fractions.			
	1st.	2nd.	3rd.	4th.	1st.	2nd.	3rd.	4th.
1	156-160°	160-175°	175-200°	200-230°	·8562	·8571	·8689	·9415
	30%	45%	8%	12%	+ 30·4	+ 27·2	+ 21·0	+ 32·4
2	156-161°	161-165°	165-200°	200-228°	·8545	·8555	·8649	·9434
	30%	22%	37%	6%	+ 32·6	+ 32	+ 30·7	+ 33·5
3	Below 160°	160-165°	165-180°	180-225°	·8477	·8494	·8561	·9256
	27%	37%	16%	12%	+ 32·4	+ 31·6	+ 30·5	+ 34·2
4	Below 161°	161-165°	165-200°	200-225°	·8550	·8565	·8664	·9416
	27%	27%	31%	7%	+ 30·5	+ 29·3	+ 27·2	+ 32
5	Below 160°	160-165°	165-200°	200-225°	·8529	·8537	·8649	·9322
	28%	28%	32%	7%	+ 32·2	+ 31·7	+ 30·6	+ 32·5

CALLITRIS ARENOSA.

The oil from the leaves of this tree resembled ordinary pine-needle oil in odour, but had a marked suggestion of lemon. The principal constituent was limonene, dextro-limonene preponderating. Two specimens had the following characters :—

Specific Gravity, ° C.	Rotation, α_D.	Refractive Index, ° C.	Ester, per Cent.	Yield, per Cent.
0·8491 @ 23	+ 35·8°	1·4760 @ 23	5·17	0·402
0·8452 @ 26	+ 18·9°	1·4764 @ 26	3·57	0·294

CALLITRIS INTRATROPICA.

The leaf oil is practically a terpene oil, the principal constituents being pinene, dipentene, and laevo-limonene. Bornyl and geranyl esters are present in very small amount. Two samples gave the following values :—

Specific Gravity, ° C.	Rotation, α_D.	Refractive Index, ° C.	Ester, per Cent.	Yield, per Cent.
0·8481 @ 22	- 21·6	1·4768 @ 22	3·81	0·11
0·8563 @ 23	—	1·4755 @ 19	4·75	—

Callitris Gracilis.

The leaf oil of this tree contains about 12 per cent. of esters (acetic and butyric) of borneol, terpineol, and, probably, geraniol. Pinene is the principal terpene, and traces of a phenol are present. The oil has the following characters:—

Specific gravity	0·8683 at 20°
Rotation	+ 8·7°
Refractive index	1·4752 at 20°
Esters	12·1 per cent.
Yield	0·723 „

Callitris Calcarata.

Both forms of limonene are present in the leaf oil as well as dextro-pinene. The oil contains 46·8 per cent. of esters, both of geraniol and borneol, the former preponderating. The oil from the fruits is practically identical with that from the leaves, the samples examined having the following characters:—

Oils from the Leaves.

No.	Specific Gravity, ° C.	Rotation, α_D.	Refractive Index, ° C.	Ester, per Cent. by Boiling.	Ester, per Cent. in Cold.	Yield, per Cent.
1	0·8949 @ 17	+ 11·7	1·4747 @ 19	46·58	39·4	0·168
2	0·8861 @ 19	+ 14·1	1·4760 @ 19	41·33	27·08	0·162
3	0·8803 @ 23	- 4·5	1·4752 @ 19	38·6	—	0·164

Oil from the Fruits.

Specific Gravity, ° C.	Rotation, α_D.	Refractive Index, ° C.	Ester, per Cent. by Boiling.	Ester, per Cent. in Cold.	Yield, per Cent.
0·8797 @ 23	+ 2·15	1·4744 @ 23	33·37	31·18	0·229

Callitris Rhomboidea.

The esters present in the leaf oil—nearly 30 per cent.—consist almost entirely of geranyl acetate, borneol and bornyl acetate being almost entirely absent. The terpenes present appear to be laevo-limonene and pinene. The oil has the following characters :—

Specific gravity	0·8826 at 22°
Rotation	− 19·2°
Refractive index	1·4747 at 25°
Esters (boiling)	30·4 per cent.
,, (cold)	29·8 ,,
Yield	0·0335 ,,

This tree has been introduced into India, and a trial distillate of the leaves gathered in the Nilghiris yielded 0·04 per cent. of essential oil having the following characters :—

Specific gravity	0·871 at 16°
Optical rotation	− 27·6°
Refractive index	1·4695 at 18°
Ester value	51·1
Acid ,,	1·2
Saponification value (after acetylation)	59·8
Free alcohols (as geraniol)	2·1 per cent.
Total ,,	17·3 ,, ,,
Esters as geranyl acetate	17·85 ,, ,,

Callitris Tasmanica.

Pinene and limonene are the predominant terpenes in the leaf oil of this species. Traces of a phenol are present, and about 60 per cent. of geranyl acetate, which imparts a fine odour to the oil. Two samples had the following characters :—

No.	Specific Gravity, ° C.	Rotation, α_D.	Refractive Index, ° C.	Ester, per Cent. by Boiling.	Ester, per Cent. in Cold.	Yield, Per Cent.
1	0·9036 @ 22	+ 1·0	1·4738 @ 25	59·95	59·91	0·14
2	0·8976 @ 15	− 5·8	1·4739 @ 15	62·75	62·2	0·208

Callitris Drummondii.

In the leaf oil dextro-pinene, dextro-limonene, and dipentene were present. The fruit oil is practically identical with the leaf oil. They have the following characters :—

Oil from the Leaves.

Locality and Date.	Specific Gravity, ° C.	Rotation α_D.	Refractive Index, ° C.	Ester, per Cent.	Yield, per Cent.
West Australia, 26/6/03	0·8591 @ 17	+ 42·2	1·4739 @ 19	1·85	0·547

Oil from the Fruits.

Locality and Date.	Specific Gravity, ° C.	Rotation, α_D.	Refractive Index, ° C.	Ester, per Cent.	Yield, per Cent.
West Australia, 26/6/03	0·8663 @ 15	+ 45·1	1·4798 @ 19	2·4	0·3

Callitris Muelleri.

The leaf oil resembles turpentine in odour and consists almost entirely of terpenes, amongst which both forms of pinene and limonene predominate. The oil has the following characters:—

Specific gravity	0·8582 at 24°
Rotation	− 4·7°
Refractive index	1·4749 at 20°
Esters	2·76 per cent.
Yield	0·103 ,,

Callitris Oblonga.

The principal constituent of the leaf oil is pinene, but both forms of limonene are present in small amount. The esters were principally geranyl acetate, with traces of bornyl acetate, and a high boiling constituent, probably a sesquiterpene, was present in small amount. The oil had the following characters:—

Specific gravity	0·8735 at 16°
Rotation	+ 38·1°
Refractive index	1·4783 at 16°
Esters (boiling)	6·05 per cent.
,, (cold)	5·6 ,,
Yield	0·054 ,,

Callitris Macleayana.

Dextro-pinene, dextro-limonene, dipentene (?), and a hydrocarbon which is probably menthene were identified in the leaf oil. A sesquiterpene, probably cadinene, and traces of esters are present. The oil has the following characters:—

Specific gravity	0·8484 at 20°
Rotation	+ 42·5°
Refractive index	1·4791
Esters (boiling)	3·5 per cent,
,, (cold)	3·2 ,,
Yield	0·172 ,,

Oils of Dacrydium Franklini.

This tree is known as the Huon pine and is found chiefly in Tasmania.

The results of the investigation[1] of the oil from its leaves, and also of those of the oil of the timber, are interesting. The principal constituent occurring in the leaf oil is apparently a previously undetected terpene of the formula $C_{10}H_{16}$, for which the name Dacrydene is proposed. This terpene readily forms a nitrosochloride, melting sharply, and with decomposition at 120° to 121° C. (cor.), which is far away from the melting-point of any nitrosochloride formed with a previously known terpene. The boiling-point of dacrydene appears to be 165° to 166° C. (cor.); the specific gravity at 22° C. = 0·8524; the refractive index at 22° C. = 1·4749; and the rotation a_D = + 12·3°, or a specific rotation $[a]_D$ + 14·48°. It is a colourless mobile oil, with a turpentine-like odour, but slightly more aromatic and less pungent than pinene. It is very volatile, and quickly and entirely evaporated from a watch glass without leaving any residue.

As it occurs in this oil together with a small quantity of laevo-rotatory pinene and dextro-rotatory limonene, it was, of course, impossible

[1] Baker and Smith, *The Pines of Australia*, 397.

to obtain it pure by fractional distillation; but by continued redistillations 10 per cent. of the oil was obtained, boiling between 165° to 166° C., which gave the results recorded above.

The presence of the small amount of dacrydene still remaining with the pinene fraction, raised the melting-point of the nitrosochloride pre-

FIG. 9.—"Huon Pine." *Dacrydium Franklini*, Hook of Tasmania.

pared from that substance several degrees, and no melting-point less than 110° C. was obtained.

Dacrydene forms a liquid bromide, and no crystalline product was formed when the oil was saturated with dry hydrochloric acid. Scarcely any colour was produced when concentrated sulphuric acid was added to a solution of the terpene in acetic anhydride, but when treated with nitric acid a yellow nitro-compound was obtained. When dissolved in light petroleum and treated with sodium nitrite and acetic acid, no crystalline product separated, but after some hours, a thick, dark-coloured mass formed at the junction of the liquids. After two days this was washed in ether and then dissolved in ether-alcohol, from which solution on evaporation a yellow-coloured substance separated, and after drying on a porous plate an ochre-yellow powder was left. This darkened much at about 130° C. and melted with decomposition at about 150° C. It readily dissolved in nitrobenzene, but did not become blue on heating. On further investigation it may become possible, perhaps, to prepare a nitrosonitrite more definite in character with this terpene.

The higher boiling portion of the leaf oil contained the methyl ether of eugenol, and veratric acid was prepared from it by oxidation. This methyl ether is the main constituent of the oil from the timber of this tree.

The crude oil was of a very light amber colour, and the odour somewhat resembled that given by the oil from the wood, thus indicating the presence of the methyl ether of eugenol, as the constituents of the wood oil had previously been determined. The leaf oil was very mobile, and had a low specific gravity. As it was mostly a terpene oil, it was but little soluble in alcohol, and it required 1 volume of absolute alcohol to form a clear solution, but it was soluble in all proportions afterwards.

The constituents identified were *d*-limonene, *l*-pinene, dacrydene, and the methyl ether of eugenol.

An oil was also obtained, to the extent of 0·56 per cent., from the timber of this tree. It had a specific gravity 1·035 at 18°, optical rotation + 1·4°, and refractive index 1·5373 at 23°. Its chief constituent is the methyl ether of eugenol, with a small amount of a sesquiterpene, possibly cadinene.

Leaf Oil of Pherosphæra Fitzgeraldi.

This tree grows freely in New South Wales, and its leaf oil has been investigated by Baker and Smith.[1]

The specific gravity of the crude oil at $\frac{22°}{15}$ C. = 0·8705; rotation a_D = + 15·1°; refractive index at 23° C. = 1·4841. The saponification number is 2·4, equal to 0·84 per cent. of ester as bornyl or geranyl acetate.

Pinene and cadinene were detected in the oil, and possibly limonene.

Leaf Oil of Phyllocladus Rhomboidalis.

This Tasmanian tree is known locally as the Celery Top Pine. Its essential oil, which has been examined by Baker and Smith,[2] is of particular interest, as it contains the only solid crystalline diterpene so far identified. The oil contains *l*-pinene, probably mixed with *d*-pinene, the diterpene referred to, and possibly a sesquiterpene.

The diterpene was readily prepared in a perfectly pure condition, so that it was possible to determine satisfactorily its composition and physical properties. This well-crystallised body is thus one of the very few members of this class of plant substances which can be prepared from natural sources in a perfectly pure condition. The oil was obtained by steam distillation, and contained about 3 per cent. of the solid diterpene.

[1] *The Pines of Australia*, 412. [2] *Ibid.*, 419.

The oil was almost free from compounds containing oxygen, and esters, alcohols, aldehydes, and similar bodies were practically absent, only a very small amount (about 1 per cent.) of an alcohol being determined by acetylating the oil. The higher boiling liquid portion showed no tendency to resinify, so that when the semi-solid crystalline mass, which contained the diterpene, was spread upon porous plates for a few days, the whole of the liquid portions were absorbed, the diterpene remaining in a perfectly white, and even at this stage, almost pure condition. It was slightly soluble in cold alcohol, but more readily in hot alcohol, and dissolved easily in chloroform, ether, petroleum ether, and benzene. The best method for purification, after the first separation from alcohol, was to dissolve it in chloroform and precipitate by the addition of alcohol. If the chloroform was in excess, so that on the addition of alcohol no precipitate was formed, then, on slow evaporation, crystals readily separated. These crystals were microscopic needles, but were not well defined. When only a small amount of chloroform was used as solvent, then on addition of the alcohol the solid substance at once crystallised out. When this was dried it had more of a tabular structure, was pure white, of a nacreous lustre, and was practically without odour. It was dextro-rotatory, and the determination of the specific rotation was made with both benzene and chloroform, the specific rotation, $[\alpha]_D = + 16{\cdot}06°$, being identical with both solvents. Its ready solubility in benzene enabled the molecular weight to be determined by the cryoscopic method, and this, together with the results of the analyses, showed it to have the formula, $C_{20}H_{32}$. Its melting-point was 95° C. (cor.), and it did not matter what the solvent had been. The fused substance also melted again at the same temperature. This diterpene has been named phyllocladene.

The Pine-needle Oils.

It will be convenient to deal with the oils obtained from the leaves of various species of Pinus, Larix, Abies, and certain closely allied plants, under the general description of pine-needle oils. A considerable number of these have been examined, but only a few are commercial articles.

1. *Oil of Pinus sylvestris.*—The oil distilled from the leaves of this tree varies in character to some extent, according to the locality in which the tree is grown.

English distilled oil is laevo-rotatory, whilst most other varieties are dextro-rotatory. It has the following characters :—

Specific gravity	0·885 to 0·890
Optical rotation	− 7° ,, − 22°
Esters as bornyl acetate	2·5 to 4 per cent.

Umney[1] examined samples distilled in summer and winter, and found that the two specimens had the following characters :—

	Distilled in June.	Distilled in December.
Yield	0·5 per cent.	0·133 per cent.
Specific gravity 15° C.	0·885	0·889
Optical rotation (100 mm.)	− 19°	− 7·75°
Percentage of bornyl acetate	3·5	2·9

[1] *Pharm. Jour.*, 55 **(1895)**, 161, 542.

On fractionation the two oils gave the following results :—

	June Oil.	December Oil.
157° to 167°	8 per cent.	13 per cent.
167° ,, 177°	27 ,,	24 ,,
177° ,, 187°	20 ,,	9 ,,
187° ,, 197°	3 ,,	6 ,,
197° ,, 240°	7 ,,	7 ,,
240° ,, 252°	6 ,,	4 ,,
Residue	29 ,,	37 ,,

The oil contains *l*-α-pinene, dipentene, sylvestrene, and a small amount of bornyl acetate.

German distilled oil, which is obtained to the extent of 0·4 to 0·6 per cent., has the following characters :—

Specific gravity	0·865 to 0·888
Optical rotation	+ 5° ,, + 10°
Esters as bornyl acetate	1 to 4 per cent.

The oil is soluble in 7 to 10 volumes of 90 per cent. alcohol. On fractional distillation 10 per cent. is obtained between 160° and 170°, and about 45 per cent. between 170° and 185°. The ester value of the acetylated oil is about 15. (Austrian oils are sometimes slightly laevo-rotatory.)

The oil contains α-pinene, principally the dextro-rotatory variety, *d*-sylvestrene, cadinene, the acetic ester of either borneol or terpineol and probably dipentene. A small amount of free alcohol (borneol or terpineol) is also present.

Swedish oil agrees in its characters with the German distillate.

2. *Oil of Abies Sibirica.*—This oil has of late years come into prominence on account of its high content in bornyl acetate, and its extremely low price. It is known commercially as Siberian "Pine" oil, although distilled from the leaves of *Abies Sibirica* (*Abies pichta.* Forb.). It is produced in North-East Russia, principally in the Wjatka district, very little being distilled in Siberia itself on account of the high cost of transport. The oil has a most fragrant odour and is highly esteemed as a cheap perfume.

A pure oil has the following characters :—

Specific gravity	0·900 to 0·928
Optical rotation	− 30° ,, − 43°
Refractive index	1·4700 ,, 1·4730
Acid value	1 ,, 4
Esters as bornyl acetate	29 to 43 per cent.

It is soluble in 1 volume of 90 per cent. alcohol or in 15 volumes of 80 per cent. alcohol.

Umney[1] has examined an apparently authentic sample of this oil, which had an exceedingly fine odour, and which had characters falling outside the above-mentioned limits. Its physical and chemical characters were as follows :—

Specific gravity	0·932
Optical rotation	− 41·2°
Refractive index	1·4676
Esters	53·4 per cent.

This oil contains a hydrocarbon,[2] distilling below 145°, of the formula C_9H_4, to the extent of from 3 to 4 per cent., which has been termed santene. This hydrocarbon boils at 140° and has a specific gravity =

[1] *P. and E.O.R.*, **1912**, 289. [2] Aschan, *Bericht*, 40 (**1907**), 4918.

0·8698, refractive index 1·4696, and yields a nitrosochloride melting at 109° to 111°. Pinene is present in its various stereo-isomeric forms, as well as camphene and phellandrene. The chief constituent, from the odour point of view, is bornyl acetate. The sesquiterpene bisabolene is present to a small extent in the high boiling fractions of the oil. Esters of terpineol are also present to a small extent.

3. *Oil of Pinus longifolia.*—The leaves of the Indian pine tree, *Pinus longifolia*—the so-called "Chir" pine—yield an essential oil having the following characters:—

Specific gravity at 20°	0·874
Optical rotation	− 6° 15′
Acid value	1·03
Ester ,,	14·51
Saponification value	15·54
Iodine value	271·2
Distils from 160° to 165°	23·8 per cent.
,, ,, 165° ,, 170°	29·0 ,,
,, ,, 170° ,, 180°	19·8 ,,
,, ,, 180° ,, 200°	15·1 ,,
,, ,, 200° ,, 215°	6·7 ,,
Residue above 215°	5·6 ,,

These five fractions had the following physical characters:—

Fraction.	Specific Gravity at 20°	Rotation.
160° to 165°	0·860	− 12° 12′
165° ,, 170°	0·861	− 9° 15′
170° ,, 180°	0·863	− 6°
180° ,, 200°	0·868	− 3°
200° ,, 215°	0·909	− 2°

The ester value of this oil is 14·5, which corresponds to about 5 per cent. of esters calculated as bornyl acetate, which is, of course, to be expected with an oil with the range of distillation temperatures indicated above; for bornyl acetate is sought for in essential oils in the fraction boiling at 220° to 230°, which is practically non-existent in the oil in question.

4. *Oil of Pinus pumilio.*—This oil, although usually classed as a pine-needle oil, is usually distilled from the leaves mixed with small twigs and woody portions of the branches of *Pinus pumilio* (*Pinus montana* Mill.). It is principally distilled in the Austrian Alps, largely in the Tyrol, and in several other districts. The yield of oil is, as would be expected, very variable, as low as 0·25 per cent. being obtained, or as much as 0·75 per cent. The oil has the following characters:—

Specific gravity	0·863 to 0·876
Optical rotation	− 5° to − 10° (rarely to − 16°)
Refractive index	1·4740 to 1·4800
Esters as bornyl acetate	3 to 10 per cent.

It is soluble in 5 to 10 volumes of 90 per cent. alcohol.

These limits may be exceeded if the distillation material is not fresh, and oils with a specific gravity up to 0·892 have been recorded. The following fractions were obtained on distilling an authentic specimen of the oil:—[1]

155° to 165°	2 per cent.
165° ,, 180°	59 ,,
180° ,, 200°	21 ,,
above 200°	18 ,,

[1] Umney, *Pharm. Jour.*, 55 **(1895)**, 163.

The oil contains *l*-α-pinene, *l*-phellandrene, sylvestrene, cadinene, and bornyl acetate. Böcker and Hahn have recently isolated from the oil an aldehyde of the formula $C_{15}H_{26}O$,[1] an oxygenated body of the formula $C_{15}H_{24}O$, the nature of which has not yet been determined, and a body of the formula $C_8H_{14}O$, having the characteristic odour of the oil. This body is a cyclic compound, and has been named pumilone, having the general characters of a ketone. It has the following characters:—

Specific gravity	0·9314
Boiling-point	216° to 217°
Optical rotation	– 15°
Refractive index	1·4646

The principal adulterant of this oil is ordinary turpentine, which, of course, reduces the ester value, and yields a large amount of distillate below 165°.

5. *Oil of Abies excelsa.*—This oil is distilled from the leaves and twigs of *Abies excelsa* (*Abies pectinata*, De Candolle; *Abies alba* Mill.). It has the following characters:—

Specific gravity	0·865 to 0·888
Optical rotation	– 30° ,, – 65°
Refractive index	1·4730 ,, 1·4755
Esters as bornyl acetate	4 to 12 per cent.

It contains *l*-α-pinene, *l*-limonene, bornyl acetate, lauric aldehyde, probably traces of decyl aldehyde, and the hydrocarbon santene, C_9H_{14}.[2] An oil is also distilled from the young cones of this tree. It contains more terpenes than the leaf oil, and has the following characters:—

Specific gravity	0·850 to 0·890
Optical rotation	– 60° ,, – 85°
Ester value	1 ,, 6

It consists chiefly of *l*-pinene and *l*-limonene, with a small amount of bornyl acetate.

6. *Douglas fir-needle oil.*—This oil is distilled from the leaves of *Pseudotsuga taxifolia*, the so-called Douglas pine, a native of North America. The needles, twigs, and very young wood are used for the distillation, and the yield is from 0·8 to 1 per cent. The oil has been examined by Brandel and Sweet and by A. W. Schorger. The oil investigated by the first-named was distilled in Washington, whilst that examined by Schorger was distilled in California. The two oils had the following characters:—

	Brandel and Sweet.	Schorger.
Specific gravity	0·868	0·873 to 0·876
Optical rotation	– 62·5°	– 17° ,, – 22°
Refractive index	—	1·4758 ,, 14780
Acid value	0	0·65 ,, 1·1
Ester ,,	86·6	11 ,, 25
Bornyl acetate	30·3 per cent.	3·9 per cent.
Free borneol	—	4·5 to 8·5 per cent.

Whether the difference in locality alone is responsible for the difference in the oils is not clear. Six hundred grams of the oil were fractionated by Schorger.

The following compounds were identified in the oil: (1) Furfural.—This was isolated by shaking the first fraction distilled with water. (2) α-Pinene.—By repeated fractionation 151 grms. of oil, boiling at 156° to

[1] *Jour. prakt. Chem.*, ii. 83 (**1911**), 489. [2] Aschan, *Bericht*, 40 (**1907**), 4919.

160° were obtained. On refractionation the oil boiled at 156° to 157°, had a specific gravity 0·8682, and optical rotation − 11·94°. It yielded a nitrosochloride melting at 103° and a nitrolpiperidine melting at 118°. Camphene could not be detected. (3) β-Pinene.—A fraction was obtained which boiled at 164° to 166°, and had specific gravity 0·872, and optical rotation − 17·19°. This was easily oxidised by means of alkaline permanganate to nopinic acid, which melted at 126°. This fraction was therefore β-pinene. (4) Dipentene and Limonene.—Two fractions were obtained, having the following characters:—

	1.	2.
Weight	2·9 per cent.	2·6 per cent.
Boiling-point	170° to 172°	172·6° to 178·2°
Specific gravity	0·8628	0·8616
Optical rotation	− 28·12°	− 26·24°

The former fraction yielded a dihydrochloride, melting sharply at 50° after several recrystallisations from alcohol. The latter fraction yielded a tetrabromide, which melted at 117° to 119°, or after further recrystallisation at 121° to 122°. It is probable that this fraction contains both dipentene and laevo-limonene. (5) Borneol.—The oil recovered from the saponified ester fraction was distilled. The fraction 208° to 213° had a rotation of − 19·42°. It was treated with phthalic anhydride for eight hours. The resulting acid phthalic ester was converted into the sodium salt and exhausted repeatedly with ether. After saponification the oil was oxidised with chromic acid, and camphor, melting at 174°, recovered by steam distillation. This establishes the identity of borneol. (6) Acetic Acid.—This acid was found to be the principal acid constituent of the esters. (7) A green oil having a specific gravity 0·9370 was isolated, but not identified. The approximate amounts of the above compounds present are as follows: α-pinene, 25 per cent.; β-pinene, 48 per cent.; dipentene and limonene, 6 per cent.; furfural, traces; esters, 6·1 per cent.; free borneol, 6·5 per cent., green oil, 3·0 per cent.

Pseudotsuga taxifolia is also known as *P. Douglasii*. According to Henry and Flood[1] the Douglas fir growing in Oregon is rightly described under this name, whilst the Colorado or blue Douglas fir is *P. glauca*. C. T. Bennet has examined the oils distilled from material grown in England, of both plants, and reports upon them as follows:—

"Oil from Leaves of the Oregon Douglas Fir.

"1. 50 lbs. of leaves of young trees growing at Avondale, sent in August, 1919. When distilled less than 0·01 per cent. of an oil with a very aromatic odour was obtained—a quantity too small for examination.

"2. 50 lbs. of leaves of old trees growing at Buckhold, Berks, sent in November, 1919, when distilled yielded 0·11 per cent. of oil having the following characters:—

Specific gravity	0·876
Optical rotation	−7°
Refractive index	1·4835
Esters as bornyl acetate	12·4 per cent.

"The ester content is much lower than that of the Colorado Douglas fir oil, but the odour is more fragrant. The oil contains dipentene or limonene, but if pinene is present, the quantity is very small, as practically nothing distils below 175°.

[1] *Proc. Roy. Irish Acad.*, May, 1920.

"3. A few days later another 50 lbs. of leaves from the Oregon Douglas fir from Buckhold were distilled, and enough oil was then available for further investigation. On fractionating the oil, an appreciable quantity of geraniol was separated, and this appears to be the chief odorous constituent. The proportion of total alcohols by acetylation calculated as geraniol is 31·5 per cent. The presence of bornyl acetate somewhat masks the odour in the original oil. There is also a small trace of citral, but the proportion is too small for determination. Geraniol occurs in the oils of some species of *Callitris* in Australia, but has apparently not been recorded as a constituent of the oils distilled from other conifers. Geraniol is the chief constituent of Indian palmarosa oil obtained from the fragrant grass *Andropogon Schoenanthus* Linn. (*Cymbopogon Martini*), and occurs in citronella oil, otto of roses, lemon oil, etc.

"OIL FROM THE LEAVES OF THE COLORADO DOUGLAS FIR.

"50 lbs. of leaves of moderate-sized trees growing at East Liss, Hants, sent in October, 1919, yielded on distillation 0·31 per cent. of oil.

Specific gravity	0·905
Optical rotation	− 46°
Refractive index (20°)	1·4717
Esters as bornyl acetate	34·5 per cent.

"The terpenes consist principally of pinene. The odour is chiefly due to the bornyl acetate present.

"It would appear from these analyses that the strong odour of Colorado Douglas fir, which is like that of turpentine mixed with camphor, is due to the large percentage of pinene and bornyl acetate.

"In the Oregon Douglas fir, pinene is not present, and the bornyl acetate is much less in percentage. The peculiar fragrance is chiefly due to the presence of the highly odoriferous substance geraniol, slightly modified by the small amount of bornyl acetate present."

7. The leaves and twigs of *Pinus Sabiniana*, the so-called digger pine, yield from 0·078 to 0·102 per cent. of essential oil having the following characters :—

Specific gravity	0·851 to 0·857
Optical rotation	− 20° ,, 39°
Refractive index	1·4670 to 1·4671 at 15°
Acid value	1·47 to 2·05
Ester ,,	6·77 ,, 11·98
,, ,, (after acetylation)	25·86 ,, 37·16

Schorger[1] has identified the following constituents in the oil :—

n-heptane	3 per cent.
l-α-pinene	58 to 59 ,,
l-limonene	18 ,,
Bornyl esters	3·5 ,,
Borneol	6 ,,

8. The oil from the leaves and twigs of *Pinus contorta*, the lodgepole pine, has the following characters :—

Specific gravity	0·869
Optical rotation	− 17·84°
Refractive index	1·4831
Acid value	0·9
Ester ,,	6·02
,, ,, (after acetylation)	32·3

[1] *Jour. Ind. Eng. Chem.* **(1915)**, 7, 24.

The oil contains *l*-camphene, *l*-β-pinene, l-α-pinene, *l*-phellandrene, dipentene, borneol, bornyl acetate, and cadinene.

9. The oil from the leaves and twigs of *Abies magnifica* has the following characters :—

Specific gravity	0·8665
Optical rotation	− 16·7°
Refractive index	1·4861
Acid value	0·75
Ester ,,	9·93
,, ,, (after acetylation)	36·22

Traces of furfural were found in the oil, as well as *l*-β-pinene, *l*-phellandrene, bornyl esters and free borneol.

10. Schorger[1] has examined the oil distilled from the leaves and twigs, and also that from the cones of *Pinus ponderosa*, the Western yellow pine, and found them to have the following characters :—

	Leaves and Twigs.	Cones.
Yield per cent.	0·04 to 0·126	0·063
Specific gravity (15°)	0·8718 ,, 0·8849	0·8757
Optical rotation (20°)	− 15·73 ,, − 19·59	− 11·48
Refractive index (15°)	1·4789 ,, 1·4838	1·4789
Acid value	0·67 ,, 2·36	1·27
Ester ,,	3·88 ,, 8·10	7·20
,, ,, (after acetylation)	24·11 ,, 35·68	22·41

The constituents identified are as follows :—

	Leaves and Twigs.	Cones.
Furfural	—	trace
l-α-pinene	2	6
l-camphene	—	—
l-β-pinene	75	60
Dipentene	6	12 to 13
Esters as bornyl acetate	2	2·5
Free alcohols (*l*-borneol)	7	4
"Green oil"	3	3 to 4
Sesquiterpene	—	—

11. The oils from the leaves and twigs, and from the cones of *Pinus Lambertiana* have also been examined by Schorger[2] with the following results :—

	Leaves and Twigs.	Cones.
Yield per cent.	0·045 to 0·120	0·32
Specific gravity (15°)	0·8676 ,, 0·8738	0·8692
Optical rotation (20°)	− 11·07 ,, − 16·50	− 23·18
Refractive index (15°)	1·4777 ,, 1·4794	1·4771
Acid value	0·68 ,, 2·38	0·63
Ester ,,	2·22 ,, 5·91	3·75
,, ,, (after acetylation)	23·25 ,, 32·04	17·04

The percentage composition of the oils is as follows :—

	Leaves and Twigs.	Cones.
Furfural	trace	trace
l-α-pinene	21	22
l-camphene	—	21
l-β-pinene	51	39 to 40
Dipentene	12	4 ,, 5
Esters	1·5	1·5
Borneol	8	3·5
"Green oil"	1	—
Sesquiterpene	—	1

[1] *Jour. Ind. Eng. Chem.* **(1914)**, 6, 893. [2] *Ibid.*

12. The leaves and twigs of *Abies concolor* yield 0·03 to 0·27 per cent. of essential oil which has the following characters:—

Specific gravity	0·872 to 0·878
Optical rotation	− 20° „ − 28°
Refractive index	1·4760 „ 1·4778
Acid value	1 „ 1·8
Ester „	12·5 „ 27·4

The bark of this tree yields an oil having the following characters:—

Specific gravity	0·870 to 0·877
Optical rotation	− 20° „ − 21°
Refractive index	1·4809 „ 1·4833
Acid value	0·87 „ 1·22
Ester „	6·4 „ 6·9

The approximate composition of these oils is as follows:—

	Leaf and Twig Oil.	Bark Oil.
Furfural	trace	trace
l-α-pinene	12 per cent.	9 per cent.
l-camphene	8 „	—
l-β-pinene	42 „	60 per cent.
l-phellandrene	15 „	Dipentene 12 to 13 per cent.
Esters as bornyl acetate	6·5 „	2·5 „
Free borneol	9·5 „	4·5 „
"Green oil"	3 „	5 „
Loss	1 „	7 „

The remaining "pine-needle oils" may be summarised as to their characters and constituents in the table on page 61, which includes most of any importance which have been examined.

N. O. GRAMINEÆ.

(The Grass Oils.)

The principal oils of this natural order are derived from the family formerly known as *Andropogon*. Much confusion had crept into the nomenclature of these oils which has been cleared up finally by the excellent work of Dr. Otto Stapf.[1] This authority has discarded the old names and has made the following classification, which has been universally accepted:—

Cymbopogon genus, with 10 species.
Vetiveria genus, with 1 species.
Andropogon genus, with 1 species.

The following are the salient features of these grasses as classified by Stapf:—

1. *Cymbopogon Schoenanthus* Spreng (*Andropogon Schoenanthus* L.). This is known locally as Camel grass. It is a characteristic desert plant able to exist on the smallest amount of water, and is distributed over Northern Africa and Arabia, and is also found in the Persian Province of Kirman where it grows at altitudes of 6000 ft. and in Afghanistan, Beluchistan, and the Punjab.

2. *Cymbopogon Iwarancusa* Schult. This grass is believed to be the *Nardus Indica* of the Ancients. The natives call it *Terankus* (fever-healer). The plant is highly aromatic. At great altitudes it becomes small and narrow-leaved and then represents a transition form towards *Cymbopogon Schoenanthus*.

[1] *Kew Bulletin* (1906), 8, 297.

Oil.	Specific Gravity.	Rotation.	Refractive Index.	Esters as Bornyl Acetate.	Constituents.
Abies balsamea	0·890 to 0·892	− 30°	—	17 to 20°/。	*l*-pinene; bornyl acetate.
,, *canadensis*	0·907 ,, 0·913	− 18° to − 26°	—	35 ,, 40°/。	*l*-pinene; bornyl acetate, cadinene.
,, *magnifica*	0·8665	− 16·7°	1·4861	3·5°/。	*l*-β-pinene; *l*-phellandrene; furfural; borneol; bornyl acetate.
Larix decidua	0·878	+ 0° 22′	—	8°/。	Borneol; bornyl acetate.
,, *pendula*	0·881	—	—	15°/。	α-pinene; bornyl acetate.
Picea canadensis	0·922	—	—	25°/。	Limonene; dipentene, bornyl acetate.
,, *excelsa* (*P. vulgaris*)	0·875 to 0·890	− 20° to − 40°	—	6 to 10°/。	*l*-pinene; dipentene; l-phellandrene; cadinene; bornyl acetate.
,, *nigra*	0·913	− 23° 50′	—	38°/。	Bornyl acetate.
,, *rubens*	0·954	—	—	8°/。	Borneol; bornyl acetate.
Pinus cembra	0·880	+ 29°	—	—	Pinene.
,, *contorta*	0·869	− 18°	1·4831	2°/。	Furfural; *l*-pinene; *l*-camphene; *l*-phellandrene; dipentene; cadinene; bornyl acetate and methyl-chavicol(?).
,, *cubensis*	0·868	+ 9° 40′	—	—	
,, *halapensis*	0·880 to 0·896	− 20° to − 25°	—	2·5 to 3°/。	*l*-pinene; limonene; borneol; bornyl acetate and free fatty acids.
,, *heterophylla*	0·887 ,, 0·889	− 32° ,, − 36°	1·4845 to 1·4869	3·5°/。	Furfural; *l*-pinene; l-camphene; dipentene; borneol; cadinene; bornyl acetate and free fatty acids.
,, *laricio*	0·865 ,, 0·870	+ 3° ,, + 8°	—	1 to 3·5°/。	Bornyl acetate.
,, *palustris*	0·883 ,, 0·885	− 26° ,, − 31°	1·4810 to 1·4830	1·5 to 2°/。	Furfural; *l*-pinene; *l*-camphene; dipentene; cadinene; borneol and its esters.
,, *strobus*	0·901	− 39·7°	1·4827	8·6°/。	*l*-pinene; bornyl acetate.

3. *Cymbopogon Nardus* Rendle (*Andropogon Nardus* L.), citronella grass. In this grass also various botanists suspected the *Nardus Indica* of the Ancients, and for this reason Linnæus gave it the name *Andropogon Nardus*. Grimm, a physician who lived at the end of the seventeenth century in Colombo, called the grass *Arundo Indica odorata*. He stated that it grows in large quantities in the neighbourhood of Colombo (probably cultivated), and that an essential oil is distilled from it.

Citronella grass is only found cultivated; it is grown particularly in the southern part of Ceylon, and also in the Malacca peninsula and Java; in most recent times attempts have also been made to cultivate it in the West Indies. The mother-plant is very probably the wild mana grass, *Cymbopogon confertiflorus* Stapf, found in Ceylon.

It is well known that citronella grass is cultivated in two varieties: "Maha pengiri" (Maha pangiri), and "Lenabatu" (Lana batu). The first-named is also called "old citronella grass" or "Winter's grass," as it was only cultivated by Winter & Son, whilst it is chiefly grown in the Malacca peninsula and in Java. It has fairly broad leaves, larger tufts than "Lenabatu," and is characterised by a rapid growth, but it is soon exhausted and dies after ten to fifteen years. As for this reason it frequently has to be planted out afresh, it has gradually been replaced by the Lenabatu variety. It yields an oil of lower specific gravity, readily soluble, and rich in acetylisable constituents.

The variety known as "Lenabatu" ("new citronella grass"), which forms the bulk of Ceylon citronella grass, yields an oil of less value, which differs from the other by a higher specific gravity, lower geraniol-content, and contains methyl eugenol.

Morphological differences between the two varieties do not exist, as Stapf proved with two specimens which, however, were not very well preserved.

4. *Cymbopogon confertiflorus* Stapf (*Andropogon confertiflorus* Steud.). This grass is found in India in the district between the Nilghiris and Ceylon, and also in Ceylon itself. It grows to a height of 5 ft. and more. It is possibly the wild mother-plant of citronella grass, and it has also the same odour as the latter, but it only gives a small yield of oil. The Singhalese name for it is "Mana," whilst in the Nilghiris it is known as "Bambe".

5. *Cymbopogon flexuosus* Stapf (*Andropogon flexuosus* Nees ex Steud.), Malabar or Cochin grass, is distributed in the Tinnivelli district and in Travancore. During the last few years large areas have been planted there with this grass. According to Stapf, the oil, which is known in commerce as lemon-grass oil, is obtained from two wholly different plants, the oil coming from the Malabar coast being produced from *C. flexuosus* Stapf, whilst the other species of lemon-grass, designated as *C. citratus* Stapf, yields an oil which, though very similar, is not identical with the former.

6. *Cymbopogon coloratus* Stapf (*Andropogon coloratus* Nees) is found from the Tinnivelli district to the Anamalai mountains, and also in the district Karnatik (Madras Presidency), and belongs to the lemon-grasses of the Malabar district. It differs from *C. flexuosus* in that it is much smaller.

7. *Cymbopogon citratus* Stapf (*Andropogon citratus* D. C.) only occurs in the cultivated state. It is found in most tropical countries,

especially in Ceylon and the Straits Settlements, also in lower Burma and Canton, Java, Tonquin, Africa, Mexico, Brazil, the West Indies, French Guiana, Mauritius, Madagascar, New Guinea, etc.; in the Malay peninsula, and especially near Singapore, the grass is cultivated on a large scale.

The oil obtained from it differs from the Malabar lemon-grass oil by its lesser solubility in alcohol and the frequently lower citral-content, for which reason it is usually less valuable than the Malabar oil.

8. *Cymbopogon Martini* Stapf (*C. Martinianus* Schult.), Rusa grass, geranium grass. The vernacular name "rusa," which is largely used for this grass, is probably derived from the brown-red coloration of the panicles in autumn.

Rusa grass is found from the Rajmahal mountains (on the bend of the Ganges) to the Afghan frontier, and from the sub-tropical zone of the Himalayas to the twelfth degree latitude, with the exception of the desert and the steppe region of the Punjab, the external slopes of the Westghat, and apparently a large part of Northern Karnatik.

A distinction is made between two varieties of the grass: "sofia" and "motia," but it is for the present undecided whether they are botanical varieties, or only different conditions of maturity of the same species.

9. *Cymbopogon caesius* Stapf (*Andropogon caesius*, α et β Nees), Kamakshi grass.

This grass is very closely allied to *C. Martini*, but yet distinctly different; only where the two species meet, transition-forms occur which unite the two species completely. *C. caesius* grows in the greater part of Karnatik, where it is fairly common. It replaces *C. Martini* in South-Eastern India.

In the Administration Report of the Government Botanic Gardens and Parks, the Nilghiris, for 1901, p. 5, an article appears on this oil. The grass used for the distillation originated from Arni, in the North Arcot district, Madras Presidency, where it is found in large quantity. The oil-yield from the grass freshly cut at the end of December was 0·431 per cent. Another parcel, distilled at the end of April in an entirely dry state, yielded 0·711 per cent. oil. With regard to the properties of the oil no data are available.

10. *Cymbopogon polyneuros* Stapf (*Andropogon polyneuros* Steud.) is a fairly strong grass which has a great tendency to form numerous ramifications upwards, so as to form thick clusters. It is found in the South-Western parts of India, especially in the Nilghiris, and here replaces *C. Martini*. It is also found in Ceylon, preferably at greater altitudes (up to 4500 ft.), and further in the island Delft in Adams' Strait, where it is called "Delft grass" and forms good food for horses. The odour of the rubbed leaves is said to remind one of fennel or anise. The oil was distilled in 1902 in Utakamand, and obtained in a yield 0·25 per cent.; with regard to its properties nothing is known.

11. *Vetiveria zizanioides* Stapf (*Andropogon muricatus* Retz.) is vetiver-grass, Anglo-Indian "Khas khas," or "cus-cus," a name which is probably of Hindu origin and means "aromatic root".

Vetiver grass occurs both in the wild state and cultivated, and is much esteemed on account of its roots which are used for many purposes. In the wild state it is distributed all over British India and Ceylon, being chiefly found on the banks of rivers and on rich marshy soil up to an elevation of 1800 ft. Occasionally it is cultivated, for example in Rajputana and Chutia-Nagpur. In the Malay districts vetiver only occurs cultivated or accidently growing wild;—the same applies to the West Indies, Brazil, and Réunion. The best sort of grass is found in the neighbourhood of Tutikorin, which is still at the present day the most important shipping port of vetiver root.

The vetiver oil prepared from the root does not appear to be distilled in India itself or in the other countries of origin with the exception of Réunion.

Jowitt,[1] however, does not consider that the so-called Mana grass, *C. confertiflorus*, is the true mother-plant of the cultivated citronella grasses. He states that *C. Nardus*, the genuine citronella grass, also grows wild in Ceylon. *C. Nardus* Rendle, which occurs in a wild state, is identical with the so-called "old citronella grass" (Winter's grass, Maha Pengiri) and forms a species distinct from Mana grass. It is possible that the third variety, known as Lenabatu grass, has resulted from hybridisation of citronella and Mana grasses. In these circumstances Jowitt considers it advisable to designate the Maha Pengiri variety as a separate species under the name of *Cymbopogon Winterianus*. In this case the name *C. Nardus* Rendle should be applied only to the Lenabatu grass.

There are numerous intermediate plants between the regular cultivated varieties and the wild mother-plants, and the oils from these have been investigated by Pickles,[2] whose results point to the fact that by continued cultivation the oils become richer, in some cases in geraniol, in other cases in citronellal, the former approaching Lenabatu oil and the latter Maha Pengiri oil.

The principal oils belonging to this important family are the following:—

Ceylon Citronella Oil.

This oil is distilled in Ceylon, from the grasses described above, on a very large scale. The oil, which is used in cheap perfumery, especially in soap manufacture, is produced in more or less primitive stills, to the extent of between 1,500,000 and 2,000,000 lb. per annum. It is largely distilled by natives, and is almost universally adulterated.

The grass is cultivated almost entirely in the south of Ceylon. About 50,000 acres are devoted to it, and but little attention is necessary. The growth is continuous, and three harvests can be obtained annually in very favourable years, but usually two are found sufficient—one in July or August and one in December, January, or February. The yield of oil is about sixteen to twenty bottles (22 oz.) per acre for the summer crop, and five to ten for the winter crop. After a lapse of fifteen years the plants become exhausted, and new plants become necessary. There are somewhere about 600 stills at work on the island, and the enormous output of the oil indicates its importance as an article of commerce. The distilleries are usually situated in small sheds and consist of a steam boiler with a safety valve and water indicator, resting upon a solid foundation; two cylindrical stills 6 to 8 ft. high and 3 to 4 ft. broad, and a spiral condenser in a tub of cold water. A still will produce from 20 to 30 lb. of oil a day.

[1] *Annals Royal Botanic Gardens, Peradeniya*, **iv.** 1908, **4**, 185.
[2] *Bull. Imp. Inst.*, 8 **(1910)**, 144.

The trade in citronella oil is in a very unsatisfactory condition, as practically all the oil exported from Ceylon is, by consent, adulterated, the amount of pure oil sold being almost negligible. This aspect of the question will be discussed latter.

A pure distillate should have the following characters :—

Specific gravity	= 0·898 to 0·920
Optical rotation	= − 7° to − 18° (rarely to − 20°)
Refractive index	= 1·4785 to 1·4900

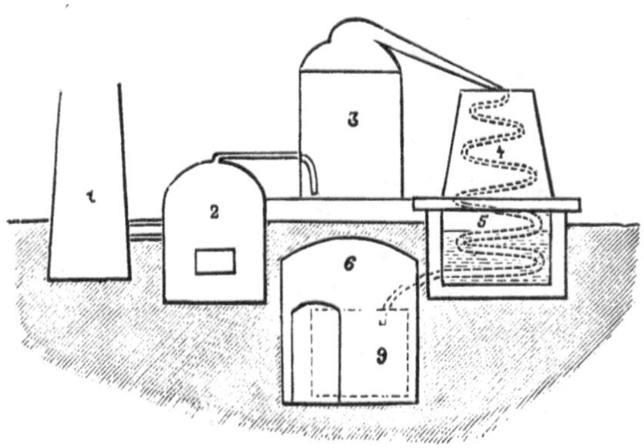

FIG. 10.—Citronella oil still. 1 chimney; 2 steam boiler; 3 still; 4 and 5 condenser; 6 underground cellar; 9 receiver.

The total acetylisable constituents, calculated as geraniol should not fall below 55 per cent., and is often above 57 per cent., rarely reaching 60 per cent. The oil should dissolve to a clear solution in 1 to 2 volumes of 80 per cent. alcohol, and on the addition of more alcohol up to 10 volumes, no turbidity—or at most a slight opalescence—should be produced. The principal adulterant of this oil is petroleum, but resin spirit has also been used, and during the past few years the lighter fractions of petroleum of the petrol type have been employed. The author and Bennett have examined the effect of adulteration on the characters of the first 10 per cent. distilled from the oil, with the results set out in the following tables :—

TABLE I.—PURE CITRONELLA OILS.

Characters of first 10 per cent. distilled at 20 to 40 mm. pressure.

	Specific Gravity at 15·5°.	Rotation in 100 mm. Tube.	Refractive Index at 19·5 C.
A	0·861	− 44°	1·4680
B	0·861	− 43°	1·4732
C	0·860	− 35°	1·4692
D	0·863	− 23°	1·4655
E[1]	0·866	− 6°	1·4590
F[2]	0·867	− 11°	1·4571

[1] E is an oil from the Straits Settlements.

[2] F is a Ceylon oil distilled by Winter, and probably prepared by fractionation.

TABLE II.—CITRONELLA OILS MIXED WITH ADULTERANT.

Characters of first 10 per cent. distilled at 20 to 40 mm. pressure.

	Specific Gravity at 15·5°	Rotation in 100 mm. Tube.	Refractive Index at 19° C.
1 E + 5 per cent. resin spirit	0·855	− 7°	1·4563
2 E + 10 ,, ,,	0·848	− 5°	1·4545
3 E + 15 ,, ,,	0·841	− 2°	1·4525
4 E + 20 ,, ,,	0·833	− 1°	1·4505
5 D + 5 ,, ,,	0·854	− 17°	1·4570
6 D + 10 ,, ,,	0·845	− 14°	1·4515
7 D + 20 ,, ,,	0·835	− 8°	1·4490
8 C + 10 ,, petroleum spirit	0·848	− 24°	1·4514

TABLE III.—ADULTERATED CITRONELLA OILS ON THE MARKET.

Characters of first 10 per cent. distilled at 20 to 40 mm. pressure.

	Specific Gravity at 15·5°	Rotation in 100 mm. Tube.	Refractive Index at 19° C.
1	0·822	− 26° 48′	1·4492
2	0·824	− 23° 50′	1·4504
3	0·833	− 18° 20′	1·4525
4	0·820	− 22°	1·4486
5	0·844	− 17° 30′	1·4540
6	0·836	− 29°	1·4495

TABLE IV.—RESIN AND PETROLEUM SPIRITS.

Characters of first 10 per cent. distilled at 20 to 40 mm. pressure:—

	Specific Gravity at 15·5°	Rotation in 100 mm. Tube.	Refractive Index at 19° C.
1 Resin spirit	0·806	+ 4°	1·4403
2 ,, ,,	0·802	+ 2° 30′	1·4425
3 Petroleum spirit[1] (turpentine substitute)	0·798	+ 1°	1·4435
4 Petroleum spirit[1] (turpentine substitute)	0·801	+ 1° 30′	1·4416

The lightest fraction which the author and Bennett obtained from the adulterated oils had a specific gravity of 0·798, an optical rotation of − 17°, and commenced to distil at 112° C. In none of the fractions has any odour of paraffin been observed, nor any fluorescence—hence the conclusion that resin spirit is the adulterant.

Suggested Standard.—From a consideration of these figures the following standards for pure citronella oil are suggested:—

1. The oil must dissolve either clear, or at most with the slightest opalescence, in 10 volumes of 80 per cent. alcohol at 20° C.

2. The first 10 per cent. distilled under reduced pressure (20 to 40 mm.) must have a specific gravity not below 0·858 and a refractive index at 20° not below 1·4570.

It may be regarded as fairly certain that no adulterated sample will pass these tests, although the limit for refractive index has been placed

[1] The actual source of these turpentine substitutes has not been ascertained, as the distillers decline to give information. They have a faint odour of paraffin, and a distinct though slight fluorescence, and probably consist only in part of petroleum spirit.

lower than the figures found for normal Ceylon oils, in order to meet the requirements of other oils, such as those from Java or the Straits Settlements, and, in general, those with low terpene-content. But any adulterated oil which might show a higher refractive index than has been indicated would fail to pass the specific gravity limit if more than traces of the adulterant were present.

For many years it was the custom to sell citronella oil "guaranteed to pass Schimmel's test". This was merely a solubility test, which required that the oil should dissolve in 10 volumes of 80 per cent. alcohol with not more than slight turbidity, without any oil globules separating. It was soon found that the oil could be adulterated down to the limits of this test, and a so-called "raised test" was suggested, but never applied to contracts commercially. This consisted in adding 5 per cent. of kerosene to the oil in question, which should in this condition pass the ordinary "Schimmel's test". To-day, however, the empirical character of these tests is fully realised, and discriminating buyers insist on valuing the oil on its acetylisable constituents calculated as geraniol. This determination is carried out in the usual manner for the determination of free alcohols, the citronellal present being converted into isopulegol and then into isopulegyl acetate, so that the result expresses the geraniol together with the citronellal. If it be desired to determine these two constituents separately, the process of Dupont and Labaune [1] may be employed.

This depends on the fact that citronellal-oxime, produced by shaking the oil in the cold with a solution of hydroxylamine, is converted, on heating with acetic anhydride, into the nitrile, which is not affected by saponification with alcoholic potash.

The difference in the molecular weight of the nitrile formed and that of citronellal is so small as to be negligible, and the calculation of the percentage of geraniol from the saponification is made by the usual formula.

The method of procedure is as follows: 10 grms. of hydroxylamine hydrochloride are dissolved in 25 c.c. of water; 10 grms. of carbonate of potash separately dissolved in 25 c.c. of water are added and the mixture filtered. With this solution 10 grms. of the oil are thoroughly shaken for two hours at 15° to 18° C. The oil is then separated, dried by means of anhydrous sodium sulphate, and acetylated with twice its volume of acetic anhydride in the usual way for 1½ hours on a sand bath under a reflux condenser. The oil is washed, dried, and neutralised, and a weighed quantity (about 2 grms.) saponified with alcoholic potash.

Working on a mixture of citronellal and geraniol, the authors obtained 55·9 per cent. geraniol as against 55 per cent. theoretically present, and when the heating with potash was prolonged for four hours the result was not appreciably altered, showing the stability of the citronellic-nitrile.

A Java oil which showed 83 per cent. of total acetylisable constituents gave 43 per cent. of geraniol and 40 per cent. citronellal, whilst a Ceylon oil containing 60·2 per cent. of total geraniol and citronellal was found to contain 43 per cent. of geraniol.

The authors are convinced, by treating considerable quantities of citronella oil, notably the Java type, that the quantity of geraniol really

[1] *Roure-Bertrand's Bulletin*, April, **1912**, 3.

contained in it is lower than that indicated by analysis. This is due to the presence of a compound, or a mixture of compounds, distilling at a higher temperature than geraniol and susceptible to acetylation, which they are further investigating.

Geraniol can also be directly estimated by the following[1] process :—

Two grms. of the oil are heated with 2 grms. of phthalic anhydride and 2 c.c. of benzene for two hours on a water-bath. The mixture is allowed to cool, and shaken for ten minutes with 60 c.c. of semi-normal aqueous caustic potash, in a stoppered flask. At the end of this time the anhydride is converted into neutral phthalate, and the acid geranyl ester into its potassium salt. The excess of alkali is titrated with semi-normal sulphuric acid. By deducting from the proportion of alkali absorbed by the quantity of phthalic anhydride used, the quantity of alkali used up in the experiment, the quantity of alkali absorbed by the geranyl ester is obtained, and hence the percentage of geraniol can be calculated. The results of this method applied to the above mixtures were as follows :—

I.	Geraniol present, 45 per cent.	Found, 47·9 per cent.
II.	,, ,, 40 ,,	,, 41·1 ,,

The slightly high results are evidently due to the fact that the citronellal contained a small proportion of alcohols which react with the phthalic anhydride.

Schimmel & Co. examined five samples of Ceylon oil by the above method, and give the following results :—

Specific Gravity.	Rotation.	Total Acetylisable Constituents.	Geraniol.	Citronellal.	
				Kleber's Method.	Dupont's Method.
0·9012	− 11° 22′	54·1 per cent.	30·7 per cent.	11·4 per cent.	7·9 per cent.
0·9016	− 11° 28′	57·2 ,,	33·2 ,,	11·0 ,,	8·0 ,,
0·9034	− 11° 40′	58·6 ,,	30·0 ,,	11·6 ,,	7·8 ,,
0·9039	− 11° 43′	56·3 ,,	29·8 ,,	10·8 ,,	6·7 ,,
0·9033	− 11° 31′	57·6 ,,	30·8 ,,	11 5 ,,	7·9 ,,

For other methods, which do not appear to be of great practical value, papers by Kleber[2] and Boulez[3] may be consulted.

Ceylon citronella oil contains about 10 to 15 per cent. of terpenes, which include camphene, *l*-limonene, and dipentene. Geraniol, borneol, citronellal, methyl-eugenol, methyl-heptenone, isoamyl alcohol, isovaleric aldehyde, valeric acid esters, *a*-terpineol, thujone (?), nerol, esters of citronellal, farnesol, a mixture of at least two sesquiterpenes (one of low and the other of high specific gravity), and a terpene of very low specific gravity have all been detected in this oil.[4]

JAVA CITRONELLA OIL.

This oil is distilled from *Cymbopogon nardus* in Java, the grass being of the Maha Pangiri type. The Singapore distilled oil is of almost exactly the same character.

[1] Schimmel's *Report*, October, **1912**, 43.
[2] *American Perfumer*, 6 (**1912**), 284.
[3] *Bull. Soc. Chim.*, iv. 11 (**1912**), 915.
[4] Schimmel's *Report*, April, **1912**, 44.

It is of a considerably higher value than the ordinary Ceylon oil, and has a quite different and much finer odour. The principal difference between it and the Ceylon oil lies in its considerably higher citronellal value, and, naturally, a much higher content of total acetylisable constituents.

De Jong has carried out a series of experiments on the oil value of Java citronella grass in various stages of development in order to decide which is the most favourable time for distilling. As the leaf ages the oil content decreases; the sheaths of the leaves and the roots contain much less oil than the leaves themselves. De Jong considers it the best plan to cut the grass as soon as four or five leaves have developed. As regards the properties of oil from different batches of old leaves, the optical rotation and total geraniol content (geraniol and citronella) show only insignificant variations. The rotation varies irregularly; values ranging from − 2° 7′ to − 7° 36′ being observed in oil from leaves of various ages. The total geraniol content first increases in an ascending scale from the newest up to the oldest leaf and afterwards decreases in a corresponding manner; the lowest value was 85·5 per cent., the highest 93·3 per cent. The results of this investigation induced de Jong to make experiments as to the time required for acetylation, the behaviour of the oil towards acetic anhydride, the amount of acetate of sodium to be added, and the time required for saponification. As a result of these inquiries de Jong considers that two hours are needed both for acetylation and for saponification, and that for every 20 c.c. of oil and acetic anhydride, 2 grms. of sodium acetate are required in order to yield accurate results. He draws these conclusions because in his experiments on citronella oils he obtained the highest values by using these precautions.

The oils examined by de Jong were soluble in three parts of 80 per cent. alcohol; when more than four parts of the solvent were used, turbidity ensued.

Ultée has carried out a series of experiments on the oil distilled from plants grown at Salatiga, which is south-east of Buitenzorg where de Jong's oils were distilled. The former place is situated at a considerably higher altitude than the latter. Ultée obtained 0·66 per cent. of oil having the following characters:—

d_{29° 0·8721, α_D − 3° 15′, total geraniol 92·75 per cent., soluble in 1·5 volume and more of 80 per cent. alcohol. The oil was distinguishable from the Buitenzorg distillates in the first place by its greater solubility, for according to de Jong the Buitenzorg oils only gave a clear solution with 3 volumes of 80 per cent. alcohol, which became cloudy when diluted to over 4 volumes. As Ultée only obtained an oil yield of 0·66 per cent., whereas in Buitenzorg the yield ranged from 0·5 to 0·9 per cent., experiments were made at Salatiga to ascertain whether the oil content of the grass could be raised by suitable fertilising. Out of four experimental fields, three were treated for this purpose with differently composed artificial manures. The grass was cut after ten weeks and equal quantities of it distilled in an exactly similar manner. The oil yields obtained ranged from 0·60 to 0·65 per cent., showing that manuring had not affected the oil content of the grass.

Java citronella oil has the following characters:—

Specific gravity	0·882 to 0·898
Optical rotation	− 2° to − 3° (rarely + 2° to − 5°)
Refractive index	1·4640 to 1·4725
Total acetylisable constituents .	80 to 94 per cent.
Geraniol	35 ,, 45 ,,
Citronellal	35 ,, 50 ,,

Schimmel & Co. have examined seven samples with especial references to their geraniol and citronellal values, and give the following results :—

Specific Gravity.	Rotation.	Total Acetylisable Constituents.	Geraniol.	Citronellal.	
				Kleber's Method.	Dupont's Method.
0·8913	- 2° 15′	85·4 per cent.	35·3 per cent.	36·0 per cent.	36·6 per cent.
0·8859	- 1° 45′	87·6 "	35·3 "	40·0 "	45·2 "
0·8866	- 1° 35′	88·7 "	36·4 "	40·0 "	46·3 "
0·8868	- 1° 22′	88·0 "	33·5 "	38·8 "	39·5 "
0·8883	- 1° 28′	87·5 "	40·1 "	38·7 "	35·4 "
0·8925	- 2° 11′	84·7 "	35·9 "	38·0 "	37·2 "
0·8881	- 1° 11′	91·0 "	37·0 "	36·8 "	40·1 "

Abnormal samples are occasionally met with, which are slightly dextro-rotatory, or whose specific gravity may rise to 0·900 to 0·905.

In the determination of acetylisable constituents in Java citronella oil it is necessary to use considerable excess of acetic anhydride. Durrans[1] prefers the following quantities and conditions :—

Ten c.c. of the oil with 20 c.c. acetic anhydride (95 to 100 per cent.) and 3 grms. anhydrous sodium acetate are boiled for two hours under a reflux condenser. The mixture is then cooled, about 50 c.c. of water added through the condenser, the flask heated on the steam-bath for fifteen minutes and then cooled, the condenser being connected to the flask during the whole of these operations. The contents of the flask are washed in a separatory funnel with cold water (or brine), then with a 1 per cent. Na_2CO_3 solution until neutral, and finally with water. The washed oil is dried with anhydrous sodium sulphate, and from 2·5 to 5 grms. saponified with 50 c.c. normal alcoholic KOH for two hours under a reflux condenser, the excess being titrated back with acid and phenolphthalein as usual.

Java citronella oil owes its odour principally to citronellal with geraniol and some citronellal. Traces of methyl-eugenol, about 1 per cent. (as against 7 or 8 per cent. in Ceylon oil), are present, and about 0·2 per cent. of citral. A sesquiterpene, isolated by Semmler and Spornitz,[2] and termed by them sesquicitronellene, is also present. This body has the following characters :—

Specific gravity at 20°	0·8489
Optical rotation	+ 0° 36′
Refractive index	1·5325
Boiling-point at 9 mm.	138° to 140°

Spornitz[3] also detected an oxide in the higher boiling fractions of the oil, which he named dicitronelloxide, a colourless liquid of the formula $C_{20}H_{34}O$, and having a specific gravity 0·9199 at 20°, and optical rotation - 4°.

Furukawa[4] finds eugenol and probably chavicol, and traces of citronellic acid in Java citronella oil.

There are a few varieties of citronella oil, distilled in various coun-

[1] *P. and E.O.R.* **(1911)**, 201; **(1912)**, 123. [2] *Bericht*, 46 **(1913)**, 4025.
[3] *Ibid.*, 47 **(1914)**, 2478. [4] *Jour. Chem. Ind. Tokyo* **(1918)**, 21, 515.

tries, which are, in most cases, obtained from practically the same grass as that grown in Java. None of them, however, are commercial articles except, to a very small extent, the Burmese oil.

Burmese oil has characters identical with those of the Java oil, except that the geraniol is usually present in rather larger amount than the citronellal. Practical tests with soap appear to show that this oil has not got so high a perfume value as the Java oil.

Citronella oil distilled in French Guiana (probably from *Cymbopogon citratus*) was obtained to the extent of 0·316 per cent. It has the following characters :—[1]

Specific gravity	0·8864
Optical rotation	− 0° 2′
Esters	5·8 per cent.
Total alcohols as geraniol	71·3 ,,

The chief portion of the "alcohols" appears to be citronellal.

Citronella oil from the Comoro Islands has been examined[2] and found to have a specific gravity 0·892, optical rotation − 0° 52′, and to yield 80 per cent. to a solution of sodium bisulphite.

The oil distilled in German New Guinea does not differ from the Java oil in any respect, and the same is true of that distilled in the Malay peninsula. Jamaica oil has been found to have the following characters :—

Specific gravity	0·895
Rotation	− 4° 16′
Refractive index	1·4710
Total alcohols, etc.	86·4 per cent.

Seychelles oil of citronella, distilled from grass derived from Ceylon, resembles the Ceylon oil in all its characters (Lenabatu oil).

The oil distilled from the wild parent grass (*C. Nardus, var. Linnaei* and *Confertiflorus*) has been examined by Pickles.[3] Those oils had the following characters :—

	Var. Linnaei.	Var. Confertiflorus.
Specific gravity	0·894 to 0·926	0·900 to 0·929
Rotation	+ 4° 54′ ,, − 6° 32′	+ 12° 12′ ,, − 2° 11′
Total "geraniol"	43·5 to 64·7 per cent.	39·1 to 64·2 per cent.

Formosan citronella oil has been examined by Furukawa,[4] who found it to have the extraordinary specific gravity 0·960, and to contain 22·15 per cent. of geraniol and 11 per cent. of citronellal. Its odour is poor. It is not clear whether the grass used is true citronella grass or not.

Lemon-grass Oil.

The lemon-grass oil of commerce is almost entirely East Indian. A certain amount was imported a few years ago from the West Indies, but it was not a successful industry and such oil is not available to-day.

Some few years ago it was regarded as essential to a pure lemon-grass oil that it should be soluble in 3 volumes of 70 per cent. alcohol. The West Indian oil was, however, found to be insoluble in 70 per cent. alcohol (although much of it appears to have been soluble when freshly distilled, but to have become insoluble during the voyage to this country). To-day most of the East Indian lemon-grass is not soluble in 3 volumes of 70 per cent. alcohol. This fact is mentioned now, as

[1] *Roure-Bertrand's Report*, April, **1910**, 62. [2] *Ibid.*, October, **1909**, 42.
[3] *Bull. Imp. Instit.*, 8 (**1910**), 144. [4] *Jour. Chem. Ind. Tokyo* (**1918**), 21, 515.

leading up to a speculative view as to the actual type of grass used in the distillation.

Stapf, in the admirable work on the Cymbopogon grasses described under Citronella Oil (p. 60), has held that the soluble lemon-grass oils are obtained from *Cymbopogon flexuosus* as the parent plant, and that the insoluble oils are derived from *Cymbopogon citratus*.

The chemists of the Imperial Institute,[1] however, hold the view that the insolubility of much of the Cochin lemon-grass oils is due to the fact that the oil is distilled by steam, which has carried over the less volatile and insoluble fractions, whilst the soluble oils are distilled over a naked fire in the old-fashioned manner.

The conclusions drawn by the Imperial Institute chemists do not seem to have any great amount of evidence to support them. The author observed, some years ago, that West Indian lemon-grass oil decreased considerably in citral value by keeping, and also became far less soluble. The freshly distilled oil from *Cymbopogon citratus* distilled in the West Indies was soluble in 3 volumes of 70 per cent. alcohol when freshly distilled. Since then he has had the opportunity of examining numerous samples of Cochin oil which had been tested before shipment, and which on arrival in Europe had lost from 3 to 5 per cent. or even more of their citral contents. In the author's opinion, there is room for the belief that the recently found insolubility of Cochin lemon-grass oil may be due to differences in the exact type of grass used, and also in the conditions of cultivation.

The oil-content of the individual parts of *Cymbopogon citratus* in their various stages of development has been the subject of researches by A. W. K. de Jong so as to ascertain which period is the most favourable one for the harvest and the distillation. According to him, the leaves contain most of the oil, of which the quantity is always largest in the leaf of the latest formation, whilst with the increasing age of the leaf, the oil-content continuously diminishes. The citral-content of the oil becomes slightly higher as the age of the leaf increases, *viz.*, from 77 to 79 per cent. in the youngest, up to 83 per cent. in the oldest leaves. The sheaths of the leaves also contain oil, but considerably less than the leaves themselves. In the roots of *Cymbopogon citratus* essential oil is, according to de Jong, also present—not in the thin, fibrous roots, but in the thick bulbs, and in the young bulbs more (about 0·5 per cent.) than in the older ones (about 0·35 per cent.). For this reason de Jong recommends that the roots should also be included in the distillation, but unfortunately he gives no information as to the properties of the root-oil. De Jong concludes from his experiments that it is inadvisable to wait longer for the cutting of the grass than until four to five leaves have been formed.

De Jong observed, that although the Java oil immediately after distillation is soluble in 2 volumes of 70 per cent. alcohol, the solubility diminishes considerably after a few days, and after a prolonged time the oil only forms a very cloudy solution from which, if left standing, part of the oil again separates. The cause of this change is, in de Jong's opinion, the polymerisation of a terpene; an oxidation process appears to him out of the question, as the oil becomes insoluble if it is kept with the air excluded. Various attempts made to prevent the change, re-

[1] *Bull. Imp. Instit.*, June, 1914.

mained unsuccessful; it was found that the change takes place rapidly when the oil is heated to 100°.

Distillation experiments have also been carried out at one of the agriculturo-chemical experimental stations in Cochin China. It was then found that the content of oil during the dry season is considerably larger than during the rainy season, and that the portion of the leaf equal to one-third its length measured from the tip, is much more aromatic than the other two-thirds.

From well-dried leaves 8 to 8·5 per cent. of oil was obtained, whilst from leaves distilled directly after harvesting, 2 per cent. was obtained in the rainy, and 5·5 per cent. in the dry season.

According to Bacon,[1] a variety of grass which, from the characteristics of the lemon-grass oil it produces, is regarded as *Cymbopogon citratus* D.C., is cultivated to a small extent in the Philippines, although it occurs everywhere throughout the Archipelago, both as a garden-plant and in the wild state, and grows in special profusion in the highlands of the province of Benguet. In the Tagal language this oil-grass is called by the name given to it in 1635 by the Spanish Jesuit, Juan Eusebius Nürnberg, who was the first to describe it, *viz.*: *tanglat* or, more accurately, *tañglad*. Other native names for the plant are *salai* and *balyoco*; its Spanish name is *Paja de Meca*. A grass five months old, distilled two days after being cut, produced 0·2 per cent. of an oil having the following characters: $d\frac{30^\circ}{4^\circ}$ 0·894, a_{D30° + 8·1°, n_{30° 1·4857, citral-content 79 per cent. The same plants, when cut again four months later, produced 0·2 per cent. oil with the following characters: $d\frac{30^\circ}{4^\circ}$ 0·8841, a_{D30° + 2·1°, n_{D30° 1·4765, citral-content 77 per cent. A grass seven months old from another plantation, distilled immediately after being cut, produced 0·21 per cent. oil, possessing the following characters: $d\frac{30^\circ}{4^\circ}$ 0·891, a_{D30° + 7·76°, n_{D30° 1·4812, and citral value 78 per cent.

Lemon-grass oil is also produced in small quantities in Burma, the

Origin.	d15°.	a_D.	Citral-content (Bisulphite Method).
Ceylon[2]	0·9058	– 0° 6′	76 °/o
India, Tyrna[2]	0·9021	– 0° 20′	72·5 °/o
,, Cochin[2]	0·9053	– 0° 39′	84·5 °/o
,, Mariani	0·9140	– 0° 5′	73 °/o
Uganda[2]	0·870 to 0·894	+ 0° 10′ to – 0° 24′	64·5 to 75 °/o
Bermuda[2]	0·8689	– 0° 21′	40 °/o
Montserrat[2]	—	—	over 74 °/o
German New Guinea[3]	0·881 to 0·892	– 0° 11′ to – 1° 15′	65 to 78 °/o
Mayotte[3]	0·893	– 0° 23′	87 °/o
Seychelles[4]	0·895 to 0·910	– 0° 15′ to – 2° 40′	76 to 81 °/o
Formosa[5]	0·865 ,, 0·914	0° to – 0° 8′	54 ,, 76 °/o
Celebes[5]	0·880	– 0° 15′	69 °/o

[1] *Philippine Journ. Sc.*, 4 **(1909)**, 111.
[2] *Bull. Imp. Instit.*, 9 **(1911)**, 334.
[3] *Schimmel's Report*, April, **1909**, 65.
[4] *Ibid.*, April, **1914**, 67, 68.
[5] *Ibid.*, April, **1913**, 72.

Seychelles Islands, Uganda, Bermuda, and various other places. There is, however, nothing particularly characteristic about these oils, the foregoing figures illustrating the general characters of a number of them.

Lemon-grass oil has the following characters if distilled in the East Indies; it may occasionally fall outside these limits, and, as will be seen from the above, may vary considerably from them when distilled elsewhere:—

Specific gravity	0·895 to 0·908
Optical rotation	+ 1° 30′ to − 5°
Refractive index	1·4825 to 1·4885
Citral (bisulphite method)	68 to 85 per cent.

The oil is somtimes soluble in 3 volumes of 70 per cent. alcohol, sometimes insoluble.

Generally speaking, the citral is determined by absorption with sodium bisulphite, but this determination will include bodies other than citral. Some chemists prefer to determine it by neutral sodium sulphite, but as this process gives lower results the method used ought always to be stated. The sulphite method gives from 3 to 6 per cent. lower results than the bisulphite method.

According to most chemists who have worked seriously on this oil, citral is the only aldehyde present in the oil, with the exception of traces of an isomeric aldehyde, and of decyl aldehyde, and citronellal. Upon this point Tiemann, Semmler, and Doebner, the three chief authorities on this oil, are completely agreed. Stiehl[1] has claimed to have separated the aldehydic constituent into three different bodies. He urges that the large amount of citral found in the oil usually is due to the isomerisation of the other aldehydes by means of the bisulphite of sodium used in separating it, which is generally too acid. If this salt be perfectly pure he claims that this isomerisation does not take place. These aldehydes he terms citral (the aldehyde usually known as such), allolemonal and citriodoraldehyde, and he gives the following as their properties:—

	Citriodoraldehyde.	Allolemonal.	Citral.
Boiling-point	228° to 229°	233° to 235°	225° to 227°
Specific gravity at 20°	·8883	·9017	·8868
Rotation	0°	− 5° 6′	0°
Melting-point of compound with naphthocinchoninic acid	204°	235°	197°

Doebner, however, has shown—in the author's opinion conclusively—that these other aldehydes are merely impure citral. By the term citral is meant a mixture of the stereoisomeric bodies, α-citral and β-citral.

Traces of an isomer of citral, having a specific gravity 0·908, have been detected, and also traces of decyl aldehyde. Methyl-heptenone is present, as well as geraniol, linalol, dipentene, and limonene.

Kafaku[2] has identified myrcene, and a new aldehyde in Formosan lemon-grass oil.

This oil is largely used for the preparation of citral for the manufacture of ionone, so that it is usually sold on its citral value.

Oil of Cymbopogon Sennaarensis.

O. D. Roberts[3] has examined at the Imperial Institute the distillate from the above grass, which is known in the British Sudan as "ma-

[1] *Jour. prakt. Chem.*, **1898**, 51. [2] *Jour. Chem. Ind. Tokyo* (**1917**), 20, 825.
[3] *Jour. Chem. Soc.*, **1915**, 1465.

hareb" grass. The material yielded 1·0 per cent. of a pale yellow oil, possessing a pleasant odour resembling that of pennyroyal herb, but more fragrant. It had a specific gravity of 0·9383, optical rotation + 34° 14′, saponification value 14, ester value after acetylation 60·2. A larger supply of material gave 1·2 per cent. of oil having a specific gravity of 0·9422, optical rotation + 29° 38′, acid value 4·2, ester value 14·5, ester value after acetylation 62·1 The results of further investigation showed that the oil has the following approximate composition:—

	Per Cent.
Terpenes, including *d*-limonene and probably pinene	18·0
Ketones, chiefly or entirely Δ'-menthenone	45·0
Sesquiterpene alcohol $C_{15}H_{26}O$	25·0
Undetermined alcohol with rose-like odour	3·0
Phenols undetermined (benzoyl derivative melting-point 70° to 72°)	0·2
Acids, probably acetic, octoic, and decoic, both in free state and as esters, with palmitic acid as esters	2·0
Residue, probably containing sesquiterpenes	11·8

Δ'-menthenone was first obtained synthetically by Wallach and Meister, who prepared it from 1 : 3 : 4 trihydroxyterpane, and its properties were fully investigated by Schimmel & Co., who isolated it from Japanese peppermint oil, which appears to be the first record of its occurrence in nature. On reduction it yields menthol.

Oil of Vetivert.

This oil is distilled from the roots of *Vetiveria zizanioides* Stapf, the plant formerly known as *Andropogon muricatus* Retz. These roots are known in Bengal under the name of khas-khas, from which the name cus-cus is easily traced. The plant is a perennial tufted grass growing to a good height, and is found all over the country near the Coromandel coast, Mysore, Bengal, Burma, and the Punjab. In contradistinction to the grasses previously described, the leaves themselves are practically odourless, whilst the roots have a strong, agreeable odour. The perfume of the roots is suggestive of myrrh. It appears to be indigenous to the East Indies; is common in the islands of the Malay Archipelago, on the Malabar coast, in Bengal, Ceylon, in the islands of the Indian Ocean, the Antilles, Brazil, Jamaica, New Caledonia, and many other localities.

The plant requires a warm, damp climate, a mean temperature of 25° C., and a firm, sandy, clayey soil. Propagation is made by means of fragments of the root nipped off and transplanted. This is generally done in the rainy season, the fragments being planted in borders along the roads or on slopes where landslips are feared, as the long roots bind the soil together.

After one year the vetivert has developed strongly, and as a hedge to plantations of sugar-cane, rice, vanilla, etc., forms a thick curtain, effectually protecting the crops against storms of wind and dust. On a commercial scale the plants are set in rows about 1½ metres apart, 65 rows to the hectare, a total of 6500 metres representing about 43,000 roots. Such a plantation from the second year will give from 30,000 to 35,000 kilos. of roughly-shaken roots, which after washing will yield about 2½ tons of cleaned product. In other words, a hectare of land gives 2½ tons of marketable vetivert roots.

On distillation, the first fractions passing over are much more volatile and lighter than those which come over with difficulty towards the end of the distillation. These "light" and "heavy" oils can be separated by using a steam-jacketed still with steam at a pressure of about 10 lb., until no more oil comes over. The heavy oil is then obtained by passing steam at a higher pressure direct into the retort. The oil distilled in Europe, however, is usually, if not always, a single distillate, and the fractionated oils are scarcely commercial articles. Réunion distilled oil is, however, of much lighter specific gravity than European distillates, and is usually found to vary from 0·980 to 0·995. The yield of oil is a matter of much disagreement. According to Piesse, 100 lb. of oil yield about 9 to 10 oz. of oil, whilst Watts states that only 2 oz. are obtained. Other observers give from ·4 to ·9 per cent. Probably about 1 per cent. is the usual average.

In order to determine the yield of cus-cus oil from the roots grown in various localities, Puran Singh examined the following seven samples from different localities in India with the following results:—

Locality.	Month of Collection.	Moisture, per Cent.	Ash (per Cent.) in Washed Roots Calculated on Dry Materials.	Oil (per Cent.) by Steam Distillation.
1. Godhra Range, Panch Mahal District, Bombay	April, 1913	8·51	2·66	0·72
2. Ajmer, Rajputana	March, 1913	10·14	3·10	1·13
3. Pilibhit, United Provinces	April, 1913	10·46	4·30	0·69
4. Ditto, and inferior sample, thin White Roots	July, 1913	9·51	4·71	0·37
5. Belgaun Forest Division, Bombay	May, 1913	10·88	3·84	0·99
6. Yeotmal Forest Division, Central Provinces	Aug., 1913	9·45	3·10	0·45
7. Delhi, from the Bazar. The best long Fibres used by Native Perfumers.	Jan. to Feb., 1914	9·81	2·97	1·14

A sample distilled in India by Puran Singh,[1] and rectified by steam distillation, had the following characters:—

Specific gravity	1·011
Optical rotation	− 30·65°
Refractive index	1·5165
Acid value	10·5
Ester "	69·6
" " (after acetylation)	132·8

The resinous matter left in the still had the following characters:—

Specific gravity at 30°	1·132
Optical rotation	+ 488°
Acid value	46·9
Ester "	64·3

Vetivert oil of commerce, however, is always dextro-rotatory, obvi-

[1] *Chemist and Druggist*, 85 **(1914)**, 225.

ously containing the high boiling constituents left behind in the above rectification.

Vetivert oil distilled in Europe has the following characters :—

Specific gravity	1·014 to 1·042
Optical rotation	+ 25° ,, + 40°
Refractive index	1·5200 ,, 1·5230
Acid value	25 ,, 65
Ester ,,	10 ,, 25
,, ,, (after acetylation)	130 ,, 160

On fractionation the oil yields results as follows :—

Boiling-point at 23 mm.	Per Cent.	Rotation.
144° to 164°	8	− 4° 10′
164° ,, 170°	10	− 3° 20′
170° ,, 180°	24	+ 2° 5′
180° ,, 185°	30	+ 31° 40′
185° ,, 200°	20	+ 47° 5′

Vetivert oil distilled in Réunion, or in the Seychelles, has the following characters :—

Specific gravity	0·982 to 1·020 (rarely, over 1·050)
Optical rotation	+ 10° to 38°
Refractive index	1·5150 ,, 1·5285
Acid value	4 ,, 20
Ester ,,	5 ,, 20
,, ,, (after acetylation)	103 ,, 150

Oil distilled in the Fiji Islands has a specific gravity 1·018 to 1·0298, acid value about 30, ester value about 17, and ester value after acetylation about 145. In all other respects it corresponds with European distilled oil.

Vetivert oil is soluble in 1 to 3 volumes of 80 per cent. alcohol, usually becoming cloudy on the addition of more alcohol.

A vetivert oil distilled from roots from the Philippines has been examined by Roure-Bertrand Fils[1] and found to have the following characters :—

Specific gravity	1·0139
Optical rotation	+ 30° 8′
Acid value	21·5
Ester ,,	6·5
Free alcohols ($C_{15}H_{24}O$)	71·5 per cent.

Seychelles distilled oil agrees generally with Réunion oil in characters. It has, however, a lower optical rotation—usually from + 10° to + 20°—and a rather lower free alcohol value, the ester value of the acetylated oil varying from 100 to 135.

Castor oil and other fatty oils are sometimes used as adulterants. These oils will raise the ester value, and lower the specific gravity and refractive index.

The chemistry of vetivert oil is in a somewhat unsettled state,

[1] *Report*, April, **1914**.

each successive investigator differing somewhat in his results from previous workers. Furfural, diacetyl, and methyl alcohol are present in the distillation waters of the oil.

Fritzsche & Co. claimed [1] to have isolated a mixture of ketones which they termed vetiverone, having the empirical formula $C_{13}H_{22}O$ and a specific gravity about 0·990. Two alcohols were also found having the following characters:—

	1.	2
Formula	$C_9H_{14}O$	$C_{11}H_{18}O$
Specific gravity	0·980	1·020
Boiling-point at 10 mm.	150° to 155°	174° to 176°

Genvresse and Langlois [2] have isolated a sesquiterpene from the oil, $C_{15}H_{24}$, which they term vetivene. It boils at 262° to 263° at 740 mm., and has a specific gravity 0·932 and optical rotation + 18° 19′. A sesquiterpene alcohol, $C_{15}H_{26}O$, was also isolated, which the discoverers term vetivenol. It boils at 169° to 170° at 15 mm., and has a specific gravity 1·011 at 20° and optical rotation + 53° 43′.

An acid, or mixture of acids of the empirical formula $C_{15}H_{24}O_4$ was also found, and the authors consider that the characteristic odour of the oil is principally due to the presence of esters of these acids with vetivenol.

Bacon [3] considers the acid or acids present in this oil to correspond with the formula $C_{15}H_{24}O_2$. He also found benzoic acid present in the oil. The most recent work on this oil is that of Semmler, Risse, and Schröter.[4] According to these chemists the formulæ given by Genvresse and Langlois are incorrect, the correct ones being, for vetivenol $C_{15}H_{24}O$ and for the acid $C_{15}H_{22}O_2$. The last-named investigators examined an oil possessing the following properties: $d_{20°}$ 1·0239; a_D + 31°; $n_{D20°}$ 1·52552. They distilled 400 grms. of this oil, at 12 mm., into the following four fractions:—

(A) Boiling-point 129° to 175°, 23 per cent.
(B) ,, ,, 170° ,, 190°, 34 ,,
(C) ,, ,, 190° ,, 250°, 8 ,,
(D) ,, ,, 250° ,, 300°, 30 ,,

Fraction (D), when subjected to further distillation, was split up into two portions, as follows:—

(D, *a*) Boiling-point 138° to 260° (13 mm.), 28 per cent.
(D, *b*) ,, ,, 260° ,, 298° (13 mm.), 62 ,,

Analysis showed that the portion (D, *b*) consisted of an ester $C_{30}H_{44}O_2$ of the acid $C_{15}H_{22}O_2$ and the alcohol $C_{15}H_{24}O$. The characters of the free alcohol are as follows: boiling-point 170° to 174° (13 mm.); $d_{20°}$ 1·0209; a_D + 34° 30′; n_D 1·52437; molecular refraction 65·94. Hence the vetivenol of this fraction can only be a tricyclic, simply unsaturated alcohol. The acid $C_{15}H_{22}O_2$ boils at 202° to 205° (13 mm.), is tricyclic and is called vetivenic acid by the authors. Its methyl ester boils at 170° to 173° (18 mm.) and has the following characters: $d_{20°}$ 1·0372; a_D + 42° 12′; n_D 1·50573; molecular refraction 71·05.

Fractions (A) and (B), when subjected to repeated fractional distillation *in vacuo* yielded a fraction boiling between 173° and 180° (13 mm.), from which vetivenol was obtainable by treatment with phthalic anhy-

[1] *D.R.P.*, 142415 (**1902**)
[2] *Comptes rendus*, 135 (**1902**), 1059.
[3] *Philipp. Jour. Sc.*, 4 (**1909**), 118.
[4] *Bericht*, 45 (**1912**), 2347.

dride. It therefore appears to be a primary alcohol. Vetivenol (tricyclic) yields an acetate, boiling-point 180° to 184° (19 mm.); $d_{20°}$ 1·0218; a_D + 28° 48′; n_D 1·50433; molecular refraction 75·91. Fraction (C), upon saponification, yielded an oil from which, under repeated distillation, there was obtained a portion boiling between 178° and 185° (19 mm.): $d_{20°}$ 1·0137; a_D + 52° 12′; n_D + 1·52822; molecular refraction 66·81. This body, therefore, represents a mixture of bi- and tricyclic vetivenol. The acid of fraction (C) was identical with vetivenic acid. The fractions (A) and (B) also contained a primary bicyclic vetivenol, which may possibly be ident cal with that from fraction (C). These portions also contained tricyclic vetivenol.

Repeated fractionation *in vacuo* of fraction (A), the last fractionation being conducted over sodium, yielded two hydrocarbons :—

I. Boiling-point 123° to 130° (16 mm.); $d_{20°}$ 0·9355; a_D + 2° 16′; n_D 1·51126; molecular refraction 65·32.

II. Boiling-point 137° to 140° (16 mm.); $d_{20°}$ 0·9321; a_D − 10° 12′; n_D 1·51896; molecular refraction 66·42.

The characteristic ester of the European distilled oil is absent from, or only present in very small amount in, Réunion distilled oil, which is very low in ester-content.

It has been suggested that vetivert oils with high specific gravities are superior in odour value to the Réunion oil. This, however, is certainly not the case when money value is taken into account, these latter being quite disproportionate from the odour values, which do not differ very materially.

Palmarosa and Ginger-grass Oils.

These oils are distilled in India from the grass *Cymbopogon Martini* (*Andropogon Schoenanthus* Linn.). There are two forms of this grass, known in India as "Motia" and "Sofia," but their botanical differences are not yet established.

The "Motia" grass yields the important commercial oil, palmarosa oil, whilst the "Sofia" grass yields the less valuable and quite different ginger-grass oil.

The common name in India for this grass is Rosha or Rusa grass and as for many years the oil found its way into Europe viâ Constantinople, it was called—and still is to some extent—Turkish geranium oil, as it is an oil rich in geraniol and having an odour something like that of geranium oil.

The plant has a wide distribution, being generally found in the drier, hot localities of India. Watt in his *Commercial Products of India* states that there are specimens in the Kew Herbarium from Kashmir, Punjab Hills, Simla, Almora, Garhwal, Singhbhum, the Central Provinces, Bombay, Rajputana, and Southern India. From a commercial point of view the more important areas in which Rosha grass is worked for the oil are the Bombay Presidency, the Central Provinces, and the Berars and certain native States in Central India.

According to Haines it is found growing from Singhbhum to the Santal Parganas in valleys near water-courses and on the northern or shady sides of hills. Blunt states that in the Berars the "Motia" variety prevails in open forests and the less valuable "Sofia" in dense forests. Pearson[1] has found "Motia" generally growing on relatively bare, hot slopes, such as the Taloda and Shahada Satpuras and "Sofia"

FIG. 11.—Distillation of palmarosa oil. [*Indian Forest Records.* R. S. Pearson, I.F.S.

on the lower ridges and in shady nullas such as are found in the Pimpalner and Nandurbar Ranges of the Khandesh District, the Yeola,

[1] *Indian Forests Records*, **V**, vii. 2.

Sinnar, and Kalwan Ranges of the Nasik District, and in the Dohad and Jhalod Ranges of the Panch Mahals.

Dunbar Brander states that "Motia" is found on bare, hot slopes

FIG. 12.—Distillation of palmarosa oil.
[*Indian Forest Records.* R. S. Pearson, I.F.S.

and "Sofia" in denser moist areas. He states that the two forms are divided by a sharp line and that there must be some factor governing the occurrence of the two varieties. Donald supports the above statement and further states that "Sofia" is often found in poorly-drained areas.

"Motia" grass is rarely, if ever, found growing gregariously over large tracts of country; it is generally found in single clumps a foot or two apart which together form open grass land, while "Sofia" on the other hand is much more gregarious in its habits and in some cases covers considerable areas as a dense crop of grass.

Pearson[1] gives the following account of the distillation of the oil:—

The local practice, which is at present the only method by which the oil is extracted from the stems, leaves, and flowers, is that of direct firing of the still, steam distillation having only been tried as an experiment.

The portion of the plant from which the oil is obtained is the flower and leaf, while only small quantities of the oil can be obtained from the stem itself. The flowers and upper third of the stem are collected in September and October, being tied into bundles or "pulas" of about $\frac{1}{4}$th of a pound each. Donald[2] in describing the process of distillation states that the inflorescences are cut twice, a second flowering occurring after the first inflorescence has been cut. The grass is either distilled in a partially green state or the flower heads and portion of stem are carefully dried in the shade before distillation, during which period they lose from about 20 to 40 per cent. of their weight according to the degree of dryness which they are allowed to attain before being treated.

The plant used by the local distillers varies little from district to district in the Central Provinces and Bombay, being in every case of a very primitive type. The still is in all cases erected within a few feet of a stream or pond, because considerable quantities of water are required for the distillation process itself and also for cooling the condenser. The still itself is of iron or copper and of cylindrical shape, having a sufficiently large opening at the top to allow of the charge being introduced; it is also often fitted with handles for transport purposes. The size of such stills varies in different localities, varying from 12 to 14 gallons.

The still is either built in and covered with mud up to the neck, or is simply supported on stones over a fire-grate by which direct heating is obtained. The opening at the top of the still is closed either with an iron or wooden disc in which a $1\frac{1}{2}$ in. to 2 in. hole is drilled in the centre, the disc being luted to the still with flour paste or mud and chaff, with the help of a rag which forms a wrapper over the joint. In the hole in the disc is fitted a bamboo 18 ins. to 2 ft. long, the joint of which is also made good with flour paste. The upper end of the bamboo is bevelled to an angle of 60° to take another section of bamboo about 8 ft. long, the joint being either pegged or made good by wrapping cloth round it, which is further secured by binding twine over the rag and the whole covered with paste or mud. In order to form the bamboo into a pipe the nodes are drilled through with the help of a hot iron, this being done before the bamboo is fixed to the still. In some cases the bamboo pipe is covered to within 6 ins. or 8 ins. of the lower end with a strip of rag or it is bound the whole of its length with twine, thus preventing the bamboo from splitting. The condenser is generally made of copper, of bulbous shape with a long neck. Donald, however,

[1] *Loc. cit.* [2] *Indian Forester*, xxxix. 3, 149.

Fig. 13.—Modern steam distillation of palmarosa oil.
[*Indian Forest Records.* R. S. Pearson, I.F.S.

states that an earthern *garrah* was in use in the distillery visited by him. It is placed in running water or in default of running water into a pond, with the lower end of the condensing tube leading to it. The capacity of the condenser varies according to the size of the still and is capable of holding anything from 8 to 15 gallons.

The still is charged with grass and cold water in proportions of one to four by weight. The fire is then put under the still and the distillate commences to come over within three-quarters to one and a half hours, according to the capacity of the still and the size of the grate.

The yield of oil may be as low as 0·15 per cent. in the case of carelessly dried material, or as high as 1 per cent. if the material is carefully dried.

Experiments on the steam distillation of the oil carried out by Messrs. Pheroze & Co. of Bombay resulted in a yield of 0·87 per cent. of oil by steam-distilling the entire plant, which consisted of 54 per cent. of stalks, 28·5 per cent. of leaves, and 17·5 per cent. of flower-heads. Each portion of the plant when distilled separately gave the following figures:—

Stalks	·04 per cent. of essential oil.
Leaves	1·32 ,, ,, ,,
Flower-heads	1·71 ,, ,, ,,

They remark that the yield might have been much more, say 1 per cent. to 1·25 per cent., had the distillation been carried out with absolutely fresh grass immediately after it was cut, instead of with grass which must have lost some of its volatile oil during transit to Bombay.

Pheroze & Co. did not report the percentage of moisture in the grass at the time of distillation. In the fresh green grass, the moisture is usually about 60 per cent., so that, assuming the moisture in the grass distilled by Pheroze & Co. to have been 50 per cent., the yield of oil calculated on completely dry grass works out to 1·74 per cent. on the entire plant and 2·94 per cent. of oil on flower-heads and leaves only, neglecting the traces of oil in the stalks.

Experiments carried out with air-dried "Sofia" grass, obtained from the Melghat Division of the Central Provinces, and containing 13 per cent. of moisture, gave 0·41 per cent. of oil by steam distillation. The entire plant consisted of 60 per cent. of flower-heads and leaves and 40 per cent. of stalks, so that, calculated on the completely dried flower-heads and leaves only, the percentage of oil obtained comes to 0·78 per cent. The yield of oil on the entire plant of "Motia" grass containing 12·88 per cent. of moisture, was 0·5 per cent. The entire plant consisted of 40 per cent. of flower-heads and leaves and 60 per cent. of stalks. Leaving out of consideration the stalks, the yield comes to 1·42 per cent. calculated on the completely dried flower-heads and leaves only. It may be noted here that the grass distilled was dry and over-ripe, while most of the flower-heads had been damaged in transit. The percentage, therefore, cannot be compared with that obtained by Pheroze & Co. The above results obtained by the distillation of dry grass have only been mentioned to show that by steam distillation not only can the green grass be worked in October and November, but that the air-dried grass containing about half the amount of oil as compared with fresh grass can be kept in stock for running the still after the crop has been reaped.

Palmarosa oil is a pale yellow liquid (sometimes rather dark on account of the crude methods of distillation used by the natives, wherein too little water is used in the stills, with the result that the grass is burned), but sometimes water-white when rectified. It has a characteristic geranium odour, and is a most popular perfume where an odour recalling roses and geranium is required at a comparatively low price.

The pure oil has the following characters :—

Specific gravity	0·886 to 0·899
Refractive index	1·4720 ,, 1·4780
Optical rotation	− 3° ,, + 5°
Acid value	0 ,, 3
Ester ,,	12 ,, 50
Total geraniol	78 to 94 per cent.

The oil is soluble in 3 volumes of 70 per cent. alcohol.

Samples distilled in Java have been examined by Schimmel & Co.[1] and found to have the following characters :—

Specific gravity	0·8906 to 0·8920
Optical rotation	+ 0° 30′ ,, + 0° 42′
Acid value	1·2 ,, 1·8
Ester ,,	37·5 ,, 51·6
,, ,, after acetylation	272·7 ,, 276·8
Total geraniol	94·3 to 96 per cent.

The oil is soluble in 3 to 3·5 volumes of 60 per cent. alcohol.

Palmarosa oil contains free geraniol as its principal constituent together with a small but varying amount of esters of the same alcohol, principally those of acetic and caproic acids. Methyl-heptenone is present in traces, as well as the terpene dipentene. Flatau and Labbé [2] claimed to have detected citronellal, but this is denied by Schimmel & Co. Elze [3] states that farnesol is present in the oil.

Ginger grass or "Sofia" grass oil differs in characters entirely from palmarosa or "Motia" grass oil. It has the following characters :—

Specific gravity	0·900 to 0·955
Optical rotation	− 30° ,, + 50°
Refractive index	1·4780 ,, 1·4950
Acid value	2 ,, 6
Ester ,,	8 to 40 (rarely to 55)
,, ,, (after acetylation)	120 to 200

The oil is usually soluble in 3 volumes of 70 per cent. alcohol, but generally becomes cloudy on the addition of more alcohol. Some samples, however, do not appear to be soluble in 3 volumes, but require considerably more 70 per cent. alcohol to effect solution.

The known constituents of ginger grass oil are the terpenes, dipentene, *d*-*a*-phellandrene and *d*-limonene, together with geraniol, and an alcohol which is identical with that obtained by the reduction of perillic aldehyde, and which is known as perillic alcohol.

Walbaum and Hüthig [4] have isolated an aldehyde from the oil, having the formula $C_{10}H_{16}O$, having the following characters :—

Boiling-point	221° to 224°
Specific gravity	0·9351
Optical rotation	± 0°
Refractive index	1·47348
Oxime melting point	115° to 116°
Semi-carbazone melting-point	169° ,, 170°
Phenylhydrazone ,,	63°

[1] *Bericht*, October, **1914**; April, **1915**, 37.
[2] *Comptes rendus*, 126 (**1898**), 1725.
[3] *Chem. Zeit.*, 34 (**1910**), 857.
[4] *Bericht*, Schimmel & Co., April, **1904**, 52; October, **1904**, 41; April, **1905**, 34.

Carvone is also present in small amount. The principal constituents of the oil, however, are geraniol and perillic alcohol, $C_{10}H_{16}O$.

Schimmel & Co.[1] give the following figures for a number of typical samples of both oils:—

No.	Place of Production.	D 15°.	α_D.	ND 20°.	Acid No.	Ester No.	Total Geraniol.	Solubility in 70 °/。Alcohol.
				Palmarosa Oils (Motia Oils).				
							Per Cent.	
1	Chitar .	0·8904	− 0° 6′	1·47225	0·8	39·8	91·5	1·5 vol.
2	Kumbi .	0·8911	− 0° 5′	1·47176	1·0	47·3	90·7	1·5 ,,
3	Mohana .	0·8906	− 0° 20′	1·47264	0·8	34·8	92·0	1·5 ,,
4	Naoli .	0·8903	− 0° 15′	1·47205	0·5	40·7	93·0	1·5 ,,
5	Ragarwal .	0·8946	− 1° 20′	1·47382	0·8	38·4	88·2	1·5 ,,
6	Udhala .	0·8906	− 0° 35′	1·47225	0·8	38·7	92·4	1·5 ,,
				Ginger-grass Oils (Sofia Oils).				
7	Astumber .	0·9396	− 19° 42′	1·48996	6·2	10·2	39·2	2·2 vols. if more cloudy.
8	Bandhera .	0·9372	− 15° 20′	1·48967	2·0	10·0	42·6	2·1 vols. if more opalescent.
9	Chanseli .	0·9392	− 38° 47′	1·49082	5·3	8·0	41·1	2·2 vols. if more cloudy.
10	Rapapur .	0·9533	− 27° 15′	1·49252	4·5	8·8	48·1	1·8 vols. if more cloudy.

Lesser-known Cymbopogon Oils.

Fiji distilled oils, probably from *Cymbopogon coloratus*,[2] have been found to have the following characters:—

Specific gravity	0·9111 to 0·920
Optical rotation	− 7° 43′ ,, − 10° 42′
Geraniol	15·6 per cent.
Citronellal	45·7 to 49·5 per cent.

Another sample, however, examined at the Imperial Institute had the following approximate composition:—

Terpenes	7·5 per cent.
Aldehydes, chiefly citral	40 ,,
Geraniol	23 ,,
Esters, chiefly geranyl acetate	10 ,,
Acetic acid	0·75 ,,
Phenols	0·75 ,,
Substances not identified	18 ,,

Four samples of the oil from *Cymbopogon polyneuros* from Ceylon were found, on examination at the Imperial Institute[3] to have the following characters:—

	1.	2.	3.	4.
Specific gravity	0·942	0·951	0·936	0·943
Rotation	—	+ 50° 39′	+ 53° 15′	+ 30° 53[1]
Total alcohols	44 per cent.	44 per cent.	38·7 per cent.	51·8 per cent.
Solubility in 80 per cent. alcohol	1 in 1	1 in 1	1 in 1	1 in 1

Cymbopogon Schoenanthus yields the so-called "camel grass" oil.

[1] *Report*, April, **1910**, 85. [2] *Bull. Imp. Instit.*, 10 (**1912**), 27. [3] *Ibid.*

According to Dymock[1] and Schimmel[2] this oil has the following characters:—

Specific gravity	0·905 at 29°
,, ,,	0·915 ,, 15°
Optical rotation	− 4° to + 34° 38′

The oil contains phellandrene.

Hofman[3] has investigated the properties of an essential oil distilled from a species of *Cymbopogon*, which, in the absence of a more precise determination, he calls *Cymbopogon javanensis*. It is a light yellow oil, with a pleasant sweet smell, resembling more the palmarosa type than the lemon-grass kind. An analysis of the oil yielded the following results:—

Specific gravity	0·9747
Optical rotation	− 2° 54′
Refractive index	1·51352
Acid number	1·25
Saponification number	30·9
Geraniol content	48·2 per cent.
Free alcohols	33·9 ,,
Esters	14·3 ,,
Methyl-isoeugenol	30·4 ,,

Among the aldehydes isolated from the oil methyl-vanillin was found, which had never previously been found in an essential oil. Citral, *l*-α-pinene, and methyl-isoeugenol are also present in the oil.

Oil of Andropogon Gryllus.

The green portions and root of *Andropogon Gryllus* (*A. Ischaemum*) yield up to 1 per cent. of essential oil, which has been examined by Schimmel & Co.[4] Two samples had the following characters:—

Specific gravity	0·968	to	0·9673
Optical rotation	− 10° 52′	,,	− 11° 4′
Refractive index	1·4995	,,	—
Acid value	1·9	,,	0·9
Ester value	8·4	,,	16·8
,, ,, (after acetylation)	—	,,	37·3

It has no value as a perfume oil. It contains about 10 per cent. of an alcoholic body calculated as $C_{10}H_{18}O$.

Oil of Elionurus Tripsacoides.

This grass is a native of Central America, extending from Mexico to the United States of Colombia and Venezuela. It is very abundant in Bolivia. It belongs to the same tribe of grasses (*Andropogoneæ*) as the ginger-grass, lemon-grass, citronella, and vetivert, but differs from the genus *Andropogon* in the spikes of flowers not bearing awns, and approaches *Rottboellia* in the slender spikes, differing from it in having hairs on the rachis of the spikelets, and on the stalks of the flowers, being, in fact, intermediate between the two genera. The odour of the root approaches that of vetivert, or cuscus, but is of a somewhat sweeter and less penetrating character.

It is interesting to find that it is not the only aromatic species of the genus, since in South Africa a species occurs which has a thyme-like odour, whence its name *Elionurus thymiodorus*. It occurs in the

[1] *Pharmacographia Indica*, vi., 564.
[2] *Bericht*, April, **1892**, 44.
[3] *Pharmb. Weekbl.*, September 6, 1919.
[4] *Bericht*, **1917**, 8; **1918**, 5.

Albany and Uitenhaage and Fort Beaufort Divisions, at an altitude of 1 to 3000 ft.

Oil from *Elionurus tripsacoides* has been examined by Umney, the yield being about 0·1 per cent. It was yellow in colour, and had an odour somewhat resembling vetivert. It gave a reddish-violet colour with ferric chloride, but only contained traces of a phenol, probably eugenol.

The characters of the oil were as follows:—

Specific gravity	0·976
Optical rotation	− 10
Ester number	10° 5′
Saponification number (after acetylation)	42

The boiling-point of the oil was very high, indicating the presence of a sesquiterpene as the chief constituent.

The following figures are the result of fractionation:—

Portion distilling below 275° C.	nil.
" " " 280° "	2 per cent.
" " " 290° "	14 "
" " " 300° "	42 "
" " " 310° "	50 "
" " " 320° "	56 "
" " " 325° "	68 "
" " above 325° "	32 "

PALMÆ.

Oil of Saw Palmetto.

The fruits of the palm, *Sabal serrulata*, which is found in the Southern United States, especially in Florida, have been stated to yield about 1·2 per cent. of essential oil having the following characters:—

Specific gravity	0·868 at 20°
Optical rotation	0°
Refractive index	1·4275

The oil is probably not a true essential oil, but the product of reaction between the free fatty acids of the fruits and the alcohol in which they had been kept. According to Shermann and Briggs [1] the oil contains free caproic, caprylic, lauric, palmitic, and oleic acids, together with about 37 per cent. of the ethyl esters of these acids.

Oil of Coco Nut.

A minute quantity of essential oil is obtained by distillation from crude fatty oil of coco nut, *Cocos nucifera*.

The oil contains [2] methyl - heptyl - carbonyl, $CH_3 . CH(OH)C_7H_{15}$, methyl - nonyl - carbinol, $CH_3 . CH(OH)C_9H_{19}$, methyl - heptyl - ketone, methyl-nonyl-ketone, and methyl-undecyl-ketone.

Oil of Elæis Guineensis.

Salway [3] has examined the essential oil of palm kernels, the fruit of *Elæis guineensis*.

The oil has the following characters:—

[1] *Pharm. Arch.*, 2 (**1899**), 101.
[2] *Comptes rendus*, 150 (**1910**), 1013; 151(**1910**), 697.
[3] *Jour. Chem. Soc.* (**1917**), 407.

Specific gravity at $\frac{20^\circ}{20^\circ}$	0·842
Acid value	30
Saponification value	15
Iodine value	93·2

Apparently portions of the fatty oil were present, having been mechanically carried over during the distillation. The only constituent of the volatile oil identified with certainty was methyl-nonyl ketone.

N. O. AROIDEÆ.

Calamus Oil.

This oil is distilled from the rhizome of the sweet flag, *Acorus calamus*. This plant is a member of a family with aquatic tendencies, and is probably a native of Asia, although now found largely in Europe. Although the essential oil obtained from the rhizome is known either as *Ol. Acori calami* or *Ol. Calami aromatici*, the plant must not, as Pereira points out, be confused with the *Calamus aromaticus* described by Royle, which is one of the *Andropogon* family. The banks of European rivers afford a plentiful supply for the markets, and it is also cultivated in damp swamps in India. The fragrance of the plant and its oil appears to increase as the climate in which it is cultivated is warmer. Commercial rhizome of European origin is obtainable in the shops, and occurs as somewhat flattened pieces of about 4 ins. long and about ¾ of an inch thick. Keeping does not improve it, nor does the process of decorticating it, as is often done for appearance sake. Many of the essential oil glands occur in the outer layers of the rhizome, so that the process of peeling naturally lessens the odour value. Dymock thus describes the Indian rhizome in his *Notes on Indian Drugs*: "The root stock occurs in somewhat tortuous, sub-cylindrical or flattened pieces, a few inches long and from ½ to 1 in. in greatest diameter. Each piece is obscurely marked on the upper surface with the scars, often hairy, of leaves, and on the under with a zig-zag line of little, elevated dot-like rings, the scars of roots. The root stock is usually rough and shrunken, varying in colour from dark brown to orange-brown, breaking easily with a short, corky fracture, and exhibiting a pale brown spongy interior. The odour is aromatic and agreeable, the taste bitterish and pungent." The powdered rhizome is used for sachet and toilet powder. Dried roots yield from 1·5 to 3·5 per cent. of essential oil, whilst the fresh root yields under 1 per cent. Japanese root which is possibly that of *Acorus spurius* yields as much as 5 per cent. The oils do not quite agree in their chemical properties, and that from the green rhizome has a finer odour than that from the dried. The chief difference noticeable is that the oil from green root is more soluble in weak alcohol than that from the dried.

According to Russell[1] all parts of the plant contain oil, the roots having the largest percentage. The oils obtained by steam distillation from the aerial portions, rhizome, and roots possess varying physical and chemical constants. The results of fractionation of these oils indicate that the components of each are present in varying amounts, also that these components vary to some extent. The plants experimented

[1] *Jour. Amer. Chem. Soc.*, 37 (**1915**), 2387.

with were cultivated at Madison, Wis., U.S.A. The oil distilled from the fresh tops of this material was straw-yellow, with a pleasant camphor-like odour and an aromatic bitter taste: specific gravity 0·9509 at 23° C.; a_D + 12·2°; $n_{D_{23°}}$ 1·5035; acid value nil; ester value 12·6; acetyl value 53·05. The oil from the rhizomes was reddish-brown; the odour aromatic and somewhat camphoraceous, the taste acrid and bitter: specific gravity 0·9547 at 23° C.; a_D + 21·7°; $n_{D_{23°}}$ 1·4990; acid value 0; ester value 15·3; acetyl value 38·4. The root oil was golden-yellow and had a pleasant camphoraceous odour; specific gravity 0·9491 at 23° C.; a_D + 18·7°; $n_{D_{23°}}$ 1·5063, and value 0; ester value 23·7; acetyl value 42. The rhizome oil was soluble in all proportions of ethyl alcohol 90 and 70 per cent. The herb oil was soluble 1:5 in ethyl alcohol with turbidity; and 1:8 in ethyl alcohol 70 per cent. The root oil was soluble 1:6 in ethyl alcohol 90 per cent. and 1:40 in ethyl alcohol 70 per cent. None of the oils contained phenols: all indicated the presence of one or more aldehydic constituents.

In the oil from the fresh tops butyric and œnanthylic acids in the form of esters were found.

A genuine calamus oil has the following characters:—

Specific gravity	0·958 to 0·970 (rarely 0·950)
Optical rotation	+ 9° to + 35°
Refractive index	1·5000 ,, 1·5080
Acid value	0 ,, 3
Ester ,,	5 ,, 20
,, ,, (after acetylation)	30 ,, 55

The oil is easily soluble in 90 per cent. alcohol.

Kurbatow [1] was the first chemist to examine this oil. He found about 5 per cent. of a terpene boiling at 158° to 159°, which was probably pinene, whose presence has since been confirmed by Semmler and Spornitz.

Thoms and Beckstroem [2] isolated *n*-heptylic acid, palmitic acid, an unrecognised unsaturated acid, eugenol, and asarylic aldehyde. There is also present a body of the formula $C_{15}H_{26}O_2$ melting at 128°, which was termed calamus camphor by Schimmel & Co., but has been named calameone by Thoms and Beckstroem. There are present two hydrocarbons of the formula $C_{15}H_{22}$ having the following characters:—

	1.	2.
Boiling-point	146° at 19 mm.	151° at 22 mm.
Specific gravity	0·933 at 18°	0·9336 at 12°
,, rotation	+ 34·83°	− 13·38°

A sesquiterpene of specific gravity 0·931 and boiling at 255° to 258° is present in the oil in small amount.

F. W. Semmler and K. E. Spornitz [3] have recently isolated several new constituents from this oil. The low-boiling portions (boiling-point 45° to 50°, 9 mm.; a_D + 59°) contained α-pinene, which was identified by converting it into the nitrosochloride (melting-point 108°). In addition to pinene, these chemists succeeded in identifying camphene by conversion into *iso*borneol (melting-point 212°). Another fraction contained camphor (melting-point of the oxime 115°). A fraction with boiling-point 130° to 135° (12 mm.), after being boiled over sodium, afforded a sesquiterpene $C_{15}H_{24}$, calamene, which possessed the following

[1] *Bericht*, 6 (**1873**), 1210. [2] *Ibid.*, 34 (**1901**), 1021.
[3] *Ibid.*, 46 (**1913**), 3700; and Schimmel's *Report*.

properties: boiling-point 123° to 126° (10·5 mm.); $d^{20°}_{19°}$ 0·9224; a_D + 5°; n_D 1·50572. By treatment with hydrogen in the presence of platinum black two double bonds were saturated, giving rise to tetrahydrocalamene, $C_{15}H_{28}$, boiling-point 123° to 125° (10 mm.); $d^{20°}_{19°}$ 0·8951; a_D ± 0°; n_D 1·48480.

They also obtained a fraction (boiling-point 150° to 160° at 13 mm.; $d_{28°}$ 0·9611; n_D 1·5098) of which the analysis pointed to the presence of a substance $C_{15}H_{24}O$. This body appears to be a tertiary alcohol which readily separates water. Water was invariably eliminated from it, when it was boiled with acetic anhydride, or treated with phthalic anhydride or with zinc dust in a sealed tube. When boiled for a short time with concentrated formic acid the body $C_{15}H_{24}O$ affords a hydrocarbon, calamenene, $C_{15}H_{22}$, boiling-point 136° to 143° (15 mm.); $d^{20°}_{19°}$ 0·9324; a_D + 6°; n_D 1·52317.

As a result of these investigations Semmler and Spornitz regard it as highly improbable that the hydrocarbon $C_{15}H_{22}$ described by Thoms and Beckstroem [1] is a natural constituent of calamus oil; they consider it to have been produced by the splitting-off of water from the body $C_{15}H_{24}O$.

A sesquiterpene alcohol, $C_{15}H_{24}O$, of specific gravity 0·9688 at 20°, optical rotation + 8°, and refractive index 1·5126, was also isolated, but has not yet been named.

Japanese calamus oil is distilled from the rhizome of the same plant according to Asahina,[2] although Holmes [3] considers the plant to be *Acorus spurius*.

At all events the Japanese oil differs from that distilled in Europe. It has the following characters:—

Specific gravity	0·970 to 0·995
Optical rotation	− 12° ,, + 25°
Refractive index	1·5095 ,, 1·5175
Acid value	0 ,, 2
Ester ,,	1 ,, 10
,, ,, (after acetylation)	16 ,, 28

It is soluble in an equal volume of 90 per cent. alcohol.

The Japanese oil appears to be free from terpenes, but it contains methyl-eugenol.

A calamus oil distilled by Dr. Carthaus in Java has been examined at the Botanical Institute in Buitenzorg. It possessed the following constants: $d_{26°}$ 1·06; a_D + 0° 52′; saponification number 9.

Schimmel & Co.[4] have examined two calamus oils from the same source which behaved very similarly. They had a yellow colour and a faint calamus odour. The constants of the one were: $d_{15°}$ 1·0783; a_D + 0° 53′; $n_{D_{20°}}$ 1·55043; ester number 12; of the other: $d_{15°}$ 1·0771; a_D + 0° 51′; $n_{D_{20°}}$ 1·55065. The oils here referred to differ from ordinary calamus oil in their greater specific gravity, their much lower rotation, and their higher index of refraction. In normal calamus oil these values range within the following limits: $d_{15°}$ 0·96 to 0·97; a_D + 9° to + 35°; $n_{D_{20°}}$ about 1·506. Moreover, the Javanese oils are soluble (with slight separation of paraffin) in 1 to 1·5 volumes of 70 per cent. alcohol, while ordinary calamus oil is only soluble in 90 per cent. alcohol (in almost every proportion).

[1] *Bericht*, 46 (**1913**), 3700.
[2] *Apotheker Zeit.*, 21 (**1906**), 987.
[3] *Ph. Jour.*, iii. 10 (**1879**), 102.
[4] *Report*, April, **1909**, 21.

N. O. LILIACEÆ.

Oil of Garlic.

This oil is distilled from the entire fresh plant *Allium sativum*. The average yield of oil, according to Schimmel, is about 1 oz. from 1 cwt. of the plants, although Wertheim obtained 3 to 4 oz. from this quantity. The oil is a liquid possessing the characteristic unpleasant odour of garlic, optically inactive, of specific gravity 1·045 to 1·060. It was examined by Wertheim,[1] who stated that it contained allyl sulphide and allyl oxide. The most modern and most trustworthy examination of the oil, however, is due to Semmler.[2] He states that the garlic yielded 0·009 per cent. of oil of specific gravity 1·0525 at 14·5°. A small quantity of crystals was deposited when the oil was cooled in a freezing mixture. The oil contains no oxygen, and therefore no allyl oxide. When fractionated under a pressure of 16 mm., four main fractions were obtained. Fraction 1 (6 per cent.) consisted of allyl-propyl disulphide, $C_6H_{12}S_2$, a bright yellow oil of specific gravity 1·0231 at 15°, and boiling at 66° to 69° at 16 mm. Its odour is that of onions. Fraction 2 (60 per cent.) consisted of diallyl disulphide, $C_6H_{10}S_2$, a light yellow oil of garlic odour, much resembling the last-named body, boiling at 78° to 80° at 16 mm., and decomposing at 150°. Its specific gravity is 1·0237 at 15°. Fraction 3 (20 per cent.), boiling at 112° to 122° at the same pressure, consisted of a body $C_6H_{10}S_3$, the exact constitution of which was not elucidated. Fraction 4 (10·5 per cent.) boiled above 122° at 16 mm. pressure. It contained still more sulphur, and corresponded to the empirical formula $C_6H_{10}S_4$. Semmler states that allyl sulphide $(C_3H_5)_2S$, does not exist in the oil, as originally stated by Wertheim. Wertheim also attributes to Beckett and Wright the statement that the oil contains a sesquiterpene, which he (Semmler) also contradicts. In this, however, Semmler is in error, as Beckett and Wright[3] examined the sesquiterpene from oil of cloves, which was mistranslated into the *Jahresbericht*[4] as *Knoblauchöl*: hence Semmler's mistake, which is reproduced in Schimmel's report.[5]

Oil of Onion.

Oil of Onion is distilled from the bulb and fresh herb *Allium cepa*. The average yield of the plant is 0·05 per cent. of an acrid oil of unpleasant odour and of dark brown colour. Its specific gravity is about 1·038 and optical rotation − 3° to − 6°. The oil has been examined by Semmler,[6] who states that he obtained 0·05 per cent. of oil, which contains no oxygen; its specific gravity was 1·041 at 9°, and its optical rotation − 5°. As it decomposes when boiled under ordinary pressure, it was fractionated at 10 mm. The chief portion of the oil consisted of a compound $C_6H_{12}S_2$, an oil of specific gravity 1·0234 at 12°, boiling at 75° to 83° at 10 mm. From the higher boiling fractions a substance was obtained, apparently identical with one of the constituents of oil of asafœtida. No allyl sulphide was found. The employment of this oil, which is somewhat disagreeable to manufacture, is quite similar to that of oil of garlic.

[1] *Annalen* 51 **(1844)**, 289.
[2] *Arch. Pharm.*, 230 **(1892)**, 434.
[3] *Jour. Chem. Soc.*, **1876**, i. 6.
[4] *Jahresbericht*, **1878**, 398.
[5] *Report*, October, **1893**.
[6] *Arch. Pharm.*, 230 **(1892)**, 443.

OIL OF HYACINTH.

The perfume of the flowers of *Hyacinthus orientalis* is usually employed in the form of an extract or pomade, but an essential oil is obtainable from them.

By extracting hyacinth flowers with benzene and distilling off the solvent under reduced pressure, precipitating the fatty matter with dilute alcohol, and again concentrating, Spalteholtz and Enklarr,[1] obtained 0·016 per cent. of crude oil which, until largely diluted, had a somewhat unpleasant odour. On washing with dilute alkali sulphuretted hydrogen was detected. After removing the fatty matter by cooling the petroleum ether solution to − 20°, filtering and removing the solvent, the oil was fractionated under reduced pressure. The first fraction contained a very volatile substance, with a disagreeable odour. In the second fraction an unknown oxygenated body was isolated. The third fraction contained benzyl benzoate, free benzyl alcohol, and a cinnamic ester. The benzoic acid liberated from an alkaline solution had an odour of vanillin. A fluorescent body free from nitrogen was also isolated, which became red with acids and yellow with alkalies. No esters of anthranilic or methyl-anthranilic acids were detected. The natural wax present crystallised from alcohol in colourless flakes, which persistently retained the odour of the flower.

OIL OF ACAROID RESIN.

The Australian "grass tree gums" *Xanthorrhœa hastilis* and *Xanthorrhœa australis* yield essential oils of highly aromatic odour. These oils smell of storax, tolu, or Peru balsam, and have the following characters :—

	Xanthorrhœa hastilis (yellow gum).	*Xanthorrhœa australis* (red gum).
Specific gravity . .	0·937	0·963
Optical rotation . .	− 3° 14′	± 0°
Acid value . . .	4·9	47·6
Ester ,, . . .	69·4	37·5

Both oils contain cinnamic acid and the red gum oil contains styrol.

Rennie, Cooke, and Finlayson [2] have investigated the essential oils of several species of Xanthorrhœa resin. They state that little evidence as to the species has been forthcoming in previous investigations. The red resin was obtained from either West or South Australia, and its botanical source is uncertain. A specimen of the leaves was submitted to Professor Osborne of Adelaide University, who stated that it would require investigation on the spot to determine whether the leaves belonged to *Xanthorrhœa tateana* (the tree which yields the yellow resin from West and South Australian trees), or to a variety of that species, or to a different species altogether. The essential oil distilled from this red resin amounted to about 0·36 per cent. of the weight of the resin, and was semi-solid at ordinary temperatures. It was found to consist almost entirely of paeonol, the peculiar acetophenone derivative found in the essential oil of *Paeonia montan.*

The essential oil obtained from the yellow resin, derived from *Xanthorrhœa tateana*, was found to contain paeonol, and a hydroxypaeonol,

[1] *Chem. Weekblad*, 7 (**1910**), 1. [2] *Jour. Chem. Soc.*, **1920**, 338.

having the constitution of 2 : 4 : 6-trihydroxy-acetophenone, in which the position of the methoxy group remains to be determined.

The oil obtained from the West Australian red resin from *Xanthorrhœa preissii* was obtained in a yield of over 1 per cent. It was found to contain paeonol and hydroxypaeonol, and a crystalline compound melting at 69·5°, of the formula (probably) $C_{13}H_{12}O_2$. The authors consider this compound to be a methoxydiphenyl ether. A small amount of laevo-citronellol was also found in the oil.

AMARYLLIDACEÆ.

Oil of Tuberose.

The flowers of the tuberose *Polianthes tuberosa* are cultivated to a considerable extent in the South of France, and are principally used for the manufacture of concretes, pomades, and similar perfume materials.

Polianthes tuberosa is a bulbous plant originally derived from Mexico. It is grown in England under glass for the fragrance of its white flowers, which is most powerful at night. The tuberous bulbs are imported from Genoa, and also from North America, but even with artificial heat the flowers do not develop such a powerful perfume as in the South of France, where the plants are cultivated in the open air.

In the districts round Grasse the tubers are planted 9 to 12 ins. apart in rows, with a distance of 2 ft. between the rows. A deep, rich soil is preferable, as the roots penetrate downwards to a considerable depth in search of moisture. If the soil is dry the plants require well watering and manuring. Under good cultivation each plant will bear 10 to 12 flowers, or even more. These are picked off as soon as they open, and the harvesting for the pomade factories commences about the first week in July and lasts until the middle of October. After that time the flowers are so deficient in perfume that they are no longer of use to the manufacturer. In November the roots are taken out of the ground and packed away in dry sand, ready for planting out in the following March or April.

The essential oil cannot be separated from the flowers by direct distillation, as apparently it is decomposed in the process, but the extract prepared by enfleurage yields a small amount of oil on steam distillation.

The oil, according to Hesse,[1] has the following characters :—

Specific gravity	1·007 to 1·043
Optical rotation	− 2° 30′ ,, − 3° 45′
Acid value	22 ,, 32·7
Ester ,,	224 ,, 243

Verley[2] isolated from the oil a body of the formula $C_{13}H_{20}O$ which he considered to be a ketone and which he named tuberone. Hesse isolated from the oil the methyl esters of benzoic and anthranilic acids, the latter being present to the extent of 1·13 per cent. Benzyl benzoate is also present, as well as free benzyl alcohol. Methyl salicylate was detected in the oil obtained by distilling the enfleurage pomade oil, but not in that obtained from the petroleum ether extract of the flowers.

[1] *Bericht*, 36 (**1903**), 1459. [2] *Bull. Soc. Chem.*, iii. 21 (**1899**), 307.

Fig. 14.—Gathering tuberose at Grasse.

[*Rouer-Bertrand Fils.*

IRIDEÆ.

Oil of Orris Root.

Three species of the Iris are used for the preparation of this oil. These are (*a*) *Iris germanica*, common in the districts near Florence, and also found in Central Europe, Northern India, and Morocco; (*b*) *Iris pallida*, found in Istria, Florence, and Lucca; (*c*) *Iris florentina*, occurring on the Macedonian coast, and near the coasts of the Black Sea, and also in the neighbourhood of Florence and Lucca. The finest roots are produced in the Tuscany district, and are known as Florentine orris root, but it must be remembered that *Florentine* root is not synonymous with the root of *Iris florentina*, but embraces the three varieties, which grow to the highest degree of perfection in this district. After the Tuscany root, the Veronese is most valued. The roots, or more correctly the rhizomes, exported from Morocco and East India are of very secondary value, and do not arrive in the market in very good condition. The plant used to be left entirely to itself to grow wild, but extensive plantations are now cultivated. The plants are grown on hills, generally on sunny open spaces, or between rows of vines. A dry calcareous soil is most suitable. The plants are left undisturbed for two or three years, when the harvesting commences. Locally the plants are known as *giaggiolo*. The fresh rhizome has very little odour, and must be carefully dried before being sent away. Certain changes appear then to set in and the odour develops.

The odour of the dried rhizomes resembles the delicate violet perfume, and the powdered substance is a usual constituent of *violet powder*. The earlier application of this delicate perfume was in the form of an alcoholic extract, often known as essence of violets. But non-odorous substances are also extracted from the root by alcohol, and the odour is not so delicate as that of the carefully distilled oil. Pure concrete orris oil melts at about 40° to 50° and has an acid value between 204 and 236. Its ester value should not exceed 10. The yield of oil is from ·1 to ·2 per cent.; but when the rhizome is treated with dilute sulphuric acid, the starch granules are broken up and liberate more essential oil, but the odour is much impaired. The essential oil obtained from the plant is of a buttery consistence, and contains certain non-volatile fatty bodies which are carried over mechanically by the steam. Flückiger, who investigated this substance, showed that it consisted chiefly of myristic acid with traces of the true volatile oil. Our present knowledge of the oil, however, is chiefly due to Tiemann and Krüger.[1] These chemists obtained the oil by extracting the roots with ether, and steam distilling the residue left on evaporation. The non-volatile portion contains a little myristic acid, whilst the volatile oil contains a large amount of myristic acid and its methyl ester, oleic acid, an oleic ester, oleic aldehyde, and a body which is responsible for the characteristic odour which they termed irone. This body was separated in a pure state by converting it into its phenylhydrazone and decomposing this with dilute sulphuric acid. Irone $C_{13}H_{20}O$ is an oil almost insoluble in water, readily soluble in alcohol, etc., boiling at 144° at 16 mm., of specific gravity ·939 at 20°. It is dextro-rotatory. In the pure form the odour is sharp and unlike violets, but when diluted it somewhat resembles

[1] *Bericht*, 26 **(1893)**, 2675.

[*Parfumerie Moderne.*

FIG. 15.—Cultivation of orris in the south-east of France.

the natural violet flowers. Tiemann and Krüger then attempted to synthesise irone, believing it to be the odorous principle of the violet flower (of which so small a yield is obtained as to render an exact examination almost impossible). Whilst they did not succeed in producing irone, they obtained an isomeric body, pseudo-ionone, by condensing citral with acetone, which on heating with dilute sulphuric acid is converted into another isomer, which they termed ionone. This body is the now well-known artificial violet perfume, so extensively used in the so-called violet scent and soaps.

Schimmel & Co.[1] have isolated from the oil furfurol; a terpene which was not identified (specific gravity 0·861; rotation + 10° 40′); decyl aldeyde; nonyl aldehyde; naphthalene (a very rare constituent of essential oils); and a ketone of the formula $C_{10}H_{18}O$. They do not agree with Tiemann and Krüger that oleic aldehyde is present in the oil.

Commercially, the aspect of orris oil is very important. Its unique delicate odour will prevent it from being replaced by the more intense but less delicate odour of ionone, and further, it forms an excellent "fixer" for this artificial perfume, so that a judicious mixture of orris oil and ionone leaves little to be desired. Its employment in fine perfumery is, and has been for many years, very extensive. A quite liquid oil, from which the inodorous solid constituents have been removed, is now on the market. According to Stead it has a specific gravity 0·949, optical rotation − 28°, and congeals at − 5° C. Schimmel & Co., however, have shown that its characters are, normally, as follows:—

Specific gravity	0·930 to 0·940
Optical rotation	+ 14° „ + 30°
Refractive index at 20°	1·4950
Acid value	1 to 8
Ester „	20 „ 40

It is soluble in from 1 to 1·5 volumes of 80 per cent. alcohol.

ZINGIBERACEÆ.

Oil of Ginger.

This oil is the product of distillation of the rhizome of *Zingiber officinale*, a native of tropical Asia, which is also cultivated in both East and West Indies, and in Africa, and to a small extent in Australia. Possibly, also, it is found in China, although the greater part of the "Chinese ginger" appears to be the product of *Alpinia Galanga*.

The rhizomes are sometimes imported in the soft juicy condition known as "green ginger," but the ordinary ginger of commerce consists of the dried rhizomes which have been picked when the stalks have withered, and are either washed, dried and scraped (*uncoated* or *white* ginger), or merely washed and dried (*coated* ginger). Sometimes it is bleached or limed before sale. The gingers of commerce are of widely different value, according to their place of origin, and the aroma and value of the oil naturally depend to some extent upon this. The Jamaica product is most highly valued. The oil, which is obtained to the extent of about 2 to 3 per cent., is a pale yellow to dark yellow liquid of characteristic aromatic odour, but lacking the pungency of the rhizome.

[1] *Report*, April, **1907**, 76.

The oil has the following characters:—

Specific gravity	0·8740 to ·886
Optical rotation	− 25° ,, − 50°
Refractive index	1·4885 ,, 1·4950
Acid value	0 ,, 2
Ester ,,	1 ,, 15
,, ,, (after acetylation)	30 ,, 45

Abnormal ginger oils are from time to time to be found. Bacon[1] reports on a sample distilled in the Philippines, having an optical rotation of + 5·9°, and a Japanese oil has been found having a specific gravity 0·894 and an optical rotation + 9° 40′. A Java oil has been recorded[2] with a rotation of + 13° 9′. Thresh has fractionally distilled a specimen of this oil prepared in England from the dried root, and obtained the following results:—

Below 150°	5 per cent.
150° to 200°	10 ,,
200° ,, 240°	8 ,,
240° ,, 265°	60 ,,
265° ,, 300°	7 ,,
Residue	10 ,,

These results, however, must be accepted with reserve, as the oil undergoes decomposition by distillation at ordinary pressure. The fraction obtained at 155° to 165° is usually dextro-rotatory.

Dextro-camphene and β-phellandrene constitute the bulk of the terpenes present in the oil. Cineol, citral, and borneol have been identified, with possibly traces of geraniol. Van Soden and Rojahn[3] have isolated a sesquiterpene which they have named zingiberene. This body has a specific gravity 0·872, optical rotation − 69°, and boils at 269° to 270°.

Dodge[4] has isolated a small quantity of an aldehyde, which appears to be decyl aldehyde. Brooks[5] has identified methyl-heptenone, nonyl aldehyde, linalol, acetic and caprylic esters, a phenol (chavicol ?), and a sesquiterpene alcohol having the characteristic odour of the oil. This body $C_{15}H_{26}O$, which has been termed zingiberol, boils at 154° to 157° at 14·5 mm. pressure.

Oil of Galangal.

This oil is obtained from the rhizome of *Alpinia officinarum*, a plant cultivated in China and Siam. The oil is obtained by moistening the ground rhizomes with water, and after some hours distilling them with steam. The oil is of a greenish-yellow colour, with a sharp taste and pungent camphoraceous odour. The yield is from ·5 to 1·5 per cent. The oil has the following characters:—

Specific gravity	0·915 to 0·925
Optical rotation	− 1° ,, − 6°
Refractive index	1·4760 ,, 1·4825
Acid value	0 ,, 2
Ester ,,	6 ,, 15
,, ,, (after acetylation)	40 ,, 50

The constituents of galangal oil so far identified, are cineol, *d*-α-pinene, cadinene, together with the following compounds described by

[1] *Philipp. Jour., Sci.*, 5 (**1910**), 259.
[2] *Jaarb. dep. Landb. Ned. Ind. Batavia* (**1912**), 57.
[3] *Pharm. Zeit.*, 45 (**1900**), 414. [4] *Chem. Zeit.*, 34 (**1912**), 1217.
[5] *Jour. Amer. Chem. Soc.*, 38 (**1916**), 430.

Fromm and Fluck.[1] These chemists isolated a small fraction of from 1 to 2 per cent. boiling at 208° to 210°, which consisted mainly of an oxygenated compound of the formula $C_{10}H_{16}O$, the exact nature of which is undetermined. Two sesquiterpenes were separated, one boiling at 138° to 140° at 12 to 15 mm., and the other yielding a dihydrochloride identical in melting-point with that of cadinene. But when mixed with pure cadinene dihydrochloride, a depression of 10° in the melting-point results, so that the sesquiterpene is not cadinene. From the highest boiling residues, a sesquiterpene hydrate $C_{15}H_{26}O$ melting at 167° was separated. Eugenol, which has been stated to be present by Horst, could not be found.

According to Dragendorff, the oil should not be quite soluble in an equal volume of 85 to 90 per cent. alcohol. He states that the chief adulterants are oils of pimento and cloves, which are much more soluble in alcohol. The ready solubility of samples of oil in all proportions of 80 per cent. alcohol is, therefore, probably indicative of adulteration. Oil of lemon and turpentine have also been used to sophisticate the pure oil. These will be indicated by the lowering of the specific gravity and the alteration in the optical rotation.

A galangal oil has also been distilled from *Alpinia Galanga* and has been examined by Ultée.[2] The oil has the following characters:—

Specific gravity	0·974 to 0·985
Optical rotation	+ 4° ,, + 6°
Refractive index	1·5164
Acid value	1 to 2
Ester ,,	145

It contains about 50 per cent. of methyl cinnamate and 20 to 30 per cent. of cineol. Pinene is also present.

OIL OF ALPINIA ALBA.

Pickles and Earl[3] have made an examination of the essential oil distilled from the fruit of this plant which is also known as *Amomum medium*. They find that the oil resides almost entirely in the seeds. By distilling the fruits with steam, about 1 per cent. of a pale yellow oil, having an odour recalling those of lemon and eucalyptus, was obtained. The oil has the following characters:—

Specific gravity	0·9366
Optical rotation	− 2° 15′

The oil was shaken, first with dilute sodium-carbonate, next with sodium hydrogen sulphite solution, then with dilute sodium hydroxide, and finally with 50 per cent. resorcinol solution to absorb cineol. From the results of this treatment and the subsequent examination of the various products, the composition of the oil was found to be approximately as follows: Cineol, 69 per cent., characterised by the crystalline additive product with iodol; aldehydes and ketones, 27·5 per cent., consisting mainly of citral, which was characterised by means of the semi-carbazone and the β-naphthacinchoninic acid; phenols, 1·5 per cent.; acids, 1 per cent. A small quantity of crystalline acid, melting-point, 46° to 48° was isolated, but not in sufficient quantity for identification. The residue, amounting to about 1 per cent., seemed, from its odour, to consist chiefly of terpenes.

[1] *Annalen*, 405 (**1914**), 175. [2] Schimmel's *Bericht*, April, **1911**, 19.
[3] *Proc. Chem. Soc.* (**415**), 164.

Oil of Alpinia Galanga.

The rhizome and roots of *Alpinia galanga* yield 0·04 per cent. of essential oil having the following characters :—

Specific gravity	0·978 to 0·985
Optical rotation	+ 4° „ + 6°
Refractive index	1·5164
Acid value	1·8
Ester „	145·6

The oil contains about 50 per cent. of methyl cinnamate, and 20 to 30 per cent. of eucalyptol. Camphor and *d*-pinene are also present.

Oil of Alpinia Malaccensis.

The fresh roots of *Alpinia malaccensis* yield, according to Schimmel & Co., about 0·25 per cent. of essential oil having the following characters :—

Specific gravity	1·039 to 1·047 at 27°
Optical rotation	+ 0° 15′ to + 1° 30′
Solidifying-point	25·5°
Saponification value	256 to 278·5
Refractive index	1·5477

It contains methyl cinnamate.

The leaves also yield an oil which has been examined by van Romburgh,[1] who obtained 0·16 per cent. from the fresh leaves. The oil has the following characters :—

Specific gravity	1·020 at 26°
Optical rotation	+ 6° 5′

It contains *a*-pinene, and about 75 per cent. of methyl cinnamate.

Oil of Alpinia Nutans.

The root of this plant yields an essential oil having a specific gravity 0·950 at 29°. It distils principally at a temperature below 230°. The higher boiling fractions contain an ester of (probably) cinnamic acid. The oil from the leaves has recently been examined by Kafuku[2] who obtained 0·053 per cent. It had the following characters :—

Specific gravity at 19°	0·9301
Optical rotation	+ 38·4°
Refractive index	1·4750
Saponification value	9·88
„ „ (after acetylation)	36·1

It contains 30 per cent. of *d*-camphor, 17 per cent. of *d*-camphene, cineol, an ester of cinnamic acid, a sesquiterpene, a high boiling phenol, and probably limonene.

Oil of Cardamoms.

The oils from different kinds of cardamoms, although similar in general properties, differ to a certain extent chemically. The seeds from which the oil is distilled are chiefly exported from South-West India and Ceylon in their pericarps, forming the cardamoms of commerce.

The following description[3] of the principal characters of the more important varieties of cardamoms is of interest :—

"*Mysores.*—Divided into rounds and longs. The former are what

[1] *Kon. Akad. Weten. Ammerdam* (**1898**), 550.
[2] *J. Chem. Ind. Tokyo*, 20 (**1917**), 349.
[3] *Chemist and Druggist Diary*, **1899**, 500.

Photo Garnier.] [Parfumerie Moderne.

FIG. 16.—Cardamoms in Tahiti; Lake Vahiria.

the *B.P.* calls 'ovoid'; they vary in length from $\frac{1}{4}$ in. to $\frac{1}{5}$ in. (the latter 1 in 10), and have a smooth pericarp of a cream colour, due to the use of bleaching agents. Their quality is judged by their weight. Sometimes the seeds are shrivelled (unripe), so that the fruit is husky. This is not so frequent in the longs, which are simply thinner than the rounds, and are not so smooth on the surface, nor so pale, as a rule. The *B.P.* description, 'longitudinally striated,' might exclude most of the rounds, as they look smooth until closely examined.

"*Malabars.*—These are smaller than Mysores, and there is a greater proportion of seed to pericarp in them. They are fat pods, with a pointed apex. Generally pale brown or pink and longitudinally striated. Rarely more than $\frac{1}{2}$ in. long. They have a full flavour.

"*Mangalores.*—These are almost globular in shape and not unlike Malabars. All three are washed or bleached before exportation.

"*Seeds.*—A goodly proportion of cardamoms come straight into the market freed from their pericarps. The *B.P.* excludes these because they are supposed to lose flavour on keeping. Wholesalers and large manufacturers use them when fresh; fine brown seed is as strong as the kind just freed from the pericarps. Grey seeds are inferior."

It is fairly certain that the plant *Elettaria cardamomum* produces most of the ordinary cardamoms, whilst the "Ceylon wilds" are the product of a variety of this plant, known as *variety β*. The majority of the cardamoms of commerce are imported from Ceylon, and may be described as *Ceylon-Malabars* or *Ceylon-Mysores*, according as they fit in with the above descriptions. The Ceylon wilds are mostly consumed in the manufacture of a certain cake made in South Germany, and are not an ordinary article of commerce. Some confusion exists as to the botanical characters of some of these seeds, and more so as to the essential oils they yield. The oil is usually distilled from Ceylon seeds, which yield from 3 to 6 per cent. of oil. According to Schimmel, the oil derived entirely from Malabar cardamoms (*Elettaria cardamomum*) is obtained to the extent of from 4 to 8 per cent. of the seeds used. Its specific gravity was ·943 and its optical rotation + 34° 52′. It dissolved in 4 parts of 70 per cent. alcohol. It was found to require 13·2 per cent. of potash for saponification, thus indicating a very high ester content. Haensel gives the specific gravity as 0·933 and the optical rotation as + 26°. Samples of oil distilled from both Malabar and Mysore (Ceylon) seeds, obtained from authentic sources by the author, showed that there was practically no difference between the two oils, although Schimmel states that Ceylon cardamom oil has a specific gravity of ·895 to ·910 and a rotation of + 12° to + 13°. It is probable that the oil from the so-called Ceylon wild cardamoms is here meant.

The Ceylon-Malabar seeds distilled by the author yielded 1·3 per cent. of oil, and the Ceylon-Mysores 2·6 per cent. Both were bright yellow liquids, whose odours were scarcely distinguishable. The specific gravities and optical rotations were as follows:—

	Specific Gravity at 15·5°.	Optical Rotation at 16° (100 mm. Tube).
Oil of Malabar Cardamoms . .	0·9418	+ 40° 41′
Oil of Mysore Cardamoms . .	0·9418	+ 46° 39′

These figures are in fair agreement with those given for Malabar oil, but in no way resemble those quoted by Schimmel for Ceylon oil, which are difficult to understand.

The oils were soluble with a slight opacity in 40 to 45 volumes of 60 per cent. alcohol.

Little difference exists between the two oils. On distillation at ordinary pressure the oil, which is very rich in esters, in both cases decomposes partially, and a considerable quantity of free acid distils over. According to Weber (*Annalen*, 238, 89), formic and acetic acids are found in the distillate. Acetic acid is undoubtedly the chief acid constituent of the esters, but the author is unable to confirm the presence of formic acid. If it is present, it is only in faint traces. On distillation under reduced pressure the earlier fractions (the boiling-point rises gradually until 50 per cent. has distilled over) contain cineol, but only to the extent of 5 to 10 per cent. of the oil. This figure is the result of an approximate estimation by means of phosphoric acid.

The following figures may be taken as covering most pure samples:—

Specific gravity	0·923 to 0·945
Optical rotation	+ 24° ,, + 48°
Refractive index	1·4620 ,, 1·4675
Acid value	1 ,, 4
Ester ,,	90 ,, 150

It is soluble in 2 to 5 volumes of 70 per cent. alcohol.

The oil contains a considerable amount of terpinyl acetate (and therefore requires at least two hours for complete saponification). The terpineol was identified as *d*-α-terpineol by Schimmel & Co.[1] The terpene terpinene has been stated to be present, but this is not confirmed by the author, nor by Schimmel & Co. Limonene is probably present in small amount. Free terpineol is also present.

The oil distilled from the Ceylon wild cardamoms, *Elettaria cardamomum β*, has the following characters:—

Specific gravity	0·895 to 0·906
Optical rotation	+ 12° ,, + 16°
Acid value	1 ,, 3
Ester ,,	25 ,, 75

This oil contains terpinene, sabinene, terpineol and its esters, and a solid crystalline substance melting at 60° to 61°.

Siamese cardamom oil is distilled from the seeds of *Amomum cardamomum*, which are known on the market as camphor seeds on account of their camphor-like odour. Schimmel obtained 2·4 per cent., which was semi-solid at the ordinary temperature, and had an odour of camphor and borneol. In order to liquefy the oil it was necessary to warm it to 42°. Its specific gravity at this temperature was ·905, and its optical rotation + 38° 4′. It required 1·88 per cent. of KOH to saponify the esters present, and contained alcohols equivalent to 22·5 per cent. of borneol. It was soluble in 1·2 volumes of 80 per cent. alcohol. From 800 grms. of the oil 100 grms. of the solid compounds were obtained by centrifugal action. These were examined and found to consist of about equal parts of dextro-borneol and dextro-camphor.

Kameroon cardamom oil is obtained from fruits which are identical with the Madagascar cardamoms. They are derived from *Amomum*

[1] *Bericht*, October, **1897**, 9.

danielli. The yield of oil was 2·33 per cent., its specific gravity was ·907 and its optical rotation − 20° 34′. The oil is soluble in 7 to 8 parts of 80 per cent. alcohol. It contains cineol and possesses an odour which prevents its being employed as a substitute for the ordinary oil.

Bengal cardamom oil is obtained from *Amomum aromaticum*, which yields about 1 to 1·2 per cent. of oil having a specific gravity 0·920 and optical rotation − 13°. This oil contains a large quantity of cineol.

The seeds of *Amomum Melegueta*, one of the Zingiberaceæ indigenous to tropical West Africa, yield the so-called "grains of paradise" oil, which is obtained to the extent of 0·3 to 0·75 per cent. The oil has a specific gravity about 0·894 and optical rotation about − 4°. It is not clearly soluble in 90 per cent. alcohol. Schimmel & Co.[1] have distilled a sample, which was found to have the following characters:—

Specific gravity	0·897
Optical rotation	− 3° 10′
Refractive index	1·49116

Its ester value was 41·2, and ester value after acetylation 63·9.

Schimmel & Co.[2] have examined the oil obtained from cardamom roots from Indo-China. They obtained 0·64 per cent. of oil having the following characters: d_{15° 0·9066; a_D − 32° 57′; n_{D20° 1·48151; acid value 3·7; ester value 87·9; ester value after acetylation 96·7. The oil was soluble in 0·5 volumes of 95 per cent. alcohol; when more alcohol was added the mixture rapidly turned turbid, and did not become clear again until the solvent had been increased to 4 volumes. Fractional distillation under diminished pressure (5 mm.) gave the following result:—

1.	to 35°	5·4 per cent.	a_D − 0° 10′
2.	35° ,, 40°	8·7 ,,	a_D − 0° 32′
3.	40° ,, 100°	5·4 ,,	a_D − 17° 5′
4.	100° ,, 110°	10·6 ,,	a_D − 31° 10′
5.	110° ,, 115°	44·2 ,,	a_D − 45°
6.	115° ,, 145°	6·4 ,,	a_D − 33° 14′
7.	residue	19·3 ,,	a_D − 39° 15′

Fractions 1 and 2 contained cineol, which was identified from the double compound it gave with resorcinol. From fractions 4 and 5 a portion boiling between 117° and 120° (5 mm.) was separated by repeated fractionation. Saturated with hydrochloric acid gas in dry ethereal solution at − 18°, this fraction yielded a hydrogen chloride compound which, when recrystallised from methyl alcohol, had melting-point 79° to 80°. The hydrochloride was inactive; the chlorine-determination gave the following result:—

0·4306 subst.: 0·5928 g. AgCl

Cl Found	Calc. for $C_{15}H_{24}$, 3HCl.
34·06 per cent.	33·9 per cent.

The probability of this body being bisabolene hydrochloride was confirmed by further examination. With sodium acetate and glacial acetic acid a sesquiterpene was eliminated from it which, when twice fractionated, was found to possess the following constants: 265° to 267° (757 mm.); d_{15° 0·8748; a_D ± 0°; n_{D20° 1·49063. With hydrochloric acid this hydrocarbon yielded again the trihydrochloride melting between 79° and 80°.

The residue of distillation of cardamom-root oil solidified at about

[1] *Report*, October, **1914**, 32. [2] *Ibid.*, **1911**, 105.

15° with separation of paraffin; the last-named body, when recrystallised from alcohol, melted at 62° to 63°.

Korarima cardamom oil is the product of *Amomum Korarima*, the yield of oil being from 1 to 2 per cent. It has the following characters:—

Specific gravity	0·903
Optical rotation	− 3° to − 7°
Acid value	3 „ 5
Ester „	22 „ 50
„ „ (after acetylation)	100 „ 110

Oil of Aframomum angustifolium (*Amomum angustifolium*) is an oil obtained from the seed of a species of cardamoms indigenous to (German) East Africa from which Schimmel & Co. obtained 4·5 per cent. of a colourless oil possessing the following characters: $d_{15°}$ 0·9017; a_D − 16° 50′; $n_{D20°}$ 1·46911; acid value 0·4; ester value 4·2; soluble in 6 volumes of 80 per cent. alcohol. So far as aroma is concerned it cannot compare with Ceylon cardamom oil, and owing to its high cineol-content it resembles cajuput oil. Unfortunately, the quantity of oil was too small to enable its composition to be determined with any exactitude. Botanical examination of the entire fruit showed that the seed was derived from *Aframomum angustifolium* K. Schum., a plant which occurs also in Madagascar and in the Seychelles Islands.

The wild cardamoms of Indo-China, derived from *Amomum globorum*, yield about 4 per cent. of essential oil having the following characters:—

Specific gravity	0·9455
Optical rotation	+ 43° 54′
Refractive index	1·4714
Acid value	0·8
Ester „	128·4

The oil probably contains camphor.

The leaves, roots, and stalks of *Amomum Walang* (*Achasma Walang*) yield essential oils, which have been examined.[1] They have the following characters:—

	Oil from Leaves.	Oil from Stalks.	Oil from Roots.
Specific Gravity at 16°	0·850	0·860	0·856
Optical Rotation	− 0° 34′	− 0° 58′	− 0° 30′
Acid Value	10·1	26	18·9
Saponification Value	70	75·6	70
Sap. Value after Acetylation	—	285·6	285·6
Aldehyde-content (Sulphite Method)	97 per cent.	96 p. cent.	96 p. cent.
Oil-yield from Green Material	0·25 „	0·2 „	0·15 „

Oil of Turmeric.

Turmeric oil is distilled from the roots of *Curcuma longa*, one of the Zingiberaceæ indigenous to India, and cultivated also in China and Southern Asia, etc. The oil, which is not of commercial importance, is obtained to the extent of about 5 per cent. as a thick yellow oil having the following characters:—

[1] *Jaarb. dep. Landb. Ned. Ind. Batavia* (**1911**), 45.

Specific gravity	0·934 to 0·962
Optical rotation	− 25° ,, + 35°
Acid value	1·5 ,, 3·5
Ester ,,	8 ,, 16
,, ,, (after acetylation)	30 ,, 55

Bacon[1] has examined an oil distilled in the Philippines and found it to have a specific gravity 0·930 (at $\frac{30°}{30°}$); optical rotation + 8·6°; refractive index 1·5030; and ester value 81.

According to Jackson and Menke[2] it is separable by fractional distillation under a pressure of 60 mm. into three portions, the first boiling below 193°, the second from 193° to 198°, and the third forming a semi-solid residue. The chief ingredient was the fraction 193° to 198° which consisted of a nearly pure alcohol, a pale yellow oil with an aromatic smell, of specific gravity 0·956 at $\frac{24°}{4°}$, and optical activity $[\alpha]_D = + 24·58°$. At ordinary pressure it boils at 285° to 290°. It appears to be an alcohol of the formula $C_{13}H_{18}O$ or $C_{14}H_{20}O$, which has been called turmerol. Valeric and caproic acids have also been isolated. According to Suida, the principal ingredient is a body isomeric with carvone, $C_{10}H_{14}O$, but this is probably incorrect. The terpene phellandrene has been found in the oil.

The most recent work on this oil is that of Luksch,[3] who isolated from it a ketone, curcumone $C_{13}H_{18}O$, probably isomeric with turmerol, It is a colourless oil of sharp ginger-like taste, having a specific gravity 0·9566 at 20°, optical rotation + 80·55°, and refractive index 1·50526. It boils at 119° to 122° at 8 to 11 mm. Its oxime boils at 159° and its phenylhydrazone melts at 92°.

Oil of Zedoary.

The roots of *Curcuma Zedoaria* yield from 1 to 1·5 per cent. of an essential oil having the following characters:—

Specific gravity	0·982 to 1·012
Optical rotation	+ 8° ,, + 18°
Refractive index	1·5020 ,, 1·5060
Acid value	0 ,, 2·5
Ester ,,	15 ,, 25
,, ,, (after acetylation)	55 ,, 68

It is soluble in 1·5 to 2 volumes of 80 per cent. alcohol.

Cineol has been identified as a constituent of this oil, and Bacon[4] has isolated a sesquiterpene alcohol, which forms crystals melting at 67°. It is optically inactive and has a specific gravity 1·010 at $\frac{30°}{30°}$.

Oil of Kæmpferia Galanga.

The roots of *Kæmpferia Galanga* yield an essential oil having a specific gravity about 1·025, optical rotation − 16°, and refractive index 1·5428. It contains the ethyl ester of *p*-methoxy-cinnamic acid melting at 50°, and the ethyl ester of cinnamic acid. It also contains pentadecane. The roots of *Kæmpferia rotunda* yield 0·2 per cent. of an oil of specific gravity 0·890 to 0·900, and optical rotation about + 12°. It contains cineol.

[1] *Philipp. Jour. Sci.*, 5 (**1910**), 262.
[2] *Amer. Chem. Jour.*, 4 (**1882**) 368.
[3] *Inaug. Dissert. Basle*, 1906.
[4] *Philipp. Jour. Sci.*, 5 (**1910**), 261.

Oil of Kæmpferia Ethelæ.

The oil distilled from the tubers of *Kæmpferia ethelæ* has been examined by Goulding and Roberts,[1] who obtained a total yield of 1·5 per cent., equivalent to 2·0 per cent. on the dry material. The plant grows in large quantities in the extreme north-east of the Transvaal, and could easily be cultivated. The oil has the following characters:—

Specific gravity	0·923 to 0·9437
Optical rotation	+ 19° 47′ ,, + 30·4°
Acid value	1·0 ,, 2·3
Ester ,,	5 ,, 11·5
,, ,, (after acetylation)	33·6 ,, 47·6

The oil is of pale yellowish-brown colour, having a rather pleasant odour somewhat resembling that of orange flower oil, combined with a less agreeable smell, recalling that of crushed ivy leaves. On fractional distillation under low pressure in an atmosphere of carbon dioxide two fractions were obtained: (1) (42 per cent.) distilling at 160° to 195°; specific gravity ·901; optical rotation − 4° 30′. (2) (25 per cent.) distilling at 195° to 270°; specific gravity ·988; optical rotation + 26° 30′. The residue (33 per cent.) deposited colourless crystals amounting to about 8 per cent. of the original oil. The first fraction consisted chiefly of terpenes and cineol. Dipentene was identified by means of the tetrabromide (melting-point 124°), and cineol by its iodol additive compound (melting-point 112°). An estimation of the cineol by the resorcin method showed that fraction 1 contained 44 per cent., equivalent to 18·5 per cent. of the original oil. Acetic acid was found to be present, together with a phenolic body having a creosote-like odour, amounting to 0·6 per cent. of the original oil. From the second fraction linalol was isolated and identified by conversion into citrylidene-cyanoacetic acid, melting at 120° to 121°. A small quantity of methylanthranilic acid was also separated, melting at 179°, probably present in the form of its methyl ester. The crystalline substance separated from the portion boiling above 270°, after repeated crystallisation from alcohol, had a melting-point of 102° and optical rotation + 198° 20′ in chloroform solution. In dilute alcoholic solution it possesses the characteristic odour resembling crushed ivy leaves, which was noticed in the oil itself. It is highly unsaturated, and readily combines with about twice its weight of bromine. It forms a hydroxylamine-oxime melting at 184°, an oxime melting at 166°, and a benzoyl derivative melting at 260° with decomposition. The heavy portion of the oil, amounting to 7 per cent., appeared to consist of a solution of the ketone compound in liquid constituents of high boiling-point, probably sesquiterpenes. The composition of the oil is summarised as follows:—

	Per Cent.
Terpenes (dipentenes and probably pinene)	21·8
Cineol	17·2
Ketonic compound (melting-point 102°)	13·0
Alcohols (including linalol)	11·2
Esters (chiefly methyl methyl-anthranilate)	1·3
Phenols	0·5
Acids (chiefly or entirely acetic acid)	0·1
Residue (probably chiefly sesquiterpene)	34·9

[1] *Jour. Chem. Soc.* **(1915)**, 314.

PIPERACEÆ.

Cubeb Oil.

This oil, which is official in the *British Pharmacopœia*, is distilled from the fruit of *Piper cubeba* (*Cubeba officinalis* Miq.). The cubebs of commerce are the dried unripe fruits, which resemble ordinary black pepper in appearance, except that they are rather lighter in colour; they are chiefly imported from Batavia and Singapore.

The berries are coarsely ground and distilled with steam, yielding from 10 to 18 per cent. of volatile oil. This is of a faint greenish or greenish-blue colour, and of very characteristic and somewhat aromatic odour. The colour has been stated to be due to copper, but this is erroneous, as the higher fractions contain a blue oil in greater or less quantity, which determines the tint of the liquid.

The characters of normal cubeb oil are as follows:—

Specific gravity at 15°	0·912 to 0·930
Optical rotation	− 25° „ − 42°
Refractive index at 20°	1·4930 „ 1·4970

The solubility in 90 per cent. alcohol is very variable, some samples requiring only 1 to 2 volumes, others up to 10 volumes, to effect solution.

A fractional distillation of the oil yields valuable results, from 65 to 80 per cent. distilling between 250° and 280°. Four pure samples examined in this manner by Messrs. Oranje[1] gave the following results:—

1.

Yield, 14 per cent. oil.
Specific gravity (15°), 0·9268; optical rotation − 33·70°.
Fractionation of 50 grms.:—

	grms.	per cent.		per cent.
Below 250°	0·0	0·0	Below 250°	0·0
250° to 260°	11·7	23·4	250° to 280°	86·4
260° „ 270°	26·8	53·6		
270° „ 280°	4·7	9·4		
Residue	6·8	13·6	Residue	13·6
	50·0	100·0		100·0

2.

Yield, 9·5 per cent. oil.
Specific gravity (15°), 0·9258; optical rotation − 27·40°.
Fractionation of 40 grms.:—

	grms.	per cent.		per cent.
Below 250°	0·0	0·0	Below 250°	0·0
250° to 260°	10·9	27·3	250 to 280°	82·8
260° „ 270°	15·8	39·5		
270° „ 280°	6·4	16·0		
Residue	6·9	17·2	Residue	17·2
	40·0	100·0		100·0

3.

Yield, 15·0 per cent. oil.
Specific gravity (15°), 0·9226; optical rotation − 34·60°
Fractionation of 50 grms.:—

	grms.	per cent.		per cent.
Below 230°	0·0	0·0	Below 250°	18·0
230° to 240°	4·0	8·0		
240° „ 250°	5·0	10·0		
250° „ 260°	27·4	54·8	250° to 280°	72·2
260° „ 270°	7·0	14·0		
270° „ 280°	1·7	3·4		
Residue	4·9	9·8	Residue	9·8
	50·0	100·0		100·0

[1] *P. and E.O.R.* (1914), 372.

4.

Yield, 14·0 per cent. oil.
Specific gravity (15°), 0·9233 ; optical rotation − 37·60°.
Fractionation of 50 grms. :—

Below 220°	0·0 grms.	= 0·0 per cent.	Below 250°	12·8 per cent.
220° to 230°	2·3 ,,	= 4·6 ,,		
230° ,, 240°	1·5 ,,	= 3·0 ,,		
240° ,, 250°	2·6 ,,	= 5·2 ,,		
250° ,, 260°	20·3 ,,	= 40·6 ,,	250° to 280°	79·0 ,,
260° ,, 270°	15·5 ,,	= 31·0 ,,		
270° ,, 280°	3·7 ,,	= 7·4 ,,		
Residue	4·1 ,,	= 8·2 ,,	Residue	8·2 ,,
	50·0	100·0		100·0

The chemistry of this oil is in need of further elucidation, as it is clear that there are several constituents present which have, so far, escaped identification. The terpene dipentene is present, and probably small quantities of pinene or camphene; cadinene is present, as well as a second sesquiterpene not yet identified. There is also present, especially in the oil distilled from old fruit, which has doubtless become partially oxidised, a small amount of the so-called cubeb-camphor. This body, $C_{15}H_{25}OH$, appears to be a crystalline sesquiterpene alcohol derived from the sesquiterpenes (or one of them) by oxidation. From a mixture of ether and alcohol it crystallises in rhombs, melting at 65° [1] and is laevo-rotatory. It boils at 248°, with decomposition. The nature of the blue oil found in the higher fractions is unknown.

Umney and Potter [2] have reported on some "false cubebs," the fruit of an unknown species of pepper. These false cubebs may, and apparently do, find their way into commerce, and if distilled with genuine cubebs produce a quite abnormal oil. The oil from these false cubebs has a specific gravity about 0·894 and an optical rotation + 16°. The following results were obtained on fractional distillation, and are compared with those of a genuine oil under identical conditions :—

FRACTIONATION.

False Cubeb Oil.		Pure Cubeb Oil.	
Below 160°	12 per cent.	—	
160° to 170°	8 ,,	—	
170° ,, 180°	20 ,,	—	
180° ,, 190°	6 ,,	—	
190° ,, 200°	4 ,,	Below 200°	5 per cent. only.
200° ,, 230°	8 ,,	200° to 230°	11 ,,
230° ,, 240°	2 ,,	230° ,, 240°	3 ,,
240° ,, 250°	4 ,,	240° ,, 250°	15 ,,
250° ,, 260°	6 ,,	250° ,, 255°	31 ,,
260° ,, 270°	10 ,,	255° ,, 257°	25 ,,

The fractionation figures, gravity, and optical rotation indicate that this oil has an entirely different composition from that of normal cubeb oil. The oil has been further investigated as regards its ester and alcohol values which are :—

[1] Schmidt, *Arch. der Pharm.*, 191 (**1870**), 23 ; Schaer and Wyss, *Arch. der Pharm.*, 206 (**1875**), 316, give 67° ; and Winckler, *Annalen*, 8 (**1833**), 203, gives 70°.

[2] *P. and E.O.R.* (**1912**), 64.

Esters . . . . None
Alcohols . . . . Saponification number after acetylation 56·1

As a result of their observations on these essential oils they come to the conclusion that certain of the abnormal oils of commerce have been produced from mixtures of the genuine berries with either this hitherto unidentified one which has been described, or possibly another variety. It is perfectly clear that a mixture of one part of the normal cubebs with one part of this variety would yield an oil having approximately the characters which have been found in these abnormal oils, namely, a rotation of − 15° or − 16° and a specific gravity of about 0·910. It must not be forgotten that, in taking a mixture in equal parts of the two fruits, the yield of essential oil of the genuine is practically twice as much as that of the substitution fruit.

A sample of false cubeb oil examined in the laboratory of Evans Sons, Lescher & Webb, Ltd.,[1] had the following characters:—

Specific gravity	0·916
Optical rotation	− 9° 30′
Refractive index	+ ·4921

On fractionation it behaved as follows (sample *a*) compared with two pure oils (*b* and *c*):—

VACUUM-DISTILLED FRACTIONS OF CUBEB OILS.

Flask	Fraction.		Optical Rotation.	Boiling-Point.	Refractive Index.	Specific Gravity.
4 pear sill head.	5 per cent.	(*a*)	+ 27° 20′	52° to 59°	1·4675	0·854
		(*b*)	− 31° 5′	57° „ 69°	1·465	0·851
		(*c*)	− 35° 47′	53° „ 109°	1·4735	0·868
	5 per cent.	(*a*)	+ 24° 30′	to 63°	1·469	—
		(*b*)	− 37°	„ 86°	1·47	0·855
		(*c*)	− 29°	„ 129°	1·485	—
Ordinary Classen flask.	20 per cent.	(*a*)	+ 10°	„ 89°	1·4753	—
		(*b*)	− 32°	122°	1·49	0·9078
		(*c*)	− 26° 30′	133°	1·496	—
	20 per cent.	(*a*)	− 20° 20′	119°	1·4953	0·918
		(*b*)	− 31° 40′	127·5°	1·4957	0·9198
		(*c*)	− 25° 40′	139·5°	1·498	0·930
	10 per cent.	(*a*)	− 20° 5′	126°	1·5005	—
		(*b*)	− 32° 30′	130°	1·4965	0·927
		(*c*)	− 24° 45′	140·5°	1·4993	—
	40 per cent. (residual)	(*a*)	− 18°	over 126°	1·5041	—
		(*b*)	− 30° 30′	„ 130°	1·498	0·9438
		(*c*)	− 26° 20′	„ 140·5°	1·5012	—

The following investigation, which was carried out by Dr. J. Dekker,[2] Director of the Colonial Museum at Haarlem, throws considerable light on the question of the abnormalities observed in certain oils of cubeb.

In a quantity of 500 grms. of cubebs were principally found four different kinds of fruit, which have been sorted out and classified as follows:—

Class 1.—Real cubebs, *i.e.* those showing both macroscopically and microscopically the features of piper-cubebs, producing, if brought into touch with concentrated sulphuric acid, the well-known red colouring.

[1] Evans' *Annual Report* (**1913**), 29. [2] *P. and E.O.R.* (**1913**), 89.

Class 2.—Much resembling real cubebs, but outwardly shrivelled, with longer and broader tail than of real cubebs.

This kind produced, with concentrated sulphuric acid, a yellow colouring.

Class 3.—Not resembling real cubebs at all—without tail and showing a totally irregular microscopic figure—colouring yellow with concentrated sulphuric acid.

Class 4.—Much resembling the real cubebs, but larger and with a different design of the net-work. With concentrated sulphuric acid colouring yellow.

In order to check the assortment, 100 fruits of the No. 1 class have been treated with concentrated sulphuric acid one by one; it then appeared that 98 gave a red and 2 a yellow colouring. These two proved to belong to the No. 2 class, and have been taken out, so that at last it could be accepted that the Class No. 1 consisted of real cubebs exclusively.

The various isolated kinds have then been distilled, and also the branches and the original (unassorted) sample:—

1. *Original Sample.*

Production of oil	14·44 per cent.
Colour of oil	light green
Refraction 15° C.	1·4955
Optical rotation 15° C.	− 29·12 (length of tube 10 cm.)
Specific gravity 15° C.	0·922

2. *Branches of Original Sample.*

Production of oil	1·67 per cent.
Colour of oil	dark green
Refraction 15° C.	1·5011
Optical rotation 15° C.	− 21·76 (length of tube 10 cm.)
Specific gravity 15° C.	0·938

3. *Assortment of Same Sample.—Class* 1.

Production of oil	17·78 per cent.
Colour of oil	light blue-green
Refraction 15° C.	1·4964
Optical rotation 15° C.	− 29·60 (length of tube 10 cm.)
Specific gravity 15° C.	0·924

4. *Assortment of Same Sample, Broad-tailed.—Class* 2.

Production of oil	11·56 per cent.
Colour of oil	yellow-green
Refraction 15° C.	1·4928
Rotation 15° C.	− 11·24 (length of tube 10 cm.)
Specific gravity 15° C.	0·912

5. *Assortment of Same Sample, Without Tail.—Class* 3.

Production of oil	7·21 per cent.
Colour of oil	slight yellow
Refraction 15° C.	1·4970
Optical rotation 15° C.	+ 14·24 (length of tube 10 cm.)
Specific gravity 15° C.	0·922

6. *Assortment of Same Sample, Larger Kind.—Class* 4.

Production of oil	13·3 per cent.
Colour of oil	slight light green
Refraction 15° C.	1·4924
Optical rotation 15° C.	− 33·6 (length of tube 10 cm.)
Specific gravity 15° C.	0·923

From these figures there results:—

1. That the sample under review did not consist entirely of real cubebs, and that fruits must have been present that produced an oil, the physical qualities of which are very different from those of genuine cubeb oil, even so much that there are fruits among them which produce instead of the usual laevo-rotation a strong dextro-rotation.

2. That the reaction with concentrated sulphuric acid may be regarded an excellent expedient to recognise different kinds from irregular cubebs, for all kinds colouring yellow with H_2SO_4 also show important irregularities of the physical constants of the oil.

Oil of Matico.

This oil is distilled from the leaves and spikes of *Piper angustifolium* (*Artanthe elongata* Miq.), a plant indigenous to tropical America. Probably other species are sometimes used for distillation, as the imported leaves show considerable variation from time to time, and the following species have been recognised in commercial samples :—

Piper angustifolium.
,, *camphoricum.*
,, *lineatum.*
,, *angustifolium* v. *Ossanuno.*
,, *acutifolium.*
,, *mollicomum.*
,, *asperifolium.*

The yield of oil is very variable, from less than ½ per cent. to 6 per cent. The oil is a liquid of somewhat penetrating odour, and has the following characters :—

Specific gravity	0·938 to 1·135
Optical rotation	− 30° ,, + 6°
Refractive index	1·4950 ,, 1·5300
Acid value	0 ,, 4
Ester ,,	2 ,, 8
,, ,, (after acetylation)	25 ,, 50

It is usually soluble in 10 parts of 80 per cent. alcohol, but sometimes 25 parts are required. Very little is known of the chemistry of the oil. There is present a small quantity of "matico camphor," a crystalline odourless solid of specific gravity 1·080, melting at 94° and having a specific rotation in chloroform solution − 28·7°. Its formula is $C_{15}H_{26}O$, and it has been suggested that it is a derivative of ordinary camphor, but there is no experimental evidence in favour of this, and it appears to be a sesquiterpene alcohol. This body seems to be generally absent from the oil distilled at the present time. Schimmel & Co. state that they have found asarone $C_{12}H_{16}O_3$ in one sample of the oil, which had the specific gravity 1·077, and was slightly laevo-rotatory.

An oil has recently been examined by Schimmel which contained much asarone. A mixture of terpenes and cineol was also identified. A sesquiterpene alcohol also appeared to be present in this oil, which is possibly identical with the matico camphor which is now not always present in normal distillates.

Fromm and v. Emster[1] have isolated from matico oil a phenol

[1] *Bericht*, 35 (1902), 4347.

ether, which appears to be a methyl-butenyl-dimethoxy-methylene-dioxy-benzene, but Thoms[1] considers this ether to be a mixture of parsley apiol with a little dill apiol. A sample distilled from Peruvian leaves was found by Thoms[2] to contain limonene, dill apiol, and a sesquiterpene having a specific gravity 0·914, refractive index 1·5125, and boiling-point 138° to 139° at 17 mm. Traces of palmitic acid were also present.

The following oils, distilled solely from single species of matico have been examined:—

Source.	Specific Gravity.	Rotation.	Constituents.
Piper camphorificum[3] .	0·9465	+ 19° 21′	Camphor, borneol, terpenes, free acids, phenols, sesquiterpene alcohol.
,, *lineatum*[4] . .	0·958	+ 8° 45′	—
,, *acutifolium*[5] .	1·0965	+ 0° 24′	Pinene, dill apiol, sesquiterpene.
,, ,, [6]. .	0·9355	+ 0° 24′	Pinene, dill apiol, sesquiterpene, an alcohol $C_{10}H_{16}O$.

Oil of Betel.

Betel oil is distilled from the leaves, either fresh or dried, of *Piper betle* (*Chavica betle* Miq.). The yield is from ·5 to 1 per cent., but if the leaves are bleached by being kept in a warm shady place, the yield may reach 4 per cent. The oil varies considerably in characters according to the nature of the raw material, and the following results[7] have been obtained by various investigators:—

Origin.	Specific Gravity.	Rotation.	Constituents.
Siam . .	1·024	—	Betel-phenol and cadinene.
	1·020	—	—
Manila . .	1·044	—	Betel-phenol.
Java . .	0·958	− 1° 45′	Chavicol, betel-phenol, a sesquiterpene.
	0·959	+ 2° 53′	—
Bombay .	0·940 at 28°	—	—
Manila . .	1·0566	− 1° 10′	—
	1·055	+ 0° 40′	—
Java . .	1·0325	− 1° 55′	Betel-phenol, allyl-pyrocatechol, cineol eugenol methyl ether, caryophyllene.

The refractive index of betel oil exceeds 1·5100, usually about 1·5200.

Eykman investigated the oil some years ago and stated that it contained the phenol, chavicol (*q.v.*), several terpenes, a sesquiterpene, and possibly cymene and cineol. He described it as a yellowish-green oil, with a burning taste, and peculiar, pleasant odour. It was stated by him to be laevo-rotatory. Bertram and Gildemeister, however, in 1890, distilling the oil from the dried leaves, stated that the oil contained 70

[1] *Arch. Pharm.*, 242 (**1904**), 328.
[2] *Ibid.*, 247 (**1909**, 591.
[3] Thoms, *Apotheker Zeit.*, 24 (**1909**), 411.
[4] Thoms.
[5] *Ibid.* (leaves mixed with some of *P. mollicomum* and *P. asperifolium*).
[6] *Ibid.*
[7] Gildemeister and Hoffmann, 1913, ii. 222.

to 75 per cent. of betel-phenol, which was found to be a methoxy-chavicol. They identified the sesquiterpene as cadinene. Schimmel states that the oil distilled from leaves from all sources contained this betel-phenol, and in addition the oil from Java leaves contained several terpenes and chavicol, that from dried Siam leaves contained a sesquiterpene, and that from Manila contained betel-phenol, but no chavicol.

Eykman [1] has found a body of peppermint odour in the Java oil (menthone or menthol?).

Oil of Pepper.

This oil is distilled from the unripe berries of *Piper nigrum*. The yield of oil is from 1 to 2·5 per cent. The oil is often of a greenish colour, somewhat resembling oil of cubebs, but with a much more pungent odour. The green colour is due to a high boiling constituent, possibly identical with that in oil of cubebs.

Pepper oil has the following characters:—

Specific gravity	0·870 to 0·915
Optical rotation	– 10° ,, + 4°
Refractive index	1·4875 ,, 1·500

The oil consists largely of phellandrene, principally the laevo-modification. Dipentene is also present, and a small quantity of a sesquiterpene which is probably caryophyllene. Haensel [2] has isolated from the high boiling fractions a body melting at 138°.

Other "pepper oils" of interest are as follows:—

The so-called "long pepper," a mixture of *Piper longum* and *Piper officinarum*, yields about 1 per cent. of an oil of specific gravity 0·861, distilling mainly between 250° and 260°.

Piper Lowong, known as a kind of "false cubebs," yields about 12 per cent. of oil having a specific gravity about 0·865. It contains a crystalline substance of the formula $C_{10}H_{16}, 2H_2O$, melting at 164°.

Oil of Ashanti Pepper.

According to Schimmel & Co. the fruit of *Piper guineense*, or Ashanti pepper, yields an essential oil which has not previously been examined. The yields in two cases were 11·5 per cent. and 10·96 per cent. The oil was colourless or pale yellow, and had a peppery odour. The two samples gave the following figures on analysis:—

	1.	2.
Specific gravity	0·8733	0·8788
Optical rotation	– 3° 43′	– 5° 34′
Refractive index	1·48905	1·48847
Acid value	0·6	0·9
Ester ,,	5·5	4·2
,, ,, (after acetylation)	—	22·1
Solubility in 90 per cent. alcohol	1 in 8	1 in 9

Oil of Piper Aduncum.

The fruits of *Piper Aduncum*, known in Java as Daon Sisirihan, yield an essential oil having the following characters:—

[1] *Bericht*, 22 (1899), 2736. [2] *Pharm. Zeit.*, 50 (1905), 412.
[3] *Arch. Pharm.*, 234 (1896), 238.

Specific gravity at 27°	0·856
Optical rotation	+ 5° 40′
Acid value	1·2
Saponification value	8·4
,, ,, (after acetylation)	48·7

CANNABINACEÆ.

Oil of Hops.

Oil of hops is distilled from the flowers of the ordinary hop, *Humulus lupulus,* which yield from ·3 to 1 per cent. The oil glands are situated under the scales of the flowers (catkins); these grains or glands can be separated from the flowers, and are known in commerce as lupulin. The quality of the oil is, of course, dependent on the state of the flowers, and when the crops have had to be "sulphured," the oil obtained from the flowers is less valuable and contains traces of sulphur. The pure oil has a pale yellow to faint green tint, but can be obtained colourless by rectification. It has a penetrating hop odour. Its characters are as follows:—

Specific gravity	0·850 to 0·893
Optical rotation	− 2° ,, + 2°
Refractive index	1·4850 ,, 1·4920
Acid value	1 ,, 10
Ester ,,	15 ,, 40
,, ,, (after acetylation)	68 ,, 80

It is not soluble in 90 per cent. alcohol as a rule, and often not in 95 per cent. alcohol. In 1822 Payen and Chevalier stated that there were at least two bodies present. Wagner in 1853 obtained a terpene boiling at 175° and an oxygenated body which yielded valeric acid on oxidation. Personne claimed to have obtained a dextro-rotatory terpene of specific gravity ·888. The most modern and reliable investigation, however, is due to Chapman.[1] Four samples of authentic origin gave the following figures:—

	Specific Gravity at 15°.	Rotation $[\alpha]_D$.
1	·8802	+ 0·41′
2	·8662	+ 0·58′
3	·8771	+ 0·50′
4	·8743	—

All these samples were free from sulphur and neutral to litmus. No ketones or aldehydes were detected in the oil. A prolonged series of fractionations at 60 mm. pressure yielded the following fractions: (1) 89° to 91°; (2) 145° to 150°; (3) 163° to 168°; (4) 168° to 173°. The first and fourth were the main fractions, the second and third being very small. Fraction No. 1 was a colourless oil, practically unacted upon by sodium. When distilled from this metal, it boiled at 86° to 89° at the same pressure. Under ordinary pressure it boiled at 166° to 171° almost entirely, but towards the end, the temperature went up to 250°, due to polymerisation of the original substance. The specific gravity at 20° was found to be ·799 and the rotation − 0 56′. It is clear that this body is not a normal terpene, and Chapman's experiments lead him to consider it as a mixture of tetrahydrocymene, and one of the so-called "olefenic terpenes," bodies as yet but little understood. Frac-

[1] *Jour. Chem. Soc.*, 83 (**1903**), 505; 67 (**1895**), 54 and 780.

tion No. 2 was a liquid of specific gravity ·885, and may be geraniol. Fraction No. 3 was a mixture of 2 and 4. Fraction No. 4 was found to be a sesquiterpene, to which the name humulene (*q.v.*) was given. The same observer has also published figures for hop oils obtained from different sources as follows:—

	Specific Gravity.	Rotation.
Bavarian hops, 1894	0·8676	+ 0° 30′
,, ,, 1901	0·8403	− 0° 08′
Californian ,, 1899	0·8639	− 0° 20′

On distilling one of these oils he again obtained the "olefenic" terpene above referred to, which is probably myrcene, and linalol, together with an acid $C_9H_{18}O_2$ (probably isononylic acid) in the form of esters. Dipentene was also detected and probably esters of geraniol. The sesquiterpene "humulene" and the "olefenic" terpene make up more than 80 per cent. of the oil. Schimmel & Co. have detected gurjun oil as an adulterant of this oil.

Deussen,[1] however, has since suggested that the sesquiterpene "humulene" is in fact *i*-α-caryophyllene, with a small amount of β-caryophyllene.

Rabak[2] has also isolated free formic, valerianic and heptylic acids, and also formic, acetic, octylic, nonylic, decylic, œnanthylic, and butyric acids in the form of esters. Myrcenol was also found in the oil. Rabak considers that hop oil contains 15 to 20 per cent. of sesquiterpenes and 20 to 40 per cent. of esters of myrcenol. This latter figure must, however, be accepted with reserve, as the ester value is rarely high enough to account for this quantity of esters.

Rabak gives the following table of characters of a selected number of samples of hop oil:—

Origin.	Year.	Yield per Cent.	Rotation.	Refractive Index.	Acid No.	Ester No.	Solubility in 94 per Cent. Alcohol.
California .	1907	0·24	0·823	1·4856	2·3	44·4	5·4 Vols.
	1908	0·35	0·836	1·4738	1·7	45·5	4 ,,
	1909	0·38	0·839	1·4730	1·8	46·8	3·6 ,,
Oregon . .	1907	0·20	0·8343	1·4802	1·6	57·0	4·6 ,,
	1908	0·32	0·838	1·4730	1·0	50·2	3·3 ,,
	1909	0·30	0·8433	1·4705	2·8	56·0	3 ,,
New York .	1907	0·16	0·862	1·4804	3·6	61·0	3·5 ,,
	1908	0·14	0·837	1·4756	2·1	47·0	3·5 ,,
	1909	0·15	0·8777	1·4800	2·5	51·8	3·7 ,,
Washington .	1908	0·36	0·850	1·4763	1·0	51·8	3·5 ,,
	1909	0·38	0·8464	1·4734	1·5	53·8	3·3 ,,
Bohemia .	1907	0·32	0·855	1·4905	1·5	52·8	4 ,,
	1908	0·23	0·824	1·4852	1·0	52·8	6 ,,
	1909	0·24	0·861	1·4829	3·0	55·3	4 ,,

Oil of Hemp.

Hemp Oil.—The flowers of *Cannabis sativa*, the ordinary hemp, yield about $\frac{1}{10}$ per cent. of volatile oil of specific gravity about ·930. It is laevo-rotatory to the extent of about − 10° to − 12°. The oil

[1] *Jour. prakt. Chem.*, ii. 83 (**1911**), 483.
[2] *Jour. Agricult. Research Dept., Agric., Washington*, 2 (**1914**), **1**15.

appears to consist entirely, or nearly entirely, of sesquiterpenes and terpenes. The principal sesquiterpene present has a specific gravity 0·919 and optical rotation − 10·8°. It boils at 256° to 258° and forms a solid hydrochloride.

Indian Hemp Oil is obtained by subjecting the flowering twigs of *Cannabis Indica* to steam distillation. According to Valenta the resulting oil is of thin consistency, and possesses an agreeable aromatic odour. It boils between 248° and 268°. Its specific gravity is about ·930, and optical rotation − 10°, and it appears to consist chiefly of a sesquiterpene, which has been named cannibene.

Vignolo obtained an oil by distilling the female flowering plant, of similar boiling-point, and of optical rotation − 18°. Wood, Spivey, and Easterfield [1] isolated from the resinous extract an oil of the formula $C_{18}H_{24}O_2$, which is resinous at ordinary temperatures, and boils at 265°. Its specific gravity is 1·0424 at 18°. It has been termed cannabinol. They also found a slightly laevo-rotatory terpene (?), of specific gravity 0·819 and boiling at 160° to 180°. They found that the sesquiterpene boiled at 258° to 259° and had a specific gravity 0·898 at 18° and optical rotation − 8·6°. This body (cannibene) was also isolated by Vignolo,[2] who found its characters to be practically identical with those described by Wood, Spivey, and Easterfield.

A substance of the formula $C_{21}H_{29}O \cdot OH$ has been described by Frenkel and Czerkis [3] under the name of cannabinol, but does not appear to be the same body as that described above under the same name.

JUGLANDACEÆ.

Walnut-leaf Oil.

The essential oil of the leaves of *Juglans regia* has been examined by Schimmel & Co.[4] Two samples were distilled, one at Miltitz, and the other at Barrême. The distillates were of an olive-brown colour and possessed the characteristic odour of walnut leaves. At about 10° they formed semi-solid, butter-like masses, which melted at about 20°. From the solutions in 90 per cent. alcohol large quantities of paraffin separated out, which after repeated recrystallisation from alcohol had a melting-point 61° to 62°. The constants of the oil are shown in the table below :—

Distilled at	Yield.	$d_{30°}$.	a_D.	$n_{D_{25°}}$.	Acid v.	Ester v.	Ester v. after Acet.
	Per Cent.						
Miltitz	0·014	0·9137	±0°	1·49657	9·3	27·0	—
Barrême							
83 p. cent. Principal Oil	0·0072	0·9174	− 17° 36′	1·49177	3·7	9·3	—
17 ,, Water Oil .	0·0015	0·9231	− 16° 12′	1·49366	4·7	9·7	—
100 ,, Total Oil .	0·0087	0·9185	− 17° 0′	1·49215	—	—	98·5

[1] *Jour. Chem. Soc.* **(1896)**, 543.
[2] *Gazzetta* **(1895)**, 110.
[3] *Annalen*, 351 **(1907)**, 467.
[4] *Report*, April, **1912**, 131.

MYRICACEÆ.

Oil of Bayberry.

The leaves of the Bayberry, or wax myrtle, *Myrica cerifera*, yield about 0·02 to 0·05 per cent. of an essential oil having the following characters :—[1]

Specific gravity	0·9168
Optical rotation	− 1° 30′
Refractive index	1·4945 at 25°
Acid value	3·5
Ester ,,	21
,, ,, (after acetylation)	58

The oil was soluble in 0·5 volume of 90 per cent. alcohol, the solution becoming turbid upon the addition of 2 volumes or more.

This oil had previously been described as having a specific gravity of about 0·885 and an optical rotation − 5°. It is doubtful, however, whether these figures were obtained on a pure sample.

The leaves of the so-called Bog Myrtle, *Myrica gale*, which (as is the case with *Myrica cerifera*) must not be confused with the true Myrtle (*q.v.*), also yield an essential oil in small quantity. According to an early investigation by Rabenhorst,[2] the oil had a specific gravity of 0·876 and was a solid mass at 12·5°. The more recent investigations of Laloue,[3] however, would indicate that, as is true of many of the early investigations of essential oils, the material investigated was impure. Laloue obtained about 0·045 per cent. of oil having the following characters :—

Specific gravity	0·8984 at 25°
Optical rotation	− 5° 16′
Acid value	3·5
Ester value	15·5
,, ,, (after acetylation)	50·2

The oil was soluble in half its volume of 90 per cent. alcohol. On cooling it becomes turbid at 5°, but does not completely solidify at − 17°.

The oil has also been investigated by Pickles.[4] He obtained 0·076 per cent. of oil from a mixture of fresh leaves and twigs, and 0·203 per cent. from the dried leaves. The oils had the following characters :—

	Fresh Leaves.	Dried Leaves.
Specific gravity	0·915	0·912
Optical rotation	− 5° 17′	− 11° 26′
Acid value	7·0	4·0
Ester ,,	24·7	19·2
,, ,, (after acetylation	—	56·4

Enklaar[5] has examined the oil from the flower spikes of *Myrica Gale*. He obtained about 0·5 per cent. of oil having a specific gravity 0·899 and optical rotation − 5° 36′. The lowest boiling fractions contained pinene ; but whether α- or β-pinene, Enklaar was unable to distinguish. It would appear that the oil contains *d*-α-phellandrene

[1] Rabak, *Midland Drug and Pharm. Rev.*, 45 (**1911**), 484.
[2] *Report Pharm.*, 60 (**1837**), 214.
[3] *Bulletin*, Roure-Bertrand Fils, April, **1910**, 54.
[4] *Jour. Chem. Soc.*, 99 (**1911**), 1764.
[5] *Chem. Weekblad*, 9 (**1912**), 1

(melting-point of the nitrite 112° to 113°). The presence of cineol was proved by the preparation of the hydrobromide and the iodol compound. The highest boiling fractions of the oil contain a sesquiterpene (boiling-point 150° to 152° [17 mm.]; 263° to 265° [760 mm.]; $d_{15°}$ 0·928; a_D – 4° 30′), possibly caryophyllene. The high boiling portions also contain a solid body which crystallises from alcohol in needles and possesses the odour of Dutch myrtle.

The oil contains about 0·75 per cent. of paraffin hydrocarbons, of the empirical formula $C_{29}H_{60}$ melting at 63° to 64°. Free palmitic acid was also found, as well as cineol, dipentene, and high boiling sesquiterpenes and alcohols not yet identified.

Myrica asplenifolia, a North American plant, known as sweet fern, yields about 0·1 per cent. of an essential oil having a specific gravity about 0·926, which solidifies on cooling. It probably contains cinnamic aldehyde.

SALICINEÆ.

Oil of Poplar Buds.

The young buds of the black poplar, *Populus nigra*, yield from 0·3 to 0·6 per cent. of an essential oil of a pale yellow colour, with an odour somewhat recalling that of chamomiles. It has the following characters:—

Specific gravity	0·890 to 0·907
Optical rotation	+ 1° „ + 6°
Acid value	2 „ 12
Ester value	8 „ 14
„ „ (after acetylation)	18 „ 58

According to Piccard,[1] the principal constituent is a hydrocarbon of the empirical formula $(C_5H_8)x$, which from a vapour density determination appears to be $C_{20}H_{32}$.

Fichter and Katz,[2] however, consider that the hydrocarbon is a sesquiterpene $C_{15}H_{24}$, which is probably α-caryophyllene. A small quantity of a mixture of paraffin hydrocarbons melting at 53° to 68° is also present in the oil.

URTICACEÆ.

Oil of Pilea.

An oil distilled in Réunion from an unnamed species of *Pilea* has been examined by Schimmel & Co.[3] and found to have the following characters:—

Specific gravity	0·852 to 0·8533
Optical rotation	+ 33° 53′ „ + 58° 20′
Refractive index	1·4686 „ 1·4690
Ester value	5·1 „ 7·7
„ „ (after acetylation	24·2 „ 34·4

On fractionation the oil behaved as follows:—

[1] *Bericht*, 6 (**1873**), 890; 7 (**1874**), 1486. [2] *Ibid.*, 32 (**1899**), 3183.
[3] *Ibid.*, October, **1906**, 84; April, **1907**, 113.

	Distilled at 748 mm. Pressure.
158° to 159°	6 per cent.
159° „ 160°	35 „
160° „ 161°	10 „
161° „ 161·5°	10 „
161·5 „ 163°	10 „
163° „ 165°	10 „
165° „ 168°	7 „
168° „ 174°	2 „
174° „ 194°	8 „
Residue	2 „

The oil contains α-pinene and sabinene.

CHENOPODIACEÆ.

Oil of Wormseed.

This oil, known as American wormseed oil, is distilled from the herb *Chenopodium ambrosioides*, var. *anthelminticum*, in the United States, principally in Maryland. A certain amount of oil is distilled from the fruit only. Normal oils have the following characters:—

Specific gravity	0·958 to 0·990
Optical rotation	− 4° „ − 12°

The oil is soluble in 3 to 10 volumes of 70 per cent. alcohol.

No portion of the oil distils below 170°, and any distillate obtained below that temperature indicates adulteration with turpentine.

The oil has been very exhaustively examined recently by Nelson, to whom our knowledge of the chemistry of this oil is in the main due.

The lowest boiling fraction consists of *p*-cymene; *l*-limonen, sylvestrene and camphor are also present, but the principal constituent of the oil is the body ascaridol. Nelson[1] has fully investigated this body, $C_{10}H_{16}O_2$, which was originally isolated by Schimmel & Co. He examined four samples of the oil which had the following characters: d_{25° 0·955 to 0·9691; α_D − 5·4 to 8·8°; n_{D25° 1·4723 to 1·4726; soluble in from 3 to 7 volumes 70 per cent. alcohol. A fifth sample, at least one year old, which was used for the purposes of the examination, had the following characters: d_{25° 0·9694; α_D − 0·35°; n_{D20° 1·4780; soluble in 3 volumes 70 per cent. alcohol; acid value 0; ester value 5·0. Ascaridol has the following characters:—

Specific gravity	1·008
Optical rotation	− 4° 14′
Refractive index	1·4731
Boiling-point at 5 mm.	83°

The results obtained by shaking ascaridol with saturated solution of ferrous sulphate are, however, of particular interest. At ordinary temperature this peculiar reaction proceeds with such violence as to decompose the ascaridol, the temperature rising considerably with evolution of a combustible gas. *Iso*propyl alcohol is also formed as a product of reaction and was isolated from the syrupy product of conversion by steam-distillation and identified as acetone by oxidising it with chromic acid. Below 35° the reaction proceeds normally and without decomposition. In these conditions ascaridol adds the elements of the water and

[1] *Jour. Amer. Chem. Soc.*, 33 (**1911**), 1405.

is converted into a glycol, $C_{10}H_{18}O_3$, which, when benzoated by Schotten-Baumann's method, yields a solid benzoate, melting-point 136° to 137°.

According to Schimmel & Co., normal American worm-seed oil contains from 62 to 65 per cent. of ascaridol, and about 22 per cent. of cymene. Light oils of inferior quality, due probably to the quality of the material distilled, having a specific gravity about 0·942, contain from 45 to 50 per cent. of ascaridol and about 38 per cent. of hydrocarbons.

The oil is used as an anthelmintic.

E. K. Nelson[1] has recently investigated the changes taking place in this oil on keeping. He points out that this oil increases in specific gravity and decreases in optical rotation after being kept at ordinary temperatures for a year. If kept in an ice-chest the changes are not so marked. The following table shows the changes in four samples of the oil kept for a year at room temperature, and in a refrigerator:—

At 25°.	1.	2.	3.	4.
	Fresh.			
Specific gravity	0·9691	0·9700	0·9550	0·9584
Rotation	− 5·4°	− 6·2°	− 8·8°	− 6·3°
Refractive index	1·4726	1·4723	1·4726	1·4725
	Refrigerator.			
Specific gravity	0·9700	0·9711	0·9590	0·9595
Rotation	− 5·3°	− 5·9°	− 8·6°	− 6·1°
Refractive index	1·4740	1·4740	1·4743	1·4740
	Room Temperature.			
Specific gravity	0·9703	0·9804	0·9838	0·9600
Rotation	− 5·2°	− 5·5°	− 6·6°	− 6·1°
Refractive index	1·4745	1·4750	1·4770	1·4746

Nelson finds that when ascaridol suffers rearrangement, either by heating or by treatment with ferrous sulphate, there are two other bodies formed besides the glycol described by Schimmel. One of these is also a glycol, which he terms β-glycol, to distinguish it from the other one. It has the formula $C_{10}H_{18}O_3$, and is decomposed with the formation of thymol on warming with dilute sulphuric acid. The third substance is an erythrite melting at 128° to 130°, of the formula $C_{10}H_{20}O_4$, which when boiled with dilute sulphuric acid yields a ketone with a strong menthone odour, and a phenolic body melting at 80° to 81°. The ketone yields a semi-carbazone melting at 182° to 184°. It is probably an isomeric menthenone, or a mixture of several isomers.[2]

LAURACEÆ.

Oil of Cassia.

This oil is distilled from the leaves, twigs, and other parts of the plant *Cinnamomum cassia*, which is probably a native of Cochin China, but which is chiefly cultivated in China proper, the three chief districts being Taiwu (Lat. 23° 34′ N., Long. 110° 18′ E.), in the Kwangsi province; Lukpo (Lat. 23° 6′ N., Long. 112° 24′ E.), and Loting (Lat. 22° 52′ N., Long. 111° 8′ E.), both in the Kwangtung province. Much confusion and mystery existed till recently in regard to the exact botanical source of the plant and the nature of the oil and the methods of its distillation, etc., which were largely caused by the shameful adulteration by the Chinese. The following details furnished by actual

[1] *U.S.A. Dept. of Agriculture, Bulletin*, No. 109.

[2] In a still later publication (*P. and E. O. R.* (**1920**), *from J. Amer. Chem. Soc.*), Nelson deals with the distillation of this oil. An abstract of this paper will be found in the Appendix at the end of this volume.

observers give the most reliable information on much affecting these questions. In a *Report on a Journey to Kwangsi*, by H. Schroeter, 1887, the writer says:—

"The shrubs destined for the production of the *Cassia lignea* proper are partly stripped during the summer months of their minor branches and exceptionally juicy leaves. They are then conveyed in huge bundles into the valley, where they are boiled in large vessels. From the aromatic juice thus obtained the esteemed cassia oil is recovered by means of a most primitive distilling apparatus. As the Li-kin stations on the road to Canton levy an excessive duty upon the oil, in addition to that exacted by the Imperial Customs, the oil is carried in tins across the hills to Pakhoi and thence transported to Hong-Kong *viâ* Macao, instead of reaching Canton by the waterway intended by nature for its conveyance."

These statements are in agreement with those contained in a report presented to the Hong-Kong Government by Mr. Charles Ford, which will be found reprinted in the *Journal of the Linnæan Society*, December, 1882, and in which much valuable information is given. More recently (1895) Messrs. Siemssen & Co. of Hong-Kong, on whose behalf Herr Schroeter's visit was made, sent a representative, Mr. Struckmeyer, to the Loting-Chow districts, in company with the German Consul, Dr. Knappe. The report of these travellers is sufficiently interesting to warrant its reproduction here, as this oil is of great commercial importance, and reliable information as to its production is very scarce. It read as follows:—

"We embarked on the right bank of the West river, opposite Tack-Hing-Chow, at six a.m. on 15th December, 1895. We left there our large boat and went to a place with extensive matting factories, located on the Loting-Chow or Lintau river, arriving there about half-past two o'clock in the afternoon. Next morning we continued our march to Loting-Chow, under the guidance of a Chinese, a manager of one of the largest matting establishments in Lintau. We were attended by a servant of the Consul, who also served as interpreter, and by several Chinese coolies. After a brisk march we took up our quarters in a small Chinese temple, and made strolls through the city, which, although small, seems to be a busy trading place. Loting-Chow is pre-eminently a centre for all the products of the cassia tree, and for firewood to be shipped thence to Canton and Macao.

"The cassia plantations are still further inland, and we had to defer visiting them until the next day, because they were too far off, according to our informants in Loting-Chow. They proposed, however, to accompany us the following day to cassia plantations as well as to a distillery. During the afternoon we visited a dealer in cassia oil, the only one in the place, he pretended. Subsequently, on our return, we ascertained this statement to be false. At first he was very reticent, but subsequently yielded to some pressure. He possessed a separating funnel, a syphon, test-tubes, etc., and pretended to examine all oils offered to him for sale, because he had been condemned, as he said, to $900 damages for inferior cassia oil sold by him some five years ago. He assorts the commercial oil into four or five grades, but deals only in the three best qualities, unadulterated in his opinion. He had in stock but a small quantity of the best oil, and gave us small specimens of three kinds of his oil, of which I shall treat more fully further on.

"Although everything was arranged for the next day with our Chinese guides, we had to face considerable difficulties. During the evening we were informed that a certain 'Mr. Wong' (who never appeared in person) had arrived from the interior, and reported that a visit to the cassia plantations would be of no use, and that it would not be possible to inspect any distillery, because some children had been instructed by thieves to steal the leaves from the cassia trees, and that in consequence a brisk contest between the owners and the thieves was in progress. Besides, it was claimed that only one distillery existed, at a distance of 20 Lis (about 8½ miles), and that this was not in operation in consequence of the disturbances. This wondrous story, and the circumstances under which it was imposed upon us, convinced us that it was an impudent lie, or, at best, an absurd exaggeration. The more they tried to dissuade us from continuing our trip, the more we persisted upon the realisation of our plan. Finally the Lintau Chinaman volunteered to accompany us to the plantations, but would not make any promise in regard to visiting any distillery. The departure was fixed for six o'clock the next morning. We were in a peculiar position, since our interpreter also appeared to antagonise our aims; nor had we any time to lose in order to comply with all other arrangements of our expedition.

"The next morning, 17th December, we and our men were ready at six o'clock, but the two Chinese failed to appear, although we had requested them several times to hurry. We were waiting with our attendants in front of the temple when messengers brought the idle excuse that the two Chinese had not yet received their breakfast, and that no carriers for our baggage could be obtained. At a quarter to eight we sent word that we would start without guides, and when they caught up with us at ten o'clock we paid no attention to them.

"We finally succeeded in inducing one of the many onlookers to guide us to one of the distilleries. The march went on very slowly, and the actions of our interpreter led us to foresee further intrigues. As on the previous day, our route passed through a hilly country with numerous side valleys and little ravines. After having passed a number of hills covered with young cassia trees, we reached a distillery by ten o'clock, but found out that it was not in operation. We were again told that this was the only distillery in the neighbourhood. But as luck would have it, during our brief stay a small trader appeared upon the scene and offered some thick cassia bark for sale. By small purchases and promises we succeeded in inducing the man to lead us to another distillery, and one in operation.

"We returned the same way by which we had come, and at noontime arrived at a distillery in full operation. It was a little off the road, but was much nearer to our starting-point in the morning than the distillery to which we had been led at first. We met with a very kind reception by the owner and his employees, and they readily gave us any desired information. We ascertained that about twenty distilleries existed in this immediate neighbourhood, and that most of them at the time were in full operation. We subsequently found out on our return to Canton that a few slight disturbances had occurred in the cassia districts, but the fact remained that we had been treated to absurd stories and gross lies, probably at the instigation of some dealers in Loting-Chow.

"The distillery, like all others, is located in a ravine abounding in springs which furnish an ample supply of water for coolers. The distillery with all its arrangements closely resembles the one which some time ago was described and illustrated in one of the reports of Messrs. Schimmel & Co.

"An iron pan is securely placed in a brick hearth with a large fireplace; upon this pan a large wooden cylinder lined with sheet iron is placed, upon whose upper rim a large cover of strong sheet iron rests. The space between the cylinder and the cover, around the rim, is made tight by strips of moist cloth or rags. Around the lower side of the cover there runs an outer groove for gathering and drawing off the water serving for cooling, and a narrow interior groove for collecting the distillate containing the essential oil. This distillate passes into tin cans and the oil collects at the bottom.

"At each charge of the cylinder, about 1 picul of leaves and twigs and 250 catties (= 2½ piculs) of water are employed; the water used is mostly from preceding distillates. The distillation lasts about two and a half hours.

"The yield of one such charge, when consisting of leaves only, is 1½ to 2 taels, and when consisting of 70 per cent. of leaves and 30 per cent. of twigs, 2½ to 3 taels of oil.[1] The oil obtained from leaves alone is in general the superior one.

"With a still of this capacity, as much as 50 catties of oil are obtained per month, or 3 to 5 piculs per year. The festivities connected with the Chinese New Year cause an interruption of the distillation for one or even two months. The quality of the oil depends upon the material used for distillation. Both too old and too young trees furnish leaves less rich in oil; large leaves are better than small young ones; this explains the fact that the twigs and leaves gathered in spring and in winter yield an inferior oil to those gathered in midsummer and in autumn.

"The distiller who furnishes us this information claims to produce but one grade of oil, using 70 per cent. of leaves and 30 per cent. of twigs. A specimen, marked A, was taken by us from the apparatus. The tin can contained but a small amount of oil resting under the water, and the sample, collected in a rather primitive way, contained some water. On standing, some drops of water separated at the top of the slightly yellow oil. It was subsequently examined by Mr. E. Neidhardt in Hong-Kong, with the following result:—

"Specimen A (1895). A turbid liquid which on standing separated into two layers, the upper one water, the lower one cassia oil of a light Rhine wine colour.

Consistency: very limpid.
Odour: the true oil of cassia, somewhat smoky.
Taste: the same.
Specific gravity: 1·059.
Residue of distillation: 3·75 per cent.; resinous, sticky.
Cinnamic aldehyde: 86 per cent.
Solubility in alcohol 80 per cent.: complete.
Solubility in solution of sodium salicylate: complete.
Reaction with lead subacetate: no turbidity.

"The result of this investigation indicates that a rather large proportion of twigs seems to be required to materially reduce the quality

[1] 16 taels = 1 catty = 605 grms.; 1 picul = 60 kilos 500 grms.

of the oil. Perhaps twigs too old and too thick have a detrimental effect upon the quality of the distillate.

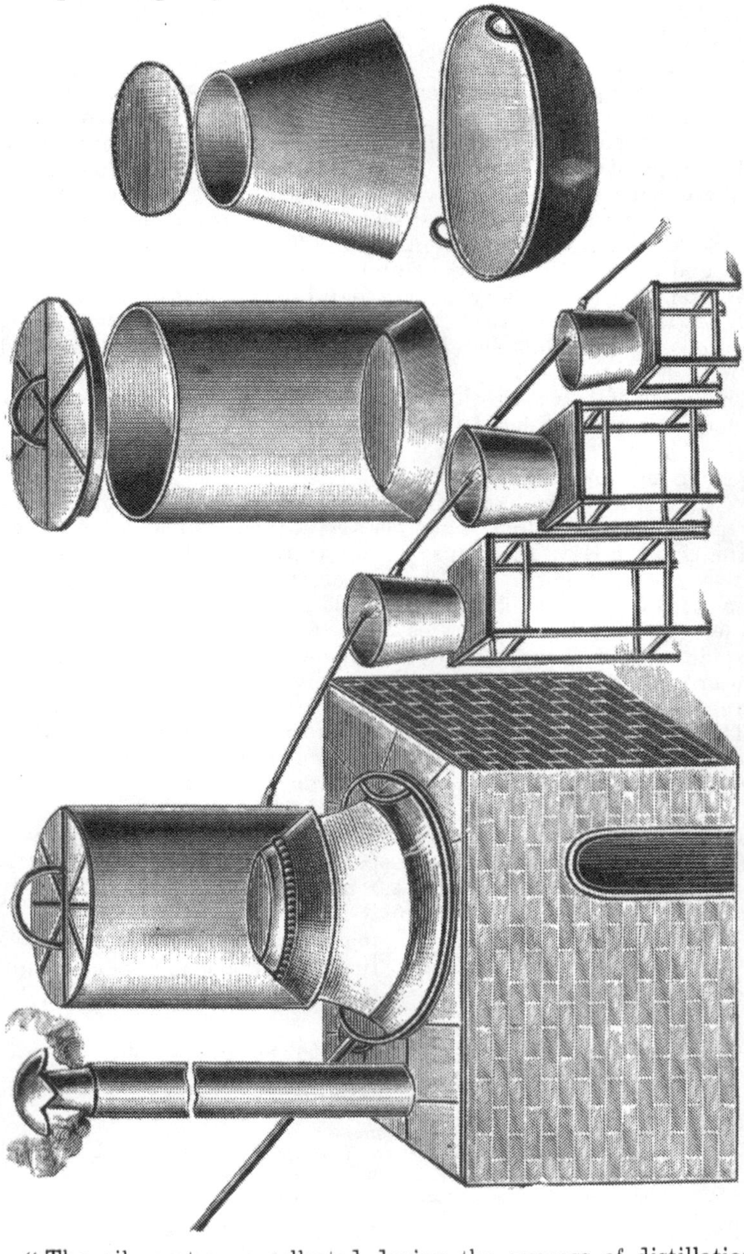

FIG. 17.—Cassia still.

"The oily water, as collected during the process of distillation, is milky. Specimen B represents this water as separated from the oil and mixed with Specimen A.

"Specimen C represents the ready oil as our distiller sends it to Loting-Chow for sale. The specimen was taken from a tin can by ourselves.

"For these samples we were charged in proportion to the regular price of $600 per picul.

"The distillers generally purchase the leaves and twigs and carry on the distillation on their own account; occasionally they rent their stills to the owners of the raw material, mostly planters, or they run their stills on the latter's account in return for a compensation.

"In Loting-Chow we obtained two more samples of commercial oil. Specimen D was said to be a good oil of prime quality obtained from old leaves and commanding a price of $600 per picul. It did not change when brought into contact with water, but separated into smaller or larger globular masses, showing a few oil drops on the surface. It was of a light colour.

"Specimen E, said to be distilled from young leaves; price $580 per picul. It showed in water a more compact mass, with a thin, transparent surface, and more oil drops on top of the water.

"Specimen F was said to have been obtained from old and young leaves and to have been adulterated with kerosene, and therefore to be of no use to the dealer; price $550. When brought into water, this oil formed a compact mass with a thick surface, partly of a whitish and striped appearance. Colour rather dark. The dealer stated that he had showed us this oil only as a specimen of adulterated oil, that he kept it only as a sample, and that he had only a small amount of it. Consequently he could only sell us such a small sample.

"The adulteration is charged by every one to somebody else. We found no suspicious material in the distilleries; there is an abundance of pine leaves growing in the region, but their admixture with the cassia leaves was peremptorily denied; nor did we find any such admixture in the contents of the still examined by us. Although a few distillers may be guilty of dishonest practices, there is no reason to assume that the majority resort to deliberate adulteration by the admixture of foreign matter with the cassia leaves and twigs. On the other hand, we may conclude from the existence of adulterated oil in Loting-Chow, and from our examination made there, that the sophistication ought not to be imputed exclusively to the Chinese of Macao and Hong-Kong, if at all. The fact is, before the oil is sold to the foreign exporters, it comes from an extensive district and passes through so many hands that it is impossible to determine where and how it is actually adulterated. In Loting-Chow it was alleged that most adulterations are committed in the province of Kwangsi, where ten or even more qualities of cassia oil can be had.

"The question may here be raised whether the present methods of examination suffice to detect the most artful and subtle adulterations, as also whether the Chinese really possess the requisite knowledge and skill for such refined forms of sophistication.

"Sample A bears sufficient evidence that excellent oil can be, and is, produced. This detailed account, however, may indicate what difficulties the foreigner has to meet with when trying to get at the bottom of facts.

"Incidentally it may be added that the residue of the leaves and twigs, after having been dried, is used as burning material. The remaining ashes are subsequently extracted by hot water, and a kind of potash called Kamalni is obtained therefrom.

"Consul Dr. Knappe has forwarded a collection of the samples A, B, C, D, E, and F and an elaborate report to the Foreign Office in Berlin. Sample A is the only one which we had examined in Hong-Kong."

Schimmel & Co. have obtained specimens of raw materials of various parts of the plant and examined the oils yielded by them. The results of their investigations are summarised as follows :—

"*Cassia Budsticks.*—Small, thin stalks, about 1 mm. thick and 5 mm. long, which possess a strong and pure cinnamon taste. Six hundred and fourteen kilos distilled by us yielded 10 kilos 120 grms., or 1·64 per cent., of a beautifully bright oil with excellent sweet taste, indicating a high percentage of aldehyde.

"We determined the specific gravity to be 1·0463, the aldehydic contents 92 per cent.

"*Cassia Leaves.*—The dried, leather-like leaves of the cassia plant with the strong petioles and small twigs. On mastication of leaves and petioles separately, both communicated a faint but pure taste of cinnamon, whilst the twigs exhibited strong cinnamon taste.

"In order to obtain exact results we caused a certain quantity of the material to be sorted into leaves and stalks. One hundred kilos material yielded—

Leaves	85 kilos.
Small twigs	15 ,,

"On distillation the leaves yielded a beautiful, sweet-tasting oil, with specific gravity 1·056 at 15° C. The aldehydic contents were determined as 93 per cent.

"The thin twigs showed a less percentage of essential oil than the leaves, but this possessed a similar sweet taste to the oil from the leaves. The specific gravity was 1·045 at 15° C. The aldehyde amounted to 90 per cent.

"The following parts of the cassia plant have been distilled by us with the appended results :—

1. The Cassia bark (the *Cassia lignea* of commerce).
   Yield of essential oil : 1·5 per cent.
   Aldehyde in oil : 88·9 per cent. Specific gravity 1·035.
2. The Cassia buds (the *Flores cassiae* of commerce).
   Yield of essential oil : 1·550 per cent.
   Aldehyde in oil : 80·4 per cent. Specific gravity 1·026.
3. Cassia budsticks.
   Yield of essential oil : 1·64 per cent.
   Aldehyde in oil : 92 per cent. Specific gravity 1·046.
4. Cassia leaves, leaf-stalks and young twigs mixed.
   Yield of essential oil : 0·77 per cent.
   Aldehyde in oil : 93 per cent. Specific gravity 1·055.

"Nos. 1 and 2 are completely excluded as raw materials for the preparation of cassia oil on account of the price.

"No. 3 is also out of the question, because the value, about $4 per picul, is still higher than that of broken cassia and because the yield would not be sufficient.

"It can, therefore, be assumed with safety, that the cassia oil of commerce is distilled in China from the leaves, leaf-stalks, and young twigs of the cassia plant, probably together with various refuse products worthless for other purposes."

Some interesting information has been published recently by the United States Consul at Swatow [1] on recent developments in the cassia-producing industry.

He states that in 1910 Huang Hsi-ch'uan, a member of the Chinese provincial assembly, instituted a series of experiments in cassia-culture at Kaying, in the course of which it was found that the soil and climate there were very suitable for the growth of this plant. Huang Hsi-ch'uan had previously carefully studied the corresponding conditions at Wuchow, on the West river, where there are large cassia forests, and accordingly he started a model plantation, not only, it is said, with a view of making money from the undertaking, but especially in order to promote industrial enterprise and to counteract the tide of emigration among the people of the region. Two species of cassia are grown in the new plantations, one large and the other small. The former flowers when ten years old, but it is only cut down at the age of twenty years for the purpose of harvesting the bark. Between the ages of ten and twenty years only the seed, of which the value is small, is harvested. The smaller variety yields leaves for the manufacture of oil at three to four years of age, and its bark is only harvested at the age of six or seven years. As soon as the trees (and this applies to both varieties) are cut down, new shoots immediately spring up around the stumps, so that such a plantation, once started, lasts practically indefinitely. The wood is used for fuel, the wood-ashes and the small twigs are employed for medicinal purposes, the bark yields the "cassia" of commerce, and from the leaves oil is distilled.

Cassia oil is imported, principally from Hong-Kong, in leaden vessels holding about 16½ lb. It is always sold as of a definite content of cinnamic aldehyde, the lowest commercial quality being 70 to 75 per cent., and followed by 75 to 80 per cent., 80 to 85 per cent., and 85 to 90 per cent. The lower grades are usually adulterated, the usual adulterant being common rosin, and if a really pure oil is required, one containing 85 to 90 per cent. of cinnamic aldehyde should be insisted upon.

A pure cassia oil should have the following characters:—

Specific gravity	1·055 to 1·072.
Refractive index	1·6000 ,, 1·6060.
Optical rotation	− 1° ,, + 6°.
Acid value	6 to 16 (rarely to 20),
Cinnamic aldehyde	rarely below 85 per cent.

The pure oil is completely soluble in 2 volumes of 80 per cent. alcohol, and usually in 3 to 4 volumes of 70 per cent. alcohol. The solubility is impaired by most adulterants, such as colophony, heavy petroleum oil or gurjun oil.

Colophony is indicated by a high acid value, and may be detected and approximately determined by the lead acetate test. If to a solution of 1 c.c. of the oil in 3 c.c. of 70 per cent. alcohol, there be added, drop by drop, half a c.c. of a saturated solution of lead acetate in 70 per cent. alcohol, no precipitate should be produced in the absence of colophony. A precipitate indicates this adulterant. For an approximately quantitative determination 5 grms. of the oil are dissolved in 20 c.c. of 70 per cent. alcohol, and sufficient lead acetate solution as above added for complete precipitation, 10 c.c. being the average quantity. The precipitate

[1] *Daily Consular and Trade Reports*, 3 July, 1911.

is collected on an ignited asbestos filter, washed with 70 per cent. alcohol, and dried at 100°. The weight of the precipitate is approximately equal to that of the colophony present.

The following rough method gives an approximate estimate of the amount of adulteration, if this is colophony, heavy petroleum oil, or a fixed oil. Weigh 50 grms. of the oil into a small Wurtz flask, and connect in the usual way to a condenser, and gradually heat with a naked flame. The oil commences to boil at about 210°, and the bulk distils over between 240° and 260°. When the thermometer rises to 280° and white vapours rise the distillation is stopped.

On cooling the flask (whose weight is known) its contents are weighed; the residue should not exceed 10 per cent.—it is seldom more than 8 per cent.—and should not solidify. Over 10 per cent. residue indicates adulteration, and if this residue solidifies colophony is present.

The most important method of examination, however, is the determination of the percentage of cinnamic aldehyde, and upon this percentage commercial oils are always sold on the market. In this process the following details should be observed. Ten c.c. of the oil are run into a Hirschsohn flask (capacity about 100 to 150 c.c., with a neck about 5 ins. long and $\frac{1}{4}$ in. in diameter, graduated in $\frac{1}{10}$th c.c.). The flask is then filled about three-quarters full with a 30 per cent. solution of sodium bisulphite, and the whole well shaken. The flask is then placed in the water-bath for several hours with occasional shaking, until the precipitated compound of the aldehyde and bisulphite is completely dissolved, and only a clear oil floats on the surface. Bisulphite solution is then carefully poured in until the oil is driven up into the neck, and when it has attained the temperature at which the oil was measured, the amount is read off. This gives the percentage of non-aldehydic constituents, the difference being returned as cinnamic aldehyde. Pure oils should not give much less than 85 per cent. of aldehyde, the best oils yielding 85 to 90 per cent., or occasionally even higher. Strictly speaking,, these percentages are by volume, but the errors of reading the result, and those due to solubility of the non-aldehydes in the aqueous liquid render any correction for the specific gravity of the constituents unnecessary in practice. Care must be taken that every particle of the aldehyde compound is dissolved, as otherwise the reading of the oily layer will be obscured, and a serious error may be introduced. If the precipitate does not dissolve after standing three hours in the water-bath with repeated shaking, the presence of solid resin may be inferred, and the estimation of the aldehyde in this way, with accuracy, is then impossible. Hirschsohn recommends shaking the oil in a graduated tube with three times its volume of light petroleum ether. A diminution in volume of the oil indicates petroleum, resin or fatty oils (except castor oil). An increase in volume of the oil indicates castor oil. The petroleum layer is then shaken with copper hydroxide. A blue solution indicates copaiba or resin.

Hanus[1] publishes a new method for the determination of cinnamic aldehyde in cassia and cinnamon oils depending on the combination of the aldehyde with semioxamazide. Ten grms. of finely powdered hydrazine sulphate are dissolved in a solution of 9 grms. of caustic soda

[1] *Pharm. Central.*, **1904**, 37.

in 100 c.c. of water, and the alkaline sulphate produced is precipitated by the addition of 100 c.c. of alcohol. After filtration the solution is warmed, 9 grms. of oxamæthane are added in small portions, the whole warmed for half an hour and allowed to cool. The azide separates in crystalline tables, and these are separated and recrystallised. To estimate the aldehyde by means of this reagent, a small quantity, not more than 0·2 grm., of the oil is well shaken in 85 c.c. of water, and about 0·35 grm. of semioxamazide in 15 c.c. of hot water is added and the whole well shaken. After five or ten minutes the compound begins to be precipitated, and after standing twenty-four hours can be collected on a Gooch filter, washed with cold water, and dried for a few minutes at 105°. The amount of the precipitate is multiplied by 0·6083 to obtain the amount of aldehyde. The constitution of the semioxamazone of cinnamic aldehyde is $NH_2 . CO . CO . NH . N : CH . CH : CH . C_6H_5$.

The main constituent to which the odour of this oil, and of oil of cinnamon, is chiefly due, is cinnamic aldehyde. Small quantities of cinnamic acid, its oxidation product, are naturally found in the oil, the amount varying according to its age. A terpene, and the acetic esters of cinnamyl and phenyl-propyl alcohols, also exist in very small quantities. A stearoptene was also found in this oil in 1850 by Rochleder. According to this chemist its formula is $C_{28}H_{29}O_5$. But a recent investigation by Bertram and Kursten[1] shows that it is *ortho*-methyl-coumaric aldehyde $C_6H_4(OCH_3)(CH : CH . COH)$.

Dodge and Sherndal[2] have recently shown that cassia oil yields at least 0·5 per cent. of constituents removable by shaking out with 2 per cent. sodium hydroxide solution. The oily liquid obtained by liberating these alkali-soluble constituents in the usual manner was found to consist of approximately 25 per cent. of salicylic aldehyde; 60 per cent. of coumarin; from 8 to 10 per cent. of cinnamic acid; with small amounts of salicylic and benzoic acids, and of a liquid acid not yet identified. Although the amounts of salicylic aldehyde and of coumarin are small, they are sufficient to contribute to the composite aroma of the oil. In fact, the presence of salicylic aldehyde may be detected by odour alone in the first fractions of the redistilled oil. It is to this constituent that the purple colour reaction is due, which is obtained when cassia oil is shaken with dilute alkali and the acidified aqueous portion is treated with ferric chloride.

The same chemists have since[3] shown that very small amounts of benzaldehyde and methyl-salicylic aldehyde are also present in the oil.

Cassia oil, being packed in lead containers, usually contains traces of lead, and in some quarters a demand has arisen for a lead-free cassia oil. For the determination of the lead the method proposed by Lubetti[4] yields useful results.

5 c.c. of the cassia oil to be tested is diluted to 20 c.c. with 90 per cent. alcohol; ·5 or 2·5 c.c. of this solution is taken according as the lead content is more or less than 0·025 per cent., which will be determined by a preliminary test. This amount is pipetted into a 50 c.c. Nessler glass of small bore (2·5 cm.) and diluted to about 20 c.c. with 90 per cent. alcohol. A blank from lead-free cassia oil, obtained by distilling the commercial oil, is prepared in a similar way; 1 c.c. of ammonium sulphide solution is added to both glasses and standard lead solution (1

[1] *Jour. Prakt. Chem.*, ii., 51 (**1895**), 316. [2] *Jour. Eng. Ind. Chem.* (**1915**), 1055.
[3] *Jour. Ind. Eng. Chem.* (**1918**), 10, *12*, 1005. [4] *Jour. Soc. Chem. Ind.*, **1920**, 35.

c.c. = 0·0001 grm. Pb.) added to the blank until the brown coloration produced matches the sample under examination, the liquid being stirred three times after each addition of standard lead solution by means of a piston-shaped glass rod.

The results are calculated by means of the formula $x = n \times 0{\cdot}04 \div m$, where x is the percentage of lead sought, n the number of c.c. of standard lead solution required, and m the amount of original solution taken. The tint of the solutions can be compared by looking through the Nessler glasses resting on a white tile. But better results are obtained by using a Duboscq colorimeter. The values for lead found by this method are slightly high, the average excess found being 0·00018 grm. This correction may be applied if strictly accurate results are required. The following were the results obtained on five oils in comparison with the results by gravimetric analysis :—

Oil	% found by Gravimetric Method.	% found by Colorimetric Method.	Difference.
1	0·0566	0·0536	+ 0·0030
2	0·0524	0·0520	+ 0·0004
3	0·0494	0·0512	− 0·0018
4	0·0546	0·0520	+ 0·0026
5	0·0611	0·0592	+ 0·0019

Experiments were made on the lead-dissolving power of cassia oil. 10 c.c. of the lead-free cassia oil was introduced into each of 22 test tubes, together with a strip of bright lead about 1 cm. broad and 6 cm. long. The tubes were then closed with clean corks. shaken at intervals, and tested at intervals. The amount of lead absorbed increased with the time, and reached a maximum of 0·074 per cent. after one month, and then seemed to remain constant. It is noteworthy that after that period the lead content exceeds the percentage found in commercial samples, which is, as a rule, from 0·04 to 0·06 per cent.

Oil of Cinnamon.

The cinnamon oil (*i.e.* the bark oil) of commerce is understood to be the product of distillation of the bark of *Cinnamomum zeylanicum*, the Ceylon cinnamon, a native of that island. The leaves and root bark also yield essential oils, the former of which is of considerable importance commercially. The cinnamon tree is a small evergreen tree with shining leaves and panicles of greenish flowers, of somewhat unpleasant odour. It is extensively cultivated in Ceylon, from which nearly all the bark of commerce is obtained. It is also grown in India, Mauritius, the Seychelles and Jamaica, but Ceylon is responsible for practically the whole of the cinnamon used. Several varieties are cultivated, but whether they are merely the same plant slightly modified by the influence of soil, etc., or not, is uncertain. The Colombo, Negumbo, and Matura districts in the south-west of the island produce the majority of the bark, and the famous "cinnamon gardens" of Colombo, which occupy a considerable space, form one of the great attractions of the island. The age of the tree, the care devoted to its cultivation, and the time at which the bark is gathered largely determine the value of the product. The appearance and size of the "quills," together with their odour and flavour, determine their value on the market. The small fragments or chips and the bark of inferior quality are used in Ceylon for distillation, and also form the chief material from which the oil is distilled in England. The oil distilled in Ceylon is usually obtained by macerating the fragments of bark in sea-water or strong brine for two or three days and then subjecting them to distillation. The yield is from ·5 to 1 per cent. The majority of that exported, however, is not genuine. Either the leaves are added to the bark when distilled, or cinnamon leaf oil is added to the oil after distillation. The important difference between the two oils is that the bark oil owes its characteristic odour to the cinnamic aldehyde it contains, whilst the leaf oil contains only traces of that body; the chief constituent of the latter oil is eugenol, the characteristic phenol of the oils of cloves and pimento.

The more delicate odour of cinnamon oil causes it to be preferred to cassia oil in England, whilst the stronger and coarser odour of the latter, together with its much lower price, gains for it preference in many parts of the continent.

In ancient literature there appear many references to cinnamon, but as these do not appear to mean the cinnamon bark as we know it to-day, and the latter is so largely a British product, the following interesting details, due to E. M. Holmes,[1] may be quoted :—

"Some kind of spice, bearing the name of cinnamon, appears to have been known from the earliest times, but it was apparently not the Ceylon cinnamon of the present day, since, according to Tennent (*Ceylon*, 1859, p. 575) 'in the pages of no author from the earliest ages to the close of the thirteenth century is there the remotest allusion to cinnamon as an indigenous production or even as an article of commerce in Ceylon'. Hanbury (*Pharmacographia*, second edition, p. 521) regards the cinnamon of the ancients as the bark now known as Chinese cassia, and the cassia of the ancients as perhaps one of the thicker and less aromatic barks of the same group, such as are still found in commerce.

"Although cinnamon seems to have been the first spice sought after in all the Oriental voyages in ancient times, and although both cinnamon and cassia are mentioned by Theophrastus, Herodotus, Galen, Dioscorides, Pliny, Strabo, and many other ancient writers on plant products, it is evident that the barks distinguished by those names were extremely analogous, since Galen remarks that the finest cassia differs so little from the lowest quality of cinnamon that the first may be substituted for the second provided a double weight of it be used. That there was a distinct difference in flavour of the barks called cinnamon and cassia recognised by ancient writers is quite clear.

"In 1511 Barbose distinguished the fine cinnamon of Ceylon from the inferior *Canella trista* of Malabar ('Canella' being the Portuguese name for cinnamon). About fifty years later Garcia D'Orta stated that Ceylon cinnamon was forty times as dear as that of Malabar, and in 1571 he saw branches of the tree at Bristol, in Holland. Cinnamon was then cut from trees growing wild in the forests in the interior of Ceylon, the bark being exacted as tribute from the Kings of Ceylon by the Portuguese. The bark appears to have been collected by the *Chalias*, a peculiar caste, who are said to have emigrated from India to Ceylon in the twelfth century, and who possibly brought the knowledge of the value of the tree with them from the Malabar coast; in aftertimes they

[1] *P. and E.O.R.* (**1916**), 41.

became the regular cinnamon peelers. The bark previous to shipment was examined by special officers to guard against frauds on the part of the *Chalias*. The cinnamon trade fell into the hands of the Dutch in 1636, but it was not until 1770 that the idea of De Koke of attempting the cultivation of the tree was carried out, and with such success that the Dutch were able, independently of the kingdom of Kandy, to obtain 400,000 lb. of cinnamon annually, and supply the whole of Europe, and so ruled the trade that they would even burn the cinnamon in Holland lest an excessive supply should reduce the price.

"In 1796, when Ceylon was taken from the Dutch by the English, the cinnamon trade became a monopoly of the East India Company, and when the kingdom of Kandy fell under British control in 1815, the wild cinnamon produced there was added to the cultivated kind, but the annual export did not usually exceed 500,000 lb. The monopoly granted to the company was abolished in 1833, and the merchants of Galle and Colombo were allowed a share in the trade. But a heavy export duty of one-third or one-half its value, and the competition of cinnamon grown in Java, and of cassia from China and other countries, decreased the cultivation in Ceylon. This duty was not removed until A.D. 1853.

"The *Cinnamomum zeylanicum* is a very variable tree, and, as already pointed out, it is difficult to classify the varieties that occur, some botanists regarding them as distinct species. Dr. Beddome, who was Conservator of Forests in Madras, states that there are seven or eight well-marked varieties of *Cinnamomum zeylanicum* in the moist forests of South-Western India, which might easily be regarded as so many distinct species, but for the fact that they are so connected by intermediate forms that it is impossible to find constant characters worthy of specific distinction. He was, therefore, inclined to regard them as forms of *Cinnamomum zeylanicum*, especially since they grew from the sea level up to the highest elevation, and might owe their differences to their local environment. They are, perhaps, better regarded from a horticultural and commercial point of view as deserving a distinct varietal name, in so far as their products possess a different odour, or a different chemical composition. There can be little doubt that special strains might be obtained by careful selection and cultivation, yielding barks richer in oil, or oil of better quality than others.

"According to Thwaites the best cinnamon is produced from a cultivated form (var. *a*), having large leaves varying somewhat in breadth and length on the same tree. But the shape of the foliage does not afford any guide to the quality of the bark, so far as uncultivated trees are concerned, so that the collectors usually taste the bark before commencing to peel the wild trees. Thus the var. *b. multiflorum* and *y. ovalifolium* yield a very inferior bark, which is only collected for purposes of adulteration.

"In the Herbarium of the Pharmaceutical Society four varietes of the wild Ceylon Cinnamon tree are represented, collected near Peradeniya, which were presented by the late Dr. H. Trimen, at one time director of the Botanical Gardens in Ceylon. The first variety has oblong-lanceolate leaves, tapering above, and has a cinnamon taste with a slight flavour resembling that of *Calamus aromaticus*. This is the best kind of the collector, but, of course, not equal to the cultivated plant.

"The second variety has rather broad and obtuse leaves, and, in addition to the cinnamon taste, a slight flavour resembling that of pepper.

"The third variety has smaller, rather ovate-lanceolate leaves, and the fourth kind narrow elongate-lanceolate leaves only half the diameter of those of the first variety. The extent to which the bark of these varieties enters into commerce is difficult to estimate, but, as a rule, the most slender and longest quills are obtained from the finest varieties, so that the size of the quill may be taken as some guide, but the final judgment must always be determined by the sense of taste. The conditions under which the best variety of cinnamon is grown are a very sandy clay soil, or fine white quartz with a good subsoil, and free exposure to sun and rain on an elevation up to 1500 ft.

"The seeds are sown in seed beds and planted out 4 to 6 ft. apart, and when about 2 or 3 ins. high the tops are cut off, so as to induce them to form stools, on which four or five shoots are allowed to grow for about two years, or until the bark turns brown by the formation of a corky layer. They are not all cut at the same time, but only as they arrive at the proper stage of development of the bark; they are then about 6 to 10 ft. high and ½ to 2 ins. thick. The bark is most easily separated in May and June, and again in November and December, after the heavy rains have caused an increased flow of sap, so that there are two harvests, one in the spring and another in the latter part of the year. A long sickle-shaped hook called a 'catty' is used to cut off the shoots and strip off the leaves, and the shoots are then slightly trimmed with a knife, the slender parts thus removed being kept separate and sold as 'cinnamon chips'. The bark is then cut transversely at distances of about a foot and slit lengthwise, and is taken off the shoots by insertion of a peculiar knife called a 'mama,' the separation being assisted, if necessary, by strongly rubbing the bark with the handle. The quills thus obtained are carefully put one into the other, and the compound sticks thus formed are firmly bound together into bundles, and are left for twenty-four hours, during which a heating or fermenting process goes on, which facilitates the subsequent removal of the corky layer. This is done by placing each quill on a stick of wood of suitable thickness and carefully scraping off the outer and middle cortical layers with a knife. After a few hours the operator places the smaller quills within the larger, also inserting the small pieces so as to make up an almost solid stick about 40 ins. in length. These sticks are kept one day in the shade and then placed on wicker trays in the sun to dry. When dry the sticks are arranged in bundles of about 30 lb. each.

"The cinnamon that comes into commerce from Ceylon occurs in three forms, viz.: (1) The quills above described; (2) the trimmings or 'cinnamon chips,' which are very aromatic; and (3) coarse bark as much as an inch thick, the pieces of which are slightly curved longitudinally, and are very deficient in aroma. This bark is derived from old stools, when they are rooted up, and occurs only in small quantity. The chips or trimmings, which have only been exported since 1867, sometimes form nearly a third of the export of bark, and are used for distilling oil of cinnamon, and are also probably powdered for use in mixed spice.

"There are other commercial varieties of the bark, such as the Tellicherry cinnamon, which is almost as good as the Ceylon product, and

Malabar or Tinnevelly cinnamon, which is also a product of Southern India.

"Cinnamon has also been cultivated in Java since 1825, the variety cultivated having large leaves often 8 ins. long by 5 ins. broad. It is also grown in Brazil, and French Guiana and the Seychelles.

"But the cinnamon of all these countries is appreciably different from Ceylon cinnamon, partly, perhaps, from want of careful cultivation, and the absence of the acquired skill of the regular cinnamon peeler, and partly from the age of the shoots or branches used, and partly also from climatic differences, or the altitude at which the tree is cultivated. Cinnamon bark in powder, especially in the cheaper qualities, requires careful examination. Bark that has been exhausted by distillation is apparently sometimes used; in other cases powdered cassia is sometimes substituted for it.

"The oil of cinnamon imported from Ceylon usually contains 10 to 20 or even 50 per cent. of leaf oil, possibly due to the distillation of the leaves along with the bark trimmings or chips. The oil distilled in this country from the imported cinnamon chips, which are not mixed with the leaves, would naturally be free from leaf oil. The presence of leaf oil in oil of cinnamon bark can be detected easily by dissolving one drop of the suspected oil in five of alcohol and adding ferric chloride, which will produce a pale green colour with the pure bark oil, whereas leaf oil, or bark oil adulterated with leaf oil, produces a deep blue colour."

There is some difference of opinion as to the limit values in specific gravity and cinnamic aldehyde-content in pure cinnamon bark oils, and it is also certain that the characters of the oil will differ materially according to the origin of the bark.

The oil distilled on the continent usually has a higher specific gravity and aldehyde-content than English distilled oil. Much of the so-called cinnamon oil of continental origin is undoubtedly adulterated with artificial cinnamic aldehyde, which would explain these characters, but oils, probably genuine, with these high figures, appear to owe them to a different method of distillation from that practised in England. With these reservations, the characters of pure cinnamon bark oil may be taken as follows:—

	English Distilled.	Continental Distilled.	Seychelles Oil.
Specific gravity . .	0·995 to 1·040	1·020 to 1·040	0·943 to 0·975
Optical rotation . .	0° ,, − 1°	0° ,, − 1°	− 1° ,, − 3°
Refractive index. .	1·5700 ,, 1·5850	1·5850 ,, 1·5910	1·5280 ,, 1·5335
Aldehydes (bisulphite method) . . .	58 to 70 per cent.	63 to 76 per cent.	25 to 36 per cent.
Phenols . . .	5 ,, 10 ,,	4 ,, 10 ,,	6 ,, 16 ,,

The oil is usually soluble in 3 volumes of 70 per cent. alcohol, except in the case of Seychelles oil, which requires up to 10 volumes.

Umney and Bennett, in a paper read before the British Pharmaceutical Conference in 1910,[1] give the following figures for a series of cinnamon oils, calling attention to the absence of light fractions in the oils distilled on the Continent:—

No. 1.—Very fragrant, very sweet, but not a normal oil, imported direct from Ceylon:—

[1] *P. and E.O.R.* (**1910**), 169.

	Specific Gravity.	Refractive Index at 25°.
Original oil	·944	1·5178
Fraction 1. 20 per cent.	·867	1·4790
2. 20 ,,	·883	1·4880
3. 20 ,,	·936	1·5115
4. 20 ,,	1·008	1·5564
Residue 5. 20 ,,	1·005	1·5748

No. 2.—Distilled in England, from broken quills :—

	Specific Gravity.	Refractive Index at 25°.
Original oil	1·016	1·5760
Fraction 1. 20 per cent.	·890	1·5002
2. 20 ,,	·994	1·5568
3. 20 ,,	1·032	1·5864
4. 20 ,,	1·048	1·5928
Residue 5. 20 ,,	1·051	1·5746

No. 3.—Distilled in England, from Ceylon "chips" :—

	Specific Gravity.	Refractive Index.
Original oil	·996	1·5602
Fraction 1. 20 per cent.	·886	1·4934
2. 20 ,,	·984	1·5442
3. 20 ,,	1·027	1·5750
4. 20 ,,	1·046	1·5912
Residue 5. 20 ,,	1·051	1·5736

No. 4.—Distilled in England :—

	Specific Gravity.	Refractive Index.
Original oil	1·004	1·5615
Fraction 1. 20 per cent.	·884	1·4925
2. 20 ,,	·993	1·5565
3. 20 ,,	1·029	1·5826
4. 20 ,,	1·048	1·5934
Residue 5. 20 ,,	1·052	1·5729

The following two samples were of Continental distillation, guaranteed pure :—

No. 5.—

	Specific Gravity.	Refractive Index.
Original oil	1·021	1·5840
Fraction 1. 20 per cent.	·918	1·5142
2. 20 ,,	1·024	1·5878
3. 20 ,,	1·042	1·6038
4. 20 ,,	1·048	1·6078
Residue 5. 20 ,,	1·053	1·5955

No. 6.—

	Specific Gravity.	Refractive Index.
Original oil	1·030	1·5920
Fraction 1. 20 per cent.	·963	1·5428
2. 20 ,,	1·026	1·5883
3. 20 ,,	1·040	1·6002
4. 20 ,,	1·050	1·6068
Residue 5. 20 ,,	1·058	1·6070

Three samples of cinnamon oil from bark grown in the Gold Coast have been examined recently at the Imperial Institute.[1] The characters of these oils, together with those of one previously examined, were as follows :—

[1] *Bulletin*, **1919**, 17, 189.

	Sample No. 1 from Assuantsi.	Sample No. 2 from Coomassie.	Sample No. 3 from Aburi.	Previous Sample from Tarquah.
Specific gravity at 15°	1·038	1·042	1·041	1·042
Refractive index	1·594	1·605	1·603	1·603
Aldehydes, per cent.	74 (approx.)	88 (approx.)	86 (approx.)	86 (approx.)
Soluble in 70 per cent. alcohol	Soluble in 2·5 vols.	Soluble in 2·4 vols.	Soluble in 2·4 vols.	Soluble in 2·4 vols.

Adulteration with cinnamon leaf oil is indicated by a higher percentage of eugenol, and the deep blue colour yielded with ferric chloride solution. Artificial cinnamic aldehyde is frequently used as an adulterant. The oil should be tested for traces of chlorine as described under oil of almonds, which, if present, is due to the artificial aldehyde. If this adulterant, free from chlorine, has been used, it will be indicated by a high specific gravity, refractive index, and aldehyde-content.

Ceylon cinnamon bark oil contains, as its principal constituent, cinnamic aldehyde; it also contains eugenol as well as the following bodies which were isolated by Schimmel & Co.,[2] who have published the following details in reference to them:—

Methyl-n-amyl ketone: $CH_3 . CO . CH_2 . CH_2 . CH_2 . CH_2 . CH_3$.

The first three fractions, distilling from 32° to 163°, were treated with bisulphite, and from the crystalline double compound a ketone was isolated, which was recognised as methyl-n-amyl ketone. Its semi-carbazone melts at 122° to 123°. Two combustions of this body gave figures corresponding to the formula $C_8H_{17}ON_3$:—

Furfurol: $C_4H_3O . CHO$.

The mixture of methyl-amyl ketone and benzaldehyde isolated from the first fractions distilled in the following three fractions:—

1. 54° to 95°. 2. 95° to 165°. 3. 165° to 180°.

Of these No. 2 and 3 gave an intense furfurol reaction with a solution of aniline hydrochloride in aniline.

Pinene: $C_{10}H_{16}$.

Fraction 160° to 165° had a pronounced terpene odour. Specific gravity 0·8695 at 15°; optical rotation $\alpha_D = -15° 10'$. The nitrosochloride of the melting-point 102° to 103°, and the nitrol benzylamine melting at 122° to 123°, obtained from the former, proved the presence of pinene.

Phellandrene: $C_{10}H_{16}$.

Fraction 170° to 174° (42° to 43° at 4 to 5 mm. pressure) had the specific gravity 0·8614 at 15°; the optical rotation α_D was $-5° 4'$.

The presence of phellandrene was proved by the nitrite, whose melting-point 102°, after recrystallisation from acetic ether and cold methyl alcohol, rose to 103° to 104°. $(\alpha_D) = +11° 39'$ in chloroform solution.

Analysis gave the following values:—

[1] *Report*, April, 1902, 14.

0·1517 grm. of the substance yielded 0·1019 grm. H_2O and 0·3154 grm. CO_2.
0·1445 „ „ „ „ 17·4 cc. N at 19°, at 750 mm. pressure.

Calculated for $C_{10}H_{16}N_2O_3$:—	Found :—
C 56·60 per cent.	56·70 per cent.
H 7·55 „	7·51 „
N 13·21 „	13·65 „

Cymene : $C_{10}H_{14}$.

For the identification of this body, a fraction boiling at 48°, at 4 to 5 mm. pressure, was first oxidised in the cold with dilute permanganate solution, for the purpose of decomposing terpenes present, and the oil driven over with water vapour was boiled for some time with sodium in order to remove possible oxygenated bodies. The oil, thus purified, boiling at 175° to 177°, yielded on oxidation with 1 to 2 per cent. permanganate solution, with the application of heat, an acid of the melting-point 154° to 156°, *p*-oxyisopropyl benzoic acid, whose mother liquors, heated with concentrated hydrochloric acid, yielded propenyl benzoic acid of the melting-point 161° to 162°, which dissolves with difficulty in water. There was further obtained terephthalic acid, which also occurs in the oxidation of cymene.

Benzaldehyde : $C_6H_5 . CHO$.

Some of the oil boiling at 168° to 171° (45° at 4 to 5 mm. pressure) was treated with bisulphite. The aldehyde regenerated from the bisulphite compound distilled at 174° to 179°, and was found to be heavier than water. The body was identified as benzaldehyde by the phenylhydrazone melting at 156°, and the semi-carbazone of the melting-point 213° to 214°.

Nonylic aldehyde: $C_9H_{18}O$.

In the fractions boiling above 180° (62° to 92° at 6 to 7 mm. pressure), nonylic aldehyde was detected, in addition to benzaldehyde, cinnamic aldehyde, and the aldehydes mentioned below. This body, recently isolated from rose oil, was identified by the nonylic acid (pelargonic acid), obtained by oxidation of the aldehyde with moist silver oxide. Three analyses of its silver salt gave the following figures :—

1. 0·3129 grm. of the substance left 0·1271 grm. Ag.
2. 0·3990 „ „ „ „ 0·1613 „ „
3. 0·2524 „ „ „ „ 0·1040 „ „

Calculated for $C_9H_{17}O_2Ag$:—	Found :— 1.	2.	3.
Ag 40·75 per cent.	40·62 per cent. ;	40·53 per cent. ;	41·20 per cent.

Hydrocinnamic aldehyde : $C_6H_5 . CH_2 . CH_2 . CHO$.

The semi-carbazone melting at 116° to 118° gave the following values on analysis :—

1. 0·1501 grm. of the substance yielded 0·0956 grm. H_2O and 0·3438 grm. CO_2.
2. 0·1572 „ „ „ „ 0·0954 „ „ 0·3617 „ „

Calculated for $C_{10}H_{13}ON_3$:—	Found :— 1.	2.
C 62·83 per cent.	62·56 per cent. ;	62·75 per cent.
H 6·81 „	7·06 „	6·74 „

Repeated recrystallisation from hot ether raised the melting-point to 126°. The semi-carbazone prepared for comparison from synthetic hydrocinnamic aldehyde melted at 130° to 131°. Paucity of material rendered a definite proof of the identity of the two products impossible.

The semi-carbazone of the melting-point 126°, heated with dilute sulphuric acid, gave the characteristic odour of hydrocinnamic aldehyde.

Cumic aldehyde: $C_{10}H_{12}O$.

From the bisulphite liquors from which the double compounds of the aldehydes mentioned above had separated as solid products, a small quantity of an aldehyde was obtained which had the odour of cumic aldehyde. Its semi-carbazone melted at 201° to 202°, and yielded on combustion values corresponding to those calculated for the semi-carbazone of cumic aldehyde.

1. 0·1481 grm. of the substance yielded 0·1002 grm. H_2O and 0·3494 grm. CO_2.
2. 0·1475 „ „ „ „ 0·0991 „ „ 0·3457 „ „

Calculated for $C_{11}H_{15}ON_3$:—	Found:— 1.	2.
C 64·39 per cent.	64·35 per cent.;	63·93 per cent.
H 7·32 „	7·52 „	7·47 „

The semi-carbazone obtained from cuminol had the same melting-point; the mixture of the two products also melted at 201° to 202°. On oxidation of the aldehyde with moist silver oxide, cumic acid of the melting-point 114° to 116° was obtained; a mixture of this acid with pure cumic acid melted at the same temperature.

Linalol: $C_{10}H_{18}O$.

The fractions of the boiling-point 80° to 92° (at 6 to 7 mm. pressure), freed from aldehydes by bisulphite solution, were rectified in vacuo, and the portions distilling at ordinary pressure at 195° to 205°, which had a distinct linalol-like odour, were oxidised in the cold with chromic acid mixture. The specific gravity of this fraction was 0·8772, the optical rotation − 8° 32′. The resulting aldehyde purified by the bisulphite compound had the odour of citral; when heated with pyruvic acid and β-naphthylamine, it yielded citryl-β-naphthocinchoninic acid of the melting-point 197°.

Linalyl isobutyrate: $C_{10}H_{17} . OCOC_3H_7$.

Linalyl isobutyrate appears also to be present. When saponifying the fractions boiling between 80° and 111° (at 6 to 7 mm. pressure) and which had been treated with bisulphite, which on the average gave the saponification number 20·4, there was isolated, in addition to an acid with a pungent odour (possibly formic acid), another acid with a distinct odour like St. John's bread. The silver content of its silver salt did not, however, agree with the value calculated for silver isobutyrate, which may have been caused by the presence of formate or acetate.

Eugenol: $C_{10}H_{12}O_2$.

In order to detect eugenol, "heavy Ceylon cinnamon oil" was shaken with 2 to 3 per cent. potash solution; the oil separated from the alkaline solution by acid gave the blue eugenol reaction with ferric chloride, and yielded with benzoyl chloride and soda solution, benzoyl eugenol of the melting-point 69° to 70°.

Caryophyllene: $C_{15}H_{24}$.

This sesquiterpene could be detected in one of the last fractions, boiling at 104° (at 5 mm. pressure).

The specific gravity of the portion distilling at 260° to 261° was 0·9047, the rotatory power a_D = − 7° 20′.

Two analyses gave the following values:—

1. 0·1904 grm. of the substance yielded 0·2012 grm. H_2O and 0·6133 grm. CO_2.
2. 0·1240 ,, ,, ,, ,, 0·1284 ,, ,, 0·3984 ,, ,,

Calculated for $C_{15}H_{24}$:—	Found :— 1.	2.
C 88·23 per cent.	87·85 per cent. ;	87·63 per cent.
H 11·76 ,,	11·74 ,,	11·51 ,,

For further identification caryophyllenic alcohol was produced from it, of which the melting-point was found at 95°. Two combustions yielded values corresponding with those calculated for an alcohol $C_{15}H_{26}O$:—

1. 0·2277 grm. of the substance yielded 0·2366 grm. H_2O and 0·6727 grm. CO_2.
2. 0·1819 ,, ,, ,, ,, 0·1923 ,, ,, 0·5399 ,, ,,

Calculated for $C_{15}H_{25}OH$:—	Found :— 1.	2.
C 81·08 per cent.	80·57 per cent. ;	80·95 per cent.
H 11·71 ,,	11·55 ,,	11·75 ,,

Seychelles cinnamon bark oil [1] contains cinnamic aldehyde, eugenol, caryophyllene, phellandrene, cymene, camphor, camphene, β-pinene, *l*-limonene, benzaldehyde, linalol, and nonylic aldehyde (?).

Cinnamon leaf oil differs from the bark oil in consisting principally of eugenol. The oil has the following characters :—

Specific gravity	1·043 to 1·066
Optical rotation	− 0° 10′ ,, + 2° 35′
Refractive index	1·5300 ,, 1·5400
Eugenol	70 to 95 per cent. (rarely lower)
Aldehydes	0 to 3 per cent.

It is soluble in 2 to 3 volumes of 70 per cent. alcohol, becoming turbid on the addition of more alcohol.

This oil contains eugenol, cinnamic aldehyde, safrol, benzaldehyde, linalol, terpenes, and traces of benzoic acid.

The oil distilled from the bark of the root, according to Pilgrim, has a specific gravity 0·994 and optical rotation + 50°. It contains pinene, dipentene, phellandrene, cineol, camphor, eugenol, safrol, borneol, and caryophyllene.

A Japanese cinnamon oil is obtained from various parts of the plant, *Cinnamomum Loureirii*, known locally as Nikkei. The oil from the leaves of young shoots has the following characters :—

Specific gravity	0·900
Optical rotation	− 8° 45′
Acid value	3
Ester ,,	18·6
Aldehydes	27 per cent.

It contains cinnamic aldehyde, citral cineol, and linalol. Oil from the trunk wood contains cinnamic aldehyde and eugenol, and oil from the root bark contains cinnamic aldehyde, camphene, linalol, and cineol,

Other cinnamon oils of less importance are as follows :—

The oil from the leaves of *Cinnamomum glanduliferum* (a Nepal species) has been examined and found to have the following characters :—[2]

Specific gravity	0·9031 to 0·9058
Optical rotation	− 24° 57′ ,, − 26° 12′
Acid value	0·34 ,, 0·9
Ester ,,	8·8 ,, 18·4
,, ,, (after acetylation)	46·9 ,, 55·3

[1] Schimmel's *Report*, November, **1908**, 41 ; April, **1913**, 42.
[2] Schimmel's *Bericht*, April (**1905**), 84.

The so-called Culilavan oil is obtained from the bark of *Cinnamomum Culilawan*, which yields about 4 per cent. of oil. This has a specific gravity 1·050 to 1·055. It contains about 60 per cent. of eugenol, together with some methyl-eugenol and terpineol. It is soluble in 3 volumes of 70 per cent. alcohol.

The bark of *Cinnamomum Kiamis* yields about 0·5 per cent. of a deep golden essential oil having an odour resembling, but less delicate than, ordinary cinnamon bark oil. It has the following characters:—

Specific gravity	1·0198
Optical rotation	− 1° 50′
Refractive index	1·5828
Aldehydes	80 per cent. (bisulphite method)

It is soluble in 1 volume of 80 per cent. alcohol. It contains about 11 per cent. of phenols (eugenol).

The bark of *Cinnamomum oliveri*, the so-called Brisbane "White Sassafras" tree, yields from 1 per cent. to 2·4 per cent. of essential oil. This has been examined by Hargreaves.[1]

The plant is a handsome tree with grey rough bark, growing in the scrubs on the North Coast Railway, Queensland. When distilled in a current of steam the bark yielded approximately 2·4 per cent. of oil, having a specific gravity of 1·030 and refractive index 1·5165 at 23°. On fractionation it was found to contain pinene (12 to 15 per cent.), *d*-camphor (18 to 20 per cent.), safrol (25 to 27 per cent.), and eugenyl methyl ether (40 to 45 per cent.). The oil from the leaves contained about 25 per cent. of a mixture of terpenes, one of which was identified as pinene, and the other, which formed a nitrosite melting at 103°, was probably phellandrene. The leaf oil contained about 60 per cent. of *d*-camphor and 15 per cent. of phenols and other substances. The phenol gave a green coloration with ferric chloride in alcohol, and had the formula $C_{10}H_{15}O_2$. No safrol was present, and the author could not detect the presence of eugenol, cinnamic aldehyde nor cineol, which had been previously suspected by Baker and Smith.

It contains cineol, eugenol, and cinnamic aldehyde, but the principal constituent is probably safrol.

The bark of *Cinnamomum peditinervium*, a tree indigenous to the Fiji Islands, yields about 1 per cent. of oil, having the following characters:—

Optical rotation	− 5°
Refractive index	1·4963
Saponification value	4 4
,, ,, (after acetylation)	115·8

The principal constituent is safrol, but linalol, eugenol, and eugenol-methyl ether are also present in the oil.

The leaves of *Cinnamomum Tamala*, a South-Asiatic tree, yield an essential oil having the following characters:—

Specific gravity	1·0257
Optical rotation	+ 16° 37′
Refractive index	1·5259
Phenols	78 per cent.

It is soluble in 1 to 2 volumes of 70 per cent. alcohol. The oil consists chiefly of eugenol, together with phellandrene and a little camphor.

[1] *Jour. Chem. Soc.*, **1916**, 751.

The wood of *Cinnamomum Parthenoxylum*, a Javanese tree, yields about 0·8 per cent. of oil, having the following characters:—

Specific gravity	1·0799
Optical rotation	+ 1° 22′
Refractive index	1·5323

It is soluble in 2·5 to 3 volumes of 90 per cent. alcohol. Its principal constituent is safrol.

Schimmel & Co.[1] have examined the oil from the Japanese bark known as Yama-nikkei, a wild so-called cinnamon tree. The oil was of a pale yellow colour, with an odour reminding one of camphor and of ginger. It had the following characters:—

Specific gravity	0·9245
Optical rotation	+ 8° 34′
Refractive index	1·47779
Acid value	0·6
Ester ,,	14·8

It contains camphor, terpenes, and, probably, cymene.

The bark of the Japanese *Cinnamomum pedunculatum* yields an oil having the following values:—

Specific gravity	0·917 to 0·932
Optical rotation	− 4° 40′ ,, − 14° 32′
Acid value	0
Ester ,,	0
,, ,, (after acetylation)	84·6

The oil contains eugenol, methyl-eugenol, phellandrene, and probably linalol.

The leaves of this tree, which is known as the Yabunikkei tree, yield 1·4 per cent. of a slightly lævorotatory oil having a specific gravity 1·0665. It contains eucalyptol, eugenol, safrol, and terpenes.[2]

The so-called Nepal sassafras, or Nepal camphor tree, is probably *Cinnamomum Ceciodaphne*. The essential oil from the wood of this tree has been examined by Pickles.[3]

The material used for distillation consisted of billets of heart wood, free from bark. Its odour was rather that of safrol than camphor. The yield was 2·95 per cent. on the original wood, or 4·16 per cent. on the wood after grinding. The oil was of a pale yellow colour, and had the following characters:—

Specific gravity	1·1033
Optical rotation	− 0° 4′
Saponification value	2·8
,, ,, (after acetylation)	7

The oil consists almost entirely of ethers, the well-defined ones being safrol, myristicin, and elemicin.

No fraction was obtained below 230°, so that terpenes were clearly absent. No aldehydes were present. A methoxyl determination showed the presence of 16·16 per cent. of CH_3O. As safrol contains no methoxyl it is obvious that other ethers are present. A complete fractionation resolved the oil into the following fractions:—

(1) First fraction from which a crystalline body separated, (2) 233° to 238° = 7·4 per cent., (3) 238° to 243° = 11 per cent., (4) 243° to 248° = 5·8 per cent., (5) 248° to 253° = 4 per cent., (6) 253° to 258° = 3·6 per cent., (7) 132° to 137° at 12 mm. = 3·2 per cent., (8) 137° to 142° at

[1] *Report*, October, **1914**, 54. [2] *Jour. Chem. Ind. Jap.*, 18 (**1915**), 913.
[3] *Jour. Chem. Soc.* (**1912**), 1433.

12 mm. = 4·4 per cent., (9) 142° to 147° at 12 mm. = 5 per cent., (10) 147° to 152° at 12 mm. = 11·4 per cent., (11) 152° to 157° at 12 mm. = 17·8 per cent., (12) 157° to 162° at 12 mm. = 11·4 per cent., (13) 162° to 172° at 12 mm. = 2·2 per cent.: residue, 1·4 per cent.

The first fraction had the following characters: Specific gravity 1·1059; optical activity 0°. Combustion figures agreed with the formula $C_{10}H_{10}O_2$, and the solidified oil melted at 9°. This was practically pure safrol. It was definitely identified by preparing from it specimens of piperonal and safrol-α-nitrosite. The piperonal melted sharply at 37°, and the nitrosite at 130° to 131°. Fractions 2 and 3 consisted chiefly of safrol also.

Myristicin was identified in the higher boiling fractions by dissolving a small quantity in petroleum ether and adding bromine, also dissolved in petroleum ether to slight excess. A heavy oily layer separated, which afterwards solidified. By recrystallisation from methyl alcohol, crystals were obtained of the formula $C_{11}H_{10}O_3Br_4$, and melting at 128°. This agrees with the characters of dibromomyristicin dibromide, and this was confirmed by converting it into dibromomyristicin. This was achieved by dissolving it in glacial acetic acid and adding zinc dust. After washing away the acetic acid, the residue was extracted with ether, and the ethereal solution evaporated. The residue solidified, and was recrystallised from methyl alcohol, when it was found to melt at 52°. It was, therefore, proved that myristicin was present in the oil. Further proof was obtained by oxidising this fraction of the oil, when myristinic aldehyde and myristinic acid were obtained.

Myristicin, however, only contains one methoxyl group, whilst the methoxyl determination showed that more methoxyl was present than is indicated by the amount of myristicin found. A trimethoxy compound was suspected to be present, but, unfortunately, could not be isolated in a sufficiently pure state for identification. Trimethyl-gallic acid was found in the oxidation products, which agreed with the presence of iso-elemicin in the oil, which had first been treated with alcoholic potash. Therefore, elemicin was probably present in the original oil. This was confirmed by preparing the crystalline dibromide of iso-elemicin, which melted sharply at 88° to 89°. Elemicin, which was isolated from oil of elemi by Semmler, has the formula $C_6H_2(OCH_3)_3(CH_2 . CH : CH_2)$.

Oil of Sassafras.

Oil of sassafras is distilled from the root of *Sassafras officinale* (*Laurus sassafras* of Linnæus), a native of North America. The tree is one of the most widely distributed in the continent, being found in Canada, in all the States east of the prairies beyond the Mississippi, and in Mexico. In Canada it seldom exceeds 30 ft. in height, but in the south, especially in Virginia and the Carolinas, it reaches nearly 100 ft. in height. The root bark (except the outer layers) is the most aromatic portion of the plant, and from this, as well as the wood, the essential oil is distilled. The roots may be dug out at any season, but those dug when the sap is not rising yield the highest percentage of oil. Some of the large roots weigh nearly half a ton, but the smaller roots are most valued on account of their higher yield of oil. The typical distilleries in the neighbourhood of Baltimore use wooden tanks as stills. These are made of kiln-dried pine and consist of wooden boxes built up of staves, with a pierced false bottom. The head is provided with a trap-door for filling by, and a large door is fitted to the bottom of the tank, for the purposes of emptying. To the upper part of the tank is fitted a copper head, leading to a condensing coil immersed in cold water. Steam at from 40 to 60 lb. pressure is admitted between the true and false bottoms. The stream of condensed products issuing from the pipe is caught in a copper funnel with a very long spout which reaches nearly to the bottom of a 20-gallon copper vessel. The oil being heavier than water collects at the bottom, leaving the water to flow away at the top of the vessel. A charge of 10 tons requires about two days to exhaust, and yields about 10 gallons of oil. In parts of Pennsylvania and New Jersey a number of small isolated distillers, often coloured men, carry on a remunerative business in this oil. A most primitive still is used, consisting merely of a barrel with its head luted on, and its bottom pierced and standing on a steam generator. A natural elbow-shaped branch, perforated throughout, serves to carry the distillation products to a metal tube placed in a trough of running water. The condensed products are then separated in the usual way. Although the yield of oil is much lower that when distillation is effected in modern apparatus, the process is a payable one, as the capital necessary for such primitive distilleries is very small. Properly treated, about 1 per cent. of oil can be obtained from the root wood, the bark yielding from 5 to 9 per cent.

Pure sassafras oil has the following characters:—

Specific gravity	1·068 to 1·082
Optical rotation	+ 1° 30′ to + 4°
Refractive index	1·5280 to 1·5310
Acid value	0
Ester ,,	1 to 2

It is soluble in 2 volumes of 90 per cent. alcohol.

A sample fractionated by Schimmel & Co.[1] gave the following results:—

	Specific Gravity.	Rotation.
208° to 216°	1·0066	+ 5° 40′
216° ,, 221°	1·0546	+ 4° 36′
221° ,, 225°	1·0764	+ 3° 55′
225° ,, 225°	1·0830	+ 2° 57′
225° ,, 226°	1·0877	+ 2° 5′
226° ,, 227°	1·0916	+ 1° 30′
227° ,, 228°	1·0930	+ 0° 45′
228° ,, 229°	1·0942	+ 0° 12′
229° ,, 234°	1·0905	− 0° 19′
Residue	1·0770	—

The principal constituent of sassafras oil is safrol, which is also an important constituent of camphor oil. Safrol is obtained in large quantities from the latter oil, and is sold in a more or less pure state, usually mixed with the lower boiling fractions of camphor oil as "artificial sassafras oil".

The earliest chemical investigation of the oil was by Grimaux,[2] who stated that it consisted of 90 per cent. safrol (*q.v.*) and 10 per cent. of a terpene which he termed safrene, with traces of a phenol, which was afterwards identified as eugenol. A recent investigation by Power and Kleber[3] has shown that "safrene" is a mixture of the terpenes pinene

[1] Schimmel's *Bericht*, April, **1906**, 62. [2] *Comptes Rendus* **(1869)**, 928.
[3] *Pharm. Review*, 14 **(1896)**, 101.

and phellandrene; dextro-camphor was also found, and a body which was probably cadinene. The average composition of the oil is:—

Safrol	80 per cent.
Pinene . . . . . . Phellandrene	} 10 "
Dextro-camphor	7 "
Eugenol	0·5 "
Cadinene (?)	2·5 "

This oil is used on an enormous scale for soap perfumery. Its strong odour and low price enable it to be used for the very cheapest of soaps, not only to give them an actual perfume, but also to cover up the bad odour of poor quality fats. But its use has been very largely diminished by the discovery that its active constituent exists in notable quantity in crude camphor oil. By means of fractional distillation and freezing (safrol melts at about 8°), this body can be extracted in a state of almost absolute purity as a water-white liquid of specific gravity over 1·100. In this form it is, in the author's opinion, preferable in every way for cheap perfumery to the natural oil of sassafras, and its manufacture is now an enormous industry. There are on the market, however, many samples of so-called "artificial sassafras oil" which are merely fractionated camphor oil, of specific gravity about 1·070. It should be remembered that these are often of low specific gravity and contain not more than 50 per cent. of safrol, and are far less valuable than the pure safrol. This body can, of course, be frozen out from ordinary sassafras oil, but it is much cheaper to obtain it from camphor oil. Safrol, which is identical with "shikimol" from oil of *Ilicium religiosum*, also finds an extensive use in the manufacture of heliotropine (piperonal), which is one of its oxidation products.

Sassafras leaves also yield a small quantity (·03 per cent.) of an essential oil, differing entirely from the oil from the wood. This has also been investigated by Power and Kleber.[1] The oil obtained from 8000 lb. of leaves was only about 2½ lb., and was of a light yellow colour and agreeable lemon odour. Its specific gravity was ·872, and its optical rotation + 6° 25′. The constituents identified were citral, pinene, phellandrene, a hydrocarbon of the paraffin series, a hydrocarbon also found in oil of bay and called by the discoverers myrcene, and the acetic and valerianic (isovalerianic ?) esters of linalol and probably of geraniol, and, possibly, cadinene. No safrol could be detected.

OIL OF LINALOE (CAYENNE).

Cayenne linaloe oil, also known as oil of "Bois de Rose Femelle," and sometimes as oil of Azelia, is distilled in French Guiana.

The botanical origin of this oil has been a matter of considerable uncertainty. The following account of the question is due to E. M. Holmes[2]:—

"The fragrant wood from which this oil is derived was apparently known in Europe in the early part of the last century, although it was evidently confused with other woods. It has been named by various writers as Bois de Citron, Bois de Rose male, Bois de Rose femelle, Bois de Cedre jaune, Bois de Citron de Cayenne, etc. The older authors, however, distinguished between the woods bearing these names.

[1] *Pharm. Review*, 14 (1896), 101. [2] *P. and E.O.R.* (1910), 32.

Guibort has pointed out that the name Bois de Citron has been applied to three distinct woods.

"1. The Bois de Citron of Domingo, which was also called Hispanille, and which he doubtfully identified as that of *Amyris balsamifera*, Linn. It occurred in commerce in large blocks 6 to 12 ft. long, 12 to 18 ins. wide, and 6 to 8 ins. thick. This he believed to have been sold in Pomet's time as Bois de Santal Citron. (This is the tree that now yields the West Indian Oil of Sandalwood.)

"2. A wood in smaller pieces, 4 to 7 ins. in thickness, harder, heavier, and of a deeper yellow colour, with a weak odour which disappeared on exposure to the air.

"3. The Bois de Citron de Cayenne, which he had also seen sold under the name of Sassafras. This appears to have been the 'Bois

FIG. 18.—Bois de Rose. Preparation of the wood.
[*Roure Bertrand Fils.*

de Rose' of Aublet, and the one which now yields the Cayenne Linaloe Oil. Planchon states that he has seen the last-mentioned wood sold in France under the name of Bois jaune de Cayenne, and Bois de Citron de Cayenne, but prefers to retain Aublet's name of Bois de Rose de Cayenne, or Bois de Licari, to prevent confusion (Licari kanali being the native Indian name of the tree in French Guiana). He states, however, that there are two woods known in Cayenne under the name of Bois de Rose, the one called by the French inhabitants Bois de Rose male, and the other Bois de Rose femelle. The former he identifies with the Licari kanali of Aublet. He describes it as hard and heavy, formed of interlacing woody layers, as having the flavour and odour of roses, and a bitter taste. He remarks that in fissures and exposed surfaces, the wood shows an efflorescence of fine white needle-shaped crystals. It has a yellow tint darkening with age. The other wood, Bois de Rose femelle, he identifies with Bois de Cedre blanc, which according to Aublet is derived from *Icica altissima*, now called *Protium altissimum*, March. This wood is soft and very light, white or greenish when fresh, but becomes yellowish on exposure. It possesses quite a different odour, resembling that of citron or bergamot, so that it would be better distinguished by the name of Bois de Citron.

"The only definite information concerning the botanical sources of the Cayenne linaloe wood is that given by Aublet in the *Plantes de Guiane Francaise*, page 313.

"He names the tree *Licaria guianensis*, and gives an illustration of a leafy twig of the natural size, a portion of which is reproduced. But he had not seen either flowers or fruit, although he had met with the tree in different districts, and in consequence, for many years, botanists were

FIG. 19.—Bois de Rose Stills at Cayenne.
[*Roure-Bertrand Fils.*

unable to identify it. He does not apply the name male or female to the tree, but simply states that it is called Licari kanali by the Galibi Indians, and Bois de Rose by the French inhabitants. He described it as a large tree 50 or 60 ft. high, with a trunk 3 ft. or more in diameter, having a cracked and wrinkled bark of a reddish hue, with a yellowish wood a little dense (*peu compacte*).

"This tree, the *Licaria guianensis* of Aublet, or Cayenne Linaloe, remained unidentified until 1889 when it was referred by Berlin botanists in the *Jahrbuch des Konigl. Bot. Gard. und des Bot. Museum zu Berlin*, 1889, page 378, to *Ocotea caudata* Mez., a Lauraceous tree of which the female flowers are still unknown, the plant being diœcious. The wood of the Cayenne Linaloe, as met with in commerce in Europe, was submitted to histological examination by Dr. Josef Moeller in 1898. He describes the wood as heavy, hard, and easily split, the fresh surface yellowish, but redder when older, with the vessels and medullary rays visible to the unaided eye, but with the annual rings of wood not perceptible. The odour is that of rose and citron. Under the microscope the sinuous medullary rays are seen to consist of one or two rows of cells. The wood consists of strongly thickened woody fibres in which the vessels occur in groups of 1 to 3, often surrounded by tracheids. The chambered parenchymatous fibres are here and there widened out into oil cavities, containing lemon yellow drops of oil. The cells of the medullary rays are mostly filled with violet-coloured amorphous masses, but in some there are yellowish or greenish globules which are soluble in alcohol, and probably contain essential oil. A characteristic feature of the wood is that the vessels often have internal cells or tyloses with unusually thick sclerotised walls. The wood resembles that of the Mexican lignaloes in the grouping and form of the vessels, and in the scattered parenchyma containing volatile oil, but in the structure of the bark, the character of the cork cells, the form and distribution of the bast fibres, and the occurrence of the oil cells and raphides, it is that of the *Lauraceæ*.

"Dr. Moeller examined the wood of several genera of the *Lauraceæ* and found that it came nearest to that of the *Ocotea*.

"It thus appears from the researches made by the German botanists, and Dr. J. Moeller, that there can be little doubt that the Licari kanali tree is the *Ocotea caudata*, Mez. It should be noted, however, that Moeller is in error in using the name Bois de Rose femelle for *Ocotea caudata*, and that it should be Bois de Rose male according to Guibourt; the Bois de Rose femelle being that of *Protium altissimum*, March."

The average yield of oil from the wood is 1 per cent., and the pure oil has the following characters:—

Specific gravity	0·870 to 0·880
Optical rotation	− 10° ,, − 20°
Refractive index	1·4610 ,, 1·4635
Acid value	0 ,, 1·5
Ester ,,	3 ,, 7

On acetylation for the prolonged period of 4 to 6 hours in three to four times its volume of xylene (since linalol decomposes by acetylation without such dilution), the oil shows a total linalol value of 60 to 90 per cent. or over. It is soluble in 2 volumes of 70 per cent. alcohol.

The principal constituent of the oil is *l*-linalol, which was originally described under the name licareol.

Schimmel & Co.[1] have shown that geraniol and *d*-terpineol are present in the oil as well as small quantities of cineol, dipentene, furfurol, and (probably) isovaleric aldehyde, and Roure-Bertrand Fils[2] have isolated methyl-heptenone and nerol from it. The last-named give the following as the composition of the oil:—

Methyl-heptenone	traces
l-linalol	90·5 per cent.
d-terpineol	5·3 ,,
Geraniol	2·4 ,,
Nerol	1·2 ,,

On fractionation the oil behaves as follows:—

[1] *Report*, April, **1909**, 68. [2] *Bulletin*, October, **1909**, 40.

	Per Cent.	Specific Gravity.	Rotation.
To 100°	0·5	—	—
194° to 195°	38·0	0·8625	− 15° 48′
195° ,, 196°	39·0	0·8638	− 16° 20′
196° ,, 200°	16·0	0·8660	− 14° 32′
200° ,, 206°	4·0	—	− 11° 50′
Residue	2·5	—	—

A sample having the following characters has been fractionated by J. C. Umney[1]:—

Specific gravity	0·875
Optical rotation	− 13°
Refractive index	1·4630
Esters	2·6 per cent.
Total alcohols	53·7 ,,

The fractionation results were as follows:—

No.	Per Cent.	Specific Gravity.	Rotation.	Refractive Index.
1	10	0·869	− 15°	1·4615
2	10	0·870	− 15° 30′	1·4620
3	10	0·870	− 15° 30′	1·4624
4	20	0·870	− 16°	1·4630
5	20	0·870	− 15°	1·4632
6	20	0·870	− 14° 30′	1·4642
7	Residue	0·873	—	1·4726

Oil of Massoi Bark.

This oil is distilled from the bark of *Massoia aromatica*, a tree found in New Guinea, in which it occurs to the extent of 6 to 8 per cent. Woy,[2] who investigated the oil, states that it contains a considerable quantity of eugenol, with some safrol, a body resembling creosote, and traces of free acetic acid. It is however, doubtful, whether the so-called massoi bark of commerce ever consists of one kind of bark only. It is possible that the principal compound of the commercial bark is that of a species of cinnamon. He also claimed that it contains a terpene not identical with any already known hydrocarbon. Wallach[3] has, however, shown that this body is a mixture of the terpenes pinene and limonene, with perhaps, some dipentene. The oil is a clear yellow liquid of clove-like odour of specific gravity 1·040 to 1·065, and refractive index about 1·5400.

Schimmel & Co.[4] obtained 8·2 per cent. of oil from the bark having the following characters:—

Specific gravity	1·0169[5]
Optical rotation	+ 1° 45′
Refractive index	1·52535

[1] *P. and E.O.R.* (1910), 33.
[2] *Arch. der Pharm.*, 223 (1890), 22, 687.
[3] *Annalen*, 258 (1890), 340.
[4] *Bericht*, 1917, 33.
[5] This is the figure given in Messrs. Schimmel & Co.'s *Bericht*, but it appears incompatible with the presence of 85 per cent. of eugenol, and is possibly a misprint for 1·0469.

It contained 85 per cent. of eugenol, the non-phenolic portion of the oil having a specific gravity 0·9343. Safrol could not be detected.

The essential oil distilled from a bark shipped from the Dutch East Indies, and known there as "Lawang" bark, has been examined by E. W. Mann. According to Holmes the bark is one of those passing under the name of massoi bark, and is derived from *Cinnamomum Litrea* or an allied genus.

The oil has the following characters:—

Specific gravity	1·0104
Rotation at 20°	− 6·97°
Refractive index at 20°	1·5095
Acid value	1·15
Saponification value	43·02
Ester value	41·87
Saponification value of acetylated oil	121·91

A portion of the oil was distilled under reduced pressure (35 mm.), three fractions being collected:—

	Specific Gravity.	Refractive Index at 15·5
Fraction (i) below 180° 35·5 per cent.	0·9335	1·4828
" (ii) 180° to 220° 37·0 "	1·0103	1·5118
" (iii) 220° " 228° 21·0 "	1·0934	1·5402

A crystalline acid, melting-point 51° to 52°, was separated.

Oil of Kuromoji.

The leaves and young twigs of *Lindera sericea*, one of the Japanese *Lauraceæ*, yield the oil known in commerce under this name. It is a dark yellow oil with a strong aromatic and balsamic odour, of specific gravity from ·890 to ·915, usually about ·900, and faintly laevo-rotary. It has been examined by Kwasnik,[1] who states that it contains dextro-limonene, dipentene, inactive terpineol, and laevo-carvone. The oil, on account of its comparatively low price, has recently come into some favour as an aromatic.

Schimmel & Co. have examined a sample distilled by themselves and found it to contain 9·5 per cent. of esters calculated as geranyl acetate. Distilled at 4 mm. pressure, it yielded the following fractions:—

Temperature.	Rotation.
35° to 40°	− 22° 26′
40° " 45°	− 15° 44′
45° " 50°	− 9° 38′
55° " 66°	− 1° 56′
66° " 70°	+ 1°
70° " 78°	− 3° 56′
78° " 81°	− 12° 37′
Residues	− 6° 10′

Linalol was detected in the oil, as well as geranyl acetate.

According to Shinohara[2] this oil contains 11·41 per cent. of esters; 20·21 per cent. of free alcohols; 7·71 per cent. of cineol; 50·98 per cent. of terpenes, with traces of a phenolic substance and free acids; and aldehydes and ketones 9·17 per cent. The alcohols consisted chiefly of linalol 11·43 per cent. and geraniol 7·16 per cent.

Schimmel & Co.[3] have examined samples of Kuromoji oil distilled

[1] *Arch. der Pharm.*, 230 (**1892**). 265. [2] *Chem. Ind. Japan* (**1915**), 18, 417.
[3] Schimmel's *Bericht*, April, **1904**, 98; and (**1907**), 67.

from other parts of the plant than the leaves and found them to have the following characters :—

Specific gravity	0·8942 to 0·8947
Optical rotation	– 14° 29′ „ – 22° 26′
Ester value	27·3 „ 29·9

They found present in the oil cineol, linalol, geraniol, and geranyl esters.

Oils of Persea.

Persea gratissima is a tree indigenous to tropical America. The bark yields about 3·5 per cent. of essential oil having the following characters :—

Specific gravity	0·969
Optical rotation	– 0° 46′

It contains anethol and methyl-chavicol (hence its original description as "anise bark" oil).

The leaves yield an oil of tarragon-like odour having the following characters :—

Specific gravity	0·955 to 0·961
Optical rotation	+ 1° 50′ „ + 2° 30′
Refractive index	1·5120 „ 1·5150
Acid value	0
Ester „	2 to 4
„ „ (after acetylation)	18 „ 22

The principal constituent is methyl-chavicol, with a little pinene and a small amount of a solid paraffin.

Rabak[1] has examined the essential oil distilled from the leaves of the so-called "swamp bay," *Persea pubescens,* a North American tree. He obtained 0·2 per cent. of oil of the following characters :—

Specific gravity	0·9272 at 25°
Optical rotation	+ 22° 4′
Refractive index	1·4695 at 25°
Acid value	2·8
Ester „	14·5
„ „ (after acetylation)	64

He found it to contain butyric acid in the free state, and butyric, valerianic, and œnanthylic acids as esters. Camphor, cineol, and probably borneol are also present, as well as traces of formaldehyde.

Oil of Pichurim.

The cotyledons of *Nectandra Puchury major* and *Nectandra Puchury minor*, two species of *Lauraceæ* indigenous to Brazil, and commonly known as the Para nutmeg, yield about ·5 to 1 per cent. of a yellowish-green oil. According to Müller, the oil contains several oxygenated constituents not completely identified. Esters of lauric and valerianic acids are probably present. It distils between 180° and 270° and contains a terpene and probably a little safrol. The fraction boiling at 256° has a deep blue colour.

Roure-Bertrand Fils[2] have examined the oil distilled from the Puchury-major bean alone, which they obtained in a yield of 2·3 per cent. It has a characteristic odour of sassafras, with a suggestion of nutmeg and cloves. Its characters are as follows :—

[1] *U.S. Dept. of Agriculture, Bureau of Plant Industry*, 235 **(1912)**, 29.
[2] *Bulletin*, April **(1920)**, 4, **1**, 35.

Specific gravity at 17°	1·0396
Optical rotation	−4° 54′
Refractive index at 16°	1·5180
Acid value	2·8
Ester value	0
Ester value after acetylation	35·5

It contains isoeugenol, safrol, and possibly cineol.

Oil of Caparrapi.

This oil is yielded by the so-called "canelo' tree of Colombia, *Nectandra caparrapi*. It varies in colour from pale yellow to dark brownish-red, and is known in commerce under the names of "white" and "black". It is a thick liquid of specific gravity about ·915 to ·935, and rotary power − 3°. On exposure to prolonged cold, crystals of an acid are deposited. The black oil probably owes its colour to overheating. From a sample of the white oil, Tapia[1] has extracted an acid of the formula $C_{15}H_{26}O_3$, which crystallises in white needles, melts at 84·5°, and has a rotary power $[a]_d$ = + 3°. In addition to this body the oil contains a sesquiterpene alcohol, $C_{15}H_{26}O$, which the same chemist terms caparrapiol. It has a specific gravity ·9146, and a rotary power $[a]^d$ = − 18·6°, and boils at 260° at atmospheric pressure. By abstracting water by means of dehydrating agents, a sesquiterpene $C_{15}H_{24}$ is formed, termed caparrapene. This is described as a colourless liquid, boiling at 240° to 250°, and of specific gravity ·9019 and a rotary power $[a]_d$ = − 2·2°. These compounds require further examination before they can be regarded as new compounds. The oil is used for the same purposes as balsam of copaiba.

Oil of Spicewood.

The various parts of the plant *Laurus benzoin (Benzoin odoriferum)*, commonly known as spice wood or wild allspice, yield aromatic volatile oils. The plant is a North American shrub, inhabiting damp, shady woods in localities extending from Canada to Florida. Schimmel & Co. state that all parts of the shrub possess pleasant aromatic odours, which, however, are very different from one another. They obtained: (1) From the bark, ·43 per cent. of an oil smelling like wintergreen. Its specific gravity was ·923, and its constituents boiled between 170° and 300°; (2) from the berries, 5 per cent. of an oil of a spicy aromatic odour, of specific gravity ·855, with constituents boiling between 160° and 270°; (3) from the twigs, ·3 per cent. of an oil smelling like camphor and calamus, specific gravity ·925; (4) from the leaves, ·3 per cent. of an oil of pleasant lavender odour, specific gravity ·888.

The oil from the bark appears to consist largely of hydrocarbons, with about 10 per cent. of methyl salicylate. It also probably contains cinnamic compounds.

Oils of Tetranthera.

Most parts of the plant *Tetranthera polyantha* v. *citrata*, a plant found in Northern India, the Indian Archipelago and Java, yield an essential oil. The fruit yields from 4 to 5 per cent. of an oil of the following characters:—

[1] *Bull. Soc. Chem.* (**1898**), 638.

Specific gravity . . .	0·885 to 0·898
Optical rotation . . .	+ 6° „ + 13°
Aldehydes	79 to 86 per cent. (bisulphite method).

The principal constituent is citral, with some geraniol, geranyl esters, and a terpene.

The leaves yield about 5 per cent. of oil having the following characters :—

Specific gravity	0·899 to 0·917
Optical rotation	– 12° „ – 16°
Refractive index	1·4644

It contains from 20 to 35 per cent. of cineol, citral, and an alcohol which is probably geraniol.

The bark yields from 0·1 to 0·8 per cent. of oil of the following characters :—

Specific gravity	0·866 to 0·906
Optical rotation	+ 10° „ + 21°
Ester value after acetylation.	230° „ 252°

The oil contains, according to Charabot and Laloue, 8 per cent. of citrol, 10 per cent. of citronellal, 56·5 per cent. of alcohols (geraniol ?), and 2·4 per cent. of esters.

Oil of Mespilodaphne Pretiosa.

This tree is one of the Brazilian *Lauraceæ* (also known as *Ocotea pretiosa*), and all parts of the plant appear to yield an essential oil. Laloue[1] has examined the oil obtained from the twigs and from the wood. From 1·600 grm. of the dry twigs he obtained 7·9 grms. of oil (0·5 per cent.), having a pronounced odour of linalol. From 25 kilos of the wood he obtained 173·2 grms. of oil (0·693 per cent.), which had an odour recalling that of rosewood and cinnamic alcohol. The oil (4) was easily separated into a light (2) and a heavy (3) oil. The benzoic esters of linalol and geraniol were found present. The following are the general characters of the oils, No. 1 being the twig oil :—

	1.	2.	3.	4.
Specific gravity	0·891	0·954	1·055	0·981
Rotation	+ 7° 20′	+ 8° 48′	+ 3° 8′	+ 7° 12′
Refractive index	1·469	1·501	1·545	1·519
Acid value	1·4	0·8	3·5	2·1
Saponification value . .	13·3	100·7	203·7	128·1
Esters as linalyl acetate . .	4·6%	35·3	—	—
Acetyl ester value . . .	165	205	248	220
Alcohols as linalol . . .	51·8	66·6	—	—

The oil from the bark has been examined by Schimmel & Co.[2] The bark yielded 0·83 per cent. of a brown oil of cinnamon-like odour. Its specific gravity was 1·1200 (15°) and its refractive index 1·52712 (20°); owing to the dark colour it was not possible to determine the rotation. It was soluble in 6·5 volumes of 80 per cent. alcohol. It is probable that its constituents do not include esters, but lactone-like compounds, for when 1·5 grm. of the oil was saponified, 5 c.c. seminormal potash

[1] *Bull. Soc. Chem.*, 11, 602. [2] *Report*, April, 1913, 76.

solution was first used up, but after neutralisation a fresh separation of alkali immediately took place and in the end the saponification value 0 was obtained. Acetylation could not be carried out, as with acetic anhydride a violent reaction ensued, gas being given off and carbonisation taking place. The presence of lactones and perhaps also of phenols is furthermore indicated by the fact that the oil is partly absorbed by caustic soda, about 65 per cent. of the oil going into solution when shaken with 5 per cent. solution. Judging by its odour, the portion of the oil which did not react with sodium hydroxide contains caryophyllene.

Oil of Ocotea Usambarensis.

Ocotea Usambarensis, an East African tree known as the Ibean camphor tree, yields an essential oil varying according to the part of the tree from which it has been obtained. The oils have been examined at the Imperial Institute,[1] and found to have the following characters :—

	Oil from		
	Sapwood.	Twigs.	Branches.
Yield	0·4 per cent.	0·52 per cent.	0·14 per cent.
Specific gravity	0·9641	0·9681	0·[illegible]327
Optical rotation	− 7° 30′	− 7° 30′	− 0° 28′
Saponification value	30·1	30·1	13·3

The oils from the sapwood and the twigs contained a considerable amount of cineol. Schmidt and Weilinger[2] have examined an oil, from the bark of the tree, which had the following characters :—

Specific gravity	0·913 at 20°
Optical rotation	− 1 ° 12′
Refractive index	1·4760
Acid value	1·2
Ester value	12·5

He found present in the oil myristic aldehyde, *l*-terpineol, and a sesquiterpene.

Oil of Litsea Odorifera.

The leaves of this plant, which are known in Java as Trawas leaves, yield an essential oil, which has been examined by Romburgh. The oil has the following characters :—

Specific gravity	0·836 to 0·846
Optical rotation	− 0° 10′ to − 7°

The oil contains nonylene-methyl-ketone, methyl-nonyl-carbinol, and undecenol.

Oil of Camphor.

Ordinary camphor oil is a by-product in the manufacture of the ordinary or "Japan" camphor. The tree yielding this is perhaps best known as *Cinnamomum camphora*, but it has been known under several other names, such as *Camphora officinalis* and *Laurus camphora*. The tree is distributed throughout the eastern provinces of Central China, on the Island of Hainan, and to a very great extent in Formosa, which is the principal seat of the industry. The Japanese islands Kinshû and Shikokû also produce it very abundantly. According to Yoshida, the older trees contain more camphor than the young

[1] *Bull. Imp. Inst.*, 9 (**1911**), 340. [2] *Bericht*, 39 (**1903**), 652.

ones, and those over a hundred years of age are exceedingly rich. The importance of a regular supply of both camphor and camphor oil is enormous, and the utter neglect which the Chinese gatherers have shown to the cultivation of the tree is very surprising. They have largely succeeded in exterminating it along the seaboard of Formosa, and are continually making fresh inroads into the inland forests, without any serious attempt to replace the trees destroyed. Attempts, more or less successful, to cultivate the camphor tree have been made in many parts of the world, such as Ceylon, the Malay States, Further India, East Africa, North America, and Italy, but as the world's supply of camphor and camphor oil practically come from Japan, these attempts have little interest from the essential oil point of view. Chinese camphor and camphor oil are found in the market to a small extent, but this trade is steadily declining.

The following details, which give in summarised form a number of experimental results, are from a report by B. J. Eaton[1] published at the instance of the Government of the Federated Malay States:—

The first experiments in the cultivation of the camphor tree in the Malay States were made in 1904 at Batu-Tiga, Selangor, seed having been obtained from Yokohama for the purpose. The plants flourished excellently and in 1909 the first camphor was distilled. The raw material consisted of shoots from five-year-old trees and the result of the experiment is shown below:—

Material.	Yield per cent.
Cut leaves	1·17 to 1·22
Small stems	0·06 ,, 0·45
Mouldy leaves	1·25 ,, 1·47
Leaves and stems	1·25 ,, 1·58
Air-dried leaves	1·10 ,, 1·16
,, mouldy leaves	1·54

In each case the distillate consisted of camphor with very little oil.

These experiments were repeated upon a large scale with a larger distilling apparatus. The raw material consisted of the parts of an entire five-year-old tree.

Material.	Yield per cent.
Leaves	1·00
Stems under ½ in. diameter	0·22
Woody stems over ½ in. diameter	0·61
Roots	1·10

The distillates consisted principally of camphor; the roots alone yielded an oil, which possessed an odour reminiscent at the same time of camphor and of lemons. Afterwards numerous experiments were carried out with various distilling apparatus which had been partly modified.

Finally, the author gives a review of similar experiments in other countries, the results of which are shown in the table on opposite page.

The crude method of preparing camphor from the trees is as follows. The tree is felled and the young branches and twigs are chopped up and packed in perforated jars, and heated over a crude steam-bath. The steam enters the jars, saturates the chips, and causes the crude camphor to sublime and condense in earthenware pots placed over the jars. The crude camphor is sent to the port, and a certain amount of

[1] *Dept. of Agriculture, Bull.* No. 15, February, 1912.

oil exudes from it which is collected and is known as oil of camphor. The majority of the oil is, however, produced by distilling the chips with water in crude stills. The crude product amounts to about 3 per

Distilled in	Distilled by	Yield per Cent.	Material.	Properties of the Distillate.
Ceylon	Willis & Bamber	1	Leaves and small Stems	—
India	Hooper	1	—	Contained from 10 to 75 per Cent. Camphor
German E. Africa	Lommel	0·06 to 0·1	Branches and thin Twigs	Camphor and Oil
		0·22	Young shoots	
Jamaica	Duncan	0·61	Wood	Crude Camphor
		0·05	Twigs	
		2·37	Green Leaves	
		2·52	Dried ,,	
		1·39	Dead ,,	
West Indies	Watts & Tempany	0·5	Wood	Camphor and Oil
		0·5	Leaves and Twigs	
		0·7	,, ,, ,,	
Italy	Giglioli	1·2 to 1·5	Green Leaves	Camphor
		2·4 ,, 3·0	Dry ,,	
		0·02 ,, 0·25	Twigs	
America	Hood & True	2	Shed Leaves	Camphor and Oil
		0·7	Trees from Shady places	
		2·77	Trees from sterile, shady places	

cent. of the wood used. The oil is drained from the crystalline camphor, of which it retains a considerable amount in solution. This is transferred to a still, and about two-thirds is distilled off, leaving the bulk of the camphor in the residue, which is cooled and pressed to separate more camphor. This process is repeated so long as it pays, and the residue forms the camphor oil of commerce. From a scientific point of view, one must regard the whole of the products of distillation of the wood as camphor oil. This, however, contains so much solid camphor, or stearoptene, that it is commercially regarded as a separate product, and the liquid portion of the oil is thus known as oil of camphor. Hence the great variability in its properties and composition.

The crude camphor arrives in England in various states of purity, and is refined by sublimation, generally with quicklime and charcoal. Formerly camphor oil was regarded as having no value. To-day, however, it is used to an enormous extent in the preparation of safrol, which is used as a cheap perfume, for the manufacture of artificial oil of sassafras, and for the synthesis of heliotropin.

Camphor oil appears on the market in various forms, and as it is a by-product, varying in characters according to the substances which have been removed from it, and the degree of fractionation which it has undergone, it is obvious that scientific standards for it cannot be set up. It is frequently sold on the basis of a guaranteed specific

FIG. 20.—The "Kame" system of distillation (Camphor Distillery No. 111, Toröyen).
[*By Permission of Monopoly Bureau of the Government of Formosa*

gravity, which gives a good indication of its general quality. The following may be regarded as the general types of oil to be found in commerce :—

1. Crude camphor oil, from which the crystallisable camphor has been removed as completely as practicable at the distillery, This is a

FIG. 21.—Full view of Camphor Distillery No. 5, Shinkôkô.
[*By Permission of Monopoly Bureau of the Government of Formosa.*

deep yellow to brownish oil of specific gravity from about 0·950 to 0·998.

2. Heavy camphor oil, black camphor oil, or brown camphor oil is

understood to be the highest boiling portion of the oil, rich in safrol and containing sesquiterpenes and a small amount of phenols. Its specific gravity usually lies between 1·000 and 1·040.

3. Light camphor oil is the lowest boiling portion of the oil, and consists principally of terpenes. Its specific gravity usually lies between 0·870 and 0·885, rarely up to 0·910. It is water white and resembles turpentine in general characters.

4. The very high boiling fractions, rendered nearly colourless by redistillation, of specific gravity from 1·060 to 1·075, are sold as "artificial sassafras oil".

These details show that camphor oil must be bought for what it is worth. No definite standards can be laid down, and only a complete analysis will show the value of a given sample.

The following figures, however, will be of assistance to the analyst. The oil is always dextro-rotary, varying from + 12° in high gravity samples to + 20° to + 32° in lighter oils. Light camphor oil, of specific gravity 0·870 to 0·910, will usually yield the following fractions on distillation:—

175° to 180°	about 25 per cent. to 35 per cent.
180° ,, 185°	30 ,, ,, 45 ,,
185° ,, 190°	10 ,, ,, 15 ,,

Camphor oil has an exceedingly complex composition. The earliest scientific investigation of this oil was by Beckett and Wright,[1] but the only conclusions they came to were that it was a complex mixture probably containing a hydrocarbon of the terpene series, a body having the composition $C_{10}H_{18}O$, a liquid containing less hydrogen than camphor, and much ordinary camphor. Oishi[2] described it as a crude oil, of specific gravity ·959, which yielded about 25 per cent. of camphor. The rectified oil was a colourless liquid of specific gravity ·895. He concluded that the oil was a mixture of terpenes, camphors, and some of the oxidised hydrocarbons. Yoshida[3] separated the oil into four chief portions, which appeared to be (1) laevo-α-pinene boiling at 150°; (2) a terpene (limonene) boiling at 172°; (3) camphor; (4) an oxygenated body which he termed camphorogenol. A sample examined by Trimble and Schroeter[4] was separated into ten fractions, each of which the authors claimed to be pure compounds, but which were clearly mixtures in several cases. No fewer than five terpenes were given, and five oxygenated bodies. Bertram and Walbaum[5] and Schimmel & Co.[6] have examined the oil since, and to them is due the discovery of the presence of safrol, which is now manufactured on an enormous scale. They also showed that Yoshida's camphorogenol had no existence. The result of their researches shows that the following bodies are present in the oil: β-pinene, phellandrene, camphene, dipentene, dextro-fenchene, dextro-limonene, and bisabolene, cineol, camphor, safrol, eugenol, terpineol, citronellol, cumic alcohol, borneol, α-terpineol, and cadinene, with traces of carvacrol, aldehydes and acids, including caprylic acid, and one of the oleic series, with the formula $C_9H_{16}O_2$, has been isolated.

The latest investigation of camphor oil is that of Semmler and Rosen-

[1] *Jour. Chem. Soc.* (**1876**), 7.
[2] *Chem. News*, **1**., 275.
[3] *Jour. Chem. Soc.* (**1885**), 782.
[4] *Pharm. Jour.*, **xx**., 145.
[5] *Jour. prakt. Chem.*, ii., 49 (**1894**), 19.
[6] *Reports, passim.*

berg[1] who have isolated a fraction boiling at 130° to 150° at 10 mm., which contained a new sesquiterpene which the authors have termed sesquicamphene and which has the following characters:—

Boiling-point at 8 mm.	129° to 133°
Specific gravity	0·9015 at 20°
Optical rotation	+ 3°
Refractive index	1·50058

It is a bicyclic compound.

They also isolated a sesquiterpene alcohol from the fraction boiling at 150° to 170° at 10 mm., which on dehydration yielded a hydrocarbon of specific gravity 0·9138 at 20°, optical rotation + 50° and refractive index 1·50895. This alcohol has been named sesquicamphenol.

The highest-boiling fractions of the oil, after being heated over sodium, yielded a hydrocarbon fraction with boiling-point 180° to 190° (11 mm.). Hydrochloric acid gas in ethereal solution was passed through this fraction, when it afforded a tetrahydrochloride melting-point 129° to 131°. The hydrocarbon which was regenerated from the hydrochloride possessed the following characters: boiling-point 177° to 178° (6 mm.); $d_{20°}$ 0·8870; a_D ± 0; n 1·50339. Combustion showed the formula to be $C_{20}H_{32}$; the value (90·6) found for the molecular refraction also agreed with the formula $C_{20}H_{32}/\overset{=}{_4}$ (90·48). This body, therefore, is a monocyclic diterpene, which has been called α-camphorene by the authors. Regenerated, α-camphorene again formed a hydrochloride melting-point 129° to 131°. Upon being reduced with platinum and hydrogen by the usual method, α-camphorene yielded octahydro-α-camphorene, $C_{20}H_{40}$, boiling-point 174° to 176° (9 mm.); n_D 1·46470. In addition to α-camphorene the highest-boiling camphor-oil fractions contain a second diterpene, to which Semmler and Rosenberg have given the name of β-camphorene. This camphorene can be regenerated from the oily residues which are formed in the process of preparing α-camphorene tetrahydrochloride. This diterpene has the following constants: boiling-point 170° to 180° (7 mm.); $d_{20°}$ 0·930 $[a]_D$ ± 0; molecular refraction found 88·61; calculation for $C_{20}H_{32}/\overset{=}{_3}$ 88·77. Its hydrochloride remains liquid.

It is a surprising fact that α-camphorene is capable of being built up synthetically with comparative ease. In order to accomplish the synthesis, Semmler and K. G. Jonas[2] heated myrcene for four hours in a sealed tube to 250° to 260° by which they obtained a yield of 50 per cent. of a fraction with boiling-point 175° to 195° (8 mm.) which afforded a solid hydrogen chloride additive product, melting-point 129° to 130°. When mixed with α-camphorene tetrahydrochloride obtained from the blue camphor oil, the melting-point of this body was not depressed; hence the synthetic hydrocarbon is identical with α-camphorene.

Semmler and Jonas have carried out another interesting synthesis by heating a mixture of myrcene and isoprene in a sealed tube to 225°. The result was a yield of about 30 per cent. of a hydrocarbon, boiling-point 125° to 135° (8 mm.): $d_{18°}$ 0·8685; a_D ± 0; n^D 1·49166; molecular refraction found 68·10; calculation for $C_{15}H_{24}/\overset{=}{_3}$ 67·86. This body is without doubt a sesquiterpene and is perhaps identical with bisabolene. Its trihydrochloride melts at 83°, but when it is mixed with bisabolene trihydrochloride the melting-point is depressed by 12°. There is no

[1] *Berichte*, 46 (**1913**), 768. [2] *Ibid.*, 1566. Schimmel's *Report*, Oct., 1913.

doubt that the synthesis of *cyclo*isoprene-myrcene represents the first complete synthesis of a sesquiterpene.

Conversion of myrcene into α-camphorene by heating in a sealed tube affords an excellent method for detecting the presence of myrcene, especially when only a small quantity of the terpene is available.

Camphor oil distilled from trees growing in the Federated Malay States and in Mauritius have been examined by the chemists to the Imperial Institute.[1]

The camphor tree was first grown experimentally by the Department of Agriculture in the Federated Malay States in 1904, when seeds obtained from Japan were sown at Batu Tiga, Selangor. The seedlings were planted out when about six months old at a distance of 10 ft. by 10 ft. The conditions at this place appear to be well suited to the tree, and the growth of the plants is stated to have been equal to that seen in Japan.

Distillation was first carried out experimentally in 1909, and has been continued at intervals ever since. In 1911, by using a wooden still of the Japanese type, a yield of 0·13 to 0·5 per cent. of camphor and camphor oil was obtained from fresh green prunings. In subsequent experiments, using a galvanised iron condenser, fresh prunings from four-year-old trees gave yields from 0·53 to 0·77 per cent., whilst air-dry material from two-year-old trees gave 0·60 to 0·80 per cent. with a metal condenser. The leaves and twigs distilled separately gave the following results:—

	Leaves.	Twigs
Two-year-old trees . .	1·4 to 1·6 per cent.	0·10 to 0·25 per cent.
Four-year-old trees . .	1·3 ,, 1·7 ,,	0·13 ,, 0·22 ,,

The distillate in all cases consisted principally of camphor, with only a small quantity of camphor oil.

In 1912 specimens of camphor oil and camphor produced in the Federated Malay States were examined at the Imperial Institute, and the oil was found to contain 19·3 per cent. of camphor, whilst safrol, the characteristic and valuable constituent of the Japanese oil, was absent.

In 1913 a further sample of camphor oil was received at the Imperial Institute which consisted of a pale yellow oil containing a considerable deposit of camphor. By cooling the oil to − 12°, a yield of 14·9 per cent. of crystalline camphor was obtained, and the residual oil possessed the following constants compared with corresponding figures for the previous sample:—

	1913 Sample.	Previous Sample.
Specific gravity (15°)	0·920	0·913
Optical rotation (100 mm. tube)	+ 38° 23′	+ 41° 1′
Acid value	1·1	—
Saponification value	3·6	—
,, ,, of acetylated oil	25·7	—

On fractionating the residual oil about 4·3 per cent. of cineol was obtained, and a further yield of 15·7 per cent. of camphor, giving a total yield of 30·5 per cent. No safrol was detected, its absence being probably due to the fact that the oil was obtained from the prunings of young trees. The camphor oil from the Federated Malay States differs,

[1] *Bull. Imp. Inst.*, **xiv.**, 4; and *P. and E.O.R.* (**1917**), 173.

therefore, from the Japanese oil in that the camphor has not been separated and that it contains no safrol.

Camphor Oil from India.

Simonsen and Ghose[1] report on experiments carried out on the distillation of camphor and camphor oil from the leaves of young trees grown in various parts of India. They obtained from 0·25 to 2·01 per cent. of camphor, and a varying amount of camphor leaf oil. This oil, after separation of the camphor, contained d-*alpha*-pinene, dipentene, cineol, terpineol, caryophyllene, and a trace of cadinene, but no safrol.

Camphor Oil from Mauritius.

The camphor tree is cultivated in the various botanic gardens and forest plantations, but so far camphor has not been produced on a commercial scale on the island. Distillation experiments have been conducted in recent years, and specimens of the oil obtained have been examined at the Imperial Institute. The samples contained a considerable quantity of cineol, and differed greatly from the camphor oil of commerce. They had the following characters:—

	1.	2.	3.
Specific gravity (15°)	0·907	0·906	0·925
Optical rotation (100 mm. tube)	− 20° 4′	− 21° 5′	− 6° 20′
Cineol by resorcinol method	69 per cent.	65 per cent.	72 per cent.

Five samples received in 1911 were described as follows:—

No. 1.—Camphor leaf oil.
,, 2.— ,, root ,,
,, 2.— ,, twig ,,
,, 4.— ,, ,, ,,
,, 5.— ,, wood ,,

These had the following characters:—

	No. 1.	No. 2.	No. 3.	No. 4.	No. 5.
Specific gravity at 15°	0·9162	0·9508	0·9143	0·9243	0·9164
Optical rotation	− 18° 38′	+ 13° 36′	− 15° 22′	− 11° 37′	− 11° 26′
Percentage of oil absorbed by resorcinol	56 per cent.	38 per cent.	54 per cent.	not determined	

No camphor was separated on cooling to − 10°, but from No. 2 as the result of fractionation about 6 per cent. of camphor was extracted, and an appreciable amount of safrol. From the other oils no camphor could be obtained and no safrol was detected.

These oils from Mauritius differ therefore from the camphor oil of commerce in containing a considerable quantity of cineol, but no safrol, and with the exception of No. 2 all were laevo-rotatory, instead of dextro-rotatory. In view of these results, a comprehensive series of distillations were conducted in Mauritius by the chemist to the Department of Agriculture. The results confirmed the previous experiments, and showed that the camphor trees in Mauritius do not yield solid camphor on distillation, and that the small amount present is readily dissolved in the oil, and is not readily separated from it. It was thought possible that the abnormal character of the oils might be due to the trees not being genuine camphor trees, but specimens examined at Kew

[1] *P. and E.O.R.*, **1920**, 301.

were found to belong to *Cinnamomum camphora* trees. It is stated that even in Formosa varieties exist which yield oil from which no camphor separates, and it is proposed that fresh trials should be made in Mauritius, and that fresh seed should be procured from selected trees in Ceylon or the Federated Malay States which are known to yield camphor.

The essential oil of *Hernandia peltata*, a Madagascar plant known as a false camphor, has been examined by Schimmel & Co.[1] From the stem wood from 1·03 to 2·06 per cent. of oil was obtained, having the following characters:—

Specific gravity at 15°	0·958 to 0·963
Optical rotation	+ 83° 45′ „ + 104° 12′
Refractive index	0·49695 „ 1·50111

It contains from 75 to 80 per cent. of aldehydes (mostly dihydrocumic aldehyde), with traces of free acids and a small amount of esters. In addition to dihydrocumic aldehyde, the oil contains a small amount of the aldehyde myrtenal, which has a specific gravity 0·986 at 20°, specific rotation + 13·6°, and refractive index 1·50618. The root wood yielded 0·5 per cent. of oil having a specific gravity of 0·9667, optical rotation + 126° 15′, refractive index 1·50383 and aldehyde content 92·5 per cent. The whole fruit gave 0·5 per cent. of oil of specific gravity 0·9528, optical rotation + 50° 10′, refractive index 1·49554, and aldehyde-content 49 per cent. The almond-like seeds gave 1·38 per cent. of oil, having a specific gravity 1·0044, optical rotation + 87°, refractive index 1·50614, acid value 7·3, and ester value 110·4. The fruit oil does not appear to contain any dihydrocumic aldehyde. It is possible, however, that the aldehyde present is perillic aldehyde, and not dihydrocumic aldehyde.

The leaves of the Japan camphor tree also yield an essential oil. According to Hooper the yield is about 1 per cent., the oil having a specific gravity about 0·930 and optical rotation + 4° 32′. It contained 10 to 15 per cent. of camphor. Another specimen contained 75 per cent., and had a specific gravity ·9314 and optical rotation + 27°. There seems not to be much difference between this and the ordinary camphor oil. The roots yield an oil of quite similar character. A sample distilled by Schimmel & Co. had a specific gravity ·975, and consisted chiefly of camphor—the liquid portion resembling commercial camphor oil very closely.

Oils of Shô-Gyū and Yu-Ju.

There are found in Formosa two trees which closely resemble the ordinary camphor tree, which are known as *Shô-Gyū* and *Yu-Ju*, and whose actual botanical differences from the true camphor tree are not yet determined. Both these trees yield essential oils which have been investigated by K. Nagai,[2] who gives the following details of them:—

"'*Shô-Gyū* Oil' is the name given to an essential oil with an aromatic odour, distilled from a perennial tree popularly called by the natives '*Chiu*ⁿ*-Gû*,'[3] growing chiefly in the highland forest regions of Formosa. On account of its not producing camphor through the reg-

[1] Schimmel's *Bericht*, April, **1915**, 54.

[2] *Publication of the Monopoly Bureau, Formosa. Taihoku*, 1914.

[3] *Chiu*ⁿ or *Shô* = camphor tree; *Gû* or *Gyū* = ox.

ular process of manufacture, the *Chiuⁿ-Gû* (*Shô-Gyū*), the source of the oil has played so far a very insignificant part in the industrial world, utterly failing to draw the attention of those connected with the industry. While the *Shô-Gyū* tree exhibits some features in common with the *Shiu-Shô* tree,[1] as in the lack of solid camphor or in other superficial indications, the latter luxuriates in the camphor-tree regions and has aroused wide public interest as a highly valuable material for industrial researches: whereas the former, thriving exclusively in an isolated region beyond that of the camphor tree, was completely overlooked, and even the essential oil it produces has hardly been the subject of discussion.

"Seen at a distance, the *Shô-Gyū* tree presents a strong resemblance to the camphor tree. The natives in the locality of *Kōsempo* (*Kah-sien-po*) call it '*Gû-Chiuⁿ*,' a term presumably corrupted from the savage tongue. At the present stage of investigation, however, its scientific name and classification are still a matter of conjecture and doubt, but, following the opinion of Dr. Hayata based upon several specimens collected in *Kōsempo*, *Taiko* (*Tai-fu*), and *Arisan* (*Ali-soaⁿ*) districts, we may assume that it belongs to the *Laurineæ*. The 'black camphor' tree, as it is generally called by the manufacturers of camphor, appears to be the same tree. There is, however, no proof of the assumption, nor has any report ever been made regarding the collection of its flowers—the blooming season is said to be from February to March; hence it is evidently impossible to determine its scientific name. In view of a future study of a collection of flower specimens, the discussions are restricted here to the scope of external forms. The following are the results of observations made on the tree as it flourishes in the thick woods along the length of the old *Arisan* road (*Kî-koan-tâi*) in *Kagi* (*Ka-gï*) prefecture, and in the vicinity of *Heishana* (*Pieng-chia-nâ*)—a mountainous region extending from a locality popularly called 'Cross road' (*Chap-jî-lō*) up to the elevated region of *Arisan*, where conifers thrive.

"The trunk of the *Shô-Gyū* tree usually attains a height of 40 to 50 *shaku*,[2] the circumference of the larger ones often measuring, at a man's height, from 15 to 20 *shaku*; and the growth of the tree is, speaking generally, uniform and vigorous with no dwarfed specimens. The greater portion of the trunk is generally covered with lichens and mosses, with a few exceptions where the bark is much exposed; also, many are entwined by vines and climbers. The bark when compared with that of the camphor tree, suggests at a glance a somewhat brownish-red colour. The large branches have smooth longitudinal wrinkles and furrows upon the greyish-white surface. A piece of *Shô-Gyū* wood from *Tebutegai* was found equally covered with lichens and other parasites in spots upon the bark, displaying identical features in its external peculiarities with specimens from *Arisan*. The longitudinal section of the branch shows that the heartwood seems rather white in comparison with that of the camphor, *Shiu-Shô*, or *Yu-Ju* tree. The wood is soft and sappy, seemingly containing much moisture, and the solidity naturally failing to equal that of the camphor tree, it can be sawed or chopped with comparatively little exertion. Despite the extreme difficulty of distinguishing the tree from the camphor tree by mere observation

[1] *Shiu* = stinking. [2] 1 *shaku* = 0·994 ft. = 0·303 m.

of external appearances in the living states, the test can readily be accomplished by shaving small pieces from the sapwood portion—particularly the root—of the two trees, and judging from their inherent odour.

FIG. 22.—Distillation of Shô-Gyū at Heishana.
[By Permission of Monopoly Bureau of the Government of Formosa.

This test has incidentally proved to be the most reliable method of detecting the difference between the two. A simpler method, however, is the examination of the leaves. The leaves of the *Shô-Gyū* tree have a bright deep-green colour upon the surface and are much thicker than those of the camphor tree (a large one often measuring about 13 cm. in length, and 4·5 cm. in width), and are slender in form with a particularly pointed apex which frequently reaches a length of 1 cm. On examination of a specimen collected in November, a large terminal leaf-bud about 1 cm. in length and 0·5 cm. in diameter was found on the upper end of every young branch, by observation of which the tree may also be distinguished from the camphor tree. The latter test, however, being only practicable when the shoots with leaves attached are within reach, it is highly advisable to employ the former test at the same time, in order to ensure a correct determination."

The following are the physical characters of a number of samples of this oil, distilled from the wood of trees collected in various localities.

Place of Origin.	No.	Colour.	d at $t°$.		α_D at $t°$.		n_D at $t°$.		Odour.	Colour-Reaction with Mercuric Sulphate.
Shiuchiku	1	Faint yellow	0·908	9	+ 29·50	5	1·47753	20½	Slightly resembling oil of savin	Trace ?
,,	2	Colourless	0·910	,,	+ 23·25	,,	1·47478	19½	Sweetish, and as that of No. 1	–
,,	3	Light yellow	0·911	,,	+ 15·88	,,	1·47468	20	Same as No. 1	Trace ?
,,	4	,,	0·951	,,	+ 14·90	,,	1·48713	18	,,	+
,,	5	Colourless	0·912	,,	+ 24·55	,,	1·47567	20	Recalling oil of nutmeg	+
,,	6	Faint yellow	0·900	,,	+ 23·50	,,	1·47665	19	—	+
,,	7	,,	0·927	,,	+ 22·30	,,	1·48133	20	—	+
,,	8	Colourless	0·945	,,	+ 18·50	,,	1·48471	20½	—	–
,,	9	Faint yellow	0·933	,,	+ 15·25	,,	1·48094	20	—	Trace ?
,,	10	,,	0·914	,,	+ 22·20	,,	1·47694	19	—	+
Kōsempo	—	Yellow	0·910	11	+ 34·45	11	1·48326	,,	Reminding of oil of nutmeg	+
Arisan	1	Light yellow	0·968	25	+ 12·83	25	1·49466	26	—	Trace ?
,,	2	Faint ,,	0·939	,,	+ 14·40	,,	1·48490	25½	—	(?)
,,	3	Light ,,	0·993	26	+ 8·55	26	1·50088	25	—	Trace ?
,,	4	Almost colourless	0·949	,,	+ 16·65	25	1·48971	,,	—	,,
,,	5	,,	0·993	,,	+ 13·25	,,	1·50227	,,	—	+
,,	6	Bright red	0·925	25	+ 12·40	,,	1·47912	,,	Recalling oil of sweet marjoram	(?)
Tebutegai	1	Faint yellow	0·943	,,	+ 19·31	25½	1·48971	,,	—	–
,,	3	Light ,,	0·970	26	+ 16·25	,,	1·49693	25½	—	(?)
,,	4	Faint ,,	1·030	,,	+ 7·75	26	1·51302	25	—	+
,,	5	Light ,,	0·993	,,	+ 11·05	,,	1·50237	,,	Reminding of oil of nutmeg	Trace ?
,,	8	,,	0·965	,,	+ 13·34	,,	1·49429	,,	—	+
,,	13	,,	0·908	,,	+ 24·25	,,	1·47987	,,	—	+

The acid value of this oil is always below 1 and the ester value does not exceed 2·5. A large amount of free alcohols, however, is present, the ester value after acetylation varying from 46 to 129.

The following constituents were identified in the oil: formaldehyde, sabinene, dipentene, α-terpinene, γ-terpinene, terpinenol-4, geraniol, citronellol, safrol, and eugenol.

"*Yu-Ju* oil" is a term applied to an essential oil which is obtained from the so-called "Oil Tree," indigenous to southern Formosa, and which is closely related to camphor oil in its properties. Its failure, however, to yield camphor in the course of production is the chief feature differing from camphor oil, and evidently indicates the reason why the natives call the mother-tree, "*Iû-chhiu*" or "Oil Tree". Similarly, the term "*Yu-bôku*" or "*Yu-bun-bôku*," as frequently applied by the Japanese to the tree, which yields only a little camphor, and a large quantity of oil, suggests the inconspicuously small percentage of camphor obtainable.

The *Yu-Ju* tree is identical, in its external forms, with the camphor tree (*Cinnamomum camphora* Nees et Eberm.), and the camphor manufacturers mainly rely, as the sole means of discriminating between them, upon the odour of shavings taken from the root, outside of which no particular method seems to be in current use.

The following are the physical characters of a number of samples of *Yu-Ju* oil :—

Place of Origin.	No.	Colour.	d at $t°$.		α_D at $t°$.		n_D at $t°$.	
Shinkôkô	1	Light golden-yellow	0·945	13	+ 29·22°	13	1·47596	19
,,	2	,,	0·942	,,	+ 18·88°	,,	1·47457	,,
,,	3	,,	0·951	12	+ 19·50°	12	1·47869	20
,,	4	,,	0·967	13½	+ 22·48°	13½	1·48191	19
,,	5	,,	0·966	,,	+ 24·42°	,,	1·48568	20
,,	6	Golden-yellow	0·961	14	+ 29·85°	14	1·48045	20½
,,	7	,,	0·961	,,	+ 26·52°	,,	1·48326	21
,,	8	Light golden-yellow	0·963	15	+ 26·95°	15	1·48326	,,
,,	9	Golden-yellow	0·955	,,	+ 25·35°	,,	1·47850	20
,,	10	Light golden-yellow	0·947	,,	+ 27·30°	,,	1·47567	19½
Daipanriau	11 (*a*)	Slightly greenish-yellow	0·954	16	+ 21·00°	16	1·47694	19
,,	11 (*b*)	Slightly greenish yellow	0·953	,,	+ 21·30°	,,	1·47694	,,
,,	12 (*a*)	Slightly greenish-yellow	0·972	,,	+ 30·80°	,,	1·47746	,,
,,	12 (*b*)	Light golden-yellow	0·951	,,	+ 28·30°	,,	1·48278	,,

This oil contains furfurol, α-pinene, camphene, β-pinene, cineol, camphor, α-terpineol, and safrol.

Oil of Shiu.

The *Shiu-Shô* tree, which thrives in the camphor regions of Formosa, in association with the camphor tree, but which was formerly considered as of no value, has recently been shown by scientific investigations to be capable of playing a prominent part as material for the production of camphor or linalol, furnishing a fair example of possible development.

The tree is closely related to the camphor tree, but its exact botanical relationship is not known.

The oil distilled from the wood of this tree is rich in linalol, and has an odour recalling that of linaloe oil. It has the following characters :—

Specific gravity	0·870 to 0·895
Optical rotation	− 1° ,, − 16°
Acid value	0 ,, 1
Ester ,,	0·5 ,, 30
Total alcohols as linalol	65 to 90 per cent.

An oil reported on by Keimazu[1] is described as having a specific gravity 0·9279, and optical rotation + 17°, but this appears to have been mixed with camphor oil, so that his researches as to the constituents present must be accepted with reserve. The principal body present in the oil is the alcohol linalol (Keimazu found formaldehyde, *d-a*-pinene, cineol, dipentene, apopinol (impure linalol ?), camphor, eugenol, and safrol).

Oil of Guiana Sandal-wood.

The wood of a tree growing in French Guiana, and known there as sandal-wood, yields an oil which has been investigated by Jeancard and Satie.[2] The tree is, in all probability, one of the *Lauraceæ*, but its botanical relationships are unsettled.

By steam distillation the wood yields oils very variable in character, as will be seen from the following tables :—

	Sample A.	Sample B.	General Limits.
Specific Gravity at 15° C.	0·9630	1·0122	0·963 to 1·0122
Rotatory Power	− 2° 10′	− 0° 30′	+ 0° 30′ to − 6°
Solubility { 80 per Cent. Alcohol	1·1	0·8	0·8 ,, 1·1
Solubility { 75 ,, ,,	10	1·6	1·5 ,, 12
Saponification Value	58·8	14·7	13 ,, 65
,, ,, after Acetylation	114·1	72·1	65 ,, 117

On fractionation *in vacuo* the following results were obtained :—

	A.	B.
Below 155° C. at 20 mm.	27·4 per cent.	18 per cent.
155° to 160° C.	59 ,,	80 ,,
Above 160°	13·6 ,,	2 ,,

The 155 to 160° fraction has the following constants :—

	A.	B.
Specific Gravity at 26°	1·024	1·037
Rotatory Power	− 4° 20′	− 6°
Solubility in 70 per Cent. Alcohol	1·8	1·8

The oils appear to contain a common constituent in varying proportion, which would explain the observed variations. To render this more evident, the authors classified a large parcel of the wood into three portions as follows :—

Wood No. I. consisting of branches.

Wood No. II. consisting of trunks split into two or three pieces according to length.

Wood No. III. consisting of logs from which the bark had been removed, and having the appearance of Indian sandal-wood.

[1] *Jour. Pharm. Soc. Japan* (**1903**), 253 and 258. [2] *P. and E.O.R.* (**1911**), 79.

The constants of the oils distilled from these several lots were :—

	No. I.	No. II.	No. III.
Specific Gravity at 15°	0·9665	0·9806	0·9968
Rotation	− 6° 16′	0	+ 0° 30′
Solubility in 85 per Cent. Alcohol	0·4	0·4	0·3
,, 80 ,, ,,	1	1	0·9
,, 75 ,, ,,	1·7	10·8	20
Saponification Value	44·1	46·9	13·3
,, ,, after Acetylation	92·4	96·6	65·8

By careful fractionation they separated a body having the following characters :—

Boiling-point	158° to 159° at 20 mm.
Specific gravity at 23°	1·0378
Rotatory power	− 6°
Solubility in 70 per cent. alcohol	1 in. 1·6
,, 65 ,, ,,	,, 2·7
,, 60 ,, ,,	,, 6·5

This body, which is termed marionol, combines with sodium with evolution of hydrogen, and is probably a tertiary alcohol. On adding to 10 grms. marionol dissolved in 10 grms. petroleum spirit, 1 grm. of sodium, a brisk evolution of hydrogen takes place in the cold. On warming over a water-bath solution is complete in the course of a day.

Oil of Californian Bay.

The leaves of the so-called Californian Bay, a large forest tree common on the mountains of California and the Pacific slope, yield from 2·5 to 5·5 per cent. of an essential oil. The tree is also known as the mountain laurel, the sassafras laurel, and the Californian olive. Botanically it is *Umbellularia Californica* (*Oreodaphne Californica*). The oil has the following characters :—

Specific gravity	0·935 to 0·950
Optical rotation	− 22° ,, − 25°
Acid value	3 ,, 5
Ester ,,	5 ,, 7
,, ,, (after acetylation)	50

It is soluble in 1·5 to 2·5 volumes of 70 per cent. alcohol.

Heaney[1] states that the yield of oil from the leaves is 4 per cent. He claimed to have found about 35 per cent. of a hydrocarbon and 65 per cent. of an oxygenated body, which he termed *oreodaphnol*, in the oil. This body, however, has never received confirmation, and it is very doubtful whether the hydrocarbon he obtained was pure. He assigned to it a specific gravity ·894 and boiling-point 175°. Stillman,[2] on the other hand, claims that it contains terpineol and an oxygenated body of the formula $C_8H_{12}O$, which he named umbellol.

Modern investigations by Power and Lees, however, have not confirmed this statement. These chemists have isolated from the oil the following constituents: (1) Traces of free fatty acids including formic acid; (2) 1·7 per cent. of eugenol; (3) 6 per cent. of pinene; (4) 20 per cent. of cineol; (5) 60 per cent. of a new unsaturated cyclic ketone of a

[1] *Amer. Jour. Pharm.*, 47 (**1875**), 105. [2] *Berichte*, 13 (**1880**), 630.

mint-like odour boiling at 220°, of specific gravity 0·858 and optical rotation − 37°. Its formula is $C_{10}H_{14}O$, and Power has named it umbellulone; (6) a small quantity of safrol; (7) 10 per cent. of eugenol methyl ether.

The ketone umbellulone is a liquid, which when purified by regeneration from its semicarbazide-semicarbazone has the following characters:—

Specific gravity	0·950 at 20°
Optical rotation	− 36° 30′
Refractive index	1·48325
Boiling-point	219° to 220° at 749 mm.

It forms a semicarbazide-semicarbazone and a normal semicarbazone the former melting at over 200° and the latter at 240° to 243°. It forms a dibromide melting at 119°.

Oil of Laurel Leaves.

The leaves of the sweet bay, *Laurus nobilis*, yield from 1 to 3 per cent. of essential oil, having an odour somewhat recalling that of cajuput. The oil has the following characters:—

Specific gravity	0·915 to 0·945 (usually 0·915 to 0·930)
Optical rotation	− 15° to − 22°
Refractive index	1·4670 ,, 1·4775
Acid value	1 ,, 3
Ester ,,	20 ,, 50
,, ,, (after acetylation)	36 ,, 96

The oil is soluble in 1 to 3 volumes of 80 per cent. alcohol. Some oils, French oils especially, are soluble in 3 to 10 volumes of 70 per cent. alcohol.

The following table, showing the characters of oils distilled in various districts, is due to Schimmel & Co.:—

	$d_{15°}$.	a_D.	$n_{D20°}$.	Acid No.	Ester No.	Ester No. after Acet.	Solubility.
German and French	0·915 to 0·932	− 15° to − 18°	1·469 to 1·477	up to 2·0	28 to 43	58·3 to 77·7	in 1·3 vol. 80 p.c. alcohol, some (esp. French) samples in 3 to 10 vols. 70 p. c. alcohol.
Fiume	0·9281	− 13° 52′	1·47156	1·0	31·9	—	in 2·5 vols. 70 p. c. alcohol and more.
Corfu	0·9177 to 0·9211	− 16° 40′ to − 21° 40′	1·46862 to 1·47107	up to 1·5	29·8 to 43·8	36·2 to 82·5	in 1 vol. 80 p. c. alcohol, mostly in 3 to 6 vols. 70 p. c. alcohol.
Cyprus	0·9405	− 5° 40′	1·47372	1·6	25·3	95·6	in 2 to 3 vols. 70 p. c. alcohol and more.
Palestine	0·916 to 0·924	− 14° to − 21°	1·46516 to 1·46575	up to 2·2	21 to 49	43·2 to 81·4	in 1 to 2 vols. 80 p. c. alcohol, some in 4 to 7 vols. 70 p. c. alcohol.
Asia Minor	0·9[illegible]68	− 15° 50′	1·46575	0·8	34·8	60·1	in 1 vol. 80 p. c. alcohol and more.
Dalmatia	0·9268	− 14° 36′	—	—	—	—	in 2·5 vols. 70 p. c. alcohol.
Syria	0·9161	− 14° 20′	—	—	—	—	in 1 vol. 80 p.c. alcohol.

This oil contains α-pinene (?), principally as the laevo-rotatory variety, and phellandrene. The oxygenated constituents include cineol, *l*-linalol, geraniol, eugenol, and methyl-eugenol. The highest boiling fractions contain a sesquiterpene and a sesquiterpene alcohol.

Molle [1] has investigated the oil and found present acetic, isovalerianic, and isobutyric acid in the free states. He also found a small quantity of esterified eugenol. Acetic, valerianic, and caproic acids were found in the form of esters, as well as a monobasic acid of the formula $C_{10}H_{14}O_2$ melting at 146° to 147°. Molle does not confirm the statement of Wallach, that pinene is present in the oil.

The leaves of a Java laurel tree, known as Trawas leaves, yield, according to van Romburgh [2] an essential oil having the following characters :—

Specific gravity	0·836 to 0·846
Optical rotation	– 0° 10′ ,, – 7°

The principal constituent of the oil is nonylene-methyl-ketone, which forms a semicarbazone melting at 116°. The oil also contains methyl-nonyl-carbinol and a second alcohol undecenol-1-ol-10 (specific gravity 0·835 ; boiling-point 233°).

PROTEACEÆ.

Oil of Stirlingia Latifolia.

H. V. Marr [3] has examined the oil distilled from the whole of the shrub, *Stirlingia latifolia*, and finds it to have the following characters :—

Specific gravity at 15°	1·031
Boiling-point	195° to 205°
Optical rotation	+ 0·2°

The oil solidifies in crystals almost completely at 12°.

The oil was found to consist almost entirely of acetophenone.

The shrub grows freely in low-lying, sandy country from Geraldton to Bunbury in Western Australia, and should form a useful source for acetophenone.

MYRISTICEÆ.

The only oils belonging to this family of the slightest importance are those distilled from the nutmeg and mace, which are the products of *Myristica fragrans*, the seed of which consists of the well-known nutmeg enclosed in a testa, which is covered by an arillus (according to Planchon, this latter is only an extension of the exostome, and therefore a false arillus, or arillode). The arillus is known as mace, and from both this and the nutmeg essential oils are distilled.

This natural order is confined to the tropics, and *Myristica fragrans* is the typical plant yielding the ordinary nutmeg. It is an inhabitant of the Moluccas, being specially abundant in the Banda Isles, three of which, Lontar, Pulo, Ai, and Pulo Nera, are often known as the "Nutmeg Islands ". At one time the Dutch attempted to confine the cultivation to these islands, and when supplies were rather more abundant than usual the excess was said to have been burned in order to keep up prices. The seed was however conveyed elsewhere by the "nutmeg

[1] *Inaugural Thesis, Basle*, 1903.
[2] *Konink Akad. Weten. Amsterdam*, 28, 10 (**1911**), p. 325.
[3] *Private communication to the author.*

bird," who in swallowing the fruit digested the mace, but voided the nutmeg uninjured. The tree has been successfully introduced into numerous other places, for example, Java, Penang, Bourbon, Zanzibar, and Singapore.

The following interesting account of this spice is acknowledged to the *Chemist and Druggist* :—[1]

"The plant *Myristica fragrans* Houttuyn (N.O. *Myristicaceæ*), is a member of the single genus containing over a hundred species which comprises this order. It is, however, the only species of *Myristica* worth cultivating, the seeds of other species having little or no aroma. The true nutmeg-plant is a somewhat bushy, medium-sized tree of pyramidal habit, attaining a height of 30 to 60 ft. In Penang the trees are somewhat more compact, and not more than 20 or 25 ft. high. The spreading branches are usually produced nearly to the base of the trunk. The lanceloate leaves, which have a rather long, pointed tip, are arranged alternately. They are about 4 ins. long and 2¼ ins. wide, somewhat leathery in texture, and with a dark green shiny upper surface. The trees are usually unisexual, but occasionally they bear both staminate and pistillate flowers. The pendulous, shortly stalked male flowers are produced in small cymes, arising a little above the leaf. The female flowers are borne in similar positions, but usually not more than one to three occur in a single influorescence. The light yellow perianth of the flower is bell-shaped, with three lobes, and encloses either a column of about twelve united stamens or a conical green ovary terminating in a pair of white stigmas. The beautiful fruit is a fleshy drupe, somewhat variable in form, resembling an apricot. The smooth husk is about half an inch thick and of a pale orange-yellow colour. It bears a groove down one side, along which it splits open when the fruit is quite ripe. When the pericarp splits open it discloses a bright crimson arillus, closely enwrapping the seed. This, which on drying yields the mace of commerce, is somewhat leathery in texture, cut into narrow finger-like portions of irregular size. The seed-coat is deep brown, polished, and marked with shallow grooves corresponding to the flaps of the mace. Inside this woody and brittle shell is the kernel, or nutmeg itself, which is about an inch in length. When fresh, the nutmeg practically fills its testa, but shrinks on drying and then rattles. Our first photograph gives an excellent idea of the structure of the fruits and the shape of the leaves.

"Nutmegs are cultivated in plantations from fresh seed sown in the shell, as they soon lose their germinating power. Seeds that rattle are rejected. The plants are generally cultivated near the sea, below the 1000 to 1500 foot level. Nutmegs have been grown on soil of widely different types with remarkable success. In the Moluccas the rich volcanic soil is a sandy friable loam, with a large proportion of vegetable matter, while in Penang and Wellesley Province the trees grow in a yellow loamy clay on the steep slopes of the granite hills; sodden or excessively dry ground is, however, fatal to nutmegs. The temperature of Banda ranges from 76° to 92° F., that of Trinidad being somewhat similar; but in the Straits Settlements it is rather less regular. A rainfall of 90 to 100 ins. per annum, well spread over the different months, with dry spells of not more than four or five days' duration, is

[1] *C. and D.* **(1914)**, 160.

the ideal for yielding heavy crops. Drought has a marked effect in lessening the yield, while excessive spells of wet weather render the plants liable to attacks of parasitic fungi. In the Banda Islands the trees are shaded by canary-nut trees (*Canarium edule*), but in the Straits Settlements no shade-trees are planted.

"The seeds are usually planted in nursery beds of well-dug and manured soil in rows from 12 to 18 ins. apart, at a depth of about 2½ ins., with some form of shading, and frequent watering. The seeds germinate in about a month or six weeks. They are planted out some six months later, when about 6 ins. high, at a distance of 26 to 33 ft. apart, in holes which have been dug some 4 to 6 ft. wide and 3 ft. deep. Some planters prefer germinating the seed in bamboo pots, as this facilitates their transport when they are big enough for planting out. The sex of the plant is not known until it flowers, some seven or eight years later. Male trees are usually exterminated with the exception of one male to ten or twelve female trees, which is sufficient for pollination purposes. Manuring is practised on the poor clay soils of the Malay Peninsula, but in the rich ground of the Banda Islands this does not appear to be necessary. The trees become productive when about eight years old, but it may be earlier in the Straits Settlements, where the whole life-history is shorter than in the Moluccas. They are at their best when about thirty years old, and continue to fruit well for twenty years more; but trees over seventy years have been known to give good crops. The fruits ripen in about six months from flowering. The fruit may be allowed to fall, but it is more usual to gather it by hand. For this purpose a long rod bearing a small basket, open at one side near its apex, on which are two downward pointing prongs, is used to collect the fruit in the Banda group. The nutmeg is caught between the forked prongs, and as it is pulled off it falls into the basket. In the Straits Settlements, where the tree is low, a hooked stick is used to knock off the fruits. The operation of collecting nutmegs is shown in our second photograph. The trees blossom and fruit throughout the year, but it is customary to gather two crops a year, usually in May and June, and again in August and September. The spring crop is said to yield the best nuts. Only quite ripe fruits are collected. A labourer in a full season can gather 1000 to 1500 nuts a day, and in the Banda Islands two or three trees keep a man employed all day, while in Malaya he may pluck from forty to fifty trees a day. In Grenada the fruit is allowed to drop from the trees as it ripens. Good trees should average from 1500 to 2000 nuts a year.

"The husk is usually stripped off on the spot by the collectors, and the nuts forwarded to the drying-sheds. There the mace is carefully removed either by hand or with the aid of a knife. It is dried in one piece ('double-blade') or after separating into two halves ('single-blade'), the former being preferred owing to its better appearance. The mace after removal is flattened by hand or sometimes between boards. In Banda, coolies tread upon it to flatten it, and it is then dried in the sun in flat baskets or on mats or trays of bamboo, this process taking from two to four days. The brilliant red colour changes to a duller orange and finally to golden-yellow, while at the same time it becomes horny and brittle. At nightfall the mace is placed in a drying-shed to avoid wetting by dew. In rainy weather the drying is done by artificial heat, usually over a smokeless fire of charcoal. The practice of sprinkling the mace with sea-water before putting it to dry seems to have been discontinued; only in cases of mildew is it occasionally washed with sea-water. It was claimed that this process kept the mace in a more pliant condition.

"The nutmegs, after removal of the husk and mace, are dried in their shells, as otherwise they run great risk of being attacked by beetles. In the Banda Islands, where nutmegs are handled in large quantities, it is usual to dry them over a charcoal fire in a drying-house. The stands have four or five shelves arranged about 8 ft. from a slow charcoal fire. The layers of seed spread on these are constantly turned and eventually shaken on to the next lower shelf. Too much heat results in shrivelled seed of less value. The drying takes from three to six weeks, when the nutmeg rattles in its shell. In Malaya the seeds are often merely dried by exposure to the sun in trays of basket-work. Sometimes the seed is exported in the shell (*e.g.* from Minahassa and Amboyna), but it is customary to remove the testa, which amounts to about one-quarter of the weight of the seed, before packing for export. The cracking of the shell is done with a wooden truncheon or hammer, or by striking several seeds spread on a kind of drumhead with a flat board. The seeds must be struck on end, otherwise oil-cells are ruptured, and a black bruise formed. The seeds are then sorted, the broken nuts being first removed, then thin nuts, and finally the large sizes are separated from the medium. In Grenada the shelling is done by a machine. This consists of a high, narrow box in which revolves a wooden wheel whose rim forms a series of deep sockets. The nutmegs are fed into the sockets, and the force with which they strike the bottom cracks the shell. The shelled seed is liable to attack by beetles, and in the Moluccas it is still usual to dip nutmegs in milk-of-lime or sprinkle them with powdered lime to protect them. In the Banda Island the nutmegs are packed for about three months in wooden bins filled with lime and water of the consistency of mortar. Tradition states that this operation was intended by the Dutch to prevent the seed (which is killed by the drying alone) from germinating. Mace is usually packed in teak chests, containing about 280 lb., the mace being trodden in. The graded nutmegs are packed in casks or in chests which have been slightly smoked inside and coated with lime and water."

The mace is detached from the seed with a knife, and is easily separated from it. When fresh it is of a brilliant red colour, tough and leathery, and possessing a flavour recalling that of turpentine. It is then dried in the sun.

Nutmegs are often imported limed, *i.e.* treated externally with lime, with the object of preventing their germination and keeping them protected from insects. Although quite unnecessary, this practice is approved of in certain markets, so much so that Penang nutmegs, the most valued in the London market, which are imported unlimed, are frequently limed in London for re-exportation. The nutmegs of other species are often used, both as nuts, mace, and oil, as a substituent or adulteration of the ordinary product. Amongst these are the Bombay nutmegs (from *M. Malabarica*) and New Guinea nutmegs (from *M. Argentea*).

The mace from the Bombay nutmegs is sometimes used to adulterate ordinary mace. It is of very faint flavour and of no pungency.

Nutmegs yield from 6 to 15 per cent. of essential oil, having the following characters :—

Specific gravity	·865 to 0·9250
Optical rotation	+ 9° ,, + 30°
Refractive index	1·4780 ,, 1·4895

The oil is soluble in 1 to 3 volumes of 90 per cent. alcohol, and about 60 per cent. should distil below 180°. The oil usually carries over a little of the fixed oil of nutmeg on distillation, but this should be present in traces only. Good oils, evaporated on the water-bath for twelve hours should not leave more than from 1 to 2 per cent. of non-volatile residue.

The oil distilled from the mace very closely resembles that from the nutmegs, and distinction between the two is often impossible. Oil of mace, however, rarely has a specific gravity below 0·890, and sometimes it may reach 0·932.

The two oils are practically identical in their composition, and the following details may be taken as referring to both oils. The early researches on this oil may be disregarded, the chemistry of the oil having now been carefully worked out by Power and Salway[1] and by Semmler,[2] and to a lesser extent by other chemists.

The terpenes, α-pinene, camphene, β-pinene, and dipentene have all been isolated from the oil, as well as a small amount of cymene. The body termed myristicol by Gladstone and Wright has been shown by Power and Salway to be a mixture of alcohols in which terpineol, linalol, geraniol, and borneol have been identified. A small quantity of terpineol-4 is present, and also a minute amount of safrol. An aldehyde smelling of citral, and yielding a β-naphthocinchoninic acid melting at 248° has been isolated. Eugenol, isoeugenol, and formic, acetic, butyric, and caprylic acid in the form of esters are all present in small amount, as well as a non-volatile monocarboxylic acid of the formula $C_{12}H_{17}COOH$. A body has been described by several chemists under the name myristicin, but this name should be retained for the body isolated by Semmler[3] and examined by Thoms.[4] This body has the formula $C_{11}H_{12}O_3$ and is, chemically, 4-allyl-6-methoxy-1 . 2-methylene dioxybenzene, of the constitution

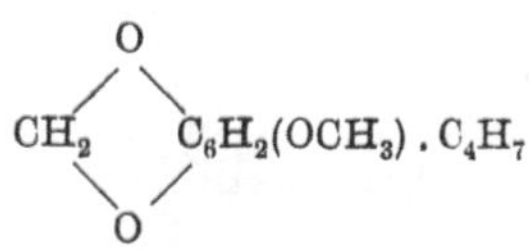

According to Power and Salway the quantities of the various constituents present are as follows :—

Eugenol and isoeugenol	0·2 per cent.
Pinene and camphene	80 ,,
Dipentene	8 ,,
Alcohols	about 6 ,,
Safrol	0·6 ,,
Myristicin	about 4 ,,

The other bodies are present in traces only.

[1] *Jour. Chem. Soc.*, 91 (**1907**), 2037.
[2] *Berichte*, 23 (**1890**), 1803; 24 (**1891**), 3818.
[3] *Ibid.*
[4] *Ibid.*, 36 (**1903**), 3446.

MONIMIACEÆ.

OIL OF PARACOTO BARK.

Paracoto bark is derived from an unknown species of *Monimiaceæ*, which is found in Bolivia. Its essential oil has a specific gravity 0·9275 and optical rotation − 2° 8′.

The oil has been investigated by Jobst and Hesse [1] who isolated two hydrocarbons which they termed α-paracotene and β-paracotene, to which they assigned the improbable formulæ $C_{12}H_{18}$ and $C_{11}H_{18}$ respectively. Cadinene and methyl-eugenol are present in the oil. Wallach and Rheindorff [2] consider that the two hydrocarbons above mentioned were not pure individuals, but mixtures of terpenes. The three bodies described by Jobst and Hesse as α-, β-, and γ-paracotol have also been found to be mixtures of methyl-eugenol and cadinene.

OIL OF ATHEROSPERMA MOSCHATUM.

The leaves of *Atherosperma moschatum* yield from 1·7 to 2·7 per cent. of essential oil, which has been investigated by Miss M. E. Scott.[3] The tree flourishes in the gullies among the Victorian ranges, and a yield of from 1·7 to 2·65 per cent. of oil is obtained from the leaves. During the steam distillation of the oil it was noticed that the first portion of the oil (about 30 per cent.) was lighter than water, whilst that which subsequently passed over was heavier than water and sank to the bottom of the vessel. The oil has the following characters :—

Specific gravity	1·0270
Optical rotation	+ 7·5°
Refractive index	1·5211

The oil was distilled at 100 mm. pressure and divided into 16 fractions boiling from 99° to 183°. The optical rotation of fractions 1 to 9 ranged from + 61·7° to 1·7°, the remaining fractions being optically inactive. The specific gravity of the first fraction was 0·881 and that of the last 1·034, the corresponding refractive indices being 1·4687 and 1·5322.

The sixteen fractions were subsequently subjected to repeated fractional distillation at ordinary pressure and the following compounds identified :—

(*a*) Pinene $C_{10}H_{16}$. A fraction was obtained boiling at 157° to 158°. Its combustion figures agreed very closely with the formula $C_{10}H_{16}$, and it yielded a hydrochloride melting at 130° and a nitrosochloride melting at 103°.

(*b*) Camphor $C_{10}H_{16}O$. Five successive fractions yielded a solid substance, which, when freed from adherent oil, was white and crystalline, and possessed the characteristic odour of camphor. It melted at 174° to 176°, and had a specific rotation of + 40·66° in 10 per cent. alcoholic solution. Its combustion figures agreed with those for $C_{10}H_{16}O$.

(*c*) Eugenol methyl ether $C_{11}H_{14}O_2$. The greater portion of the oil, after repeated fractionation was a pale yellow liquid distilling at 251° to 252° at 755 mm. pressure. Its combustion figures agreed with the formula $C_{11}H_{14}O_2$, and it formed a bromo derivative melting at 75°. Its identity with eugenol methyl ether was therefore established.

[1] *Annalen*, 199 (**1879**), 75. [2] *Ibid.*, 271 (**1892**), 300.
[3] *Jour. Chem. Soc.* (**1912**), 1612.

(*d*) Safrol $C_{10}H_{10}O_2$. The fractions of the oil with the highest boiling-points were found to be a mixture of eugenol methyl ether with another compound, the separation of which was effected by repeated fractional distillation. A nearly colourless liquid was obtained which boiled at 233°, and had the characteristic odour of safrol. Its combustion figures agreed well with those for $C_{10}H_{10}O_2$. On exposing the oil to the temperature of boiling ammonia, it congealed to form white silky needles, melting at 8° to 12°. It was thus definitely identified as safrol.

The composition of the oil is therefore as follows: Eugenol methyl ether, 50 to 60 per cent.; pinene, 15 to 20 per cent.; camphor, 15 to 20 per cent.; and safrol, 5 to 10 per cent.

Boldo Leaf Oil.

The dried leaves of *Pneumus Boldo* yield about 2 per cent. of an oil resembling wormseed oil in odour. The oil has the following characters:—

Specific gravity	0·915 to 0·960
Optical rotation	− 2° ,, + 1°
Refractive index	1·4793
Acid value	1 to 3
Ester ,,	11·2

According to Schimmel & Co.[1] this oil contains *p*-cymene, cineol, and ascaridol, $C_{10}H_{16}O_2$. Traces of terpenes and phenols are also present. Tardy has reported on an oil under this name which had a specific gravity 0·876, and optical rotation − 6° 30′. It contained *a*-pinene, dipentene, terpineol, cumic aldehyde, eugenol, sesquiterpenes, and acetic esters.

EUPHORBIACEÆ.

Oil of Cascarilla.

The barks of *Croton Eluteria* and *Croton Cascarilla* form the Cascarilla bark of commerce. These trees are found in the Bahamas, and the bark on distillation yields from 1 to 3 per cent. of essential oil. The oil has the following characters:—

Specific gravity	0·898 to 0·928
Optical rotation	+ 1° ,, + 14°
Refractive index	1·4910 ,, 1·4960
Acid value	3 ,, 9
Ester ,,	5 ,, 12
,, ,, (after acetylation)	67 ,, 72

The oil is soluble in all proportions in 90 per cent. alcohol. The older researches on this oil are not to be depended on. According to Fendler[2] the oil contains about 2 per cent. of free acids, of which the principal is cascarillic acid, $C_{11}H_{20}O_2$, melting at − 18°, and isomeric with undecylenic acid. Palmitic and stearic acid were also found in traces. Traces of a phenol, *l*-limonene, an unidentified terpene, cymene, and two sesquiterpenes were also found present in the oil. A sesquiterpene alcohol $C_{15}H_{23}OH$ was identified in the high boiling fractions.

Oil of Cathetus Fasciculata.

Cathetus fasciculata Lour. (*Phyllanthus cochin-chinensis* Muell.) one of the Euphorbiaceæ which is found in Southern China and in

[1] Schimmel's *Bericht*, April, **1888**, 43. [2] *Arch. der Pharm.*, 238 (**1900**), 671.

Cochin China, contains an essential oil of which a small sample from Annam, where the plant is known by the name of *Bruyère d'Annam* has been examined by Schimmel & Co.[1] The oil was of a bluish green colour, and possessed an odour reminding somewhat of cajuput oil. Its constants were as follows: $d_{15°}$ 0·8897; α_D − 4° 34′; $n_{D20°}$; 1·47544; acid value 1·9; ester value 3·7; ester value after acetylation 44·8 = 12·7 per cent. $C_{10}H_{18}O$; soluble in 0·5 volume a.m. of 90 per cent. alcohol. Among the constituents is cineol, which was identified from the iodol-compound (melting-point 112°). Judging by the odour of the oil, cymene and linalol are also present.

A sample of oil previously examined under the name of "Essence de Bruyère du Tonkin" possessed precisely similar characters, with the exception of the rotation: $d_{15°}$ 0·8787; α_D + 10° 32′; $n_{D20°}$ 1·47009; acid value 1·7; ester value 5·9; ester value after acetylation 45·2; soluble in 0·3 volume a.m. of 90 per cent. alcohol.

According to Murat, the plant is in full flower in the month of August. Distillation commences towards the end of January and gives different results according to the localities in which the cutting is made whether the plants be fresh or dry. The essential oil obtained is then sometimes of a green colour, sometimes of a pale yellow colour. This divergence would appear to be attributable to the nature of the soil.

Roure-Bertrand Fils[2] have examined each of these two varieties of the oil of *Cathetus fasciculata*. They possess the same odour, slightly camphoraceous, but recalling at the same time that of cineol. The yellow oil is slightly less dense than the green oil; it differs from it, further, in its lower optical rotation and acidity. Below are the analytical characters of the two samples:—

	Green Oil.	Yellow Oil.
Specific gravity at 15° C.	0·8871	0·8862
Optical rotation	− 4° 40′	− 1° 22′
Rotation of the acetylated oil	− 4° 36′	− 1° 14′
Solubility in 85 per cent. alcohol	4·5 volumes and over	3 volumes and over
Acid value	4·2	0·7
Saponification number	8·4	4·2
Ester number	4·2	3·5
,, ,, (after acetylation)	42·0	49·5

About 30 per cent. of cineol was found in the oil.

SANTALACEÆ.

Oil of Sandal-wood.

The sandal-wood oil of commerce, which is largely employed in medicine and perfumery, is the product of distillation of the wood of *Santalum album*. This oil is known as East Indian sandal-wood oil. Probably the Macassar oil is the product of the same tree. *Santalum album* is a native of the mountainous parts of India, but is especially found in Mysore, Malabar, and Coimbatore. It likewise grows on the Coromandel coast, in Madura and Assam. In the Madras presidency and Mysore it is now grown on Government plantations from seeds. The wood is sold at the Government auctions, and the latest classification—for the various woods are not of equal value—with which the billets are marked is as follows:—

[1] *Report*, April, **1914**, 102. [2] *Bulletin*, April, **1914**, 3, **9**, 5.

1. *First Class Billets* (or *Vilayat Budh*). Consists of thoroughly sound billets weighing not less than 20 lb., and not exceeding 112 pieces to the ton.
2. *Second Class Billets* (or *China Budh*). Consists of slightly inferior billets weighing not less than 10 lb., and not exceeding 224 pieces to the ton.
3. *Third Class Billets* (or *Panjam*). Consists of billets having small knots, cracks, and hollows, weighing not less than 5 lb., and not exceeding 448 pieces to the ton.
4. *Chotla* (or billets of short length). Consists of short sound pieces, no limit as to weight or number.
5. *Chat Badala.* Consists of billets with knots, cracks, and small hollows at either end weighing not less than 10 lb., and not exceeding 240 pieces to the ton.
6. *Bagaradad.* Consists of solid pieces, without limit as regards dimensions, weight, or number.

   N B.—Billets of Classes 5 and 6 are not planed nor are their edges rounded off.
7. *Roots* (First Class). Consists of pieces weighing not less than 15 lb., and not exceeding 150 pieces to the ton.
8. *Roots* (Second Class). Consists of pieces weighing not less than 5 lb., and not exceeding 448 pieces to the ton.
9. *Roots* (Third Class). Consists of small and side roots below 5 lb. in weight.
10. *Jugpokal* (First Class) (or *Badala*). Consists of hollow pieces weighing not less than 7 lb., and not exceeding 320 pieces to the ton.
11. *Jugpokal* (Second Class). Consists of hollow pieces weighing not less than 3 lb.
12. *Ain Bagar.* Consists of solid, cracked, and hollow pieces, weighing not less than 1lb.
13. *Ceria* (or large *Chilta*). Consists of pieces and chips of heartwood weighing not less than ½ lb.
14. *Ain Chilta.* Consists of pieces and small chips of heartwood.
15. *Hatri Chilta.* Consists of heartwood chips and shavings obtained by planing.
16. *Milwa Chilta.* Consists of pieces and chips having in fair proportions heartwood and sapwood.
17. *Basola Bukni.* Consists of small heartwood and sapwood chips.
18. *Sawdust* (or *Powder*). Obtained in sawing the santal-wood.

Santalum album is a small evergreen tree, which grows to a height rarely exceeding 40 ft. It is found in the drier parts of Mysore, Coimbatore, and Salem, and as far south as Madura and as far north as Kohlapur. It flourishes best at elevations of 2000 to 3000 ft. above sea level. Being somewhat delicate, it flourishes best when protected by hedges or jungle. When cultivated outside these districts it loses much of its highly valued aromatic heartwood. The cultivation is a Government monopoly.

Much discussion has taken place as to the best method for growing sandal-wood, and the general conclusions arrived at appear to be as follows: The sandal-fruits are gathered whilst fresh, spread out to dry and then stored in a dry place until the commencement of the rains. The seed is then sowed on tilepot beds and covered with a mixture of sand and leaf mould. The seedlings are not transplanted for a year, and are then kept shaded from the sun. Transplanting commences in the following year, as soon as the ground is thoroughly moistened by the early rains. The tree is a slow grower, and in the first year or two often suffers from animal depredations. During the first year the trees in the nurseries should grow to about 10 ins. in height, and by the end of the second year should, if well attended to, attain a height of 2 ft. The tree attains its commercial maturity, that is, the age at which it pays to cut it down, in from twenty-seven to thirty years. By this time the heartwood is well developed—as a rule, at about 2 ins. below the surface; and after this time the growth of the wood is so slow that it does not repay the attention bestowed upon it, and the commercial interests involved are such that it now pays to cut the tree down. It is generally felled at the end of the year, the largest roots, which contain a very fine quality of wood, being cut out, and the tree is stripped of its bark, and the logs thus obtained are sent to various depôts, where they are cut into the usual sized billets and dressed and sorted for the auction sale. The pieces which are straight and have most heartwood fetch the highest prices, as they contain the highest percentage of essential oil.

The amount of oil obtained from the wood is very variable, generally from 1·5 to 6 per cent., but usually from 3 to 5 per cent. Briggs [1] has described the following method of determining the amount of oil in the wood and gives the results of twelve samples of wood obtained from Mysore, representing the various grades there obtainable. Sandal-wood oil contains a large amount of fixed oil as well as volatile oil, and both are extracted by ether. The essential oil cannot be separated by distillation in vacuo, unless glycerine is added to assist in carrying over the last traces of oil. The latter can be extracted from the glycerine by means of chloroform. The process adopted was as follows: 100 grams of the finely powdered drug was extracted in a Soxhlet for twelve hours, or until completely exhausted. The ether extract was transferred to a 200 c.c. distillation flask with side neck, and the ether evaporated by the aid of a current of air. Twenty-five c.c. of glycerine was then added, and the mixture distilled in vacuo until about 25 c.c. distilled over. Seventy-five c.c. of water was added to the distillate, and the mixture extracted with three successive quantities of 25 c.c. of chloroform, and the chloroform evaporated over a water-bath with the aid of a stream of air. The residual oil was then dried in a vacuum desiccator over sulphuric acid and weighed. From 3·5 to 8·3 per cent. of oil was obtained from the twelve samples of wood. In the course of this investigation it was noted that the oil obtained by this method had an optical rotation of − 18° or over, whereas the same wood, distilled with steam, yielded an oil with an optical rotation of − 14° or less. It was also noted that the oil obtained by the assay method above described was readily soluble in 5 volumes of 70 per cent. alcohol, whereas the oil obtained by the ordinary distillation method required more than 5 volumes. The results of distilling 500 grams of sandal-wood with steam, showing the characters of each day's running, are of particular interest:—

	Specific Gravity.	Refractive Index.	Optical Rotation.	Solubility in 5 Vols. 70 per Cent. Alcohol.
1	0·969	1·5017	− 12° 34′	Soluble at 40° to 50°
2	0·970	1·5020	− 12° 22′	„ 25°
3	0·972	1·5027	− 12° 46′	„ 25°
4	0·974	1·5030	− 12° 54′	„ 25°
5	0·976	1·5035	− 13° 10′	„ 25°
6	0·978	1·5037	− 12° 36′	„ 25°
7	0·978	1·5045	− 11° 8′	„ 25°
8	0·9795	1·5046	− 10° 4′	„ 26·5°
9	0·981	1·5046	− 9° 10′	„ 29°
10	0·9815	1·5046	− 8° 20′	„ 32°
11	0·982	1·5045	− 7° 45′	„ 35°
12	0·983	1·5045	− 7° 8′	„ 38°
13	0·983	1·5046	− 6° 34′	„ 44°

The refractive index and specific gravity gradually increased whilst

[1] *Jour. Ind. Eng. Chem.* (1916), 420.

the optical rotation steadily decreased after the first five days. The solubility, with the exception of the first day's running, was constant until the eighth day, after which it decreased. In order to show the effect of prolonged boiling, a sample of oil having an optical rotation of − 20° 40′ was boiled continuously for several weeks in a glass flask with reflux condenser, and another portion with a 10 per cent. solution of sodium chloride. The optical rotation steadily decreased to − 13° 16′ in three and a half weeks, and the change in the brine solution was much less, being reduced to − 16° 40′ in the same length of time. It is concluded that prolonged contact with water brings about a decided change in the composition of sandal-wood oil and greatly reduces the optical rotation.

East Indian sandal-wood oil is a thick, colourless or slightly yellow liquid, of powerful aromatic odour, and having the following characters:—

Specific gravity	0·973 to 0·985
Optical rotation	− 15° ,, − 21°
Refractive index	1·5045 ,, 1·5095
Acid value	0·5 to 2 (rarely up to 6)
Ester ,,	6 to 20
Total alcohols (as santalol, $C_{15}H_{26}O$)	90 to 97 per cent.

These values may, of course, be slightly exceeded when the oil is not a complete distillation. As it is very heavy and takes many days for the wood to be exhausted on distillation, it is not uncommon to find commercial specimens which in fact represent the principal fractions of the oil, when the optical rotation may sink to about − 11°.

The oil is soluble in 6 volumes of 70 per cent. alcohol; indeed nearly all pure samples are soluble in 5 volumes and many in 3·5 to 4 volumes.

The oil is frequently adulterated with small quantities of castor oil and other oils. These adulterants present no difficulty when present in appreciable amounts, but when present in small quantity are exceedingly difficult to detect. Oils which only just satisfy the minimum requirements as to solubility, specific gravity, optical rotation, and santalol value should be regarded with considerable suspicion. Glyceryl acetate is sometimes used as an adulterant, and is detected by the high ester value of the sample (for which see Vol. II.). Benzyl alcohol is also sometimes used, but this will be indicated by an abnormally high "santalol value" and also by the distillation results. With pure sandal oil nothing distils below 280° at ordinary pressure (or 150° at 14 mm.), whilst benzyl alcohol distils at a much lower temperature. Cedar-wood oil is indicated by its insolubility in 70 per cent. alcohol, by its low "santalol value," and by its high optical rotation. Further, a considerable distillate is obtained below 280°. The so-called West Indian sandal oil (*q.v.*) is also a somewhat common adulterant, which is easily detected by the alteration of all the constants of the oil, as well as by its decreased solubility and "santalol value".

Three samples distilled in Mauritius have been examined by Schimmel & Co.[1] Two of the samples were distilled from heartwood, the third from sapwood. As will be seen from the subjoined table, all three were remarkable for their high santalol-content, and their other characters also agreed with those of East Indian oil, except that the optical rotation was a little higher:—

[1] *Report*, April, **1914**, 89.

Colour.	d_{15°	a_D.	$_{20^\circ}$	Acid v.	Ester v.	Ester Santalol.	Ester v. after Acet.	Total Santalol.
						Per Cent.		Per Cent.
Heartwood, Pale	0·9786	– 21° 50′	1·50698	1·9	5·6	2·2	207·2	96·5
,, ,, Dark	0·9816	– 20° 55′	1·50651	1·9	5·6	2·2	207·2	96·4
Sapwood, Pale .	0·9808	– 21° 15′	1·50670	1·9	5·6	2·2	206·2	95·8

All the samples gave a clear solution with 3·5 volumes of 70 per cent. alcohol.

The author and Bennett have recently examined a number of oils offered at low prices, and although the better samples, at prices very close to that of the pure oil, have so far evaded any attempt to discover the adulteration, if any, of them, in the lowest grade there is little doubt that there is present the so-called West Indian sandal oil, or, probably, a fractionated oil with the more insoluble portions removed. Two oils of this type which were examined exhibit the characters as follows:—

	1.	2.
Specific Gravity	0·976	0·9755
Optical Rotation	– 10°	– 9° 30′
Refractive Index	1·5070	1·5068
Esters (as Santalyl Acetate)	5·7 per cent.	5 per cent.
Total Santalol	82·2 ,,	83 ,,
Solubility in 6 vols. 70 per cent. Alcohol .	Soluble	Soluble
Specific Gravity of Acetylated Oil . .	0·983	0·9846
Rotation	– 8°	– 8°
Refractive Index	1·4952	1·4963

The oils were by distillation separated into ten fractions of 10 per cent. each. The characters of these fractions show that one was dealing with a body having the properties of sesquiterpene derivatives, and the odour of the first fractions suggested West Indian oil:—

1.

Fraction.	Specific Gravity.	Optical Rotation.	Refractive Index.
1	0·963	– 7°	1·5040
2	0·969	– 6° 30′	1·5060
3	0·973	– 5° 30′	1·5070
4	0·976	– 5° 30′	1·5080
5	0·977	– 5° 40′	1·5082
6	0·977	– 7°	1·5085
7	0·978	– 8° 30′	1·5089
8	0·980	– 10° 25′	1·5090
9	0·984	– 14° 45′	1·5103

2.

Fraction.	Specific Gravity.	Optical Rotation.	Refractive Index.
1	0·965	– 5°	1·5036
2	0·967	– 5°	1·5048
3	0·972	– 4° 30′	1·5062
4	0·975	– 4° 30′	1·5075
5	0·976	– 5° 50′	1·5079
6	0·978	– 6°	1·5083
7	0·978	– 8°	1·5080
8	0·981	– 10° 10′	1·5090
9	0·983	– 14	1·5095

These figures should be compared with those obtained with genuine samples of sandal-wood oil. Four samples of authentic origin gave the following results :—

A.

Fraction.	Specific Gravity.	Optical Rotation.	Refractive Index.
1	0·970	– 19° 30′	1·5055
2	0·970	– 17° 20′	1·5060
3	0 972	– 16°	1·5060
4	0·974	– 16°	1·5065
5	0·977	– 15° 30′	1·5068
6	0·978	– 15°	1·5068
7	0·980	– 16° 40′	1·5079
8	0·980	– 18°	1·5080
9	0·984	– 21°	1·5084

B.

Fraction.	Specific Gravity.	Optical Rotation.	Refractive Index
1	0·975	– 19°	1·5000
2	0·969	– 18°	1·5044
3	0·969	– 18°	1·5068
4	0·972	– 16°	1·5070
5	0·979	– 14°	1·5072
6	0·979	– 15° 30′	1·5080
7	0·982	– 16°	1·5080
8	0·984	– 17° 20′	1·5075
9	0·982	– 21° 30′	1·5085

C.

Fraction.	Specific Gravity.	Optical Rotation.	Refractive Index.
1	—	– 21°	1·5067
2	0·970	– 17°	1·5064
3	0·970	– 15°	1·5063
4	0·974	– 15°	1·5071
5	0·977	– 15°	1·5075
6	0·978	– 15°	1·5080
7	0·978	– 15°	1·5080
8	0·979	– 17°	1·5082
9	0·980	– 20°	1·5086

D.

Fraction.	Specific Gravity.	Optical Rotation.	Refractive Index.
1	0·977	– 18°	1·5078
2	0·964	– 17°	1·5038
3	0·969	– 16°	1·5051
4	0·975	– 15°	1·5068
5	0·979	– 15°	1·5072
6	0·980	– 14°	1·5078
7	0·981	– 16°	1·5083
8	0·981	– 18°	1·5083
9	0·978	– 22°	1·5086

A comparison of these figures will prove of the utmost service to the analyst.

It is obvious that a dextro-rotatory oil is present in these adulterated samples, which makes the optical rotation of the fractions very much lower than is the case with normal pure oils. The divergence between the fractions in pure oil is but little, and in no case was the optical rotation below – 14°.

The principal constituent of santal oil is a mixture of alcohols, which, for the purposes of analytical determination, are classed as "santalol" and returned for the purposes of calculation as of the formula $C_{15}H_{24}O$.

The following are the constituents which have, so far, been isolated from the oil :—

1. *Isovaleric Aldehyde.*—Traces of this fatty aldehyde C_4H_9CHO have been isolated, and recognised by its thiosemicarbazone, melting at 49° to 53°.

2. *Santene.*—This body is a hydrocarbon of the formula C_9H_{14}. It is, therefore, apparently a lower homologue of the terpenes. The purest specimen so far prepared had the following characters :—

Specific gravity at 15°	0·8708
Optical rotation	± 0°
Refractive index at 20°	1·4678
Boiling-point	140° to 141° at 770 mm.

3. *Nortricycloeksantalane.*—This body is a hydrocarbon of the formula $C_{11}H_{18}$. A specimen purified by distillation over sodium was found to have the following characters :—

Specific gravity	0·9133 at 15°
Optical rotation	– 23° 55′
Refractive index	1·47860
Molecular refraction	46·74
Boiling-point	183°

4. *Santenone.*—This compound is a ketone of the formula $C_9H_{14}O$, isolated from the oil by Müller. It forms a semicarbazone melting at 224°. It is a solid body, with an odour recalling that of cineol and camphor, and having the following characters :—

Melting-point	58° to 61°
Boiling ,,	193° ,, 195°
Specific rotation	– 4° 40′
Oxime	liquid
Semicarbazone	melting-point 224°

5. *Santenone Alcohol.*—This alcohol, also termed norisoborneol, is, in all probability, the alcohol corresponding to the ketone santenone. It

has been isolated by means of its phthalic acid ester, which melts at 230°. It is a solid body, melting at 58° to 62° and boiling at 196° to 198°.

6. *Teresantalol.*—This is also an alcohol having the formula $C_{10}H_{16}O$. It is a solid, melting at 111° to 112° when recrystallised from petroleum ether, and boiling at 210° to 220°. When purified by sublimation it melts at 112° to 114°.

7. *Nortricycloeksantalal* is an aldehyde of the formula $C_{11}H_{16}O$. It can be isolated as its bisulphite compound. When so purified it has the following characters:—

Boiling-point	86° to 87° at 6 mm.
" "	220° to 224° " 761 mm.
Specific gravity at 20°	0·9938
Optical rotation	− 38° 48′
Refractive index	1·48393 at 20°

It forms a semicarbazone melting at 223° to 224°, and a liquid oxime boiling at 135° to 137° at 7 mm.

8. *Santalone.*—This is a ketone of the formula $C_{11}H_{16}O$, also isolated by Müller. It has the following characters:—

Boiling-point	214° to 215°
Specific gravity	0·9906
Optical rotation	− 62°
Refractive index	1·5002

It forms a semicarbazone melting at 174° to 176°, and an oxime melting at 75°.

9. A ketone of the formula $C_{11}H_{16}O$, so far unnamed. It forms a semicarbazone melting at 208° to 209° and an oxime melting at 97° to 99°.

10. *Santalene*, $C_{15}H_{24}$.—This sesquiterpene was first discovered by Von Soden and Müller,[1] but afterwards found by Guerbet to consist of two isomeric sesquiterpenes. Schimmel & Co.[2] have also investigated the mixture, and the following characters are assigned by the two authorities to the compounds in question:—

	α-Santalene.		β-Santalene.	
	Guerbet.	Schimmel.	Guerbet.	Schimmel.
Specific Gravity	0·913 at 0°	0·913 at 15°	0·914 at 0°	0·894 at 20°
Rotation	− 14°	− 3° 34′	− 28·5°	− 41° 3′
Boiling-point	—	252° at 753 mm.	—	125 at 7 mm.
Refractive Index	—	1·49205	—	1·4946

Semmler[3] gives the following figures for the isomeric santalenes:—

	α-Santalene.	β-Santalene.
Specific gravity	0·8984 at 20°	0·892 at 20°
Optical rotation	− 15°	− 35°
Refractive index	1·4910	1·4932
Boiling-point	118° to 119° at 9 mm.	125° to 127° at 9 mm.

11. *Santalol*, $C_{15}H_{26}O$ or $C_{15}H_{24}O$.—This is a mixture of isomeric alcohols, the characters of which are not clearly established. Guerbet (*loc. cit.*) gives the following as the characters of α-santalol and β-santalol and their acetates:—

[1] *Comptes Rendus*, 130 (**1900**), 417. [2] Schimmel's *Bericht*, October, **1910**, 107.
[3] *Berichte* (**1907**), 3321.

(*a*) α-santalol, boiling at 300° to 301° at 760 mm., or at 162° to 163° at 13 mm.; of specific gravity 0·9854 at 0° and optical rotation − 1° 20′.

(*b*) β-santalol, boiling at 170° to 171° at 14 mm., or 309° to 310° at 760 mm.; of specific gravity 0·9869 at 0° and optical rotation − 56°.

α-santalyl acetate boils at 308° to 310°. β-santalyl acetate boils at 316° to 317°.

Schimmel & Co., on the other hand (*loc. cit.*), give the following characters for the two bodies :—

	α-Santalol.	β-Santalol.
Specific gravity	0·9788 at 15°	0·9728
Optical rotation	+ 1° 13′	− 41° 47′
Refractive index	1·49915	1·5091
Boiling point at 5 mm.	148°	158°

β-santalene boils at 125° to 127° at 9 mm., has a specific gravity 0·892 at 20°, refractive index 1·4932, and optical rotation − 35°.

Semmler and Bode [1] have shown that santalol is a primary alcohol (or alcohols), by its oxidation to santal aldehyde. According to these chemists this aldehyde boils at 152° to 155° at 10 mm., has a specific gravity 0·995 at 20°, refractive index 1·51066, and is dextro-rotary. Its semicarbazone melts at 230°.

Santalyl chloride, obtained by the action of phosphorus pentachloride on the alcohol, has a specific gravity 1·0398. Its formula is $C_{15}H_{23}Cl$ according to Semmler, who regards santalols as $C_{15}H_{24}O$.

Semmler regards α-santalol as a tricyclic compound, which, when heated to 160° with alcoholic potash for two hours, is ruptured at the double bond in the side chain, and the tricyclic compound is converted into a dicyclic alcohol which he calls dicycloeksantalol. The formulæ for the α and β-santalols which he suggests are as follows :—

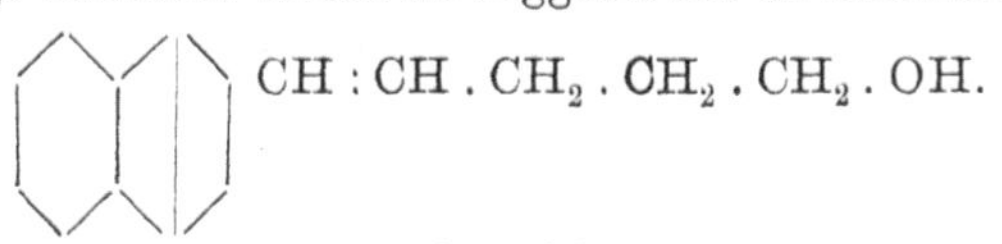

α-Santalol.

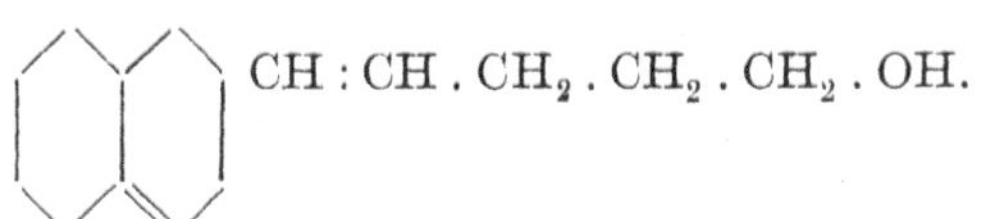

β-Santalol.

These abstruse points are dealt with fully in the following papers—*Berichte*, 1907, 40, 1120, and 1907, 40, 1124.

By means of the strychnine salts, Paolini and Divizia [2] have separated the two alcohols in a pure state. The mixture of the two alcohols as found in sandal-wood oil is first carefully fractionated and the fractions separately treated with sodium and converted into the hydrogen phthalate, in the usual manner. The strychnine salts are then formed and recrystallised until the melting-point and rotatory power are constant. The pure isomerides are then obtained by saponification. The purity is confirmed by reconverting into the strychnine salts, which should yield on saponification a santalol of identical properties. The strychnine phthalate of α-santalol melts at 155°, and has an optical rotation of

[1] *Berichte* (**1907**), 1124. [2] *Att. R. Acad. Lin.* (1914), **23**, 226.

– 1·4°. The α-santalol prepared from it boils at 159° (10 mm.), has a density of 0·979, optical rotation + 1·1°, and refractive index 1·499 (19°). The strychnine phthalate of β-santalol melts at 134° to 135°, and has an optical rotation of – 30·43°. The β-santalol prepared from it boils at 168° to 169° (10 mm.), has a density of 0·9729, optical rotation – 42°, and refractive index 1·5092 (19°).

12. *Santalal*, $C_{15}H_{22}O$, has been reported as a constituent of santal oil, but its presence is doubtful.

13. *Teresantalic Acid.*—This body exists in very small amount in santal-wood oil. It forms part of the small quantity of free acids present, and has the formula $C_{10}H_{14}O_2$. It forms well-defined prisms melting at 157°, and, according to Müller, boils at 150° at 11 mm. Its specific rotation in alcohol solution is – 70° 24′. It exists in the free state in the oil to the extent of 0·5 per cent., and can be extracted by means of dilute caustic alkali solution.

14. *Santalic Acid.*—This acid, of the formula $C_{15}H_{22}O_2$, is a second of the free acids isolated from santal-wood oil. It is, according to Guerbet, a thick liquid boiling at 210° to 212° at 20 mm. pressure.

In addition to the above-described oil, the following santal oils, or so-called santal oils, have been examined :—

New Hebrides Oil.

This is probably not distilled from *Santalum album*. It has the following characters :—

Specific gravity	0·9675
Optical rotation	– 1° 2′
Refractive index	1·5089
Acid value	1·8
Ester ,,	3·6
Santalol	92·4 per cent.

Tahiti Oil.

This oil is probably distilled from *Santalum Freycinetianum* and has the following characters :—

Specific gravity.	0·9748
Optical rotation	– 8° 29′
Refractive index	1·50848
Acid value	2·0
Ester ,,	5·1
Santalol	94·4 per cent.

New Caledonia Oil.

This oil is obtained from *Santalum austro-caledonicum*. It has the following characters :—

Specific gravity	0·9647 to 0·978
Optical rotation	+ 6° 29′ ,, – 21° 42′
Refractive index	1·5062
Acid value	0·9 to 6·4
Ester ,,	3·2 ,, 5·4
Santalol	90·5 to 96·2 per cent.

Figi Oil.

This oil is obtained from the wood of *Santalum Yasi*. It has a specific gravity 0·9768 and optical rotation – 25·5°.

West Australian Oil.

This oil is the product of one or more species of tree, being chiefly derived from the wood of *Fusanus spicatus*. Several other species are found in West Australia, but they are much less common than *Spicatus*, which alone is used to any extent for distilling. This tree was originally known as *Santalum cygnorum*. Although only a tree of small dimensions, it forms an important factor in the timber industry of Western Australia. The species, as the author has seen it growing in the interior of this colony, has a low depressed habit, and is usually very branchy and heavily topped. It is generally from 12 to 20 ft. high, and from 6 to 10 ins. in diameter. The stems usually weigh from 1 to 6 cwt. It is found fairly well distributed in the interior of the colony, except in the extreme south-west. It may be seen growing fairly freely on the most barren sandy soil, where one could not see a blade of grass nor a drop of water for miles round. The greater part of this wood is exported to China, but one or two santal oil distilleries have been started in the colony. The author has examined four samples of oil distilled from either *Fusanus spicatus* or from this tree with other species mixed, and found they had the following characteristics:—

	Specific Gravity.	KOH for Esters.	Iodine Absorption.
1 . . .	·9650	1·46	200·0
2 . . .	·9644	1·15	204·5
3 . . .	·9632	1·66	198·2
4 . . .	·9643	1·35	197·6

There was found only from 65 to 75 per cent. of alcohols in these oils, as against 90 per cent. in the East Indian oil. The optical rotation is usually about + 5° to + 8°. The distillation water of this oil contains traces of methyl alcohol, diacetyl and furfurol.

Claims that this oil can be rectified so as to yield an oil corresponding with the requirements of the British Pharmacopœia, have recently been made. They are, of course, unfounded.

South Australian Oil.

This oil is distilled from *Santalum preissianum*, but it yields an oil quite different to the ordinary santal-wood oil. It has a specific gravity of about 1·02, and becomes solid when cooled.

Berkenheim [1] has isolated from this oil a solid alcoholic constituent of the formula $C_{15}H_{24}O$, melting at 101° to 102°, and yielding crystalline esters.

West Indian Oil.

This oil is the product of a Venezuelan tree which has been identified as belonging to the N.O. Rutaceæ, but as the oil somewhat resembles true santal-wood oils, it may be conveniently described here. Its specific gravity is lower than that of the true santal oil, and it is far more insoluble in alcohol. It is dextro-rotatory and contains much less alcoholic constituents than does the oil of *Santalum album*. Five samples gave the following figures:—

[1] *Jour. Chem. Soc.* (**1893**), ii. 666.

	Specific Gravity.	Solubility in 70 per Cent. Alcohol.	Rotation.	Alcohols as Santalol.
				Per Cent.
1	·962	1 in 80	—	42
2	·966	1 ,, 55	+ 25° 30′	50·5
3	·958	Insoluble in 80	+ 8°	30·1
4	·963	,, 80	+ 30° 20′	32·1
5	·962	,, 80	+ 25° 75′	—

Mr. E. M. Holmes examined the plant yielding this oil, and named it *Schimmelia oleifera* after the firm who obtained the specimens for him, but Urban has now proved definitely that the plant belongs to the genus *Amyris*, and has identified it as *Amyris balsamifera*.

It is now well recognised that the alcoholic body or bodies of West Indian santal oil differ from the alcohols of the East Indian oil.

This alcohol has now been isolated by H. von Soden and named by him "amyrol". It forms an almost colourless, viscous liquid, which forms a clear solution with 3 to 3½ parts of 70 per cent. alcohol. Specific gravity at 15° = 0·980 to 0·982; a_D = + 27°. Boiling-point at 748 mm. pressure = 299° to 301°, at 11 mm. 151° to 152°.

Amyrol cannot be esterified with phthalic acid anhydride; quantitative acetylation is equally impossible. Amyrol is probably a sesquiterpene alcohol, $C_{15}H_{26}O$. In addition to this body, West Indian sandal-wood oil contains a sesquiterpene, whose nature has been investigated by E. Deussen. He introduced hydrochloric acid gas up to saturation into the oil dissolved in dry ether, and obtained crystals of cadinene dihydrochloride. The corresponding compounds of hydrogen bromide and iodide, of which the constants are given below, were produced in a similar manner:—

	Melting-point.	Per Cent.			$[a]d$.
		Cl.	Br.	I.	
Dihydrochloride . .	117° to 118°	25·65	—	—	− 36·65°
Dihydrobromide . .	124° ,, 125°	—	43·64	—	− 36·26°
Dihydroiodide . .	105° ,, 106°	—	—	54·96	—

It has more recently been found that amyrol, like santalol, is not a uniform body, but that by frequently repeated fractional distillation it can be separated into different alcohols. The higher boiling, very viscous compound has a faint, characteristic, fragrant odour; its formula $C_{15}H_{25}OH$. Specific gravity about 0·987; a_D = about + 36°; boiling-point 299°. The lower boiling alcohol appears to have the formula $C_{15}H_{23}OH$, and to be optically inactive.

The liquid obtained on saponifying the original oil, when acidified with sulphuric acid, yielded an oil, partly congealing in crystalline form, with an odour strongly reminding of acetic acid. This oil was shaken with carbonate of sodium solution to remove the acid. The remaining crystals, after recrystallisation from methyl alcohol, melted at 117°. This new body, to which the name amyrolin was given, contains neither nitrogen nor methoxyl groups; its composition is $C_{14}H_{12}O_3$. Amyrolin dissolves in alcoholic potash solution with a yellowish-green fluorescence, absorbs bromine, and yields with the latter an amorphous, white dibromide of the melting-point 157° to 159°. In view of its behaviour amyrolin appears to be a lactone-like body of the aromatic series.

East African Santal Oil.

A so-called santal oil has been distilled in East Africa, and examined by Haensel and by Schimmel & Co. It is the product of species of *Osyris* (probably *O. tenuifolia*) and has the following characters:—

Specific gravity	0·9477 to 0·963
Optical rotation	− 40·6° ,, − 61°
Refractive index	1·5219
Acid value	1·7
Ester ,,	8·1 to 17·9
Santalol	30·5 per cent.

ARISTOLOCHIACEÆ.

Oil of Snake-wood (European).

The ordinary European snake-wood oil is distilled from the root of *Asarum europaeum*, which yields about 1 per cent. of a thick dark brown oil of sweet aromatic odour. The specific gravity of this oil varies from 1·015 to 1·07. According to Petersen, when frozen it deposits crystals of asarone.

The oil contains pinene, methyl-isoeugenol, and a compound termed asarone. This body has the following constitution:—

$$C_6H_2\begin{cases} C_3H_5(^1) \\ OCH_3(^2) \\ OCH_3(^4) \\ OCH_3(^5) \end{cases}$$

Oil of Snake Root (Canadian).

This oil is distilled from the rhizomes of *Asarum canadense*, a plant growing in the Northern United States. The yield is from 3 to 5 per cent. of an aromatic oil. The plant is also known as Indian wild ginger. The oil has a strong nondescript aromatic odour and taste, and varies in characters. The oil distilled from the rhizomes without the rootlets has the following characters:—[1]

Specific gravity	0·9516 to 0·9520
Optical rotation	− 2° 50′ ,, − 10° 42′
Refractive index	1·4851 ,, 1·4886
Acid value	3·1 ,, 3·7
Ester ,,	86 ,, 117
,, ,, after (acetylation)	126 ,, 137

The oil distilled from the rhizomes with the rootlets has the following characters:—

Specific gravity	0·951 to 0·952
Optical rotation	− 10° 30′ ,, − 22°
Refractive index	1·4854 ,, 1·4899
Acid value	3·7 ,, 4·7
Ester ,,	74·7 ,, 115·9
,, ,, (after acetylation)	125 ,, 140

The oil from the rootlets alone, which was obtained to the extent of 1·2 per cent., had the following characters:—

[1] Schimmel's *Report*, April, **1908**, 99; April, **1909**, 85.

Specific gravity	0·9659
Optical rotation	– 39° 40′
Refractive index	1·5028
Acid value	2·2
Ester „	39·2
„ „ (after acetylation)	110·2

The oil contains a phenol of the formula $C_9H_{12}O_2$; α-pinene, linalol, borneol, *l*-terpineol, geraniol, methyl-eugenol, acetic, palmitic, and other fatty acids, and a lactone of the formula $C_{14}H_{20}O_2$.

Oil of Snake-root (Virginian).

The roots of *Aristolochia serpentaria*, the North American serpentary, or Virginian snake-root, yield from 1 to 2 per cent. of essential oil, having the following characters:—

Specific gravity	0·960 to 0·990
Optical rotation	+ 20° „ + 26°
Befractive index	1·4972 „ 1·4980
Acid value	2 „ 3
Ester „	65 „ 80
„ „ (after acetylation)	105 „ 115

Borneol has been detected as a constituent of this oil.

LABIATÆ.

Oil of Lavandula Vera (Lavender Oil).

This oil is obtained by distilling the flowers of *Lavandula vera*, the well-known lavender, a native of southern Europe. The plant is found on the stony declivities of the Provençal Alps, the lower Alps of Dauphiné and Cevennes (and even so far north as latitude 46°), in Piedmont, Switzerland, the mountains of Liguria and Etruria, and various other places. Peculiarly enough, the plants come to their greatest perfection and yield a much finer essential oil when transferred from their native soil, and carefully cultivated in certain districts in England, which produces the finest lavender oil in the world. The principal lavender plantations in England are in the neighbourhoods of Mitcham, Ampthill, Hitchin, and Canterbury.

Attempts have been made to establish plantations in Southern France from young plants taken from parent stems in England; but the plants quickly reverted to their original condition, and "English" lavender oil can so far only be produced in England. For the details of the methods of cultivation adopted, the reader is referred to the excellent monograph in Sawyer's *Odorgraphia*, vol. i. p. 361. A great deal of the finest lavender oil is distilled with water over an open fire, and, so long as care is taken that the plants themselves are not burnt, the fine quality of the resulting oil largely justifies the process. Modern distilleries, in which steam distillation replaces the old water distillation, give, however, a larger yield of oil, and, in general, these have higher ester values than the water distilled oils. This point will be referred to later. In England it is usual to allow the body of the still in which the flowers are distilled to be very shallow as compared with its width, in order to allow the oil to pass over rapidly and get away from the source of heat, which might damage its odour. If a naked fire be used, it is customary to suspend the charge of flowers in a basket of sheet copper, with perforations all over it to allow the free circulation of water. This basket, of course, holds a large charge, and is manipulated by means of a pulley or swing-crane. The basket should be about an inch from the bottom and sides of the still, and have copper legs to rest upon. Sufficient water is used with each charge to cover the top of the basket. The finest portion of the oil condenses during the first half-hour to hour of the distillation, and is usually reserved, the remainder being collected separately. Rectification of lavender oil, which renders it colourless, does not improve it. The somewhat "herby" odour of the fresh oil soon passes off, and if the oil be kept in cool dark places it improves by maturing for several years, and matured oils are often sold according to the date of their preparation. After a certain time it is said to deteriorate unless a little alcohol is added to it.

This oil, in common with several others distilled from plants of the natural order *Labiatæ*, is one of those whose characters are altered to a very remarkable degree by changes in the conditions of growth. An expert now will usually distinguish a lavender oil distilled from plants in and around Mitcham from one distilled in Cambridgeshire or Hertfordshire, and although the ester value of all English oils is fairly constant, foreign distilled oils invariably have much higher ester values. This difference in ester value is brought about by entirely different conditions from those which bring about a reduction in esters where the plants yielding the oil "*s'aspiquent*". There it is a case of cross-fertilisation of the two species of lavender.

The conditions of the lavender oil industry in France are well described by Lamothe.[1]

He gives a botanical description of the various *Lavandula* species, and discusses the conditions which are most favourable for the existence of lavender, the nature and treatment of the soil for laying out the plantations, and the manner of doing so. According to him *Lavandula* species are divided as follows:—

1. *Lavandula latifolia* Vill., *Lavandula spica*. var. β L. (grande lavande, l. mâle, l. branchue, spic, aspic).

2. *Lavandula officinalis* Chaix, *Lavandula vera* D. C., *Lavandula spica*, var. α L. (lavande véritable, l. femelle).

Jordan classifies the latter into two sub-divisions, *viz.*:—

Lavandula fragrans (lavande odorante, l. moyenne).

Lavandula Delphinensis Jordan (petite lavande, l. fine).

The first of these two varieties is widely distributed. It occurs at the lower altitudes, whilst *Lavandula Delphinensis* is found exclusively in the highest regions.

In addition to these, there occurs a cross between *Lavandula latifolia* and *Lavandula fragrans*, which Reverchon calls *Lavandula Hybrida*, and Chatenier *Lavandula fragrans* × *latifolia* (gross lavande, l. bâtarde, lavandin, spigoure).

The distillers, according to Lamothe, have the following designations for the various lavender species:—

1. Petite lavande. This yields the best oil, is more hardy, and less fastidious than the next two which occur in lower regions.

2. Lavande moyenne. Yields a less fine but still useful lavender oil.

3. Gross lavande. Bastard, yields an inferior oil. Not to be confused with the grande lavande (spike).

[1] *Lavand et Spic.*, 2nd edition, 1908.

Lavender oil is frequently judged, as to quality, entirely on its ester-content. There is, however, abundant evidence to show that this is a mistake, and lavender oils can only be so classified, provided that oils from the same district are under consideration. The fact that the true lavender and the spike lavender grow more or less together in the same district, causes a good deal of hybridisation to take place.

In dealing with the influence which the cross-fertilisation of lavender and spike on the essential oil of these plants, A. Birckenstock has also dealt with the ester question. He has pointed out that there are also good lavender oils which contain only 20 to 30 per cent. ester. Such oils are found in the Alps near the Franco-Italian frontier, and the annual production (about 5000 kilos) is in Birckenstock's opinion of sufficient importance to take the particular properties of these oils into account. They have according to him a very fine bouquet, but little "body". The content of linalol amounts to about 50 per cent., and is therefore normal, but, as already mentioned, the ester content is only 20 to 30 per cent.; they are also characterised by a low specific gravity (0·878 to 0·882), and by a considerable laevo-rotation (− 8° to − 9°); they are, moreover, very readily soluble in alcohol (in 10 to 12 volumes 60 per cent. alcohol).

According to Birckenstock, if lavender comes below 2300 to 2600 ft. it crosses with spike, "elle s'aspique"; the hybrids thus formed are known among distillers by the names of "*lavandin*" or "*spigoure*," and they represent every possible transition between lavender and spike, according to the conditions due to altitude and state of the soil.

In this connection Birckenstock's statements on the morphological differences between lavender and spike are interesting; lavender has a non-ramified stalk, which is rarely longer than 16 ins.; the blossoms are blue-violet, and have thin, heart-shaped, pointed sepals which are shorter than the bluish calyx.

The spike plant is larger than the lavender. The stalks frequently have several ramifications, and grow to a height of up to 32 and 36 ins. The sepals are linear and of the same length as the calyx, which in this case is not bluish, but whitish.

The influence of cross-fertilisation also shows itself in the properties of the essential oils, which behave entirely like mixtures of lavender oil and spike oil. Birckenstock has in the course of his observations examined a whole series of oils, from typical lavender oil to typical spike oil. He then found at the same time that as the oil approaches true lavender oil not only the ester content and alcohol content increase, but that also a gradual change takes place in the proportion of borneol to linalol and geraniol; whereas in spike oil borneol predominates, it diminishes towards lavender oil more and more as compared with linalol and geraniol. The following are the properties of two characteristic lavandin oils:—

1. d_{15° 0·9027; α_D − 0° 43′; ester-content, 6·23 per cent.; alcohol-content ($C_{10}H_{18}O$), 34·8 per cent.; soluble in 3 volumes 65 per cent. alcohol.

2. d_{15° 0·8995; α_D − 1° 35′; ester-content, 9·12 per cent.; alcohol-content, 36·5 per cent.; soluble in 3 volumes 65 per cent. alcohol.

According to Zacharewicz[1] lavender thrives best on a light, flinty-

[1] *Bull. mensuel du Synd. Agric. Vaucluse*, 23 (**1907**), 230, and Schimmel's *Report*.

clayey, stony, and sunny soil, at an elevation of about 1200 to 3700 ft. At lower altitudes lavender languishes, yields an oil less rich in ester, and has moreover a tendency to bastardise and finally to degenerate

FIG. 23.—Distillation of lavender in the environs of Castellane. [*Roure-Bertrand Fils.*

into spike, whilst at increasing altitudes the ester-content of the oil, and with it its value, increases.

By suitable cultivation at the proper elevation, both the growth of lavender and the production of blossoms can be increased. The laying-out can take place either by rearranging an existing natural lavender-field, or by starting from seed or from lay rs (*éclats*).

In the former case, in the field grown with lavender, either in the autumn or in March, furrows are ploughed, or in the more recent way, cut with a hoe, so as to leave at distances of one metre strips with lavender plants growing on them. In this manner some lavender plants are no doubt destroyed, but this loss need not be considered at all in view of the rapid development of the plants, which, when suitably fertilised, now give almost double the yield.

Besides in the foregoing manner, the cultivation can also be started with layers or with seed. Whichever is to be preferred in an individual case, depends entirely upon the circumstances. The sowing-out is no doubt cheaper, but against this, layers give a greater yield in the first four years, so that the net cost is about the same.

The planting of the layers can take place either in the autumn or in the spring; the latter is advisable (owing to the frosts) if the field is situated at a great altitude. The plants are placed in rows at distances of 60 cm., the rows being 1 metre apart, so that about 16,600 plants go to a hectare. The sowing can also be done either in the autumn or in the spring, and where the winter is not severe, the spring is preferred. The seed is sown at a depth of 2 to 3 cm. in rows 1 metre apart. One grm. of seed is calculated for an area of 1 sq. metre, which comes to 10 kilos per hectare. The cultivation requires the soil to be turned twice a year, once at the commencement of the winter, and once in March; during the latter, the manure which had previously been distributed between the rows, is worked into the soil. Zacharewicz has shown by comparative experiments that the fertilisation promotes both the growth of the lavender and the wealth of blossoms, and that the yield has thereby been raised from 2000 to 3500 kilos per hectare. Distillation showed that the blossoms from the fertilised soil were also richer in oil. Zacharewicz states that the harvest commences on 1 August, and ends in the first days of September.

Lamothe[1] has recently drawn attention to the fact that the bastard lavender "lavandin," referred to above, is being cultivated on an increasing scale.

Its flowers develop about a fortnight later in the season than those of the true lavender, which is ascribable to its hybrid character; spike, for example, flowers still later, in the beginning of September. Lavandin occurs principally in the region of the holm-oak, and even spreads beyond the boundaries of the latter, traversing in a broad belt the Departments of the Drôme, Vaucluse, Basses-Alpes, etc., where it covers the southern slopes of several mountains up to the top.

Like all hybrids, lavandin is an extraordinarily hardy plant, and in its prolific development constitutes an actual danger to the true lavender, which it robs of air and nourishment. The fact that this hybrid is gaining ground year by year gives cause for serious concern, the more so because pasturing sheep and goats shun it on account of its acrid odour and bitter taste, whereas in the true lavender these animals find an occasional welcome substitute for grass.

But in spite of this same bitter, herbaceous, and camphor-like

[1] *Parfumerie Moderne*, 5 (**1912**), 9.

flavour, which places it far behind lavender as an odoriferous plant, very considerable quantities of lavandin are cut for distilling.

Lamothe estimates that the lavandin oil which is brought to market every year amounts to about 20 per cent. of the total output of lavender oil or, in weight, to about 12,000 kilos. He regards this as most regrettable and utters an emphatic warning against the distillation of lavandin flowers, for although they yield a better return than true lavender, and with less trouble, their oil is decidedly inferior and may easily injure the present predominant position of French lavender oil. Lamothe gives a very interesting comparison of the respective outputs, which deserves to be quoted here. In the same time which is required to collect about 55 kilos true lavender flowers, nearly 400 kilos of the freer-flowering lavandin can be gathered without trouble. And whereas for the production of 1 kilo lavender oil 145 kilos of flowers are needed, from 77 to 80 kilos of flowers suffice to produce 1 kilo of lavandin oil.

With regard to the properties of lavandin oil, Lamothe states that the samples distilled by him had an average ester-content of 24 per cent.

Lavender oil is a pale yellow oil having the following characters :—

Specific gravity	0·879 to 0·900
Optical rotation	− 3° ,, − 10°
Refractive index	1·4600 ,, 1·4660

With regard to any further tests to be applied to the oil, the place of origin must be taken into consideration. An English oil will not give an ester number outside the limits 7 to 10 per cent., calculated as linalyl acetate, whilst foreign oils yield about 25 to 44 per cent., or even higher values under certain conditions. This fact is exceedingly important in discriminating between English and mixed lavender oils. The usual adulterants of lavender oil used to be spike oil and turpentine. Neither of these oils contains appreciable quantities of ester, so that the ester determination affords much information here. Spike oil being usually dextro-rotatory causes a diminution in the rotatory power of the oil. American turpentine has the same effect, whereas French turpentine increases the laevo-rotation. Mixtures of spike oil and French lavender oil can be made having the same ester content as English oil, but the optical rotation is interfered with (the specific gravity is slightly raised also, but not necessarily above the limits for normal oil). To-day, however, the most dangerous adulterant one has to guard against is spike oil, or a similar oil, enriched with artificial esters.

The esters used are, in general, glyceryl acetate, terpinyl acetate, ethyl phthalate, ethyl succinate or ethyl citrate, most of which yield a saponification value indicating a much greater amount of linalyl acetate. The specific gravity of the oil is raised by these additions, unless this is counterbalanced by the suitable addition of a low gravity oil. Schimmel gives the following method of detecting ethyl succinate. About 2 grms. of the oil are saponified and the insoluble oil washed away with ether; the aqueous solution is neutralised with hydrochloric acid and diluted to 50 c.c. Ten c.c. of cold saturated solution of barium chloride are added and the liquid warmed on the water-bath for two hours and then cooled. The formation of any crystalline precipitate indicates adulteration, since the barium salts of acetic and butyric acids (which are present as natural esters) are soluble. This test applies to most of these artificial esters, such as ethyl oxalate, etc.

But more recently Schimmel has found that the following method is more reliable than the use of barium salts: 10 grms. of the oil to be examined are saponified on the water-bath for one hour with alcoholic potash, the contents of the flask then placed in a porcelain dish, and the bulk of the alcohol evaporated. After this the liquid is washed in a separating funnel with about 100 c.c. water, the oily portions removed by extraction with ether, the aqueous solution returned to the porcelain dish, and the bulk evaporated on the water-bath. When the alkaline solution is cooled down, it is acidified with sulphuric acid, and the organic acids thus liberated absorbed with ether. The ethereal solution is placed in a dish and the ether evaporated. For the purpose of removing volatile acids (acetic acid) the residue is still heated for some time on the water-bath, then cooled, and digested with ether. In this manner a fine crystalline residue remains behind undissolved. On recrystallisation from a small quantity of alcohol, white crystals are obtained, and the melting-point can be determined. On fractional distillation the artificial esters will be found in the residues left after distilling off the more volatile portion and will be found to have a very high specific gravity and low refractive index. A comparison with similar fractions of a normal oil will at once reveal the characteristic differences.

Delphin[1] has found an artificial ester as an adulterant of lavender oil which he believes to be the ethyl ester of the fatty acids of cocoanut oil. He has also definitely identified ethyl phthalate as an adulterant.

The effects of some of the common adulterants on the characters of the oil are as follows:—

Turpentine oil lowers specific gravity and solubility; further, the rotation is affected, the laevo-rotatory French turpentine oil ($_D$ − 20° to − 40°) raising, and the dextro-rotatory American oil (a_D up to + 15°) lowering it.

Spike oil increases the specific gravity and lowers the rotation; the solubility remains the same.

Spanish lavender oil behaves like spike oil, but has a less marked influence on the rotation.

Rosemary oil also causes similar alterations to spike oil, but renders the lavender oil less soluble.

All the additions mentioned above considerably reduce the ester content.

Lavender oil is practically neutral. An acid value over about 3 indicates the addtiion cf oaids, such as benzoic or salicylic added, to bring up the apparent ester value.

Generally speaking, pure lavender oil is soluble in 2·5 to 3 volumes of 70 per cent. alcohol at or under 20°, but old oils—and rarely even freshly distilled oils, require up to 4 volumes for complete solubility, or even in rare cases up to 7 volumes.[2]

Oils are not infrequently found with a very high ester value, the following being amongst those so recorded.

[1] *Svensk. Farm. tid.* **(1908)**, 22; **(1912)**, 5.
[2] *P. and E.O.R.* **(1914)**, 130; **(1915)**, 314.

Giaconi has distilled oils in the Dalmatian Archipelago at an altitude of 900 metres, having the following characters:—

Specific gravity	0·890 to 0·899
Optical rotation	– 8° 30′ „ – 10°
Esters	57·1 to 61·6 per cent.

Schimmel & Co.[1] give the following analyses as indicating the effect of certain of these artificial esters on the analytical values of the oil. In this connection, the methods for the detection of artificial esters (Vol. II.) should be taken into consideration:—

Lavender Oil.	No. 1.	2.	3.		4.		5.	
			Original Oil.	Dist. with Steam.	Original Oil.	Shaken once with 5 % Alcohol.	15 c.c. Seminormal Potash Solution.	10 c.c. Seminormal Potash Solution.
$d_{15°}$	0·9007	0·8946	0·9072	0·8932	0·8992	—	0·8913	—
αD	– 3° 55′	– 2° 34′	– 5° 14′	– 5° 37′	– 4° 32′	—	– 8° 43′	—
Soluble in 70 per Cent. Alcohol .	2·5 vols. a.m.	2·5 vols. a.m.	1·9 vols. a.m.	2·2 vols. a.m.	2·5 vols. a.m.	—	4·5 vols. a.m.	—
Acid value . .	0·3	0·5	0·8	—	0·5	—	0·3	0·3
Ester value after 1 h. sap. with 10 cc. ½ n. Solution .	78·4	83·3	100·5	102·5	90·0	87·0	145·7	143·4
Ester-content (calc. for Linalyl Ac.) .	27·4 %	29·2 %	35·2 %	35·9 %	31·5 %	30·4 %	51 %	50·2 %
Ester value after 2 h. sap. with 20 c.c. ½ n. Solution .	82·7	87·6	104·2	104·6	91·9	—	146·0	146·0
Ester value after 1 h. sap. with 10 c.c. ½ n. Solution + 25 c.c. Alcohol . .	69·7	77·0	99·0	100·6	91·1	—	144·1	132·9
Difference . .	13·0	10·6	5·2	4·0	0·8	—	1·9	13·1
Sap. value . .	78·7	83·8	101·3	102·5	90·5	—	—	—
Acid value II. .	75·7	83·7	96·7	100·6	90·6	—	—	—
Difference . .	3·0	0·1	4·6	1·9	—	—	—	—
Terpinyl Acet. content about . .	5 %	4 %	doubtful	not ascertainable		—	—	—
Esters of spar. vol. Acids . . .	not ascertainable		doubtful	not ascertainable		—	—	—
Glyceryl Ester .	not ascertainable		doubtful	not asc.	present	—	—	—

The following remarks are applicable to the samples referred to in the table. Nos. 1 and 2 contained terpinyl acetate; No. 3 was a pure, but old resinified oil, as the analysis performed on it after steam distillation proved; No. 4 contained glyceryl acetate; No. 5 was a pure oil, but as its ester-content is so high, 10 c.c. of $\frac{N}{2}$ alkali is insufficient for a proper saponification, so that the only reliable indications are those when 15 c.c. of N/2KOH are used.

Schimmel & Co.[2] have examined a number of oils distilled by themselves at Barrême, which had the following characters:—

[1] *Report*, April, **1912**, 86. [2] Schimmel's *Bericht*, April, **1907**, 73.

Specific gravity	0·886 to 0·896
Optical rotation	$-7°$,, $-9° 33'$
Esters	47 to 56 per cent.

English oil of lavender is quite different in its odour and chemical characters from Continental lavender oil. It is generally valued at a very much higher rate than French oil and is esteemed more highly than any other type of lavender oil.

The chief difference between English and French oils of lavender lies in the fact that the former contains only about 7 to 10 per cent. of esters calculated as linalyl acetate, whereas the latter contains up to 44 per cent. and over. Messrs. Schimmel have actively endeavoured to establish this ester-content as the basis of the valuation of the oil. They maintain the superiority of fine French oil over English oil, and go so far as to say that the latter cannot compete with the former. The author, in common with most others, holds the opposite opinion, and considers that no comparison can be made between the two oils on the basis of their ester-content. This is much accentuated, if such were necessary, by the fact that linalyl acetate is not *the* odoriferous ingredient of oil of lavender. It is so much modified by the presence of other bodies, as to be regarded as only one of the odoriferous compounds in the oils. Pure linalyl acetate has a marked bergamot odour, and may be regarded as the characteristic ingredient of that oil. The fact that English oil fetches a very much higher price than French oil speaks for itself. For a comparison of oils grown in the same locality, the ester comparison may, however, be of service. The oils produced in the South of Europe are finer according as the plants are growing at greater elevations, and according to Schimmel & Co. the very finest oils are produced from the higher valleys of the Savoy Alps, yielding 44 per cent. of ester. The fine oils yielding 38 to 40 per cent. of ester are usually obtained from the Alpes Maritimes and the Basses Alpes, close to the Italian frontier. Less fine, but still excellent, oils with 28 to 32 per cent. of ester are obtained from the French Departments of the Gard, Drôme, and Hérault.

These statements, however, are based only on a comparison of the ester values, and are in direct opposition to the views and experience of several of the leading distillers of this oil. Lautier, for example, says (*Etudes sur quelques procédés et products de parfumerié*): "No one can deny that the oils from Drôme or Vaucluse are of far inferior odour to those from the Alps. Yet the former contain 35 to 40 per cent. of esters." It is also true that some of the oil distilled on the Italian frontiers with 25 to 30 per cent. of esters is of the finest grade, and Umney, who in the main agrees with Schimmel on this question, admits that with an ester value of much over 40 per cent. the oil may be rank. The "ester theory" has induced a considerable amount of adulteration with artificial esters and should be accepted only within very definite limitations.

English lavender oil, on keeping, especially in the presence of traces of moisture, alters considerably in character.

Brewis and Umney[1] have recorded the following figures for a series of samples of known authenticity :—

[1] *P. and E.O.R.* (**1912**), 5.

SOME LABORATORY SAMPLES OF OLD ENGLISH LAVENDER OILS.

Sample Marked	S. G. @ 15°.	Opt. Rot. 100 m/m. @ 15°.	Refractive Index @ 25°.	Acids as Acetic.	Apparent Esters as Linalyl Acetate.	Remarks.		
						Storage.	Colour of Oil.	Condition of Oil.
1903 A	0·8915	– 9·5	1·4667	0	10·45 %	Narrow-mouthed corked bottle	Pale yellow	Limpid.
1904 A	0·8855	– 9·0	1·4638	0	8·76 %	”	”	”
1905 MA	0·9092	– 8·0	1·4668	0·36	8·46 %	”	Dark yellow	Slightly viscid and resinous.
1906 M	0·9481	– 6·2	1·4657	1·8	14·95 %	”	”	Viscid and resinous.
1907 M & A	0·9021	– 8·3	1·4695	0·24	8·54 %	”	”	”
1907 M	0·8879	– 9·4	1·4672	0	8·34 %	”	Pale yellow	Limpid.
1908 M	0·8864	– 8·5	1·4655	0	7·20 %	”	”	”
1908 B	0·9885	– 4·6	1·4712	3·22	21·65 %	Wide-mouthed corked bottle	Dark yellow	Very viscid and resinous.
1909 M	0·9851	– 4·8	1·4715	3·22	22·7 %	”	”	Slightly less viscid and resinous than 1908 B.
1910 M	0·8859	– 9·4	1·4658	0·12	7·37 %	Narrow-mouthed corked bottle	Pale yellow	Limpid.
1911	0·8859	– 8·5	1·4638	0	7·96 %	”	Very pale yellow.	”

Those samples in which very high ester values had developed were of such acid and disagreeable odour as to be unfit for use. These observations are confirmed by the following analyses by Passmore, the original analysis being made on comparatively freshly distilled oils, and the second analysis after a period of several years on the same oil:—

		Original Analysis.	Second Analysis (1912).
A1/03 . . .	Acidity	nil	nil
	Ester Value	11·0 per cent.	11·0 per cent.
A1/04 . . .	Acidity	nil	2·00 ,,
	Ester Value	8·8 per cent.	21·2 ,,
M/05 . . .	Acidity	nil	1·31 ,,
	Ester Value	7·5 per cent.	17·7 ,,
LM/08 . . .	Acidity	nil	2·29 ,,
	Ester Value	8·7 per cent.	22·7 ,,
LM/09 . . .	Acidity	nil	2·68 ,,
	Ester Value	8·8 per cent.	20·2 ,,

The Constituents of Lavender Oil.—The earliest systematic investigation of this oil was that of Proust and Dumas,[1] who found a camphor (lauro-camphor) present. There is little doubt, however, that the oil they examined was not a genuine lavender oil, but contained other oils, possibly including spike oil.

Bertram and Walbaum[2] found that linalyl acetate was the principal constituent of French lavender oil, and that linalol was also present in the form of esters of butyric, valerianic, and caproic acids. A small amount of free linalol is also present in the oil.

Semmler and Tiemann[3] have isolated limonene and a sesquiterpene from the oil. This sesquiterpene has later been identified as caryophyllene. Very small amounts of α-pinene are present, as well as a small quantity of geraniol, probably in the form of esters.

Cumarin is present, in traces only, and furfurol, an aldehyde which is probably valeric aldehyde, amyl alcohol, ethyl-*n*-amyl-ketone, and *d*-borneol are all constituents of the oil, but in very small amount.

According to Elze[4] nerol and thymol are present in lavender oil, but Schimmel & Co. deny that thymol is a constituent of the pure oil, and consider that Elze was working on an adulterated specimen. Traces of cineol may be present, but this is doubtful.

The principal difference between English and French oil of lavender is that the former contains only a very small quantity of linalyl esters; whilst cineol appears to be a normal constituent of English oil. Semmler and Tiemann[5] have isolated from English oil the terpene limonene, linalol, and *l*-linalyl acetate. Cineol and a sesquiterpene are also normal constituents of the oil.

Oil of Lavandula Dentata.

This oil is not commercially distilled from the unmixed herb, but there is no doubt that the flowers are to some extent gathered promiscuously and distilled with other lavender species in Spain—and to a smaller extent in France.

[1] *Annalen* 6 (**1833**), 248.
[2] *Jour. prakt. Chem.*, 11, **45** (1892), 590.
[3] *Bericht*, xxv., 1880.
[4] *Chem. Zeit.*, 34 (**1910**), 1029.
[5] *Bericht*, 25 (**1892**), 1186.

The oil resembles rosemary oil rather than lavender oil in odour, having a distinctly camphoraceous smell. The dried flowers yield about 0·8 per cent. of essential oil which has the following characters :—

Specific gravity	0·942 to 0·963
Optical rotation	+ 32° „ + 44°
Refractive index	1·4680 „ 1·4800
Acid value	1 „ 8
Ester „	12 „ 25
„ „ (after acetylation)	65 „ 75

It is soluble in 2 to 3 volumes of 70 per cent. alcohol.

The oil contains *d*-camphor, *d*-fenchone, and probably fenchyl-alcohol.

Oil of Lavandula Stoechas.

Lavandula Stoechas is the herb known as *Romero santo* (holy rosemary) by the Spaniards. The essential oil is very similar in odour to that of *Lavandula dentata*, resembling rosemary rather than lavender. It has the following characters :—

Specific gravity	0·930 to 0·935
Optical rotation	− 36° „ − 40°
Refractive index	1·4700 „ 1·4730
Acid value	2 „ 4
Ester „	95 „ 110

The oil is soluble in 3 to 4 volumes of 70 per cent. alcohol.

It contains cineol, a ketone which is probably pino-camphor, and a highly laevo-rotatory alcohol.

Oil of Lavandula Pedunculata.

This oil has been obtained from lavender plants growing in Portugal. It has no commercial value. It has a specific gravity 0·939, optical rotation − 45°, and contains about 39 per cent. of esters. It is soluble in an equal volume of 80 per cent. alcohol. It contains cineol, and probably thujone.

Oil of Lavandula Burmanni.

Two samples of this oil from plants grown in India have been examined by Schimmel & Co.[1] One was distilled from the flowers only, the other from the leaves. The oils had the following characters :—

	Leaf Oil.	Flower Oil.
Specific gravity	0·9131	0·9309
Optical rotation	− 0° 40′	+ 1° 40′
Acid value	9·9	1·9
Ester „	36·3	115·7

Samples of these oils have been examined at the Indian Institute of Science at Bangalore and found to have the following characters :—

	Leaf Oil.	Flower Oil.
Specific gravity at $\frac{27°}{15°}$	0·895	0·921
Refractive index at 25°	1·4830	1·4683
Saponification value	44·25	149·5
„ „ (after acetylation)	141·5	199

[1] Schimmel's *Bericht*, October, 1913, 110.

OIL OF LAVANDULA SPICA (*L. Latifolia*).

This oil, known as spike lavender oil, is a commercial article of considerable importance, and is used very largely in cheap perfumery. Its odour is camphoraceous, and far less pleasant than that of true lavender oil. *Lavandula spica* flourishes, generally speaking, in the same districts as *Lavandula vera*, and grows to a very large extent and very freely in Spain. The yield of oil from the flowers varies from 0·5 to 1·0 per cent.

Spike lavender oil is a pale yellow or almost colourless oil having the following characters :—

	French Oil.	Spanish Oil.
Specific gravity	0·900 to 0·921	0·903 to 0·922
Optical rotation	– 4° „ + 7°	– 5° „ + 12°
Refractive index	1·4640 „ 1·4680	1·4640 „ 1·4670
Ester value	3 „ 22	4 „ 27

Dalmatian spike oil resembles Spanish oil in characters.

Pure spike oil will dissolve in 2 to 2·5 volumes of 70 per cent. or in 3 to 4 volumes of 65 per cent. alcohol to a perfectly clear solution.

The most important feature of spike oil from the perfumer's point of view, is the amount of free alcohols contained therein, which is usually returned as borneol (calculated from the acetylation results). This will, in genuine spike oils, rarely be less than 30 per cent. Spanish spike oils frequently show rather less than this—28 to 29 per cent., but it is probable that this is due to the fact that other Labiate flowers grow in close proximity with the lavender flowers, and are to some extent distilled with them, without any intention of fraud, but because it is in practice impossible to separate them. Umney [1] has recorded the analyses of about a hundred samples of spike lavender oils, with a borneol value varying from about 23 to 41 per cent. Any sample with less than 28 per cent. should be viewed with suspicion, and samples containing much less than this are undoubtedly adulterated.

If spike oil be fractionally distilled (preferably in a flask as illustrated (fig. 24), with the lowest bulb holding about 125 c.c.), using 50 c.c. for the purpose, the first 10 per cent. obtained is usually dextro-rotatory, and not differing more than about 2° in rotation from the original oil.

The author and Bennett have shown that (as is especially the case with Spanish oils) pure spike oil may be slightly laevo-rotatory up to – 3°, and also yield a laevo-rotatory fraction on distillation as above.

The following classification of spike oils grown in different districts by Birckenstock (see under *Lavender Oil*) confirms this statement. So that the statement that pure oils are always dextro-rotatory is incorrect.

Birckenstock distinguishes the following groups: Ardèche, Hérault, Drôme, Gard, Basses-Alpes, Alpes-Maritimes, and Var. Whilst, according to the author, the first-named group represents the spike type proper, and yields oils of the following properties: $d_{15°}$ 0·918 to 0·921; a_D + 7° 48′ to 9° 36′; a_D of the first 10 per cent. of the distillate + 8° to + 10°; ester-content 4 to 5 per cent.; alcohol-content 21 per cent.; soluble in 3 volumes 67 per cent. alcohol,—the "Var" group somewhat approaches the lavender type: $d_{15°}$ 0·9035 to 0·905; a_D – 1° 10′ to 0°; a_D of the first 10 per cent. of the distillate + 2°; ester-content 2 to 3 per cent.; alcohol-content 20 to 32 per cent.; soluble in 5 to 6 volumes 60 per cent. alcohol. The other groups represent intermediate

[1] *P. and E.O.R.* (**1916**), 239.

stages. Specific gravity and rotation therefore become less, whilst alcohol content and solubility increase; inasmuch as the borneol greatly preponderating in the "Ardèche" type is gradually substituted towards the "Var" type by linalool and geraniol. The laevo-rotation of the last-mentioned oil is also remarkable, but Birckenstock considers that up to $-1° 10'$ this is still normal, provided that the first 10 per cent. of the distillate rotate to the right.

Our knowledge of the chemistry of this oil is chiefly due to Bouchardat,[1] who showed that it contained pinene and cineol (eucalyptol). Later, he proved the presence of camphor, linalol, borneol, and camphene. Terpineol and geraniol are also present in small quantities. Probably a sesquiterpene also exists in this oil, but its presence has not yet been definitely proved. The similarity of the odour of this oil to true lavender and rosemary is easily explained when one considers that

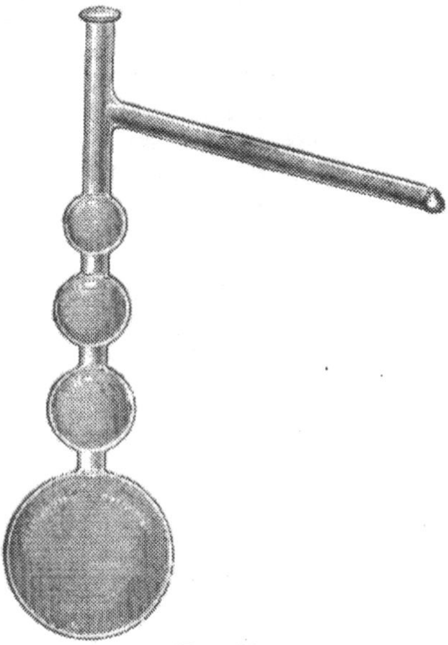

FIG. 24.

the chief ingredients of the oil are cineol, borneol, linalol, and camphor. Linalol is an important ingredient of lavender oil, whereas cineol, borneol, and camphor oil are all present in rosemary oil.

Spike oil is largely alterated with turpentine, and probably, according to Umney, with the commoner varieties of rosemary oil. The former is detected by the adulteration in specific gravity and solubility, whilst the latter interferes with the solubility and the percentage of free alcohols present. Spike oil is largely used in cheap perfumery and in the manufacture of certain kinds of lacquer for fine china manufacture. It is also largely employed as an adulterant of ordinary oil of lavender.

ROSEMARY OIL.

This oil is distilled from the flowering tops of the plant *Rosmarinus officinalis*, a native of the south of Europe. The rosemary is a plant

[1] *Comptes Rendus*, 106 (**1888**), 551; 117 (**1893**), 53 and 1094.

largely influenced, as are most of the plants of this natural order, by variations in soil and climate, and the marked differences in English and foreign oils is more than can be explained by care in the method of production. The plant is grown in a few places in England, but English rosemary oil is almost negligible as an article of commerce, the greater part of the commercial oil being produced in the south of Europe. Several countries on the Mediterranean littoral produce small quantities, but after the south of France the chief source of the oil is the Dalmatian Islands. Spain, however, now produces a quantity of rosemary oil. The Dalmatian produce is chiefly brought into commerce *viâ* Trieste, and this is the oil usually known in trade as Italian rosemary oil. The English oil is by far the most valued, and commands an infinitely higher price than the foreign oil. The Italian (Dalmatian) and Spanish oil are the least valuable. In England and the south of France rosemary is distilled as quickly as possible after it is cut, and in the same manner as lavender flowers. Care is taken to exclude the woody parts of the plant, as not only do these take up unnecessary room in the stills, but also yield a less valuable oil. The best oil is that which distils over during the first hour of the process. The finest French oil comes from the departments Gard and Hérault, and the neighbourhood. The supply of oil from the Dalmatian Islands is somewhat irregular, and, according to Schimmel & Co., the reason of this is that the rosemary woods, as they are called, are municipal property, and are leased to the peasants and distillers, but with an effective control as to harvesting. This follows a regular rotation, the general crop being gathered in one year and very sparing crops for the next two years. The richest plants grow on the island of Solta, but the successful cultivation of vineyards has nearly replaced the rosemary industry. A little is cultivated on the island of Lissa, but the majority comes from Lesina. The total production of these islands varies from 15,000 to 50,000 lb. per annum. The plants are usually sun-dried before distillation, and the process is carried out in very primitive apparatus, which facts possibly account largely for the inferior quality of the oil. The adulteration practised on this oil appears to commence at Trieste, and not on the islands. The usual adulterants are turpentine and petroleum oils, although the finer qualities are at times adulterated with spike oil.

Spanish rosemary oil is now largely employed in cheap perfumery, but its odour is not sufficiently fine for high grade work. This is, no doubt, to some extent due to the fact that other Labiate plants, especially *Lavandula spica* and *Salvia lavandufolia* grow in profusion with or near the rosemary plants, and are cut with the last named, so that the oil is frequently a mixture of Labiate oils in which rosemary predominates.

Rosemary oil of excellent quality is also produced in considerable quantity in southern Tunis, notably in the Bled. The Tunisian oil is far softer than Spanish oil, although perhaps not quite so fine as the best French distillates.

The English market takes something like 50,000 kilos of rosemary oil per annum, while America and Australia are steadily increasing their demands. French continental production, on the contrary, is restricted more and more, for the cost of collection and distillation is becoming too great for remunerative working. Whole ranges of hills in the Alpes-Maritimes remain uncultivated; in the region of Draguinan and Lorgues vast tracts of the Basses-Alpes put enormous quantities of fragrant rosemary at the disposal of the distiller, but the cost of the product would barely repay the expense of labour and fuel.

The average yield of essential oil is from 1 to 1·8 per cent.

Rosemary oil has been examined by several chemists. Bruylants[1] claimed to have isolated a terpene from the oil, but as the oil on which he worked had a specific gravity 0·885, it was probably adulterated with turpentine, so that his results cannot be accepted with confidence. Gildemeister and Stephan[2] have, however, isolated pinene from the lowest boiling fractions of the oil, which is probably a mixture of *dextro-* and *laevo-α*-pinene. Camphene has also been definitely identified as a constituent of the oil. These two terpenes have been identified, not only in French, but also in Dalmatian oil. Weber[3] has isolated cineol, and has also prepared dipentene dihydrochloride from a terpene fraction of the oil, which makes it probable that dipentene is also present. Camphor[4] and borneol[5] are important constituents to which the oil owes much of its odour. Traces of esters, probably of borneol, are also present.

Pure rosemary oil has the following characters:—

French Oil of Rosemary.

Specific gravity	0·900 to 0·920
Optical rotation	– 4° to + 13° (rarely to – 9°)
,, ,, (of 1st 10 per cent. distilled)	– 6° ,, + 14° (,, ,, – 13°)
Refractive index	1·4660 to 1·4725
Acid value	0·5 ,, 2
Ester ,,	3 ,, 14
Esters per cent.	1 ,, 5
Total alcohols (as borneol)	8 ,, 19 per cent.

The oil is usually soluble, with at most faint turbidity, in from 1 to 10 volumes of 80 per cent. alcohol.

Dalmatian Rosemary Oil.

Specific gravity	0·894 to 0·914
Optical rotation	– 1° to + 7° (rarely to + 12°)
,, ,, (of 1st 10 per cent. distilled)	– 1° to + 8°
Refractive index	1·4650 ,, 1·4700
Acid value	0·5 ,, 1·5
Ester ,,	5 ,, 20
Esters per cent.	1·8 ,, 7 per cent.
Total alcohols	8 ,, 18 ,,

The solubility is identical with that of French oil.

Spanish Rosemary Oil.

Specific gravity	0·898 to 0·922
Optical rotation	– 6° ,, + 12°
,, ,, (of 1st 10 per cent. distilled)	– 6° to + 6° (rarely a little higher)
Refractive index	1·4660 to 1·4700
Acid value	1 ,, 2
Ester ,,	2·3 ,, 17·5
Esters per cent.	0·8 ,, 6 per cent.
Total alcohols	10 ,, 20 ,,

The solubility is identical with that of French oil.

[1] *Jour. de Pharm. et de Chim.*, iv. **29** (1879), 508.
[2] *Arch. der Pharm.*, 235 (**1897**), 586. [3] *Annalen*, 238 (**1887**), 89.
[4] *Ibid.*, 114 (**1860**), 197.
[5] *Jour. de Pharm. et de Chim.*, iv., **29** (1879), 508.

[Very many samples of Spanish "rosemary" oil have figures well outside the above limits. They are usually, however, distillates from a mixture of sage with rosemary flowers—and are accepted as "commercial" Spanish rosemary oil.]

Tunisian Rosemary Oil.

Specific gravity	0·905 to 0·925
Optical rotation	− 2° ,, + 6°
Refractive index	1·4660 ,, 1·4695
Acid value	1 ,, 2
Ester ,,	5 ,, 10
Total alcohols	12 to 17 per cent.

The solubility is identical with that of French oil.

English Rosemary Oil.

Specific gravity	0·8965 to 0·924
Optical rotation	− 10° ,, + 8°
Refractive index	1·4650 ,, 1·4690
Esters	about 5 per cent.

Greek rosemary oil does not differ materially from Dalmatian or Spanish rosemary oils; but Corsican and Sardinian oil have the peculiarities of having a high optical rotation (up to + 18°), high ester value (about 10 to 13 per cent.), and a high percentage of alcohols (up to 25 per cent.).

Up till recently laevo-rotatory rosemary oils, or oils yielding a laevo-rotatory first 10 per cent. on distillation, were regarded as adulterated. The above figures indicate the incorrectness of this view. The whole question of the fractional distillation of this oil has been studied by the author and C. T. Bennett,[1] who had consigned to them authentic specimens of rosemary herb grown in Spain and France respectively (some 300 lb. in all), and the oil from these was distilled in the laboratories of Wright, Layman & Umney, Ltd., under the direct supervision of Mr. J. C. Umney. The results of the examination of these oils prove beyond doubt that a laevo-rotatory oil is consistent with purity, and also a dextro-rotatory oil giving laevo-rotatory fractions.

It is customary in France to distil the oil from herb which has been collected after the flowering period (February and March) and dried for about eight days in the sun. In Spain the distillation goes on all the year round, and some variation is therefore to be expected. Both fresh and dried herbs are employed, the latter when the supply is greater than the capacity of the stills and when it has to be carted from a long distance. In the dry state the leaves can be readily separated from the stalks by threshing, but in the fresh condition it is impossible so to separate them. The yield from the fresh herb is from 0·4 to 0·75 per cent. while the dried leaves yield about 1 per cent. of oil, which is of finer quality than that from the fresh herb. The characters of three typical samples were as follows:—

[1] *Chemist and Druggist* (**1906**), 137.

	1 (Spanish).	2 (French).	3 (French).
Source	Leaves alone	Leaves and stalks	Leaves alone
Specific Gravity . . .	0·917	0·897	0·914
Optical Rotation . . .	+ 5° 30′	8° 30′	− 3°
Esters calculated as Bornyl Acetate	3·2 per cent.	3·0 per cent.	3·6 per cent.
Total Borneol	19·7 ,,	10·9 ,,	18·5 ,,
Optical Rotation of first 10 per cent. (100 mm.) . . .	− 1°	− 12° :0′	− 10°

No. 1 was distilled from herb collected towards the end of last summer and dried in the sun. It consisted entirely of leaves which were well developed and of a fine green colour. The yield was 0·89 per cent.

No. 2 was distilled from stalky herb in the dried condition, collected in February last. The proportion of stalks amounted to nearly 60 per cent. Yield of oil 0·4 per cent.

No. 3 was distilled from the same consignment as No. 2, but the stalks were separated and the leaves alone distilled. They yielded 1·09 per cent. of oil, containing a high proportion of borneol. The leaves were somewhat discoloured, probably owing to some fermentation having taken place.

The stalks alone yielded very little on distillation, showing that they contain a much smaller percentage of oil and that of little odour value. These results show that the laevo-rotatory constituent (laevo-pinene) occurs in greater proportion when the stalks are included, and that an inferior oil is then obtained. It is evident that oils derived from carefully picked leaves yield fractions which are laevo-gyrate.

The following are the results of fractionation of the above oils :—

No 1.

Fraction.	Specific Gravity.	Rotation.	Refractive Index.	Boiling-point Commencing at
Per Cent.				
1 . . 10	0·884	− 1°	1·4676	153° C.
2 . . 10	0·890	− 1° 10′	1·4680	156° C.
3 . . 10	0·895	− 1° 20′	1·4681	159° C.
4 . . 10	0·902	+ 1°	1·4682	162° C.
5 . . 10	0·903	+ 1° 20′	1·4683	163° C.
6 . . 10	0·911	+ 2°	1·4686	165° C.
7 . . 10	0·922	+ 2° 30′	1·4700	169° C.
8 . . 10	0·940	+ 3°	1·4736	177° C.
Residue 20	Partially crystallised		1·4885	185° C.

No. 2.

Fraction.		Specific Gravity.	Rotation.	Refractive Index.
	Per Cent.			
1	10	0·874	− 12° 30′	1·4660
2	10	0·878	− 13°	1·4670
3	10	0·879	− 13° 30′	1·4670
4	10	0·883	− 12° 20′	1·4670
5	10	0·886	− 11° 20′	1·4670
6	10	0·891	− 10° 30′	1·4670
7	10	0·896	− 8° 30′	1·4678
8	10	0·909	− 5° 30′	1·4702
Residue . .	20	—	—	1·4859

No. 3.

Fraction.		Specific Gravity.	Rotation.	Refractive Index.
	Per Cent.			
1	10	0·885	− 10°	1·4660
2	10	0·888	− 10°	1·4680
3	10	0·891	− 9° 20′	1·4685
4	10	0·896	− 7° 70′	1·4686
5	10	0·900	− 6° 70′	1·4686
6	10	0·909	− 4° 50′	1·4686
7	10	0·921	− 1°	1·4686
8	10	0·938	+ 2°	1·4697
Residue . .	20	Partially crystallised		—

It is therefore quite clear that the borneol is derived principally, if not entirely, from the leaves, and a genuine laevo-rotatory oil containing a comparatively low percentage of borneol may be assumed to have been distilled from the leaves and stalks, since the leaves alone yield an oil which has a finer odour and higher borneol-content.

According to Gutkind,[1] the optical rotation of Spanish rosemary oil is very variable, even when the plants are grown in the same locality. He states that the rosemary on one hill-side frequently yields an oil of an opposite rotation to that on another hill-side, although, as a rule, the optical rotation is constant for each particular hill-side. In this respect, therefore, it differs from the observations made by Henderson on the oils of rosemary distilled at Hitchin, which he found in some instances to be dextro-rotatory and in others laevo-rotatory.

J. C. Umney[2] gives the following particulars of a sample adulterated with camphor oil, and compares it with pure French oil. Where, however, camphor oil is used as an adulterant, it is usually the light variety, free from safrol, which lowers the specific gravity and greatly diminishes the solubility.

The oil appeared to smell slightly of safrol, and was found to have a somewhat high specific gravity, ·916 being high for a natural French oil of rosemary. The oil, submitted to fractional distillation, afforded the following figures :—

[1] *Private Communication from J. C. Umney.* [2] *P. and E.O.R.*, April, **1913**, 271.

Adulterated Oil.

Specific gravity	0·916
Optical rotation	+ 14°
Total alcohols	19·3 per cent.
Esters	1·6 ,,

Solubility, 1 in 2 volumes of 80 per cent. alcohol.

Fractionation :—

Fraction.	Specific Gravity.	Rotation.	Refractive Index (25°).
Per Cent.			
1 . . 10	·883	+ 14°	1·4660
2 . . 60	·894	+ 16°	1·4675
3 . . 10	935	+ 15°	1·4750
4 . . 10	·961	+ 12°	1·4822
5 Residue 10	·978	—	1·4950

No crystallisation occurred in the higher fractions.

Pure French oil fractionated under the same conditions :—

Specific gravity	0·911
Optical rotation	+ 5°
Total alcohols	15·1 per cent.
Esters	2·2 ,,

Fractionation :—

Fraction.	Specific Gravity.	Rotation.	Refractive Index (25°).
Per Cent.			
1 . . 10	0·880	− 4°	1·4643
2 . . 60	0·897	+ 3°	1·4650
3 . . 10	0·938	+ 13°	1·4690
4 . . 10	0·953	+ 13°	1·4725
5 Residue 10	0·955	—	1·4835

Fraction 4 partially crystallised on cooling.

From a comparison of these figures it will be seen that the highest boiling fractions of the adulterated oil have a high specific gravity, a high refractive index, and, further, do not crystallise.

He was able to obtain a considerable bulk of the oil, which has been most carefully fractionated, and from which it was possible without difficulty to obtain, not only safrol, confirming the suspicion of the presence of camphor oil, but further undoubted proof of the presence of terpineol, which is a constituent of camphor oil, and not hitherto detected in rosemary oil.

Peppermint Oil.

Peppermint oil is produced in numerous countries, the principal supplies of the world being derived from the United States and Japan. English peppermint oil is, however, the most highly esteemed of all and commands by far the highest price. France, Italy, Russia, Germany, and other countries now produce appreciable quantities of peppermint oil, some of which are of very fine quality and are highly esteemed.

The oil distilled in America and Europe is derived from varieties of *Mentha piperita*, chiefly *Mentha piperita*, var. *vulgaris* or "black mint," and *Mentha piperita*, var. *officinalis* or "white mint". The former is the hardier plant and yields the greater quantity of oil, whilst the latter yields less oil, but of more delicate odour and flavour. The parent plant of Japanese peppermint is *Mentha arvensis*, var. *piperascens* Holmes, or *Mentha canadensis*, var. *piperascens* Briquet, which yields an oil much inferior to that of *Mentha piperita*.

From a commercial point of view, peppermint oil appears to have commenced its history about 1750, when cultivation was commenced at Mitcham in Surrey. To-day the English peppermint fields are chiefly located in this district, and in the neighbourhoods of Market-Deeping in Lincolnshire, Hitchin in Hertfordshire, and Wisbeach in Cambridgeshire. The oil produced by English-grown plants bears the same relation to most of the foreign oil as is the case with lavender oil, although it is claimed that certain distillates of German plants approach the English variety in fineness of odour, as they certainly do in price. Distillation of the oil in America, which is now the most important producer of peppermint oil, commenced in the early part of last century, when a small distillery was erected in Wayne County, New York State. To-day many thousand acres are under cultivation, and the principal districts producing the oil are Wayne County (New York), various portions of the State of Michigan, especially Wayne County (Michigan), Van Buren, St. Joseph's and Kalamazoo Counties; St. Joseph's County, Indiana; and to a smaller extent in some of the counties of Ohio and in some parts of the south of the Canadian province of Ontario. The whole of the peppermint cultivation in America is thus confined to the north-east portion of the United States and extreme south of Canada. In Germany, plantations have been successfully established at Miltitz, and in the neighbourhood of Leipzig, by Schimmel & Co., and a fine oil is obtained from the plants grown there. A fair amount of oil is also produced in France, in the departments of the Yonne and du Nord. Japan and, to a certain extent, China produce large quantities of peppermint oil. Mr. E. M. Holmes identified the Japanese plant as *Mentha arvensis*, and as the Chinese plant differs slightly from the Japanese, he retains, for the sake of distinction, the names *Mentha arvensis*, var. *piperascens*, for the Japanese, and *Mentha arvensis*, var. *glabra*, for the Chinese plant.

There appears to be some slight confusion in regard to the Japanese peppermint plant. This plant was brought over to England by the late Mr. Thomas Christy, and this plant was considered by Mons. Maluivaud, the French authority on mints, to be a variety of *Mentha arvensis*, and in this Mr. E. M. Holmes acquiesced, and it was named *Mentha arvensis*, var. *piperascens* Holmes. Dr. Thoms has recently carried out some cultivation experiments at Dahlem, near Berlin (which will be referred to later), with plants received from Japan and examined by Dr. Briquet, the German authority on mints, who considered it to be a variety of *Mentha canadensis*, and he named it *Mentha canadensis*, var. *piperascens* Briquet, but stated that in his opinion it was identical with Christy's plant. This was probably a misunderstanding arising from his not having seen the particular plant grown by Christy, for as grown at Sydenham, and also by Holmes at Sevenoaks, it showed a very decided difference.

The chief Japanese plantations are located North of Yokohama, and southward into the provinces of Bingo-Bitchin. The largest centre of cultivation is in the province of Uzen, which produces more oil than all the other provinces put together. Uzen is in the north-east of the island of Hondo, the chief of the Japanese islands. After Uzen, Bingo is the most important producing province. The following map (reproduced by the courtesy of the editor of the *Chemist and Druggist*) illustrates the chief peppermint districts in Japan :—

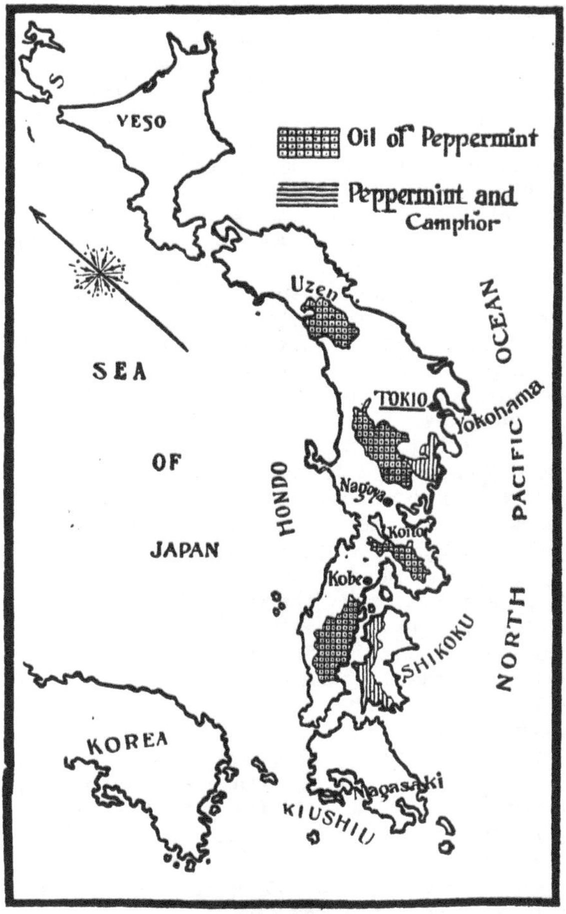

Fig. 25.

American Peppermint Oil.—According to the United States Department of Agriculture, the United States produce—mostly in Michigan, Indiana, and New York—rather under half of the world's total output of mint (peppermint and spearmint) oils, or approximately 250,000 lb. out of a universal production of 600,000 lb.

The Agricultural Department affirms that on the muck lands of Southern Michigan and Northern Indiana, where mint culture has become highly specialised, the cost of establishing an acre of new mint and caring for it during the first season is placed at about $30. In subsequent years the cost per acre would be about $15. These estimates do not make any allowance for fertilisers, taxes, interest on the land, equipment, superintendence, and other charges of this character. In Wayne County, New York, where mint is cultivated on uplands, the cost of preparing the land is somewhat less, but the tillage costs, on the other hand, are higher because of the greater quantity of labour required. The industry in the New York district, however, appears to be declining in extent. It is now carried on in rather a small way chiefly by co-operation between landowners and practical mint-growers. In Michigan and Indiana, on the other hand, many of the growers have invested considerable capital in the business, and have provided themselves with an equipment which enables them to handle the product in an economical manner.

The yield of oil per acre varies widely, ranging from almost nothing to nearly 100 lb. About 325 lb. of peppermint are required to produce a pound of oil in commercial practice.

Peppermint thrives best in deep soils which are rich in humus and retentive of moisure, but fairly open in texture and well drained. It may also be grown successfully in well-prepared upland soils, such as those suitable for corn or potatoes. The area selected should be cropped for one or two years with some plant that requires clean and frequent tillage. The tillage is also continued as long as possible during the growth of the mint, for successful mint-growing implies clean culture at all stages of progress. The crop is cut at the time of full bloom, which, with new plantings, is usually early in September. It is, as a rule, allowed to dry partially in the field before being hauled to the still. Laboratory experiments show that this results in an appreciable loss of oil, but the loss is not regarded by growers as sufficient to offset the increased cost of handling and distilling the green herb. Exposure to frost, however, must be avoided, as frozen mint yields scarcely one-half of the quantity of oil which could otherwise be secured.

Nitrate of soda, applied at the rate of 50 to 150 lb. to the acre, and potash have been widely used as fertilisers to stimulate the growth of mint, and potash is particularly useful against a form of chlorosis, due, apparently, to too much water in the soil. On the other hand, ground bone, acid phosphate, and lime do not seem to be of marked benefit. Few diseases and pests trouble peppermint. During very wet seasons a rust fungus sometimes appears, and crickets, grasshoppers, and caterpillars may always do some damage. On suitable soil, and with proper cultivation, yields of from 2 to 3 tons of mint herb per acre may be expected. The yield of oil is extremely variable, and large yields can only be expected from fields that are in the best possible condition. A fair average for well-managed commercial plantings may be said to be 30 lb. of oil per acre.

Rabak[1] has investigated the effect of cultural and climatic conditions on the yield and quality of American peppermint oil. The following tables indicate the differences in the oil distilled from the fresh and the dry herb at various stages of growth, and also those of the oils when distilled from the entire herb, the tops, and the fresh leaves at various stages of growth :—

[1] *U.S. Dept. Agric., Bull.* 454 (1916).

PHYSICAL PROPERTIES AND CHEMICAL COMPOSITION OF PEPPERMINT OIL DISTILLED FROM THE FRESH AND FROM THE DRY HERB AT VARIOUS STAGES OF GROWTH DURING SUCCESSIVE YEARS.

Year and Description of Material.	Physical Properties.					Chemical Composition (per Cent.).			
	Colour, Odour, and Taste.	Specific Gravity.	Rotation, 50 mm. Tube.	Refraction.	Solubility in 80 per Cent. Alcohol.	Free Acid (as Acetic).	Ester (Menthyl Acetate).	Menthol. Free.	Menthol. Total.
1908.									
Fresh Herb:			°C.		Vols.				
Budding	Pale yellow; pleasant menthol-like odour; cooling taste	0·912	−15·8	1·4696	1·5	0	9·55	30·56	38·08
Flowering	Pale yellowish-green; pleasant ester-like odour; very pungent taste	·924	− 4·5	1·4728	1·25	0	14·50	23·97	35·39
Fruiting	Yellowish-green; pleasant minty odour; less pungent and cooling taste	·918	−21	1·4681	1·5	0	24·40	34·33	53·55
1909.									
Fresh Herb:									
Budding	Golden-yellow; pleasant, faint odour; slightly pungent and cooling taste	·9158	− 9·2	1·4738	1	·10	6·72	30·1	35·39
Flowering	Pale yellow; agreeable menthol-like odour; pungent minty taste	·9203	−10·6	1·4662	1·2	·21	7·07	46·5	52·09
Fruiting	Straw coloured; very flowery odour; minty, pungent, taste	·9151	− 8·3	1·4676	·8	·10	12·37	33·8	43·58
Dry Herb:									
Budding	Greenish-yellow; strong minty odour; pungent, minty taste	[1] ·9337	−13·1	[2] 1·4695	[3] 1·2	·17	13·26	36·07	46·38
Flowering	Golden-yellow; minty odour; pungent, minty, slightly fatty taste	[1] ·9163	−13·5	[2] 1·4695	([4])	·16	12·73	37·37	47·44
Fruiting	Yellow; unpleasant minty odour; strong, pungent, minty taste	[1] ·9344	− 8·1	[2] 1·4718	[5] 1	·36	17·68	34·2	48·15

[1] At 22° C. [2] At 25° C. [3] Turbid in excess.
[4] Turbid in 1 volume absolute alcohol; insoluble in 80 and 90 per cent. alcohol in any proportion.
[5] Turbid in 2 volumes or more.

PHYSICAL PROPERTIES AND CHEMICAL COMPOSITION OF PEPPERMINT OIL DISTILLED FROM THE FRESH AND FROM THE DRY HERB AT VARIOUS STAGES OF GROWTH DURING SUCCESSIVE YEARS (*continued*).

Year and Description of Material.	Physical Properties.					Chemical Composition (per Cent.).			
	Colour, Odour, and Taste.	Specific Gravity.	Rotation, 50 mm. Tube.	Refraction.	Solubility in 80 per Cent. Alcohol	Free Acid (as Acetic).	Ester (Menthyl Acetate).	Menthol. Free.	Menthol. Total.
1910.									
Fresh Herb:									
Budding .	Dark golden; not fragrant, minty odour; slightly bitter, pungent, minty, taste	[3] ·9155	− 12·1	[1] 1·4718	[4] ·7	·86	16·62	43·76	56·85
Flowering.	Pale yellow; mild, agreeable odour; fatty, strong, cooling, pungent taste	[3] ·9273	− 5·7	[1] 1·4820	[2] ·6	·03	11·67	30·94	40·13
Fruiting .	Yellowish-green; very fragrant flowery odour; very bitter and slightly pungent taste	[3] ·9250	− 7	[1] 1·4712	[5] ·8	1·14	20·86	12·7	29·14
Dry Herb:									
Budding .	Light golden; herb-like, faint, minty odour; slightly bitter, pungent, minty taste	[3] ·9167	− 11	[1] 1·4724	[5] ·6	1·20	19·80	39·41	55·01
Flowering .	Pale brown; herb-like odour; pungent and cooling taste	—	—	[1] 1·4765	—	3·20	30·41	—	—
Fruiting .	Golden-yellow; fragrant, pleasant odour; very bitter and slightly pungent taste	[3] ·9383	− 3·3	[1] 1·4713	[6] ·5	·56	20·51	—	—
1911.									
Fresh Herb:									
Budding .	—	—	—	—	—	·06	12·02	32·37	41·84
Flowering .	—	—	—	—	—	·20	13·79	34·11	44·97

[1] At 22° C. [2] Turbid in 2 volumes or more. [3] At 24° C.
[4] Turbid in 4 volumes or more. [5] Turbid in $1\frac{1}{2}$ volumes or more.
[6] Turbid in $1\frac{3}{4}$ volumes or more.

COMPARISON OF THE PHYSICAL PROPERTIES AND THE CHEMICAL COMPOSITION OF PEPPERMINT OILS DISTILLED FROM THE FRESH LEAVES, FROM THE TOPS, AND FROM THE ENTIRE FRESH HERB AT THE VARIOUS STAGES OF GROWTH.

Year and Description of Material.	Physical Properties.					Chemical Composition (per Cent).			
	Colour, Odour, and Taste.	Specific Gravity.	Rotation, 50 mm. Tube.	Refraction.	Solubility in 80 per Cent. Alcohol.	Free Acid (as Acetic).	Ester (Menthyl (Acetate).	Menthol. Free.	Menthol. Total.
1909.			°C.		Vols.				
Budding stage: Herb . .	Golden-yellow; pleasant, faint odour, slightly pungent, cooling taste	[1] 0·9158	− 9·2	[2] 1·4738	[3] 1	0·10	6·72	30·1	35·39
Leaves . .	Straw coloured; fragrant but faint minty odour; fatty, pungent, slightly bitter taste	[1] ·9143	− 12·1	[2] 1·4713	[4] 1·4	·014	9·9	37·7	45·67
Tops . .	Straw coloured; very fragrant minty odour; fatty, bitter, pungent taste	[1] ·9151	− 12	[2] 1·4718	[5] 1	·116	8·84	31·8	38·76
Flowering stage: Herb . .	Pale yellow; agreeable menthol-like odour; pungent, minty taste	[1] ·9203	− 10·6	[2] 1·4662	[6] 1·2	·21	7·07	46·5	52·09
Leaves . .	Straw coloured; pleasant, minty odour; pungent taste	[1] ·9389	− 11	[2] 1·4706	[6] 1·2	·52	17·3	41·7	55·38
Tops . .	Straw coloured; flowery odour; pungent taste	[1] ·9370	− 10·3	[2] 1·4732	[6] 1	·48	15·9	37·75	50·29
Fruiting stage: Herb . .	Straw-coloured; very flowery odour; minty, pungent taste	[1] ·9151	− 8·3	[2] 1·4676	[5] ·8	·10	12·37	33·8	43·58
Leaves . .	Golden-yellow; pleasant, minty odour; minty, pungent taste	[1] ·9373	− 9·6	[2] 1·4730	[4] 1·5	·32	19·8	37·58	53·18
Tops . .	Light straw coloured; pleasant, flowery odour; minty, slightly pungent taste	[1] ·9223	− 9·5	[2] 1·4723	[5] 1	·16	15·56	34·8	47·08

[1] At 22° C. [2] At 25° C. [3] Clear in excess. [4] Turbid in 3 volumes or more. [5] Turbid in 1½ volumes or more. [6] Turbid in 2 volumes or more.

COMPARISON OF THE PHYSICAL PROPERTIES AND THE CHEMICAL COMPOSITION OF PEPPERMINT OILS DISTILLED FROM THE FRESH LEAVES, FROM THE TOPS, AND FROM THE ENTIRE FRESH HERB AT THE VARIOUS STAGES OF GROWTH (*continued*).

Year and Description of Material.	Physical Properties.					Chemical Composition (per Cent.).			
	Colour, Odour and Taste.	Specific Gravity.	Rotation, 50 mm. Tube.	Refraction.	Solubility in 80 per Cent. Alcohol.	Free Acid (as Acetic).	Ester (Menthyl Acetate).	Menthol. Free.	Menthol. Total.
1910.									
Budding stage:									
Herb . .	Dark golden; not fragrant, minty odour; slightly bitter, pungent, minty taste	[3]·9155	− 12·1	[1]1·4718	[4]·7	·86	16·62	43·76	56·85
Leaves . .	Golden-yellow; faint not especially minty odour; slightly bitter, pungent taste	[3]·9156	− 13·1	[1]1·4710	[4]·6	·90	19·80	18·79	34·39
Tops . .	Golden-yellow; agreeable, minty odour; strong, pungent, bitter, cooling taste	—	—	—	—	·13	13·44	29·19	39·78
Flowering stage:									
Herb . .	Pale yellow; mild agreeable odour fatty, strong, cooling, pungent taste	[3]·9273	− 5·7	[1]1·4820	[2]·66	·03	11·67	30·94	40·13
Leaves . .	Pale yellow: strong minty odour; slightly fatty, bitter, pungent, cooling taste	·9231	− 9·4	1·4753	·7	1·90	22·27	37·83	55·38
Tops . .	Pale yellow; fragrant, agreeable odour; fatty, bitter, pungent, cooling taste	·9242	− 6·5	1·4780	·8	1·43	22·98	48·60	66·71
Fruiting stage:									
Herb . .	Yellowish green; very fragrant flowery odour; very bitter, slightly pungent taste	·9250	− 7	1·4712	·8	1·14	20·86	12·7	29·14
Leaves . .	Golden-yellow; strong minty odour; bitter, cooling, slightly pungent taste	—	—	1·4720	—	4·80	43·40	—	—
Tops . .	Greenish-yellow; agreeable, flowery odour; bitter, slightly pungent, cooling taste	—	—	1·4738	—	3·30	28·29	—	—

[1] At 22° C. [2] Turbid in 3 volumes or more.
[3] At 24° C. [4] Turbid in 4 volumes or more.

Rabak draws the following conclusions as the results of his experiments :—

Conditions of soil and climate are influential factors in the formation of oil and its constituents in the peppermint plant. Light sandy or loamy soils appear to be most favourable for the production of an oil of high quality.

The yield of oil distilled from fresh plants apparently decreases as the plant matures. Drying the plant before distillation results in a considerable loss of oil. The largest proportion of oil is found in the leaves and flowering tops.

The percentage of esters in the oils increases as the plants approach maturity. The menthol-content of the oil bears a close relationship to the ester-content. The free acidity and ester-content of the oil distilled from dry plants is considerably higher than in the oil from fresh plants. The drying of the plants causes changes favourable to esterification, while the percentage of free and total menthol in oils distilled from dried plants is also uniformly high.

The formation of esters and menthol takes place most readily in the leaves and tops of the plant, the metabolic processes showing increased activity as the plant matures.

The effect of shade upon the peppermint plant is to decrease esterification and the formation of menthol, and is due possibly to the lessened activity of the elimination of water by the plant.

The action of frost noticeably increases esterification and the formation of menthol.

The mint is usually left in the sun in small cocks for a short time before distillation. With regard to the advisability of drying the plants before distillation, thus reducing the bulk and so increasing the charge of a still, Mr. Todd, a well-known Michigan grower, carried out experiments on a very large scale, and came to the following conclusions :—

1. No loss of essential oil of peppermint through diffusion in the atmosphere is occasioned by thorough drying of the plants and prolonged exposure to atmospheric action prior to distillation. (This conclusion, however, does not appear to be quite correct.)

2. Such exposure does not increase the crystallising tendency of the essential oil.

3. A heavy and insoluble resinoid is produced by oxidation, increasing the specific gravity of the oil and affecting the boiling-point and solubility, by raising the former and decreasing the latter.

4. To obtain the best results as to the quality of oil produced and the facility of handling and distillation, the plants should be dried as thoroughly as possible without endangering the loss of leaves and blossom; distillation should then take place as soon as convenient to prevent the formation of resin.

The earliest elaborate research on the composition of American peppermint oil is that of Power and Kleber.[1] The oil examined by these chemists was distilled from the dried plant and had the following characters :—

Specific gravity	0·9140
Optical rotation	− 32°
Menthol as esters	14·12 per cent.
Free menthol	45·5 ,,

[1] *Pharm. Rundschau*, 12 (**1894**), 157.

On fractional distillation, the following fractions were obtained:—

To 200°	2·6 per cent.
200° to 205°	2·4 ,,
205° ,, 210°	8·6 ,,
210° ,, 215°	18·8 ,,
215° ,, 220°	24 ,,
220° ,, 225°	19·6 ,,
225° ,, 230°	9 ,,
230° ,, 235°	3·6 ,,
Residue	12·2 ,,

The following compounds were identified in the oil:—

Acetic aldehyde $CH_3 . COH$ (about 0·044 per cent.).
,, acid $CH_3 . COOH$.
α-pinene (*inactive*) $C_{10}H_{16}$.
Limonene (*laevo*) $C_{10}H_{16}$.
Menthone $C_{10}H_{18}O$.
Menthyl acetate $C_{10}H_{19} . C_2H_3O_2$.
Cadinene $C_{15}H_{24}$.
A menthyl ester $C_{10}H_{19} . C_8H_{11}O_2$.
Iso-valerianic aldehyde $C_4H_9 . COH$ (about 0·048 per cent.).
,, acid $C_4H_9 . COOH$.
Phellandrene $C_{10}H_{16}$.
Cineol $C_{10}H_{18}O$.
Menthol $C_{10}H_{19} . OH$.
Menthyl-iso-valerianate $C_{10}H_{19} . C_5H_9O_2$.
A lactone $C_{10}H_{16}O_2$, melting at 23°.

Amyl acetate and dimethylsulphide have also been detected in traces.

Pure American peppermint oil has the following characters:—

Specific gravity at 15°	0·900 to 0·915 (rarely to 0·925)
Optical rotation	− 18° to − 35° (rarely below − 18°).
Refractive index	1·4600 to 1·4640
Menthol as esters	5 to 14 per cent.
Total menthol	50 ,, 62 ,,
Menthone	9 to 19 per cent. (rarely to 25 per cent.)

An examination of nine samples by Kleber[1] indicates that the menthone is usually lower as the menthol is higher. The following are his figures:—

$d_{25°}$.	α_D.	Total Menthol.	Menthone.
0·8995	− 25° 27′	49·7 per cent.	18·8 per cent.
0·8983	− 24° 20′	48·4 ,,	16·8 ,,
0·900	− 24° 14′	51·7 ,,	17·98 ,,
0·904	− 24° 28′	50·2 ,,	16·98 ,,
0·902	− 24° 30′	50·2 ,,	17·7 ,,
0·905	− 23° 18′	53·5 ,,	14·0 ,,
0·901	− 26° 12′	56·1 ,,	13·9 ,,
0·902	− 25° 30′	55·8 ,,	16·7 ,,
0·903	− 23° 49′	56·4 ,,	15·2 ,,

The following table is due to Power, and includes a few oils other than American distillates. Samples No. II. and II*a*. are so abnormal that some suspicion as to their genuineness may well be forgiven:—

[1] Schimmel's *Report*, April, **1914**, 77.

Character of Peppermint Oils (Power).

Designation of the Oil.	Specific Gravity at 15° C.	Optical Rotation in 100 mm. Tube.	Menthol in the Form of Esters.	Free Menthol.	Total Menthol.	From 50 c.cm. of Oil there Distil in c.cm. between the Temperatures noted: ? to 200° C.	200-205°.	205-210°.	210-215°.	215-220°.	220-225°.	225-230°.	230-235°.	Residue.
			Per Cent.	Per Cent.	Per Cent.									
I. F. B., from Dry Herb	0·9140	– 32° 0′	14·12	45·5	59·6	1·3	1·2	4·3	9·4	12·0	9·8	4·5	1·8	6·1
Ia. F. B., from Fresh Herb	0·9130	– 30° 0′	11·25	43·2	54·5	0·6	1·2	5·7	8·5	11·6	9·2	4·2	2·2	6·7
Ib. Wayne Co., N.Y., 1892	0·9158	– 26° 45′	9·32	40·8	50·1	1·4	1·8	6·3	10·0	12·9	7·3	2·4	1·8	6·3
Ic. " " 1893	0·9110	– 32° 30′	9·04	46·1	55·1	1·1	1·4	5·0	10·7	13·9	7·6	2·9	1·6	5·4
Id. " " Twice Rectified (F. S. & Co.)	0·9110	– 32° 45′	8·61	51·0	59·6	1·5	3·7	8·8	14·7	12·2	4·2	—	—	5·2
Ie. V. B. Co., Mich., 1893	0·9067	– 29° 20′	6·39	43·6	50·0	1·9	5·0	9·4	13·2	12·4	2·3	—	—	5·1
If. Wayne Co., Mich., 1893	0·9135	– 28° 30′	7·73	50·3	58·0	1·7	1·6	5·5	9·6	12·2	8·9	3·2	3·1	5·8
II. St. J. Co., Mich., 1893 (I.)	0·9135	– 9° 45′	3·63	28·9	32·6	1·9	2·6	12·0	14·0	11·6	4·4	—	—	3·3
IIa. " " 1893 (II.)	0·9083	– 19° 30′	4·23	29·6	35·8	3·4	7·0	7·6	13·2	10·1	3·3	1·5	—	4·1
IIb. "Rose Mitcham" } Michigan	0·9050	– 23° 35′	4·37	44·2	48·6	4·4	9·6	13·4	17·1	3·2	—	—	—	2·5
IIc. "Crystal White" } Michigan	0·9105	– 23° 55′	6·74	44·2	50·9	1·7	5·8	9·9	15·4	9·4	3·4	—	—	4·1
IId. "Redistilled Oil" } Michigan	0·9105	– 23° 30′	8·59	46·5	55·1	3·3	6·5	8·1	8·8	9·9	5·2	—	—	8·7
III. Mississippi	0·9250	– 13° 40′	11·47	24·2	35·7	0·9	0·3	1·2	6·5	16·2	11·8	4·8	2·4	6·6
IV. Japanese, Normal	0·9100	– 34° 45′	3·45	72·7	76·2	0·7	0·4	3·4	22·0	17·2	2·1	1·0	—	3·5
V. " Dementholised	0·9030	– 31° 20′	5·71	55·1	60·8	0·5	1·6	10·2	20·7	12·0	3·3	—	—	2·4
VI. Mitcham	0·9070	– 27° 55′	4·92	53·9	58·8	1·4	0·7	12·2	24·7	8·6	—	—	—	3·7
VII. Saxon	0·9100	– 26° 0′	6·38	61·2	67·6	1·7 (? to 205°)		10·9	21·3	11·6	2·1	—	—	2·1
Russian Oil	0·905 to 0·915	– 17° to – 22°	5	45	50	—		—	—	—	—	—	—	—
French Oil	0·918 " 0·921	– 6° " – 10°	5 to 10	35 to 40	45 to 50	—		—	—	—	—	—	—	—
Italian	0·912 " 0·925	– 2° " – 16°	10	46	50 " 60	—		—	—	—	—	—	—	—

Japanese Peppermint Oil.—This oil, as mentioned above, is distilled from *Mentha arvensis* (*Mentha canadensis*), and forms an important Japanese industry. The most important modern contribution to the literature of this oil is by Naojiro Inouye.[1] Nagasaki[2] describes the four following varieties of the plant which are cultivated :—

1. *Akamaru.*—The leaves of this plant are round, the stalk is reddish-violet, and the flowers of a violet colour. The yield of oil and menthol is greater in this than in the other varieties.

2. *Aomaru.*—The leaves are round, the stalk is blue, and the flowers violet.

3. *Akayanagi.*—The leaf is violet on the under surface, the stalk is blue and the flower white. The yield of oil is low.

4. *Aoyanagi.*—This variety resembles the wild plant. Its stalk is blue and flower white.

The following details[3] of peppermint cultivation are of interest; "Peppermint is cultivated in different parts of Japan, mostly on the hill-sides, although that grown on low-lying ground which can alternately be used for rice is richest in crystals. As the oil which has been freed from menthol crystals is inferior both in taste and odour to English and American oil, experiments have been made in the cultivation of English and American peppermint, but unfortunately without success. The best oil is produced in the districts of Okayama and Hiroshima, where three cuts are made yearly: in May, June, and August. The first cut yields about 47 per cent. of crystallised menthol (*i.e.*, by freezing only), the second about 53 per cent., and the third about 60 per cent. In the district of Yamagata only two, and in Hokkaido only one cut is made, but in the last-named district the area under cultivation is considerably larger than in the other, so that, in spite of a single crop and a crystallised-menthol-yield of only 45 per cent., more than one-half of the total output of menthol is produced in the province of Hokkaido.

"The method of distilling is the same everywhere. It lasts four hours, the yield from 82 lb. of dry leaf of the first cut being 14, of the second 24, and of the third 21 oz. of oil, or 1·07, 1·83, and 1·60 per cent. respectively. The value of course fluctuates, being dependent both upon the quality and the dryness of the leaves. The second cut is always the most prolific; thus, for example, a field which produces 300 lb. of leaves in the first cut will yield 800 lb. in the second and 600 lb. in the third. The production per acre is about 5000 lb. of dry leaves, yielding about 80 lb. (= 1·60 per cent.) of oil.

"The price of the leaves, like that of the oil, varies and the distillers only make small profits. The oil is collected by brokers, and resold to the large refineries, of which there are two of importance. In the districts of Odashitsuki, Okujoto, and Bingo there are so-called peppermint-guilds, who test the oil and give certificates of weight and purity. In the other districts the refiners must take the oil as it comes."

E. Marx[4] has made an elaborate report on the distillation of the oil in Japan.

From this report it appears that the industry has not benefited by the introduction of European methods, although it is a considerable one.

[1] *Communication from the Laboratory of Prof. Kremers, Madison, Wisconsin,* 1908.

[2] *Yamagata*, S. 1.

[3] *Board of Trade Journal*, 79 (**1912**), 78.

[4] *Gessell. für Natur. und Völkirkunde Ostasiens* (1896), **6**, 355.

It is Marx's opinion that the substitution of an improved method for that now followed would benefit both the quantity and quality of the oil produced. The apparatus universally adopted in Japan consists of cast-iron broad-edged boilers, A, I, H; wooden vats, B; and condenser, C. Generally there are three sets of stills, etc., combined in one battery with a common furnace. They are arranged in steps, so that the lowest is built into the ground.

The process is begun by filling the boilers, A, I, H, with water; then the vats, B, which have perforated bottoms, are placed on the broad

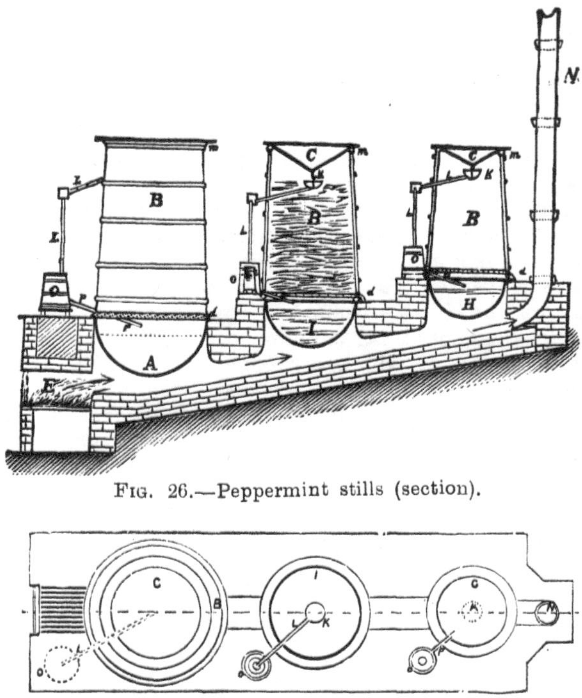

FIG. 26.—Peppermint stills (section).

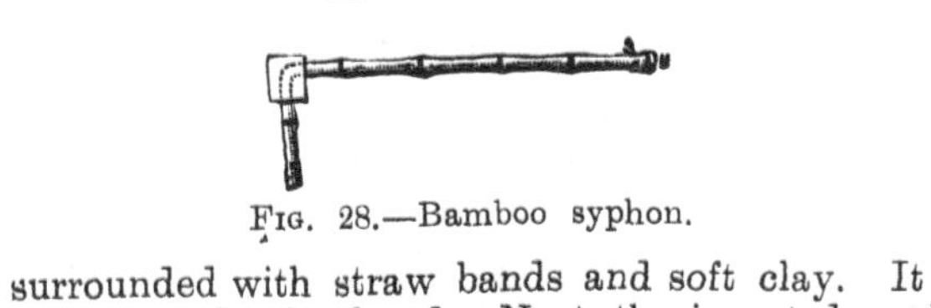

FIG. 27.—Peppermint stills (ground plan).

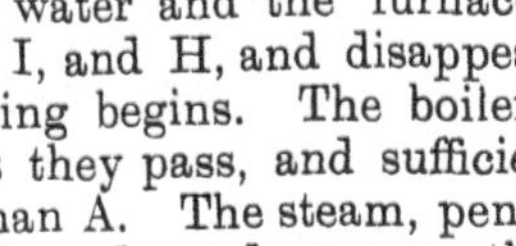

FIG. 28.—Bamboo syphon.

edges and surrounded with straw bands and soft clay. It is in these vats that the peppermint is placed. Next, the inverted condenser, C, is put on and filled with water and the furnace, E, lighted. The heat passes from E below A, I, and H, and disappears through the chimney, N, on which the distilling begins. The boilers I and H, are properly heated by the fumes as they pass, and sufficiently so, as H is smaller than I, and I smaller than A. The steam, penetrating the herb, carries with it the essential oil, and condenses on the outer side of C, being collected in the vessel, K, hanging upon it, from which the condensed water containing the oil passes through a bamboo reed, L, inserted in K to the florentine flasks, O. Each of these consists of a little wooden container in which a glass cylinder (an old bottle without bottom or neck) is inserted. In this, the water separated from the oil runs through the little pipe, P, again into the boilers, A, H, I. As soon as the water in C becomes hot, the workman replaces it with cold water. The emptying is effected by a bamboo syphon. The workman inserts the short arm in the inverted cooler, closes the lower opening and sucks the syphon until the water flows into it. According to the inborn easy-going nature of the Jap workman, who likes to pause in his work for a smoke, it often happens that the cooling water in C begins to boil and steam runs through instead of the condensed water saturated with oil, so that the atmosphere in the neighbourhood becomes so impregnated with peppermint that anyone near loses sight and sense of smell. The workman does not go unpunished, as he burns his mouth by sucking up the boiling water; that, however, does not prevent him from going on smoking, and the same thing happens again. In addition to the direct loss of oil caused by this negligence, it often happens that the water-boiler becomes dry, and then the herb in the still is burnt, in spite of the straw. This sometimes happens in spite of all precautions, as the flames frequently overheat the boilers if the water runs too low, and the products of burning straw get over with the distillate, which explains why Japanese peppermint oil sometimes has a peculiar empyreumatic flavour.

Japanese oil of peppermint contains a very high proportion of menthol, a large amount of which is frequently removed, and the dementholised oil sold as such. The term Japanese peppermint oil is understood on the market as meaning the dementholised oil. The usual manner in which the menthol is obtained is by a simple process of freezing, whereby the crystals separate, and are freed from the adherent oil either by pressure or centrifugal treatment. To purify them they should be recrystallised from a suitable solvent, such as alcohol. If it is wished to extract the maximum quantity of menthol, it is advisable to distil off the terpenes, etc., under reduced pressure, and then freeze the residual oil.

The crude distillate, containing the whole of the menthol, is known in Japan as Torioroschi.

Japanese peppermint oil was first carefully investigated by Beckett and Wright.[1] They found present in the oil besides menthol, a body isomeric with menthone but which was not identified as menthone. Other bodies found by these chemists were almost certainly mixtures and not chemical individuals.

Murayama[2] detected *l*-limonene in the oil, and Schimmel & Co.[3] isolated from it the alcohol ethyl-amyl carbinol $C_2H_5 . CH(OH)C_5H_{11}$, and[4] also the ketone Δ'-menthenone.

The identity of the ethyl-amyl carbinol found in the oil has been settled by the preparation of the same body synthetically. Schimmel & Co.[5] heated normal caproic aldehyde with magnesium methyl-iodide, and thus obtained *i*-ethyl-*n*-amyl carbinol, which was found to possess characters substantially identical with those of the natural compound.

[1] *Jour. Chem. Soc.* (**1876**), 13. [2] *Jour. de Pharm. et Chim.*, vii. 1 (**1910**), 549.
[3] *Report*, April, **1912**, 103. [4] *Ibid.*, October, **1910**, 97.
[5] *Ibid.*, April, **1913**, 82.

Pickard and Littlebury[1] have isolated from the oil an isomer of menthol, which occurs in small quantity, and which they have named neomenthol. This body is of particular interest in regard to the stereochemistry of this important body.

These chemists pointed out that a ketone of the configuration of menthone should exist in both the *cis* and *trans* forms, one of which may be termed menthone, the other isomenthone. When the keto grouping is reduced to the carbinol complex, two optically inactive menthols should exist for each of the ketones, and each of these should yield its pair of optically active menthols. It is proposed to term the optically inactive alcohols menthol and neomenthol in the case of menthone, and isomenthol and isoneomenthol in the case of isomenthone. They artificially prepared the isomers, menthol and neomenthol, and resolved each into its optically active varieties, so that the characters of four out of the eight possible optically active menthols are, to some extent, known. They have also shown that, in addition to *l*-menthol, *d*-neomenthol is present in Japanese peppermint oil.

The starting-point of the experimental work was the reduction product of thymol, by finely divided nickel. The unchanged thymol is removed by washing with a solution of soda; alcohols are transformed into hydrogen phthalic esters, which are then removed by sodium carbonate. The ketones remaining are treated with semi-carbazide in the usual manner, and a mixture of two semi-carbazones obtained, from which menthone and isomenthone can be obtained. Fractional crystallisation of the hydrogen phthalic esters from acetic acid yielded two pure products—the less soluble melting at 177° and the more soluble one at 130°. The former on hydrolysis yields pure neomenthol (inactive), melting at 51°.

The ester melting at 130° yields inactive menthol melting at 34°.

Pure menthol (inactive) is characterised by its melting-point (34°), and by those of its hydrogen phthalate (130°), and hydrogen succinate (85° to 86°). It also forms a phenyl-carbamate melting at 102° to 104°. By repeated recrystallisations of the brucine or cinchonine salt of the hydrogen phthalate ester, from acetone or alcohol, the nearly pure laevo-rotatory form can be obtained, whilst by the use of cinchonidine, the dextro-rotatory variety can be separated. *L*-menthol was found to be identical in melting-point and specific rotation with the natural menthol, which forms the principal ingredient of the peppermint oils, etc., whilst *d*-menthol has the same melting-point and an equal rotation, but in the opposite direction. Strangely enough, the dextro and laevo varieties differ considerably in odour, and each forms a crystalline benzoate, whilst the benzoate of the inactive variety is an uncrystallisable oil.

Neomenthol in the inactive condition melts at 51°, and forms a hydrogen phthalate melting at 177°, and a phenyl-carbamate melting at 114°. On oxidation it yields menthone, identical with that formed by the oxidation of menthol, thus proving that the two bodies are stereoisomers. By fractional crystallisation of brucine salt of the hydrogen phthalate, the *dextro* variety was obtained, whilst by similar treatment of the quinine salt of the hydrogen succinate *l*-neomenthol was obtained. The specific rotation of these optically active neomenthols is ± 19·6°.

[1] *Jour. Chem. Soc.*, 101 **(1912)**, 109.

Curiously enough, the menthones resulting from the oxidation of these active neomenthols are optically active in the opposite sense to the menthols themselves.

The characters, therefore, of two of the optically inactive and four of the optically active isomeric menthols have been fixed with certainty. By treating the alcoholic residue of Japanese peppermint oil, from which the bulk of the ordinary *l*-menthol had been removed, with phthalic anhydride in the usual manner, the authors obtained a mixture of phthalic acid esters, from which after twenty-eight recrystallisations from 95 per cent. acetic acid, a fraction was obtained which melted at 141° to 144°, and had a specific rotation of + 57·16° in chloroform. This substance was the phthalic acid ester of *d*-neomenthol, containing a little of the ordinary *l*-menthol ester. This was proved (1) by mixing it intimately with artificially prepared *d*-neomenthol hydrogen phthalate, when the melting-point was practically unchanged, and (2) by hydrolising the ester and oxidising the resulting menthol, when almost pure *l*-menthone was obtained, having a specific rotation − 28·33°, and yielding a semicarbazone melting at 187° to 189°. Japanese peppermint oil therefore contains *l*-menthol, together with a small amount of *d*-neomenthol.

Walbaum[1] has isolated from the fraction of the oil boiling at 250° to 310° the acetic ester of 3 hexen-I-ol, of the formula

$$CH_2(C_6H_5)CO_2 . CH_2 . CH_2 . CH : CH . CH_2 . CH_2$$

which. when pure, has the following characters :—

Boiling-point	135° to 136° at 4 mm.
,, ,,	290° ,, 760 mm.
Specific gravity at 15°	1·000
Refractive index at 20°	1·4981

It is a liquid with an onion-like odour.

Pure, natural Japanese peppermint oil, containing all the natural menthol, has the following characters :—

Specific gravity at 15°	0·900 to 0·912
Optical rotation	− 26° ,, − 42°
Refractive index	1·4600 ,, 1·4635
Acid value	0·5 ,, 2·5
Menthol as esters	3 to 8 per cent.
Total menthol	70 ,, 90 ,,
Solidifying-point	16° to 28°

The dementholised oil of commerce has the following characters :—

Specific gravity at 15°	0·894 to 0·906
Optical rotation	− 24° ,, 36°
Refractive index	1·4590 ,, 1·4650
Acid value	0·5 ,, 3
Menthol as esters	4 to 17 per cent.
Total menthol	45 to 55 per cent. (rarely higher)

Various colour reactions have been proposed with a view to distinguishing between Japanese and other peppermint oils. These are of doubtful value, but there is no question that all peppermint oils other than Japanese give the following reactions to a marked extent, whilst Japanese oil either gives no reaction, or at most a very slight one. A colour reaction, however, would be of great use if it were negative with the other peppermint oils, and positive with Japanese oil, which is the cheapest variety obtainable.

[1] *Jour. prakt. Chem.* **(1917)**, 96, 245.

Umney[1] heats 1 c.c. of the oil with 0·5 gram of a mixture of equal parts paraformaldehyde and citric acid over a water-bath. With Japanese peppermint oil there is no coloration, whereas with American, English, Italian, and Saxon oils a purple colour develops.

If five drops of oil are dissolved in 1 c.c. of glacial acetic acid a blue colour results within one hour with peppermint oils, except in the case of Japanese oil.

Many experiments have been undertaken with a view to transport peppermint plants from a country where the oil has developed a particularly fine odour and flavour, into other countries. Few plants, however, appear to alter to so great an extent under these conditions as those of the natural order Labiatæ, and when once the soil and climate suit the plant, transplanting it to different conditions causes the oil to assume different characters. Thus Mitcham plants, transplanted to France, will not yield an oil which is necessarily identical with Mitcham oil. Recent experiments, however, have been to some extent successful, and oils having characters very similar to English oil have been obtained from Mitcham plants grown in Italy. The most complete experiments, however, in this direction have been carried out with Japanese plants by Thoms,[2] with a view to determine whether oil containing as much menthol as is present in Japanese oil could be obtained in Germany or in German colonies. The Japanese plants were cultivated at Dahlem, near Berlin, and at Okahandja, South-West Africa.

At Dahlem cuttings were made, the first on 17 July and the second on 10 August. The yield of oil from the July crop was 0·74 per cent., and that from the August crop 0·624 per cent. The oils had the characters given below, which are compared with those of the oil distilled from plants harvested in the summer of 1911:—

	July Crop, 1912.	August Crop, 1912.	Summer Crop, 1911.
Specific Gravity	0·8987 at 18°	0·903 at 16°	0·8954 at 22°
Solidifying-point	13·5°	14·75°	14·5°
Optical Rotation	− 35·22°	− 35·29°	− 34·75°
Acid Value	2·66	2·73	4·01
Ester ,,	12·96	13·80	12·74
Acetylation Ester Value	283·8	280·1	283·2
Combined Menthol	3·78 %	3·56 %	3·50 %
Free Menthol	75·20 %	75·65 %	75·27 %
Total ,,	78·98 %	79·21 %	78·82 %

A sample of oil distilled from plants grown at Witzenhausen, the yield being just under 1 per cent. on the dry herb, had the following characters:—

Specific gravity at 15·5°	0·9042
Solidifying-point	14·5°
Optical rotation	− 35·3°
Acid value	3·3
Ester ,,	15·15
Acetylation ester value	291·6
Combined menthol	6·99 per cent.
Free menthol	74·16
Total ,,	81·15

[1] *P. and E.O.R.*, 2 **(1911)**, 275.

[2] H. Thoms, *Ber. deutsche pharm. Ges.*, 20 **(1910)**, 424; *Arbeit. Pharm. Instit. Univ. Berlin*, 8 **(1911)**, 93.

In dealing with the experimental cultivation at Okahandja, in German South-West Africa, the herb was divided into the leaves and the stems, which were distilled separately. The leaves yielded 1·225 per cent. of oil, and the stems 0·2 per cent. Calculated on the whole plant the yield was 0·866 per cent. Both oils were very similar, and were characterised by a very high menthol value, in each case over 84 per cent. They had the following characters :—

	Leaf Oil.	Stem Oil.
Specific Gravity at 20°	0·904	0·9000
Solidifying-point	20·75°	21°
Optical Rotation	− 35·29°	− 35·3°
Acid Value	1·56	2·77
Ester ,,	8·29	6·15
Acetylation Ester Value	306·6	304·1
Combined Menthol	2·31 %	1·71 %
Free Menthol	83·01 %	82·93 %
Total	85·32 %	84·64 %

No less than 57·5 per cent. of menthol could be separated by freezing from these oils.

Chinese Peppermint Oil.

A small quantity of peppermint oil is distilled in the neighbourhood of Hong-Kong, from a plant which is, in all probability, *Mentha canadensis*, var. *glabrata*. This oil has been examined by the Imperial Institute,[1] by Schimmel & Co.,[2] and by Shino Saki.[3] The three samples had the following characters :—

	1.	2.	3.
Specific gravity	0·918	0·9187	0·9091
Optical rotation	− 44° 40′	− 44° 2′	− 35°
Menthol as Esters	13·55 per cent.	12·6 per cent.	8·08 per cent.
Total menthol	64·24 ,,	64 ,,	78·65 ,,

French Peppermint Oil.

Peppermint oil is distilled at Grasse, Cannes, Vence, Cagnes, and Villeneuve-Loubet, as well as in the Basses-Alpes, Var, and Haute-Garonne, etc. The French varieties of *Mentha piperita* are not identical with those cultivated in England. According to A. and E. G. Camus,[4] *Mentha piperita* is not a separate species, but is a hybrid of *Mentha viridis* and *Mentha aquatica*. Two well-defined forms are cultivated in the south of France which are termed by Camus *Mentha piperita officinalis*, var. *pallescens*, and *Mentha piperita officinalis*, var. *rubescens*. These, however, do not agree in characters with the forms described by Planchon and Collin as "blanche" and "noire"—nor are they identical with our own "black" and "white" mints. These facts illustrate the very great variability of the peppermint plant. The following figures illustrate the characters of the two oils :—

[1] *Bull. Imp. Inst.*, 11 **(1913)**, 434.
[2] *Report*, October, **1910**, 96.
[3] *J. Chem. Ind. Tokio*, 1919, **22**, 458.
[4] Roure-Bertrand Fils, *Report*, October, **1911**, 3.

FIG. 29.—Cultivation of peppermint in Vaucluse.

[Parfumerie Moderne. Mus.]

	Ordinary Peppermint, *Pallescens.*		Red Peppermint, *Rubescens.*	
Specific Gravity	0·9191	0·9184	0·9170	0·9136
Rotation	− 10° 54′	− 8° 2′	− 16° 38′	− 13° 44′
Solubility in 80 °/₀ Alcohol	1 vol. afterwards clouding	—	1 vol. afterwards clouding	—
Acid Value	0·8	0·8	1·0	1·2
Ester ,,	40·5	31·7	18·9	17·5
Menthol Acetate	14·3 °/₀	11·2 °/₀	6·7 °/₀	6·2 °/₀
Ester Value after Acetylation	160·8	169·1	180·2	185·6
Total Menthol	50·9 °/₀	53·9 °/₀	58·0 °/₀	60·0 °/₀
Free ,,	39·6 °/₀	45·1 °/₀	52·8 °/₀	55·1 °/₀
Menthone	7·3 °/₀	—	16·8 °/₀	—

The limit values for French peppermint oils are, however, wider than those indicated for these two particular varieties. Most—if not all—genuine oils will have characters which fall within the following:—

Specific gravity	0·910 to 0·930
Optical rotation	− 5° ,, − 35°
Refractive index	1·4610 ,, 1·4690
Menthol as esters	6 to 20 per cent.
Total menthol	48 ,, 70 ,,

Many French oils are not soluble in 3 volumes of 70 per cent. alcohol, but will generally dissolve clearly in 4 volumes.

In rare cases where the plants have been attacked by certain insect pests, oils have been obtained which are dextro-rotatory up to + 7°.

French oil of peppermint contains,[1] in addition to menthol and its esters, isovaleric aldehyde, isoamyl alcohol, *l*-pinene, Δ^3-*p*-menthene and cineol.

English Peppermint Oil.

Two varieties of the plant are grown in England, known as "black" and "white" mint. Of these the former is the coarser plant, rarely flowering in England, and yielding more oil of somewhat less delicate aroma than the white mint. These appear to be forms of *Mentha piperita*, var. *officinalis*, but are certainly not identical with the French forms (*vide supra*) named by A. and E. G. Camus *rubescens* and *pallescens*, although these botanists believed them to be so. The herbs are distilled in the green state, although sometimes left for a day or two in heaps, after being cut. In England most of the stills used hold from 4 to 8 cwt. of herbs, and have a false bottom which is covered to the depth of 2 ft. with water, and direct heat is then applied. The distillation is carried on for about five hours. The yield varies according to the season from about ½ to 1 per cent.

According to the *Journal of the Board of Agriculture*,[2] from 500 to 1000 acres are under peppermint cultivation in England. In this country black mint yields twice as much oil as does white mint, and as a result the cultivation of the latter has been supplanted to a considerable extent. Peppermint grows best at a height not exceeding 200 ft. The principal

[1] Roure-Bertrand Fils, *Report*, April, **1909**, 33. [2] 15 (**1908**), No. 5.

production is in the districts of Ewell, Cheam, Carshalton, and Mitcham; smaller quantities are also obtained in certain parts of Sussex, Kent, Hertfordshire, and Lincolnshire. In England the plants are grown from roots, not from seed. The land having been ploughed and manured during the winter, the plants are dibbled in in May. A plantation lasts from four to five years, the best output being in the second year. After four or five years the land is not used again for mint growing for some years.

The harvest begins early in September, when a red rust or "snuff" appears on the leaves. After some slight drying the plants are distilled for about six hours. The oil-yield per acre ranges from 8 up to 25 lb.

In general, English oils have a specific gravity of about 0·900 to 0·912, and an optical rotation of − 23° to − 32°, and contain from 60 to 70 per cent. of menthol, or rather more than most American oils. The refractive index of the oil varies from 1·4600 to 1·4640. The oil is usually quite soluble in 3 to 4 volumes of 70 per cent. alcohol. Umney has examined samples of "black" and "white" mint oils comparatively, and gives the following figures for them:—

		Specific Gravity.	Rotation.	Menthol— as Esters.	Free.	Menthone.
1	Black . .	·9086	− 23·5°	3·7	59·4	11·3
2	White .	·9058	− ·33°	13·6	51·9	9·2

Fluckiger and Power[1] isolated two laevo-rotatory terpenes from English peppermint oil, but their purity is doubtful. It is probable that they were mixtures of pinene, phellandrene, and limonene. A dextro-rotatory sesquiterpene (cadinene?) is also present, as well as menthone, and menthol, both in the free state and as acetate and isovalerianate.

Italian Peppermint Oil.

A certain amount of peppermint oil is distilled in Italy from plants whose botanical origin is not yet clearly settled. This oil, which is of excellent quality, has the following characters:—

Specific gravity	0·905 to 0·926
Optical rotation	− 2° ,, − 27°
Refractive index	1·4615 ,, 1·4685
Menthol as esters	3 to 11 per cent.
Total menthol	45 ,, 65 ,,
Menthone	8 ,, 20 ,,

It is soluble in 3 volumes of 70 per cent. alcohol.

During the last few years, however, Mitcham plants have been cultivated in Italy, and the resulting oil approximates, both in odour and in general characters, much more closely to ordinary English distilled oil than the oils above described. This so-called Italo-Mitcham oil is prepared at Messina and at Pancalieri, Piedmont, Northern Italy. The oil has been examined by Umney and Bennett,[2] and has characters falling within the following limits:—

[1] *Pharm. Journ.*, iii. **11** (1880), 220.
[2] *Chemist and Druggist*, 66 (**1905**), 945; 67 (**1905**), 970.

Specific gravity	0·900 to 0·920
Optical rotation	− 12° ,, − 26°
Refractive index	1·4620 ,, 1·4645
Menthol as esters	3 to 29 per cent.
Total menthol	40 ,, 70 ,,
Menthone	12 ,, 18 ,,

Umney and Bennett found the oil distilled from July-cut plants to contain only about 40 per cent. of menthol, whilst oil from the later

FIG. 30.—Distillation of peppermint in Piedmont.
Carles.] [*Parfumerie Moderne.*

second cutting contained 70·5 per cent. But no generalisation is possible without the examination of a number of samples.

RUSSIAN PEPPERMINT OIL.

A certain amount of peppermint oil is distilled in various parts of Russia, most of which is used in the country itself; but it has recently appeared on the English market, where its reception has been favourable. It is distilled in Tambow, the Caucasus, Poltava, and the Ukraine. The dried herb yields about 1·5 to 1·7 per cent. of oil.

Maisit[1] states that peppermint oil is distilled in various parts of European Russia, generally in very primitive stills, with the result that the oil is often burnt and of poor odour. During the past few years peppermint plantations have been laid out in the Caucasus districts, which, owing to favourable climatic and soil conditions, are very suitable for the cultivation of numerous medicinal plants.

A yield of from 1·6 to 1·7 per cent. on the dried leaves and flowers is the average obtained. Maisit has examined the oil distilled from one year old and two year old plants. Both are pale yellow in colour, not very soluble in 70 per cent. alcohol, and having the following characters:—

[1] *Arch. der Pharm.*, 249 (1911), 637.

Oil from	Specific Gravity at 20°.	Rotation.	Acid Value.
One year old Plants	0·912	– 17° 42′	0·57
Two ,, ,, ...	0·913	– 17° 57′	0·56

The menthol values of the two oils are as follows: One year old plants, 42·44 per cent. free menthol, 49·17 per cent. total menthol; two year old plants, 41·33 per cent. free menthol, 50·07 per cent. total menthol.

One hundred grams of the two year old plant oil has fractionated with the following results:—

Distilled at	Per Cent.
To 200°	9·5
200° to 205°	8·0
205° ,, 210°	8·0
210° ,, 215°	17·5
215° ,, 220°	21·0
220° ,, 225°	5·5
225° ,, 230°	7·0
230° ,, 235°	4·5
Above 235°	18·0

Maisit considers that a sesquiterpene is present in the high boiling fractions.

Those districts known as the Ukraine (" Little " Russia and Ruthenia) are exceptionally favourable to the growth of peppermint, and somewhere about 20,000 lb. of oil are distilled here every year, and the output is increasing. A sample of this oil examined by Umney [1] had the following characters:—

Specific gravity	0·904
Optical rotation	– 26°
Total menthol	53·8 per cent.
Menthyl acetate	6 ,,

It was soluble in 3 volumes of 70 per cent. alcohol.

Russian peppermint oil will usually have values falling within the following limits:—

Specific gravity	0·903 to 0·920
Optical rotation	– 15° ,, – 28°
Menthol as esters	3 to 10 per cent.
Total menthol	50 ,, 60 ,,

In addition to menthol and menthyl esters, this oil contains menthone, principally in the dextro-rotatory form, *l*-limonene, dipentene, inactive α-pinene, and cineol. The esters are present as acetate and isovalerianate.

German Peppermint Oil.

A small amount of peppermint oil is distilled in Germany, successful plantations existing in Saxony especially. German oil has the following characters:—

[1] *P. and E.O.R.* (1914), 314.

Specific gravity	0·898 to 0·915 (rarely to 0·930)
Optical rotation	− 23° to − 37°
Refractive index	1·4580 ,, 1·4680
Menthol as esters	2·8 to 20·8 per cent.
Total menthol	48 ,, 81 ,,

It is soluble in 3 to 5 volumes of 70 per cent. alcohol.

Dalmatian Peppermint Oil.

French peppermint plants cultivated at Trieste yield an oil of good aroma, which has been examined by Schimmel & Co.[1] The crude and rectified oils had the following characters:—

	Crude Oil.	Rectified Oil.
Specific gravity	0·907 to 0·913	0·909 to 0·914
Optical rotation	− 15° 45′ ,, − 20° 58′	− 11° 45′ ,, − 18° 12′
Refractive index	1·4623 ,, 1·4668	1·4604 ,, 1·4678
Menthol as esters	3·4 to 4·8 per cent.	2·6 to 3·8 per cent.
Total menthol	46·3 ,, 54·2 ,,	39·6 ,, 54·1 ,,

The oil is soluble in 3 to 4 volumes of 70 per cent. alcohol.

Hungarian Peppermint Oil.

Irk[2] has described the oil of peppermint distilled in Hungary, and which resembles American oil in aroma. The dry plants yielded 1·17 per cent. of oil, having the following characters:—

Specific gravity	0·9014 to 0·9192
Optical rotation	− 26° ,, − 32·4°
Refractive index	1·4632 ,, 1·4769
Menthol as esters	8 to 12·8 per cent.
Total menthol	56·4 ,, 65·2 ,,
Menthone	7·4 ,, 13·2 ,,

It is soluble in 2 to 5 volumes of 70 per cent. alcohol.

An interesting series of experiments on the value of manuring has been carried out on Hungarian peppermint plants.[3] These experiments are summarised in the following table:—

Artificial Manure.	Farmyard Manure.	Fresh Herb (kilos per hectare).	Dried Herb (kilos per hectare).	Oil (per Cent.).	Weight of Oil (kilos per hectare).
None	Without	4·240	1·300	0·77	10·01
	With	6·760	2·000	0·88	16·40
With Sodium Nitrate	Without	6·640	1·860	0·74	18·76
	With	10·560	2·820	0·81	22·84
With Sodium Nitrate and Superphosphate	Without	7·120	1·940	0·73	14·16
	With	8·980	2·320	0·84	19·94
With Sodium Nitrate, Superphosphate and Kainit	Without	8·780	2·200	0·72	15·84
	With	13·080	3·140	0·93	29·83
Marshy ground at Admont without manure	—	7·89	2·83	0·62	17·55

It will thus be seen that the yield of oil from a given area of completely manured soil was nearly double that from the same area which

[1] *Bericht*, October, **1911**, 69. [2] *Kisérletügyi Közlemények*, 13 (**1910**), 3.
[3] *Pharm. Post* (**1912**), 2.

had received only farmyard manure, and nearly treble that from land which was not manured at all. It is interesting to note that the yield from the natural unmanured marshy ground was the lowest calculated on the amount of material employed, but was fairly good when referred to the area of land concerned. The mint grown on a plot without any manure and attacked by *Puccinium* gave oil containing the highest percentage of esters, equivalent to 13·57 per cent. of menthyl acetate. The same unaffected by the blight gave 10·73 per cent. The oil grown on the natural unmanured marshy ground had the highest percentage of total menthol, 64·25 per cent. The puccinium-attacked plant gave oil containing 62·28 of total menthol, and coming next. The lowest menthol-content, 57·69 per cent., was obtained from the plot manured with sodium nitrate and superphosphate without any farmyard manure.

Javan Peppermint Oil.

This oil is distilled from *Mentha arvensis*, var. *Javanica*. It has the following characters :—[1]

Specific gravity at 15° C.	0·9979 [2]
Rotatory power (l = 100 mm.)	+ 0° 24′
Soluble in 1½ volumes of 70 per cent. alcohol, and the solution remains clear on the subsequent addition of the same alcohol.	
Soluble in all proportions of 80 per cent. alcohol.	
Esters, as menthyl acetate	17·5 per cent.
Combined menthol	13·8 ,,
Total menthol	48·2 ,,
Free ,,	34·4 ,,
Ketones and aldehydes	none, or traces

British East African Peppermint Oil.

A sample of peppermint oil distilled from Mitcham plants grown at Molo, in the highlands of British East Africa, has been examined [3] and found to have an excellent aroma and a very high menthol value. It had the following values :—

Specific gravity	0·967
Optical rotation	− 33° 30′
Total menthol	67·5 per cent.

The most important determination to be made in the examination of peppermint oil is that of the total menthol, but the menthone value is also frequently required. This is best obtained by determining the total menthol in the oil, and then reducing the menthone, in a separate portion, to menthol by means of sodium, and determining the menthol in the reduced oil. If M_2 be the total menthol in the reduced oil and M_1 be that in the original oil the percentage of menthone present is

$$\frac{(M_2 - M_1)154}{156}.$$

The factor $\frac{154}{156}$ representing the difference in the molecular weights of menthone and menthol, may usually be neglected.

Peppermint oil is frequently adulterated. American oil is sometimes

[1] Roure-Bertrand Fils, *Bulletin*, 3 (**1910**), 1, 59.

[2] This is probably a misprint for 0·8979. A sample examined by Wielen had a specific gravity 0·921 and optical rotation + 4° 40′.

[3] *P. and E.O.R.* (**1915**), 4.

enriched by the addition of Japanese menthol, which is usually cheaper than American oil. Camphor oil is occasionally used as an adulterant. The author and Bennett have found cedar-wood oil and oil of African copaiba used as adulterants (see below). Glyceryl acetate (tri-acetin) has been found in sophisticated oils. The behaviour on distillation as compared with normal samples affords the surest indication of such adulteration. The following table represents a normal distillation of pure oil :—

PURE OIL.

(Specific Gravity, 0·911 ; Refractive Index, 1·4645).

Quantity.	Specific Gravity.	Rotation.	Refractive Index.
Per Cent.			
12½	0·898	− 10°	1·4660
12½	0·903	− 14°	1·4635
12½	0·907	− 16°	1·4645
12½	0·910	− 20°	1·4640
12½	0·912	− 23°	1·4615
12½	0·912	− 23°	1·4615
12½	0·915	− 34°	1·4630
12½	0·962	—	1·4790

An oil containing tri-acetin gave the following results (Bennett) :—

Fractions.	Quantity.	Specific Gravity.	Rotation.	Refractive Index.
	Per Cent.			
1 . .	12½	0·900	− 15°	1·4645
2 . .	12½	0·902	− 15°	1·4670
3 . .	12½	0·910	− 14°	1·4650
4 . .	12½	0·920	− 16°	1·4640
5 . .	12½	0·926	− 20°	1·4640
6 . .	12½	0·938	− 22°	1·4640
7 . .	6	—	—	1·4640
Residue	19	1·147	—	1·4450

With cedar-wood and copaiba oils, the high boiling fractions are found to have very high refractive indices (up to 1·4900 or even 1·500), and in the case of African copaiba, cadinene is found in the fraction 255° to 270°, and in the case of cedar oils, the corresponding fraction may be laevo-rotary to the extent of − 45°.

As indicating the behaviour of African copaiba oil, which has been used to a considerable extent for adulterating this oil, the following figures for the pure copaiba oil are of interest. A sample of the balsam itself was distilled by the author and Bennett and the oil collected in four equal fractions.

The figures obtained were :—

	Specific Gravity.	Rotation.	Refractive Index.
1	0·917	+ 17° 30′	1·5030
2	0·918	+ 28° 30′	1·5043
3	0·921	+ 46°	1·5061
4	0·927	+ 55°	1·5082

Other samples of African copaiba oil had the following characters :—

Specific gravity	0·919 to 0·932
Rotation	+ 30° ,, + 37°
Refractive index	1·4999 ,, 1·5040

Two of these on fractionation showed the following variations in optical rotation and refractive index :—

	I.		II.	
	Rotation.	Refractive Index.	Rotation.	Refractive Index.
25 per cent.	+ 16°	1·4960	+ 24°	1·4975
25 ,,	+ 19°	1·4965	+ 26°	1·4980
25 ,,	+ 24°	1·4980	+ 29°	1·4981
20 ,,	+ 48°	1·5089	+ 43°	1·5090
5 ,, residue	—	—	—	—

Oil of Spearmint.

There are several different essential oils known under the name spearmint oil; the botanical origin of the plant used for distillation differing with the country in which the plant is grown. In the United States and in this country several varieties of *Mentha viridis (Mentha spicata)*, are distilled. In Russia the plant distilled is *Mentha verticellata*, and in Germany either *Mentha spicata*, var. *crispata*, or *Mentha longifolia*.

The principal types of this oil have the following characters :—

American Spearmint Oil.

Specific gravity	0·920 to 0·940 (rarely higher)
Optical rotation	− 35° to − 53°
Refractive index	1·4800 ,, 1·4890
Acid value	0 ,, 2
Ester ,,	18 ,, 36
Carvone (neutral sulphite method)	35 to 66 per cent.

The oil is soluble in 1 to 1·5 volumes of 80 per cent. alcohol.

English Spearmint Oil.

Specific gravity	0·926 to 0·935
Optical rotation	− 39° ,, − 52°
Carvone	30 to 48 per cent.

Austrian Spearmint Oil.[1]

Specific gravity	0·936 to 0·952
Optical rotation	– 38° ,, – 50°
Refractive index	1·4890 ,, 1·4930
Carvone	61 to 72 per cent.

Russian Oil.

Specific gravity	0·880 to 0·890
Optical rotation	– 20° ,, – 28°
Carvone	5 to 10 per cent.
Linalol	50 ,, 60 ,,

According to E. K. Nelson, an American oil which he examined had the following characters :—

Specific gravity at 25°	0·9290
Optical rotation	– 52·16°
Refractive index at 25°	1·4866
Ester value	12·4
,, (after acetylation)	36·4

The oil contained 66 per cent. of carvone, phellandrene, and *l*-limonene, and dihydrocarveol acetate. Esters of acetic, butyric, and caproic or caprylic acids are also present.

Elze,[2] after removing the carvone from a German spearmint oil, obtained an oil with a more intense spearmint odour than the original oil. It had a specific gravity 0·917 and optical rotation – 28°. It contained 18 per cent. of esters and on fractionation yielded the following results :—

	Boiling-point (4 mm.)	Percentage.	$d_{15°}$.	Rotation.	Ester-content.
A	40° to 45°	15·0	0·860	– 27° 30′	2·8 %
B	45° ,, 75°	15·0	0·894	– 25°	
C	75° ,, 80°	24·0	0·930	– 24°	19·0 ,,
D	80° ,, 85°	7·5	0·935	– 30° 50′	
E	85° ,, 100°	35·0	0·940	– 40°	31·0 ,,

The author was able to prove the presence, in fraction A, of *l*-phellandrene, which was identified by its nitrosite (melting-point 105° to 105½°) as well as by its constants (boiling-point 175° to 176° [755 mm.]; $d_{15°}$ 0·8575°; α – 26° 0′). Fraction E, which possessed a marked odour of spearmint, was saponified. Acetic acid, together with small proportions of valeric acid (probably isovaleric acid ?) were identified. The alcoholic part consisted to the extent of 15 per cent. of dihydrocuminic alcohol ($d_{15°}$ 0·9539 ; α – 30° 15′), which yielded a naphthyl urethane melting at 146° to 147°, and also, when oxidised with Beckmann's mixture, an aldehyde (melting-point of the semi-carbazone 198° to 199°), and an acid, melting at 132°. When fraction E was saponified the typical spearmint odour disappeared ; hence, in Elze's opinion, the dihydrocuminic acetate is the carrier of the odorous principle of the oil.

F. Rabak[3] has carried out an investigation on the influence of the

[1] According to Schimmel & Co. [2] *Chem. Zeit.*, 34 **(1910)**, 1175.
[3] *Jour. Ind. Eng. Chem.* **(1918)**, 10, **4**, 275.

time of harvest, drying, and freezing of spearmint, on the yield and odorous constituents of the oil. He gives the following characters for typical oils distilled from the plant at different stages of its development. He finds the oil from the fruiting herb more fragrant than that from the budding or flowering plant. Oil distilled from the leaves and tops are richer in esters than the oil from the whole plant. Rabak concludes that the yield of oil is affected by seasonal conditions, being much higher in some seasons than in others. The maximum yield is obtained during the flowering period. Drying the plants results in a lower yield, and freezing causes the oil to contain more esters and alcohols.

Year and Description of Material.	Specific Gravity.	Optical Rotation.	Refractive Index.	Esters as Acetate of $C_{10}H_{18}O$.	Total Alcohols as $C_{10}H_{18}O$.
1908.					
Fresh Herb:					
Budding . .	0·925 at 22°	− 25·1°	1·4874 at 24°	3·8 %	9·1 %
Flowering .	0·929 ,,	− 24·5°	1·4880 ,,	4·6 %	8·54 %
Fruiting . .	0·929 ,,	− 21·7°	1·4853 ,,	8·7 %	13·4 %
1909.					
Fresh Herb:					
Budding . .	0·931 at 21°	− 16·3°	1·4891 at 21°	8·6 %	12·0 %
Flowering .	0·929 ,,	− 16·1°	1·4864 ,,	7·7 %	12·4 %
Fruiting . .	0·934 ,,	− 17·4°	1·4873 ,,	7·3 %	10·9 %
Dry Herb:					
Budding . .	0·928 ,,	− 17°	1·4888 ,,	7·0 %	11·7 %
Flowering .	0·933 ,,	− 17·5°	1·4868 ,,	10·8 %	12·7 %
Fruiting . .	0·925 ,,	− 16·5°	1·4846 ,,	9·1 %	15·6 %
1910.					
Fresh Herb:					
Budding . .	0·931 at 23°	− 22·8°	1·4895 at 23°	7·6 %	16 %
Flowering .	0·925 ,,	− 20·3°	1·4825 ,,	13·4 %	18·4 %
Fruiting . .	0·930 ,,	—	1·4800 ,,	15·9 %	21·4 %
Dried Herb:					
Budding . .	0·912 ,,	− 18·2°	1·4775 ,,	10·8 %	15·7 %
Flowering .	—	—	1·4788 ,,	31·8 %	28·2 %
Fruiting . .	0·916 ,,	—	1·4745 ,,	22·0 %	30·5 %

Russian spearmint oil contains, according to Schimmel & Co.,[1] 50 to 60 per cent. of *l*-linalol. Cineol, *l*-limonene, and carvone are also present.

Oil of Mentha Aquatica.

The dried herb, *Mentha aquatica*, the water mint, yields about 0·4 per cent. of essential oil, having an odour of pennyroyal. The characters of the oil are not well determined. An oil distilled by Schimmel & Co. had a specific gravity 0·880 and optical rotation − 2° 14′, whereas a sample examined by Tschirch had a specific gravity 0·9553, optical rotation + 64° 56′, and refractive index 1·4827.

Irk[2] found a Hungarian oil to have a specific gravity 0·9603, optical rotation + 22·73°, and refractive index 1·4900.

[1] *Bericht*, April, **1898**, 28. [2] *Pharm. Zentral.*, 52 (**1911**), 1113.

Oil of Mentha Silvestris.

The dry herb, *Mentha silvestris*, yields about 1 per cent. of essential oil, having the following characters:—

Specific gravity	0·968 to 0·971
Optical rotation	about + 30°
Refractive index	1·4954
Acid value	2·4
Ester „	20·9
„ „ (after acetylation)	170 to 175

The oil contains very little menthol, but a considerable amount of pulegone, with a small amount of a phenol (carvacrol ?).

An oil examined by Tschirch was stated by him to have the extraordinary optical rotation − 132° 52′ and refractive index 1·46856. The authenticity of this specimen must be regarded with suspicion.

Oils of Monarda.

The best known oil of the *Monarda* species is that known as America horse-mint oil, which is obtained to the extent of about 1 per cent. from the herb *Monarda punctata*. The oil has a specific gravity from 0·930 to 0·940, and is either optically inactive or slightly dextro-rotatory.

The plant is of considerable importance, as it may before long be available as a regular source of thymol, which has hitherto been manufactured principally from ajowan seeds. Experiments have been carried out by the United States Department of Agriculture,[1] with this end in view.

In 1910 seeds obtained from Volusia County were sown, and the first year's growth yielded 0·20 per cent. of oil, and the second 0·24 per cent., the oils containing 64 per cent. of phenols. The plants were observed to vary, the variations falling into two classes:—

1. Darker plants with more serrate leaves and a pronounced red colour in the stems and branches.

2. Plants with larger leaves of a light green colour and little or no red colour in the stems and branches. The lighter type also produced more leaves. The average yield of oil from the two types (thirty-four tests being made) was 0·40 per cent. of oil from the dark plants, containing 66 per cent. of phenols, and 0·42 per cent. from the light type, containing 69 per cent. of phenols.

By continuing the cultivation of the light green variety, the yield of oil was increased to 0·44 per cent. in 1914, containing 72 per cent. of phenols; and in 1915 the yield of phenols was 74 per cent., although the percentage of oil was only 0·42 per cent.

The oil has the following characters:—

Specific gravity	0·923 to 0·933
Optical rotation	+ 1° „ − 2°
Refractive index at 20°	1·5020 „ 1·5040

The oil contains about 70 to 80 per cent. of phenols, of which a small portion consists of carvacrol and the remainder of thymol, the rest of the oil being chiefly cymene, which boils at 175° to 176° C., whilst thymol boils at 232° C.

Fractional distillation in a copper retort is therefore resorted to, and

[1] *Bulletin*, No. 372; and *P. and E.O.R.* **(1916)**, 331.

the portion of the oil coming over below 215° C. is set aside. The next fraction coming over between 215° and 240° C. is then secured, leaving a thick tar-like mass in the retort, which, so far as is known at present, has no value. The first fraction may contain about 25 to 30 per cent. of thymol, the cymene acting as a solvent of it, and preventing crystallisation. This fraction is, therefore, redistilled, and the fraction redistilling between 215° and 240° C. saved.

This fraction, added to the first which was distilled at the same temperature, is cooled in a shallow dish, and a crystal of thymol added to start crystallisation. After standing overnight the crystals are separated from the mother liquor by means of a centrifuge, and the crystals washed with water whilst the centrifuge is running at full speed. If there is a large quantity of mother liquor left, it probably contains a considerable portion of the lower boiling fraction and should be redistilled, removing any that comes over between 215° and 240° C.

The oil thus treated gives a commercial yield of 66·3 per cent. of thymol of a perfectly white crystalline character. Carvacrol is also probably present in the oil, as well as cymene, and traces of *d*-limonene.

Monarda citriodora.—This plant is known in North America under the name of Lemon Mint. It yields about 1 per cent. of essential oil on distillation, which has the following characters :—

Specific gravity	0·9437 to 0·9603 at 20°
Phenols	65 ,, 80 per cent.
Aldehydes	1·2 ,, 4 ,,

The oil contains a considerable amount of carvacrol, with a little thymol. Thymo-hydroquinone is also present, and probably citral and cymene.

Monarda myristica.—The oil distilled from the seeds of this plant, known as "Owere" seeds on the Gold Coast, has been examined at the Imperial Institute.[1]

The oil has the following characters :—

	1.	2.
Specific gravity	0·849	0·8592
Optical rotation	– 44° 40′	– 57° 8′
Acid value	1·2	1·4
Ester ,, ...	1·9	6·4
,, ,, (after acetylation)	52·9	35·5

Monarda didyma.—This plant is indigenous to North America. The dried herb yields about 0·03 per cent. of essential oil, the twigs and leaves about 0·04 per cent. and the half-open flowers 0·26 to 0·32 per cent.

The oils have the following characters :—

No.	From	Specific Gravity at 15°.	Rotation.	Refractive Index.	Acid Value.	Ester Value.
1	Dry Herb . . .	0·902 at 20°	– 10°	—	—	—
2	Half-dry Herb . .	0·8786	– 24° 36′	1·4676	—	—
3	Dry Twigs and Leaves	0·8855	– 32° 38′	1·4689	5·5	20·4
4	Half-open Flowers .	0·8665	– 7° 30′	1·4689	2·4	4·4
5	,, ,, .	0·8740	– 15° 45′	1·4674	4·6	5·6

[1] *Bull. Imp. Inst.*, 13 (**1915**), 346.

Sample No. 1 was examined by Brandel,[1] the remainder by Schimmel & Co.[2] Neither thymol nor carvacrol are present in the oil.

Monarda fistulosa.—This plant is known in North America as the wild bergamot. The fresh herb yields from 0·3 to 1 per cent. of essential oil, and the dried herb, about 2·7 per cent.

The oil has a specific gravity varying from 0·915 to 0·959, according to the part of the plant distilled, and is slightly lævo-rotatory. Its phenol-content is about 50 to 70 per cent.

The oil has been exhaustively examined by E. R. Miller,[3] who finds the following substances present therein: *d*-α-pinene, *l*-α-pinene, *p*-cymene, one or two unidentified terpenes, butyric aldehyde, iso-valeric aldehyde, a third aldehyde, possibly piperonal, carvacrol, thymoquinone, hydro-thymoquinone, dihydroxy-thymoquinone, dihydrocuminic alcohol, an unidentified alcohol, acetic acid, butyric acid, valeric acid, caproic acid, basic substances, and (in the aqueous distillate) acetone, formaldehyde, methyl alcohol, formic acid, and acetic acid.

Oil of Pennyroyal.

The leaves and other parts of the plant *Mentha pulegium* yield the ordinary pennyroyal oil. This plant is a native of most parts of Europe, the Caucasus, Chili, Teneriffe, etc. The volatile oil is of a yellow or greenish-yellow colour, and possesses a strong odour of the plant.

E. M. Holmes[4] gives the following account of this herb:—

"Two forms of the plant are met with in this country, the commonest form having weak, prostrate stems, which root at the joints and form a dense green turf. This is named var. *decumbens*. The flowering stems are sparingly produced, and often lie prostrate on the leafy cushions of the plant.

"The other form, known as var. *erecta*, has stouter upright flowering stems, which end off lateral stolons or prostrate branches near the base. These root sparingly, forming new plants, which gives rise to upright flowering stems during the ensuing year. This is much rarer as a wild plant, but is the best for cultivation, as it can be reaped and tied up into bundles more readily. The stems are usually 6 to 9 ins. long, but under favourable conditions, such as moist soil and a warm climate like that of Devon and Cornwall, will grow to 15 or 18 ins. There are several other varieties of this plant found on the Continent. One variety, *eriantha*, with hairy flowers, occurs at Biarritz, and a similar form has been found near Falmouth, in this country. Another, in which the whole plant is hairy, is found in Sicily, and is named var. *tomentosa*, and a third, covered with rather stiff hairs, named var. *tomentella*, is found near Montpelier and in Hungary. A fourth, with small leaves, occurring in Eastern Europe is named var. *thymifolia*, and on the steppes of Southern Russia a hairless variety with smaller leaves and flowers, has been described as a variety, but as it is an annual plant it has probably the right to specific rank, and should be called *Mentha micrantha*. It has been called *Pulegium micranthum*.

"Of these varieties it is uncertain which are used on the Continent as sources of the oil, but the plant used in Spain, so far as can be judged

[1] *Pharm. Rev.*, 21 (**1903**), 109.

[2] *Bericht*, October, **1904**, 101; October, **1908**, 89; October, **1909**, 78.

[3] Circular 4 (**1918**), University of Wisconsin.

[4] *P. and E.O.R.* (**1911**), 254.

belongs to the variety *eriantha*. The Russian oil is stated to be derived from the *Mentha micrantha*."

It has the following characters:—

Specific gravity . .	0·930 to 0·955
Optical rotation . .	+ 15° ,, + 25°
Refractive index . .	1·4810 ,, 1·4865
Pulegone	not less than 80 per cent. (sulphite method)

The oil is soluble in 1·5 to 2·5 volumes of 70 per cent. alcohol. On fractional distillation not more than 5 per cent. distils below 212°, and from 212° to 216° at least 80 per cent. will come over.

The principal constituent of the oil is the ketone pulegone, which, in very fine specimens, may be present to the extent of over 90 per cent. According to Tétry[1] there are also present *l*-limonene, dipentene, menthol, and menthone.

A Russian pennyroyal oil, distilled from *Pulegium micranthum* has been examined by Butlerow[2] many years ago. It had a specific gravity 0·934. It commenced to boil at 202° and distilled principally at about 227°. He stated that the principal constituent had the formula $C_{10}H_{16}O$, so that it was probably pulegone.

American pennyroyal oil is distilled from a herb indigenous to North America, *Hedeoma pulegoides*. The oil is principally distilled in North Carolina, Ohio, and Tennessee. The dried leaves yield about 3 per cent., and the leaves and stalks together about 1·5 per cent. of essential oil.

The oil has the following characters:—

Specific gravity	0·925 to 0·940
Optical rotation	+ 18° ,, + 35°

It is soluble in 2 volumes of 70 per cent. alcohol.

The oil contains pulegone, up to the extent of about 30 per cent. According to Kremers[3] the oil contains two other ketones, one of which is apparently menthone, and the other he has termed hedeomol, which boils at 168° to 171° and forms an oxime melting at 41° to 43°. He also identified formic acid, acetic acid and isoheptylic acid in traces.

Barrowcliff[4] has recently examined this oil. The oil used in this investigation possessed the following characters: $d_{15°/15°D} = 0·9297$; $a_D + 25° 44'$ in a 1-dcm. tube; soluble in twice its volume of 70 per cent. alcohol. It was found to consist of: (1) an undetermined phenol, in very small amount; (2) *l*-pinene; (3) *l*-limonene; (4) dipentene, all of these terpenes being present in only small amount; (5) 1-methyl-3-*cyclo*hexanone about 8 per cent.; (6) pulegone, about 30 per cent.; (7) *l*-menthone, and (8) *d*-iso*menthone*, identical with the dextro-rotatory constituent of Beckmann's "inverted menthone"; the amount of these two menthones constituting about 50 per cent. of the oil; (9) a sesquiterpene alcohol, about 2 per cent.; (10) esters of formic, acetic, octoic, decylic, and salicylic acids, and the ester of a dibasic acid of the probable formula $C_8H_{14}O_4$, together with formic, butyric, octoic, and decyclic acids in the free state; all these esters and acids being present only in small amount.

Barrowcliff does not appear to confirm the presence of any ketone corresponding with Kremer's hedeomol.

[1] *Bull Soc. Chim.*, iii. 27 (**1902**), 186.
[2] *Jahr. d. Chem*, (**1854**). 594.
[3] *Amer. Jour. Pharm.*. 59 (**1887**, 535.
[4] *Jour. Chem. Soc.*, 91 (**1907**), 875.

Oil of Thyme.

French oil of thyme, which is the most esteemed variety of the oil known, is distilled from the fresh herb, *Thymus vulgaris*. A considerable quantity of thyme oil is also distilled in Spain, but probably from mixed species of thyme, and a certain amount is also distilled in Algeria from *Thymus algeriensis*.

Two commercial varieties are recognised, the "red" and "white" oils. The former is the crude distillate, and probably owes its red colour to the action of the phenols present on the iron of the crude stills and condensers frequently used. The red oil when properly rectified forms the colourless or "white" oil of commerce. The yield of oil is very variable—sometimes being as low as 0·2 per cent., often rising to 1 per cent. in the fresh and 2·5 per cent. in the dried herb. The French oil contains from 20 to 35 per cent. of phenols, chiefly thymol, upon which the value of the oil chiefly depends. The Spanish oil contains a much higher percentage of phenols (50 to 70 per cent.) of varying composition. A few years ago carvacrol was generally the preponderating phenol, but to-day careful selection of the species or varieties distilled has enabled a Spanish oil to be produced, so rich in thymol that this body is actually manufactured from it.

E. M. Holmes has recently shown that there are two Spanish species of thyme now known to yield thymol as the principal phenol. These are *Thymus vulgaris* and *Thymus Zygis*.[1]

French oil of thyme has the following characters:—

Specific gravity . .	0·905 to 0·935 (rarely 0·900)
Optical rotation . .	− 0° 30′ to − 4°
Refractive index . .	1·4800 ,, 1·4950
Phenols (mostly thymol)	20 to 40 per cent. (rarely below 25 per cent.)

The oil is soluble in 2 volumes of 80 per cent. alcohol.

Spanish thyme oil has the following characters:—

Specific gravity	0·928 to 0·958
Optical rotation	+ 2° ,, − 4°
Refractive index	1·5025 ,, 1·5110
Phenols	50 to 75 per cent.

It is soluble in 3 volumes of 70 per cent. alcohol. The phenols of Spanish thyme oil consist principally of carvacrol, but sometimes a fairly large proportion of thymol is present.

J. C. Umney[2] has examined three samples of thyme oil distilled in British East Africa.

The three samples of thyme oil were distilled respectively from plants grown from seeds obtained through Sutton & Co., of Reading, from young plants from the same source, and from French seeds of *Thymus vulgaris*. The following are the characters of the oils:—

No.	Yield.	Specific Gravity.	Phenols.	Refractive Index.	Remarks.
	Per Cent.		Per Cent.		
1	0·50	0·901	32	1·4900	Phenols solidify.
2	0·70	0·905	32	1·4768	,, liquid.
3	0·38	0·905	40	1·4908	,, partly crystallise.

[1] *P. and E.O.R.* **(1920)**, 339. [2] *Ibid.* **(1914)**, 423.

No. 1 was distilled from English thyme grown from Sutton's seeds.

No. 2 was distilled from English thyme grown from young plants.

No. 3 was distilled from French thyme grown from seeds.

It will be seen from the above that the percentage of phenols is highest in the case of the oil distilled from the plant grown from the French seed, and the phenols partially crystallised. In the other cases it is noteworthy that the phenols in the oils distilled from plants grown from English *seeds* consisted chiefly of thymol, whilst the phenols of the oil distilled from the English *plants* could not be induced to crystallise, and therefore appear to consist chiefly of carvacrol.

The phenols are the principal constituents of thyme oil, thymol being the most valuable for medicinal purposes, but carvacrol, an isomeric phenol, preponderating in some oils. Cymene and *l*-*a*-pinene are present in the oil, as well as a little menthone. Borneol and linalol have been detected in the high boiling fractions of the oil, and a crystalline body melting at 169° and having the formula $C_{10}H_{22}O_3$. It is probably identical with a similar body isolated from Juniper berry oil.

Admixture with the oil of *Thymus serpyllum* does not alter the specific gravity, but increases the rotation, as this oil rotates up to − 12°. Oils adulterated with French turpentine have a high laevo-rotation. A very large amount of the French "white" oil contains a considerable quantity of turpentine. The percentage of phenols is indicated by the decrease in volume on shaking with a 5 per cent. aqueous solution of potash in a cassia oil flask, and also by the amount of oil distilling above 220°, which should not fall below 25 per cent. Kremers recommends the following method of estimating the thymol :—

Five c.c. of the oil to be examined is weighed and brought into a glass-stoppered burette graduated to $\frac{1}{10}$ c.c. and is diluted with about an equal volume of petroleum ether; a 5 per cent. potassium hydroxide solution is added, and the mixture shaken for a short time, then the liquid is left standing until separation is complete. Then the alkaline solution is allowed to run into a 100 c.c. graduated flask. This operation is repeated until no further decrease in the volume of the oil takes place.

The alkaline solution of thymol is made up to 100 or 200 c.c. as the case may require, using a 5 per cent. soda solution. To 10 c.c. of this solution in a graduated 500 c.c. flask is added a $\frac{1}{10}$ normal iodine solution in slight excess, whereupon the thymol is precipitated as a dark reddish-brown iodine compound. In order to ascertain whether a sufficient quantity of iodine has been added, a few drops are transferred into a test tube and a few drops of dilute hodrochloric acid are added. When enough iodine is present, the brown colour of the solution indicates the presence of iodine, otherwise the liquid appears milky by the separation of thymol. If an excess of iodine is present, the solution is slightly acidified with dilute hydrochloric acid and diluted to 500 c.c. From this 100 c.c. are filtered off, and the excess of iodine determined by titration with $\frac{1}{10}$ normal solution of sodium thiosulphate. For calculation, the number of cubic centimetres required is deducted from the number of cubic centimetres of $\frac{1}{10}$ normal iodine solution added and the resultant figure multiplied by 5, which gives the number of cubic centimetres of iodine required by the thymol.

Every cubic centimetre of $\frac{1}{10}$ normal iodine solution equals 0·003753 grm. of thymol. Knowing the quantity of thymol in the alkaline solution, the percentage in the original oil is readily found.

The reaction taking place is represented by the equation :—

$$C_{10}H_{14}O + 4I + 2NaOH = C_{10}H_{12}I_2O + 2NaI + 2H_2O.$$

In the estimation of carvacrol a slight modification of this method must be made, because carvacrol is thrown down as a finely divided white precipitate, giving the solution a milky appearance. In order to form a precipitate the liquid is vigorously shaken after the addition of iodine solution, and is subsequently filtered. Then the liquid is acidulated with hydrochloric acid, and subsequently the same procedure is followed as was described for thymol. The calculation is also the same.

Labbe[1] has examined an authentic sample of thyme oil and found it to contain 30 per cent. of thymol, 17 per cent. of a terpene which he could not identify, 15 per cent. of menthene, 21 per cent. of cymene, 5 per cent. of linalol, 8 per cent. of borneol, and traces of carvacrol.

Thymus capitatus also yields an oil of specific gravity about ·900, and closely resembling the above-described oils. It contains pinene, cymene, dipentene, thymol, carvacrol, and bornyl acetate. Its phenol-content is about 6 per cent. A similar oil is obtained from *Thymus camphoratus*. A somewhat different oil is obtained from a variety of *Thymus serpyllum* (*citratus* or *citriodorus*). This oil has an odour recalling thyme, lemon, and geranium. Its specific gravity is 0·890 to 0·920 and its optical rotation − 10° to − 20°. It contains only a very small amount of phenols, but cymene and a sesquiterpene have been isolated from it.

The value of this oil depends so much upon the phenols it contains that it is important that these should be estimated, as the abstraction of thymol, a valuable commercial product, is by no means uncommon.

Thymus Mastichina, the so-called Spanish wood marjoram, yields an oil of thyme which has been examined recently by Dorronsoro.[2] It has a bright yellow colour, turning darker with age, and a camphoraceous odour reminding of thyme. Its constants are as follows : $d\frac{20°}{15°}$ 0·907 to 0·945 ; a_D − 0° 50′ to + 4° 40′ ; $n_{D_{23°}}$ 1·4630 to 1·4654 ; saponification value 12·7 to 18·5 ; ester-content 4·44 to 6·47 per cent. (calculated as linalyl acetate) ; ester value after acetylation 29·2 to 49·3 = 8·2 to 14·09 per cent. alcohol $C_{10}H_{18}O$.

The oil is soluble in any proportion in 94 per cent., in one-half to its own volume in 90 per cent., and in 1 to 3 volumes of 80 per cent. alcohol. It does not always dissolve readily in 70 per cent. alcohol, 30 volumes of this strength being occasionally required to make a solution.

The oil was split up into 8 fractions by distillation under reduced pressure.

Temperature (98 to 100 mm.).	Per Cent.	$n_{D22°}$.	$a_{D20°}$.
Up to 100°	2·65	1·4600	+ 3° 41′
100° ,, 110°	43·00	1·4620	+ 4°
110° ,, 120°	28·53	1·4605	+ 1° 58′
120° ,, 125	6·25	1·4617	± 0°
125° ,, 145°	10·25	1·4657	− 3°
145° ,, 160°	3·75	1·4718	− 4° 10′
160° ,, 175°	3·05	1·4784	− 1° 55′

[1] *Bull. Soc. Chim.* iii. **19**, 1009.

[2] Paper read before the Soc. Espan. Fisic. y Quimica, October, 1910.

The oil has the following composition:—

d-α-pinene	7 to 8 per cent.
Cineol	64 ,, 72 ,,
Phenols	traces
Ketones	,,
Esters	4 to 6·5 per cent. (linalyl acetate)
Free alcohols	8 to 14 per cent.

OILS OF MARJORAM AND ORIGANUM.

Some confusion has existed in regard to the botanical origin of the various origanum or marjoram oils of commerce. This confusion is well cleared up by Holmes,[1] who gives the following details concerning the plants.

The oil known as oil of sweet marjoram is derived from *Origanum majorana*:—

"The plant is known in this country as sweet, knotted, or garden marjoram, to distinguish it from the wild marjoram, *Origanum vulgare*, which differs from it both in appearance and odour. In France *Origanum majorana* is known to botanists and florists as 'Marjolaine douce,' the wild marjoram being called 'Marjolaine sauvage'. In Germany it is known as 'Meiran'. It is supposed to be a native of Mediterranean countries, and has been cultivated as a pot herb from the earliest times, being used for this purpose by the ancient Egyptians. It was introduced into this country from North Africa in A.D. 1573, but it is hardly known in a truly wild state in Europe, and its native country must be regarded as doubtful. Other species were, however, and are still, used in southern Europe.

"Although still cultivated at the present day in almost all countries for culinary purposes, the oil is chiefly distilled in and imported from Spain.

"The plant is a small bushy herb, usually about 12 or 18 ins. high, with small ovate, stalked, greyish-green, entire, somewhat hairy leaves, and small, rounded, stalkless heads of flowers forming greenish clusters at the top of the stems and upper branches, the small white blossoms being rather inconspicuous, and almost hidden among the crowded green bracts. There is a variety of the plant with woody perennial stems which was cultivated by herbalists in this country in the eighteenth century, and was distinguished by the botanist Willdenow as *Origanum majoranoides*. This is the plant that yields the oil distilled in Cyprus."

"Holmes points out that, it is generally recognised that the oil of marjoram coming from a definite port or place of manufacture, is not always identical in composition or constituents. Origanum oil from Trieste sometimes yields carvacrol and sometimes thymol, in different samples, and the same occurs in oils imported from Smyrna and from Cyprus. It would appear, therefore, that more than one species is employed in the manufacture of the oil in each district.

"The oils have been attributed to various species of *Origanum*, but the statements do not afford any definite information in cases in which the botanical authority for the name is not given. Thus there are three different species of *Origanum* bearing the name of *Origanum Creticum*, and two bearing that of *Origanum Smyrnæum*, and two bearing that of *Origanum hirtum*, so that the use of either of these names is of no

[1] *P. and E.O.R.* (1912), 322; (1913), 741.

service for identifying the plant unless the name of the botanical authority follows the name of the plant, *e.g. Origanum Creticum* Linnæus; *Origanum Creticum* Schousboe; or *Origanum Creticum* Sieber.

"When these are given it is possible for any botanist to understand which species is meant, and so far the course is clear. But the matter is further complicated when two different plants are used in the same locality, at different times, or are mixed before distillation. Failure of the crop, or the exhaustion of the wild plant from careless collection, may lead at any time to substitution of the nearest plant having a somewhat similar odour or similar appearance, and then the identification of the botanical source of the oil can only be settled by the botanical examination of a specimen of the plant actually employed for distillation.

"It is now known that the character and percentage of the phenols carvacrol and thymols differ in different species, and may possibly do so in the same species under different conditions, and that as a rule they do not both occur in the same species. But with the aid of botanical descriptions and illustrations it should be possible to select and cultivate experimentally those species which give the highest yield, and to determine the conditions of soil, or the use of particular manure under which the largest yield may be obtained. Until this is done there is little hope of obtaining anything approaching a uniform oil in commerce.

"The principal species from which the oil has actually been distilled and examined are the following:—

Origanum majoranoides Willd.
Origanum onites Linn.
Origanum maru Linn.
Origanum hirtum Link."

The oil of *Origanum majorana*, or sweet marjoram, is produced principally in Spain, where the fresh herb yields 0·3 to 0·4 per cent. of oil. This oil has the following characters:—

Specific gravity	0·895 to 0·910
Optical rotation	+ 15° ,, + 20°
Refractive index	1·4725 ,, 1·4775
Acid value	0 ,, 1·0
Ester ,,	10 ,, 30

The oil is soluble in 2 volumes of 80 per cent. alcohol.

These properties may vary somewhat, and three samples, two distilled in Cyprus and one in London from material obtained from Cyprus, were found to have the following characters:—[1]

	Cyprus.	Cyprus.	London.
Specific gravity	0·899	0·912	0·888
Optical rotation	+ 14° 2′	+ 3° 45′	+ 13° 15′
Saponification value	6·4	8·25	12·8
Solubility in 80 per cent. alcohol	1 in 1·3	1 in 1	1 in 8 to 9

This oil contains about 40 per cent. of terpenes, principally terpinene, terpineol, terpinenol-4, and small quantities of esters. Possibly borneol and camphor are present in traces.

The oil from *Origanum majoranoides* from Cyprus has been examined by Pickles [2] who gives the following results:—

Carvacrol, about 84·0 per cent.; another phenol with a creosote-like odour, 0·2 per cent.; a hydrocarbon origanene $C_{10}H_{16}$, about 2·5 per cent.; cymene and terpenes, boiling-point 170° to 180°, about 8 per

[1] *Bull. Imp. Inst.*, 11 **(1913)**, 50. [2] *Trans. Chem. Soc.* **(1908)**, 876.

cent.; terpene alcohols, boiling-point 100° to 125° C., about 3·5 per cent.; and a residue boiling above 125° C. The origanene, oxidised by peroxide of hydrogen, yielded succinic acid.

This Cyprus oil has the following characters:—

Specific gravity	0·960 to 0·970
Optical rotation	+ 0° 10′ „ + 0° 30′
Phenols (mostly carvacrol)	70 to 85 per cent.

The oil is soluble in 1 to 3 volumes of 80 per cent. alcohol.

The terpene origanene isolated by Pickles has the following characters:—

Specific gravity	0·847
Optical rotation	+ 1° 50′
Refractive index	1·4800
Melting-point of nitrosochloride	91° to 94°
„ „ nitrol benzylamine	104° „ 105°

Smyrna origanum oil is principally the product of *Origanum onites* Linn. The yield of oil is about 1·5 to 2·5 per cent. of the dried herb. The oil has the following characters:—

Specific gravity	0·895 to 0·965
Optical rotation	− 1° „ − 15°
Refractive index	1·4950 „ 1·5230
Phenols	16 to 75 per cent. (usually 25 to 50 per cent.)

The oil contains carvacrol, some thymol, linalol, cymene, an olepinic terpene, and probably α-pinene and camphor. Cedrol has been found by Schimmel & Co. in a sample of the oil, but this may have been an oil adulterated with cedar-wood oil.

Syrian origanum oil is principally derived from *Origanum maru* Linn. This oil has the following characters:—

Specific gravity	0·930 to 0·960 (rarely higher)
Optical rotation	− 0° 50′ to + 1° 35′
Phenols	55 to 75 per cent.

It is usually soluble in 3 volumes of 70 per cent. alcohol.

The phenols consist principally of carvacrol.

The principal species used in the distillation of Trieste origanum oil is *Origanum hirtum* Link. The dried herb yields from 2 to 3 per cent. of oil having the following characters:—

Specific gravity	0·940 to 0·980
Optical rotation	+ 1° „ − 1°
Phenols	60 to 85 per cent.

The oil is soluble in 3 volumes of 70 per cent. alcohols. The phenols consist almost entirely of carvacrol, and there are also present cymene and terpenes.

Origanum vulgare yields from 0·1 to 0·5 per cent. of oil having a specific gravity 0·870 to 0·910, and optical rotation about − 35°. It contains phenols which have not been carefully investigated.

Origanum Creticum yields an oil having the following characters:—

Specific gravity	0·920
Optical rotation	0°
Phenols	44 per cent., mostly carvacrol

It is soluble in 2 volumes of 80 per cent. alcohol. This oil was distilled in Sicily, and examined by Umney and Bennett.[1]

[1] *Pharm. Jour.*, 75 (**1905**), 860.

A Greek origanum oil[1] of unknown botanical origin has been found to have the following characters:—

Specific gravity	0·935
Optical rotation	0°
Refractive index	1·5020
Phenols	63 per cent.

It was soluble in 4 volumes of 70 per cent. alcohol.

The Imperial Institute has examined the oil from a species of origanum from Cyprus found near Lapithos, locally known as "rikhanon". The plant was described by E. M. Holmes as a new species under the name of *O. Bevani.*[2] In August, 1916, 49 lb. of the dried flowering plants without roots were obtained and examined at the Institute. The material, which contained 9·8 per cent. moisture, yielded 1·9 per cent. of volatile oil, which is only about half the yield of the ordinary Cyprus Origanum. The oil when first distilled was pale yellow, but soon acquired a reddish-brown colour; it had the characteristic thyme odour. It had the following characters:—

Specific gravity at 15°	0·951
Optical rotation	+ 0° 24′
Refractive index at 23° C.	1·51
Phenols per cent.	75
Solubility in 70 per cent. alcohol at 15° C.	In 2·7 vols.

Patchouli Oil.

Patchouli oil is distilled from the dried leaves of several types of the patchouli plant, of which the principal is *Pogostemon patchouli.* E. M. Holmes[3] gives the following interesting information as to its botanical source and origin:—

Patchouli has long been known and cultivated in India and the Straits Settlements and various islands of the Indian Ocean, so long indeed, that its original country was lost sight of until quite recently. The first notice of its appearance in European commerce appears to be that given by Mr. J. J. Virey in 1826,[4] who describes it as having been brought from the Isle of Bourbon, where it was evidently cultivated the flowers being unknown, and it was doubtfully referred to as *Plectranthus graveolens,* a native of New Holland. It was spoken of by Mr. Virey as an insectifuge for clothes and shawls.

The first notice of a sale of the leaves in this country is an account given by Dr. J. Pereira[5] of forty-six cases put up for sale at Garraway's Coffee House on 27 June, 1844, by Mr. Ellis, a drug broker of Fenchurch Street. It was packed in cases of two sizes, containing respectively 50 lb. and 100 lb.; 6s. per lb. was asked for it, but no bids were made. It is rather remarkable that this lot was said to have come from New York, where it was reported to have been carried from China. The dried plant as imported consisted of the leafy tops, 1 ft. or more in length, the stems being round and woody and the branches obtusely quadrangular and the leaves serrate, or crenate-serrate. But it was recognised by Dr. Pereira as an East Indian product, for he says that it had been introduced into the Botanic Gardens at Calcutta from Penang, but that up to 1841 it had not flowered there.

[1] Evans' *Analytical Notes*, 1910.
[2] *P. and E.O.R.* **(1915)**, 6, 19.
[3] *Ibid.* **(1913)**, 369.
[4] *Jour. de Pharm.* **(1826)**, xii. 61.
[5] *Pharm. Jour.* **(1)**, iv. 808.

Its introduction into Europe is accounted for by a writer in the *Gardener's Chronicle* [1] by the fact that a few years previously real Indian shawls bore an extravagant price, and purchasers could always distinguish them by their odour of patchouli. The French manufacturers, having discovered the secret of the odour, imported the dried leaves to perfume articles of their own make, and thus palm off their homespun shawls for real Indian.

The true patchouli plant was first described and illustrated in 1845 by Dr. Pelletier-Sautelet, in the *Mem. de la Soc. Roy. des Sciences d'Orleans*, tom. v., No. 6, the plant having flowered at Orleans in February of that year, and the description and figure of the plant were copied into the *Pharmaceutical Journal*.[2] In the spring of 1849 the plant flowered at Kew, and another illustration of the plant, but not quite so characteristic, was published in Hooker's *Journal of Botany*.[3] The obtuse character of the apex of the leaf, and the crenato-lobate margins of the leaf, with the base wedge-shaped and not toothed, are well shown in both.

The plant was supposed by Dr. Pelletier-Sautelet to be hitherto undescribed, in consequence of its not having previously been seen in flower, and he named it *Pogostemon patchouli*. Its native country remained unknown until March, 1896, when Holmes was able to identify it as a wild plant of the Philippine Islands, which had been described by Blanco under the name of *Mentha cablin* in the *Flore de Filipinos* in 1837. This botanical source explains why the plant was known to the Chinese, and why the first importation into this country may have gone to New York from China. One of the uses of the plant appears to be to give its characteristic odour to Chinese and Indian ink.

The true patchouli is known in the Straits Settlements as *Tilam wangi* or *Dhelum wangi* ("sweet patchouli"). A wild patchouli, known as *Tilam outan*, is sometimes added to the genuine patchouli on distillation, but its botanical origin is uncertain.

The oil distilled from Java patchouli leaves has for some years past been a commercial article, and is usually known as Dilem oil, in order to distinguish it from the oil from *Pogostemon patchouli*. Java patchouli is clearly recognised as *Pogostemon Heyneanus*, a species distinct from the true *Pogostemon patchouli*, and the following notes by E. M. Holmes [4] are sufficient to clear up any difficulties resulting from the confusion existing between the various names employed prior to 1896 :—

The typical form of this species has long been cultivated in gardens in India, from the Concan southwards to Coimbatore, under the name of patchouli, and was referred to *Pogostemon Heyneanus* by botanists, but when, in 1843, the origin of the non-flowering cultivated patchouli of the Straits Settlements was shown to be a different plant, and received the name of *Pogostemon patchouli* Pellet., the authors of the *Flora of British India* considered that in consequence of the similarity of odour, the Indian plant *Pogostemon Heyneanus* Wall, must belong to the same species, and altered its name to *Pogostemon patchouli* Dalz. and Gibbs, and that of the Straits Settlements non-flowering plant into *Pogostemon patchouli*, var. *suavis*, hence the *Pogostemon Heyneanus* became known in botanic gardens as *Pogostemon patchouli* Dalz. and

[1] **1849**, 645. [2] **1849 (1)**, viii. 574. [3] 1849, 329, pl. xi.
[4] *P. and E.O.R.* **(1913)**, 418.

Gibbs,[1] and the genuine patchouli, as its var. *suavis*, and it was not until 1896 [2] that it was shown that the true patchouli (*Pogostemon patchouli* Pellet.) was a distinct species and a native of the Philippine Islands. The native Indian patchouli was then referred back again to *Pogostemon Heyneanus* Wall.

It is practically certain that some other leaves from closely allied species are from time to time to be found in "Java patchouli leaves".

Distillation of the leaves in the Straits Settlements is carried out in stills with false-bottom cylinders, usually with steam at a pressure of about 1·3 atmospheres, when a yield of about 1·5 per cent. is obtained. By using higher pressure a larger yield of oil results, but the quality is inferior. By very prolonged distillation up to 2·6 per cent. has been obtained in the native distilleries, but in Europe, where most of the leaves are distilled for perfumery purposes, a yield of 4 per cent. is obtained from the true patchouli leaves, and of about 1 to 1·5 per cent. from the Java leaves. There appears to be no doubt that the question of fermentation of the leaves has an important bearing on the yield of oil.

A. W. K. de Jong [3] has carried out an investigation in regard to the oil obtained from plants during various stages of their development. As

No.	State of the Leaves.	Colour.	Odour.	d_{15°.	α_D.	$n_{D_{20^\circ}}$.	Acid No.	Ester No	Solubility in 90 per Cent. Alcohol.
				I. Singapore Oils.					
1	Fresh	bright yellow	faint patchouli odour	0·9655	− 51° 18′	1·50820	1·5	2·0	Soluble in 0·8 vol. if 1·5 to 5 vols. are added, the solution becomes cloudy and afterwards again clear.
2	Dried	bright yellow	faint patchouli odour	0·9587	− 50° 58′	1·50766	—	1·1	Soluble in 6 to 7 and more vols.
3	Fermented	bright yellow	faint patchouli odour	0·9628	− 52° 33′	1·50784	0·9	1·5	Soluble in 7 and more vols.
				II. Java Oils.					
4	Fresh	bright yellow	terpene-like	0·9344	− 15° 20′	1·50050	0·8	9·9	Soluble in 0·3 and more vols. (soluble in about 10 and more vols. 85 per cent. alcohol).
5	Fresh	bright yellow	faint patchouli odour	0·9450	− 15° 20′	1·50483	—	5·8	Soluble in 0·6 and more vols. (insoluble in 10 vols. 85 per cent. alcohol).
6	Dried	bright yellow	calamus-like	0·9168	+ 3° 15′	1·50030	—	6·1	Soluble in about 10 and more vols.
7	Slightly fermented	bright yellow	calamus-like	0·9229	+ 2° 32′	1·50058	0·8	4·9	Soluble in about 10 vols. with separation of paraffin.
8	Strongly fermented	bright yellow	faint patchouli odour	0·9210	− 0° 26′	1·50207	—	5·0	Soluble in 8 and more vols.

[1] *Pharm. Jour.* (4), xxvi. 349. [2] *Ibid.* (4), ii. 223. [3] *Teysmannia* (1906), No. 6.

a result he came to the conclusion that the best results are to be obtained by cutting the plants when they have five pairs of leaves, and to distil only the leaves, as the leaf-stalks contain so little oil that they do not pay to distil.

The table on opposite page shows the variation in oil distilled from leaves which were either fresh, dried, or fermented before distillation. The oil from fresh leaves appears to be practically valueless.

The above considerations indicate the difficulty in fixing standards for an oil, which varies so much by alterations in the conditions of cultivation, preparation of the leaves, and distillation.

Oils distilled locally from the dried Singapore leaves ("imported oils") will usually have characters falling within the following limits:—

Specific gravity at 15°	0·955 to 0·980
Optical rotation	− 44° ,, − 62°
Refractive index at 20°	1·5065 ,, 1·5130
Acid value	0 ,, 1
Ester	1·5 ,, 8

Soluble in 3 to 10 volumes of 90 per cent. alcohol.

The oil is sometimes less soluble than this, but any marked deviation from these figures should be regarded with suspicion, and the oil should be subjected to minute examination.

Oils distilled in Europe from imported Singapore leaves, which are of much finer odour than most of the native distilled oil, have in general the following characters:—

Specific gravity at 15°	0·965 to 0·995
Optical rotation	− 50° ,, − 68°
Refractive index at 20°	1·5060 ,, 1·5130
Acid value	1 ,, 5
Ester ,,	2 ,, 12

Most samples are soluble in 1 to 2 volumes of 90 per cent. alcohol, although some samples require 4 to 6 volumes for solution.

Dilem oil, or patchouli oil distilled from Java leaves, has an odour distinctly recalling that of calamus oil, and has characters differing considerably from those of the patchouli oil from Singapore leaves. This oil usually has characters falling within the following limits:—

Specific gravity at 15°	0·920 to 0·940
Optical rotation	+ 4° ,, − 38°
Refractive index at 20°	1·5000 ,, 1·5060
Acid value	1 ,, 5
Ester ,,	6 ,, 20

It is soluble, sometimes with turbidity, in 6 to 10 volumes of 90 per cent. alcohol. Oils are, however, from time to time found with values much outside these limits, but how far this is due to the addition of other species of patchouli leaves to the distillation material is not known. Schimmel & Co. have recorded an oil from Java leaves having a specific gravity 0·9659 and an optical rotation − 51° 16′.

A sample of patchouli oil distilled from leaves from Perak has been reported upon by the Imperial Institute, having the following characters:—

Specific gravity at 15°	0·9595
Optical rotation	− 43° 51′
Refractive index	1·5063

Soluble in 7·4 volumes of 90 per cent. alcohol.

Lehmann[1] has drawn attention to the fact that oils of recent distillation have frequently had a lower specific gravity and optical rotation than usual, and also showed a marked decrease of solubility in 90 per cent. alcohol. The amount of oil obtainable from the herb is also decidedly smaller, seldom being above 2 per cent. Two of the typically abnormal oils showed the following constants:—

Specific gravity	0·935 to 0·937
Optical rotation	− 9° „ − 34°
Refractive index	1·5050 „ 1·5070
Saponification value	4·5 „ 6·5
„ „ (after acetylation)	38 „ 42

Both were soluble in 4 to 5 volumes of 90 per cent. alcohol. The odour of the oil was especially fine and strong. The suspicion that the leaves had been adulterated proved to be unfounded. The marked reduction in the yield of the oil and the change in the nature of the constants is probably attributable to atmospheric conditions existing during the time of the growth of the herb.

Patchouli oil is adulterated to a certain extent with the oil from other leaves which are fraudulently packed with patchouli leaves, such as basil leaves and the leaves of a Malayan plant, known as "purpulut" (*Urena lobata*). The leaves of *Hyptis suaveolens* are also used as an adulterant. Sassafras (or heavy camphor oil) is sometimes found as an adulterant, as well as cubeb oil, cedar-wood oil, and—rarely—petroleum.

The earliest chemical examination of this oil was made by Dr. Gladstone in 1864, who stated that it contained a hydrocarbon analogous to that from oil of cubebs, together with a small quantity of an intensely blue colouring matter which he termed cærulein. Gal[2] and Montgolfier[3] have shown that the oil contains a solid body, which they termed patchouli camphor, and which Wallach has shown to be a sesquiterpene alcohol $C_{15}H_{26}O$. This body is probably the oxidation product of other substances in the oil, and appears to have little or no odoriferous value. It is deposited when the oil stands for long in a cold place. When recrystallised it melts at 56°. It is laevo-rotary, the specific rotation in the melted condition being − 118°, or in chloroform solution − 97° 42′. Dehydration gives rise to a hydrocarbon $C_{15}H_{24}$ patchoulene. This hydrocarbon has an odour of cedar-wood oil, boils at 254° to 256°, and has a specific gravity 0·9334 and optical rotation − 36° 52′.

Wallach claimed to have isolated cadinene from the oil, but later researches have shown that this sesquiterpene is absent from the oil, so that the oil examined by Wallach was probably adulterated.

Von Soden and Rojahn[4] have examined samples of patchouli oil of known purity, and have separated it into two main fractions by distillation under reduced pressure. The earlier fractions of specific gravity 0·984 to 1·002 were found to consist chiefly of patchouli alcohol, whilst two other fractions of specific gravity 0·946 and 0·964 were found to consist of high-boiling sesquiterpenes. A series of refractionations succeeded in separating a fraction of specific gravity 0·930 to 0·940 and of optical rotation − 50°, and one of the same specific gravity but of rather higher boiling-point. From the former, after hydrolysis, a sesquiterpene was obtained in a state of purity which boiled at 264° to 265° at 750 mm.,

[1] *Chem. Zeit.* (**1913**), 37, 1589.
[2] *Comptes rendus*, 68 (**1869**), 406.
[3] *Ibid.*, 84 (**1877**), 88.
[4] *Berichte*, 37 (**1904**), 3353.

and had a specific gravity 0·9335 and an optical rotation − 58° 45′. This latter fraction yielded a sesquiterpene of specific gravity 0·930, optical rotation + 0° 45′, and boiling-point 273° to 274° at 760 mm.

According to Schimmel & Co.[1] about 97 per cent. of patchouli oil consists of bodies of no value for odour purposes. They have detected the following constituents in patchouli oil: benzaldehyde; eugenol; cinnamic aldehyde; a terpenic alcohol of rose-like odour and of undetermined constitution; a ketone of caraway odour forming a semicarbazone melting at 134° to 135°; and a base of stupefying odour not yet identified. But as all these bodies are present in traces only it cannot be said that our knowledge of the odour bearers present in patchouli oil is complete.

Simmons[2] records two samples of commercial oil to which some artificial ester, or oil containing a high proportion of esters, had been added. These two samples had the following characters:—

	1.	2.
Specific gravity	0·9948	0·9937
Rotation	− 38° 30′	− 49° 30′
Refractive index	1·5175	1·5110
Acidity	trace	trace
Ester value	58	18·5
Solubility in 90 per cent. alcohol	1 in 0·75	1 in 0·5

Melissa Oil.

True melissa oil, or oil of balm, is the distillate of the herb *Melissa officinalis*, a plant indigenous to the Northern Mediterranean littoral and Western Asia. It is also cultivated in North America. The leaves have an odour recalling that of a mixture of lemon grass and citronella. The yield of oil is, however, so small that the melissa oil of commerce is practically invariably the result of the distillation of lemon oil with a little citronella oil (or citronellal) over the leaves of the plant.

Schimmel & Co.[3] have examined the pure oil, distilled from the fresh herb, and found two oils to have the following characters:—

	From Fresh Herb at Commencement of Flowering.	From Fresh Herb in Full Flower.
Yield	0·014 per cent.	0·104 per cent.
Specific gravity	0·924	0·894
Rotation	+ 0° 30′	± 0°

The oil contains citral, and probably a little citronellal. Flatau and Labbé[4] state that a sample examined by them contained the following:—

Geraniol	20 per cent.
Linalol	12 ,,
Citronellol	6 ,,

Oil of Sage.

Sage oil is obtained from the herb *Salvia officinalis*, a plant indigenous to the Northern Mediterranean littoral. The oil is distilled to a considerable extent in Dalmatia, and recently a fair amount of oil has been manufactured in Spain but from a different species of Salvia. The leaves and twigs of the plant yield from 1 to 3 per cent. of oil. A certain amount of oil is also distilled in Germany.

[1] *Report*, April, **1904**, 68.
[2] *Chemist and Druggist* (**1904**), 815.
[3] *Bericht*, October, **1894**, 37.
[4] *Bull. Soc. Chim.* (**1898**), ii. 636.

The oil distilled in Dalmatia and in Germany is of typically sage odour and is used for flavouring purposes, whilst that distilled in Spain is distilled from *Salvia trilobia*, with possibly other species as well, and more resembles spike lavender in odour than the ordinary sage of the garden.

Pure Dalmatian or German sage oil has the following characters :—

Specific gravity	0·910 to 0·932
Optical rotation	+ 2° „ + 25°
Refractive index	1·4575 „ 1·4690
Esters as linalyl acetate	2 to 6 per cent.
Total alcohols as borneol	20 „ 25 „

It is soluble in 2 volumes of 80 per cent. alcohol.

Spanish sage oil has the following characters :—

Specific gravity	0·912 to 0·932
Optical rotation	+ 10° to + 20° (rarely laevo-rotatory)
Esters as linalyl acetate	5 to 20 per cent.
Total alcohols as borneol	20 „ 29 „

It is soluble in 6 volumes of 70 per cent. alcohol.

On fractionating German oil of sage H. Seyler[1] obtained 1 to 2 per cent. of a first fraction boiling below 155°, which could be separated into three portions by repeated distillation over sodium. The fraction first passing over, of the boiling-point 142° to 145° ($d_{20°} = 0·80$; $n_D = 1·4438$; $a_D = + 1° 40'$), gave analyses corresponding to $C_{10}H_{18}$. This hydrocarbon, called "salvene," when oxidised with potassium permanganate, yielded an acid, whose semicarbazone $C_{10}H_{16}O_2CON_3H_3$ melted at 204°, and which possibly is identical with β-tanacetoketonic acid (β-thujaketonic acid).

Salvene has probably the following constitutional formula :—

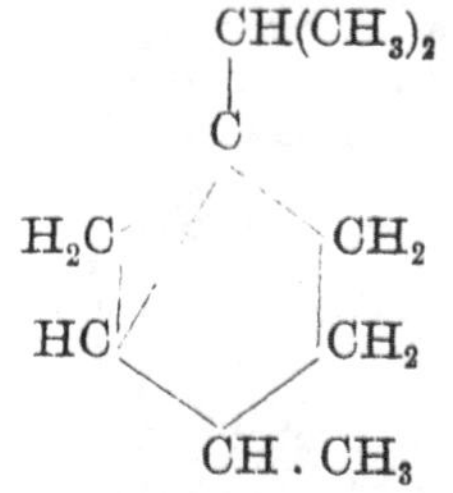

From the first runnings of a Spanish oil of sage no salvene could be isolated.

Pinene and cineol are probably present in sage oil in small amount, together with borneol, a small quantity of esters, and the ketone thujone. Muir and Sugiura isolated a body which they named salviol, which is now known to be identical with the body which has been named thujone. Dextro-camphor is also present in traces. English distilled sage oil has been said to contain cedrene, but the authenticity of the sample examined is doubtful.

Salvia cypria, a native of the Island of Cyprus, yields an essential oil having a camphoraceous odour. Two samples[2] gave the following characters :—

[1] *Berichte*, 35 **(1902)**, 550. [2] *Bull. Imp. Inst.* 11 **(1913)**, 429.

	1.	2.
Specific gravity	0·9263	0·925
Optical rotation	− 6° 31′	− 22° 23′
Saponification value	13·9	8
" " (after acetylation)	38·9	36

The oil contains about 75 per cent. of eucalyptol.

Salvia sclarea also yields an oil with a highly aromatic odour, recalling that of ambergris. It is known commercially as *clary* oil or muscatel sage.

French oil of clary has the following characters :—

Specific gravity	0·895 to 0·930
Optical rotation	− 10° " − 63°
Refractive index	1·4640 " 1·4750
Ester value	110 " 206
Esters as linalyl acetate	38 to 73 per cent.

It is soluble in 2 volumes of 90 per cent. alcohol.

German clary oil has the following characters :—

Specific gravity	0·910 to 0·960
Optical rotation	− 25° " − 48°
Ester value	18 " 160

It is soluble in 2 volumes of 90 per cent. alcohol.

Pinene, cineol, and linalol have been isolated from this oil.

Salvia mellifera (*Ramona stachyoides*) is a labiate plant found in Southern California, and known as black sage. Rabak[1] first examined the oil which he obtained to the extent of 0·75 per cent. of the green flowering plant. He found the oil to have the following characters :—

Specific gravity at 24°	0·9144
Optical rotation	+ 30·2°
Refractive index at 24°	1·4682
Acid value	2
Ester "	2·5
" " (after acetylation)	27·1

Rabak detected *d*-camphor, cineol, pinene, theyone, and borneol in the oil, as well as traces of free formic acid, and formic and acetic acids in the form of esters.

Burke and Scalione[2] have also examined the oil, obtaining 0·9 per cent. from the leaves and twigs. The oil had the following characters :—

Specific gravity at 15°	0·8979
Optical rotation	+ 24·4°
Refractive index	1·4729
Acid value	2·2
Ester "	1·6

These chemists identified the following bodies in the oil : *a*-pinene, cineol, thujone, camphor, and inactive terpenes, probably consisting of a mixture of dipentene and terpinene.

Satureja Oils.

Savory Oil.—The oil distilled from the green herb, winter savory, *Satureja montana*, has been examined by Haller,[3] who states that it is an orange-yellow oil of aromatic taste, somewhat resembling true origanum oil, of specific gravity 0·9394 at 17° and optical rotation − 3° 25′.

[1] *U.S. Dept. Agric. Bur. Plant Industry*, **235**, 14.
[2] *Jour. Ind. Eng. Chem.*, 6 (**1914**), 804. [3] *Comptes rendus*, 94 (**1882**), 132.

He found in it from 30 to 40 per cent. of carvacrol, and traces of another phenol. Two hydrocarbons at least are present, probably cymene and a terpene. A sample examined by Schimmel & Co. had a specific gravity ·939 and optical rotation − 2° 35′. It was soluble in 4·5 volumes of 70 per cent. alcohol, and contained 65 per cent. of phenols. Two samples distilled in the South of France, and examined by Schimmel & Co., had the following characters:—

	1.	2.
Specific gravity	0·908	0·919
Optical rotation	− 1° 42′	− 4° 48′
Refractive index	1·4949	—

The oil of the summer savory, *Satureja hortensis*, does not differ greatly from that of the winter savory. It has the following characters:—

Specific gravity	0·900 to 0·933
Optical rotation	+ 0° 4′ ,, − 1°
Phenols	35 to 45 per cent.

An oil examined by Jahns, having a specific gravity 0·898, was found to contain 30 per cent. of carvacrol, with traces of a second phenol, cymene, and one or more terpenes.

Satureja cuneifolia, a plant grown on the Dalmatian Islands, yields an oil which has been examined by Schimmel & Co.[1] Four samples had the following characters:—

Specific gravity.	Rotation.	Refractive Index.	Phenols.
0·9182	− 4° 45′	1·4882	28 per cent.
0·9190	− 5° 15′	1·4982	34 ,,
0·9444	− 2° 15′	1·4058	59 ,,
0·9444	− 1° 50′	1·4056	59 ,,

Carvacrol and cymene are the principal constituents of the oil.

Satureja thymbra yields an essential oil having the following characters:—

Specific gravity	0·905
Thymol	about 20 per cent.

Pinene, cymene, dipentene, and bornyl acetate are also present in the oil.

Satureja Calamintha (*Calamintha nepeta*).—A number of oils have been described under this name, but in most cases the botanical source was not traced with certainty. A sample of absolute authenticity has been examined by Roure-Bertrand Fils,[2] who reported on the oil as follows: the plants yielded 0·1426 per cent. of essential oil, and were grown in Sicily. The odour of the oil, which is brownish in colour, recalls that of pennyroyal. The oil had the following characters:—

Specific gravity at 15° C.	0·9249
Optical rotation	+ 17° 48′
,, ,, of the acetylated oil	+ 9° 40′
Acid value	1·4
Coefficient of saponification	12·6
Esters (as menthol acetate)	4·4 per cent.
Coefficient of saponification of the acetylated oil	48·5
Total alcohols (as menthol)	14·0 per cent.

Soluble in 0·5 volume and over of 80 per cent. alcohol

,, ,, 2 volumes of 75 per cent. alcohol, then slight opalescence

,, ,, 3 ,, ,, 70 ,, ,, ,, cloudy

[1] *Bericht*, October, **1911**, 108. [2] *Bulletin*, October, **1912**, 68.

The estimation of the pulegone, carried out by means of neutral sodium sulphite, shows 20 per cent. of this ketone.

Two Dalmatian oils examined by Schimmel & Co, had the following values:—

	1.	2.
Specific gravity	0·9305	0·9395
Optical rotation	+ 2° 50′	+ 6° 28′
Refractive index	1·4844	1·4892
Acid value	0·9	—
Ester „	5·4	14·6

The oil contained about 45 per cent. of pulegone.

A French distilled oil had characters very similar to those of the Dalmatian oil.

Umney & Bennett[1] have reported on an alleged Sicilian oil, which contained only a very small amount of ketones. The characters of the oil were as follows:—

Specific gravity	0·922
Optical rotation	+ 14°
Esters	4·2 per cent.
Total alcohols	18·2 per cent. as menthol
Ketones	10·8 per cent.

Satureja hortensis.—This plant yields about 0·1 per cent. of essential oil, which has the following characters:—

Specific gravity	0·898 to 0·930
Optical rotation	+ 0° 4′ „ – 0° 56′
Phenols	36 to 45 per cent.

The oil contains carvacrol, cymene, and a terpene.

Satureja calamintha subsp. *silvatica.*—This plant is known in middle Europe as Mountain Balm. It yields an essential oil having the following characters:—[2]

Specific gravity	0·8759 to 0·8771
Optical rotation	17° „ – 28°
Refractive index	1·4911 „ 1·4951
Acid value	0
Ester „	4·5 to 8·3
„ „ (after acetylation)	39

Its constituents have not been investigated.

Oil of Basil.

Sweet Basil Oil.—This oil is distilled from the fresh herb *Ocimum basilicum*, which yields about 0·05 per cent. of essential oil. This oil has a most excellent fragrance, and is used in the preparation of mignonette extract and similar perfumes.

A certain amount of Basil oil is distilled in Germany, France, Spain, and Algeria, but a fair quantity is also now distilled in Réunion, Java, and Mayotte.

E. G. and C. Camus[3] have recently investigated the botanical relations of the cultivated species of *Basilicum*. There are several varieties of *Ocimum basilicum* L. The variety *purpurascens* Benth., which is regarded by several authors as a distinct species, is cultivated under the name of violet-red basil.

[1] *Chem. and Drug.*, **67**, 970.

[2] Schimmel, *Bericht*, April, **1901**, 61; October, **1905**, 11.

[3] Roure-Bertrand Fils, *Bulletin*, October, **1910**, 23.

The var. *thyrsiflorum* Benth. is known as the common white basil. The var. *album* Benth. (*Ocimum album* L.; *O. laxum*, *O. americanum*) is cultivated under the name of lettuce-leaf basil, while the var. *crispum* (*Ocimum crispum* Thunb.?) is grown under the name of curly-leaved basil.

The authors treat in detail of the morphological and anatomical conditions of the various degenerate species, and in conclusion describe the different essential oils, of which the properties are summarised in the table below.

The variety most suitable for cultivation is *Ocimum crispum* (curly-leaved basil), because it yields the highest percentage of oil, while so far as its odour is concerned the essential oil from this variety is not inferior to the ordinary commercial product.

The oil from the lettuce-leaved basil (var. *album*) gives a slightly cloudy solution with 70 per cent. alcohol; all the other oils give clear solutions when mixed with from 2 to 3 times their volume and more of 70 per cent. alcohol. The methylchavicol (estragol) content of all the oils (acertained by the determination of the methoxyl-group) was about 55 per cent.

Constants.	*Ocimum Basilicum.*					
	Var. *Thyrsiflorum.*		Var. *Crispum.*	Var. *Album.*	Var. *Purpurascens.*	
	Crop 1910.	Crop 1909.	Crop 1910.	Crop 1910.	Crop 1910.	Crop 1909.
$d_{15°}$	0·9168	0·9125	0·9123	0·8975	0·8959	0·9100
α_D	− 10° 52′	− 11° 28′	− 10° 30′	− 12° 54′	− 13° 52′	− 10° 14′
α_D of the Acetylated Oil	+ 1° 36′	+ 1° 34′	+ 0° 30′	+ 0° 2′	+ 1° 52′	+ 3° 28′
$n_{D_{20°}}$	1·488	1·484	1·484	1·479	1·477	1·477
Acid Value	2·8	1·4	2·1	0·7	3·5	2·1
Ester Value	5·6	3·5	4·9	7·0	9·8	7·7
Ester-content calc. as Linalyl Acetate	1·96 p. c.	1·22 p. c.	1·71 p. c.	2·45 p. c.	3·43 p. c.	2·68 p. c.
Ester Value after Acetylation	116·9	114·8	116·9	130·2	126·7	117·6
Alcohol-content calc. as Linalol	35·19 p. c.	34·50 p. c.	35·19 p. c.	39·66 p. c.	38·46 p. c.	35·43 p. c.
Yield	0·0855 p. c.	—	0·1285 p. c.	0·0780 p. c.	0 0370 p. c.	—

Normal samples of German or French distilled Basil oil have the following characters:—

Specific gravity	0·900 to 0·930
Optical rotation	− 6° „ − 20°
Refractive index	1·4800 „ 1·4950
Acid value	1 „ 4
Ester „	2 „ 12

Réunion oil, which is derived from an unknown variety, has the following characters:—

Specific gravity	0·940 to 0·990
Optical rotation	+ 1° „ + 12°
Refractive index	1·5150 „ 1·5175
Acid value	1 „ 3
Ester „	8 „ 25

It is soluble in 8 volumes of 80 per cent. alcohol, sometimes with slight turbidity.

Two samples of Mayotte oil examined by Roure-Bertrand Fils [1] had the following values :—

	1.	2.
Specific gravity at 15° C.	0·9677	0·9630
Optical rotation	+ 0° 58′	+ 0° 56′
Solubility in 80 per cent. alcohol	3 volumes and over	3·2 volumes and over
Acid value	1·4	0·7
Coefficient of saponification	5·6	6·3
Esters (as linalyl acetate)	1·9 per cent.	2·2 per cent.

Two varieties of Java oil have been examined in the laboratories at Buitenzorg. One is that from a large-leaved variety known as "Selasih Mekah". The oil had a specific gravity 0·900 at 26° and an optical rotation of − 15° to − 18°. It contained about 30 to 40 per cent. of phenols, of which eugenol is stated to be the principal.

A sample of the oil distilled in the Seychelles [2] had the following characters :—

Specific gravity	0·962
Optical rotation	+ 0·82°
Refractive index at 21°	1·5140
Acid value	0·8
Ester ,,	2·5

Another variety of the plant yielded an oil of specific gravity 0·948 at 25° and containing much methyl-chavicol. This plant is known as Selasih hidjan.

Dupont and Guerlain [3] have identified methyl-chavicol (estragol) and *l*-linalol in the oil. Bertram and Walbaum [4] isolated cineol, *d*-α-pinene, and *d*-camphor.

The Java oil contains cineol, estragol, pinene, an olefinic terpene (ocimene ?), and eugenol.

Roure-Bertrand Fils [5] have examined the oils of *Ocimum canum* and *Ocimum gratissimum*, distilled at Dakalba (Ivory Coast).

The essential oil of *Ocimum canum* Sims (*O. americanum* L.) consists, when slightly heated in the hand, of a mobile liquid, highly refractive, slightly yellowish. Its odour is peculiar and rather pleasant. At the time of its preparation it collects at the bottom of the receiver. At the ordinary temperature it soon gives a deposit of large acicular crystals. The separation of crystals is sufficiently abundant to cause the whole product to become semi-solid.

The oil has the following characters :—

Specific gravity at 32° C.	1·0330
Optical rotation at 30° C.	− 2° 30′
Acid value	0
Coefficient of saponification	301·4

Soluble in all proportions in 80 per cent. alcohol.
,, ,, 2¼ volumes and over of 75 ,, ,,
Solution slightly cloudy in 5½ volumes and over of 70 per cent. alcohol.

It was found to contain about 87 per cent. of methyl cinnamate, of which ester the crystals above mentioned consist.

The oil from *Ocimum gratissimum* has a specific gravity 0·9105 and an optical rotation + 0° 58′. It contains thymol and carvacrol.

[1] *Bulletin*, October, **1912**, 71.
[2] *Bull. Imp. Inst.*, xvi., I, 30.
[3] *Comptes rendus*, 124 (**1897**), 300.
[4] *Arch. Pharm.*, 235 (**1897**), 176.
[5] *Bulletin*, October, **1913**, 18.

A sample of oil from *Ocimum gratissimum* distilled in the Seychelles [1] had a specific gravity 0·995, optical rotation − 14·1°, and refractive index 1·5260 at 21°.

The essential oil from the leaves of *Ocimum viride* from Sierra Leone has been examined by Goulding and Pelly.[2] The oil had a specific gravity 0·9115 and optical rotation + 1° 30′. By distillation at atmospheric pressure the following fractions were obtained :—

Boiling-point.	Per Cent.	Specific Gravity.	Rotation.
165° to 180°	25	0·8614	+ 0° 33′
180° ,, 200°	20	0·8804	+ 1° 40′
200° ,, 220°	20	0·9164	+ 2° 38′
220° ,, 235°	20	0·9548	+ 3° 38′
235° ,, 250°	10	0·9565	+ 5° 14′

It is obvious, however, from an examination of the average optical rotations of the fractions as compared with that of the original oil, that decomposition had gone on during the distillation. The authors found present 32 to 65 per cent. of phenols, 40 per cent. of alcohols, 2 per cent. of esters, and one or more terpenes. Two samples from the Seychelles had specific gravities 0·924 and 0·942; and contained 52 per cent. and 62 per cent. of phenols, mostly thymol.

Bacon [3] has examined the oil from *Ocimum sanctum*, which he obtained to the extent of 0·6 per cent. from the leaves, and which had the following characters :—

Specific gravity $\frac{30^\circ}{30^\circ}$	0·952
Optical rotation	0°
Refractive index at 30°	1·5070
Saponification value	2·8

Methyl-chavicol, cineol, and linalol are present in the oil.

Bhaduri [4] has examined the oil of *Ocimum pilosum*. The whole plant when fresh is fragrant, but quickly loses its odour when dried. The seeds form a gelatinous mass when steeped in water; this mucilage is employed medicinally in India. The green seeds contain more essential oil than the leaves, but the whole plant was distilled. The oil obtained was pale yellow in colour and very mobile. Its characters were as follows :—

Specific gravity	0·887 at 25°
Optical rotation	− 3·7°
Refractive index	1·4843 at 24·5°

It contains limonene, citral, citronellal, cineol, and a little thymol, as well as other undetermined phenolic constituents. Unlike other Indian basil oils, it contained no methyl-chavicol.

OILS OF MOSLA.

The Japanese plant *Mosla Japonica* yields about 2 per cent. of essential oil containing 44 per cent. of thymol. Its specific gravity is about 0·920. Muryama [5] has identified thymol, carvacrol, and pinene in the oil.

Hoshino [6] states that although thymol and carvacrol are constituents of the oil, they are never found together in the same oil. A thymol containing oil had the following characters :—

[1] *Bull. Imp. Inst.*, xvi., I, 30.
[2] *Ibid.*, 6 (**1908**), 209.
[3] *Philipp. Jour. Sci.*, 5 (**1910**), 261.
[4] *J. Amer. Chem. Soc.* (**1914**), 36, 1772.
[5] *Pharm. Zentral.* 51 (**1910**), 35.
[6] *J. Chem. Ind. Tokyo* (**1919**), 22, 557.

Specific gravity	0·9154
Refractive index	1·4996
Optical rotation	+ 1°
Ester value	4·5
Thymol	50 per cent.

He identified phellandrene, *p*-cymene, γ-terpinene, caryophyllene, cadinene, and probably sabinene in the oil.

Mosla punctata yields about 1 per cent. of oil, calculated on the air-dried herb, of a fragrant odour, having the following characters:—

Specific gravity	0·8966 at 16°
Optical rotation	− 9°
Ester value	16
,, ,, (after acetylation)	36·4

It does not contain any phenols, but owes its odour to a ketone, which has been named α-mujone. It also contains a sesquiterpene, not yet identified, having a specific gravity 0·9259, optical rotation + 2·16°, and refractive index 1·51615.

The herb, *Mosla grosserata*, yields an oil which appears to be very variable in character. It is known in Japan as oil of Himeshiso. The fresh herb, according to Furukawa and Tomizawa,[1] yields 0·24 per cent. of an oil containing 25 per cent. of carvacrol, some thymoquinone and some thymohydroquinone. It also contains *p*-cymene, phellandrene, and terpinene. Murayama, however, has examined four samples, three from wild and one from cultivated plants.[2] These oils had the following characters:—

	Wild Plants.			Cultivated Plants.
	1.	2.	3.	4.
Specific gravity	0·998	0·954	1·0488	0·9006
Optical rotation	− 3·75°	− 3·2°	− 11·75°	− 4·5°
Saponification value	33	23	—	37
,, ,, (after acetylation)	66·6	54·5	—	49·2
Phenols	20 per cent.	—	—	13 per cent.

In the oils from the wild plants the following constituents were identified: thymol, methyl-eugenol, *p*-cymol, thymohydroquinone, myristicin, a terpene and a sesquiterpene. Thymol, thymohydroquinone and a terpene were identified in the oil from the cultivated plants.

Oil of Hyptis.

Hyptis suaveolens is found in South America, the Philippines, Java, and China. Bacon [3] has obtained from it 0·0135 per cent. of essential oil, having a powerful odour of peppermint. The oil distilled in Java has the following characters:—

Optical rotation	− 16° 18′
Acid value	0·7
Ester ,,	14
,, ,, (after acetylation)	31·7

[1] *J. Chem. Ind. Tokyo* **(1919)**, 22, 382. [2] *Pharm. Jour. Japan* **(1920)**, 459, 389.
[3] *Philipp. Jour. Sci.*, 4 **(1909)**, 130.

Perilla Oils.

The leaves of *Perilla arguta*, known in Japan as Shiso, yield an essential oil of a pale yellow to greenish colour, of a peculiar hay-like odour, and having the following characters :—[1]

Specific gravity	0·9265 to 0·939
Optical rotation	− 90° to − 98°
Refractive index	1·4983
Aldehydes	50 per cent. (neutral sulphite method)

The aldehyde present in the oil has been examined by Semmler and Zaar[2] who find it to be a dihydrocumic aldehyde, $C_{10}H_{14}O$, and which has been named perillic aldehyde. It has the following characters :—

Boiling-point at 10 mm.	104° to 105°
Specific gravity at 18°	0·9617
Specific rotation	− 146°
Refractive index	1·50746

Gattefossé considers that geraniol is present in the oil in the form of esters. The saponification value is very high, due partially to the presence of esters and partially to the action of alkali on the aldehyde present.

The dry leaves of *Perilla citriodora* yield from 2 to 3 per cent. of an oil of specific gravity 0·911 to 0·913, containing about 60 per cent. of citral. From the lower boiling fractions Kondo and Yamaguchi[3] have isolated a body which they term perillene, of the formula $C_{10}H_{14}O$, and having the following characters :—

Specific gravity at 20°	0·9017
Optical rotation	± 0°
Refractive index at 21°	1·4705
Boiling-point	185° to 186°

It is a complex body containing a furane nucleus. The oil also contains a sesquiterpene boiling at 251°.

Oil of Hyssop.

Hyssopus officinalis is a plant indigenous to the Mediterranean countries and Central Asia. The fresh plant yields, on distillation, up to 0·3 per cent. of essential oil having the following characters :—

Specific gravity	0·925 to 9·945
Optical rotation	+ 1° „ − 25°
Refractive index	1·4730 „ 1·4860
Acid value	0·8 „ 2
Ester „	3 „ 15
„ „ (after acetylation)	37 „ 70

An elaborate investigation of hyssop oil has been made by Schimmel & Co.[4] The following oils have been distilled by themselves :—

I. Oil from blossoming dry herb: $d_{15°}$ 0·9377 ; α_D − 22° 30′ ; acid number 1·8 ; ester number 5·8 ; ester number after acetylation 44·0. Not soluble in 10 volumes 70 per cent. alcohol ; soluble in about 6 and more volumes 80 per cent. alcohol, with separation of paraffin.

II. Oil from blossoming withered herb : $d_{15°}$ 0·9322 ; α_D − 22° 23′ ; acid number 1·3 ; ester number 3·6 ; ester number after acetylation 37·3 ;

[1] Schimmel's *Report*, October, **1910**, 146.
[2] *Bericht*, 44 (**1911**), 52.
[3] *J. Pharm. Soc. Japan* (**1919**), 446, 263.
[4] *Report*, April, **1908**, 57.

$n_{D_{20^\circ}}$ 1·48315; not soluble in 10 volumes 70 per cent. alcohol; soluble in about 8 volumes and more 80 per cent. alcohol with slight turbidity (separation of paraffin); soluble in every proportion in 90 per cent. alcohol.

III. Oil from highly faded herb which had ceased blossoming: d_{15° 0·9336; α_D − 20° 26′; $n_{D_{20^\circ}}$ 1·48441; acid number 1·8; ester number 3·1; soluble in about 7 and more volumes 80 per cent. alcohol with slight turbidity (separation of paraffin).

For the following investigation they used oil III. After repeated fractional distillation, the lowest-boiling fractions were combined and distilled several times over sodium. The portions obtained at a pressure of 758 mm. showed the following properties:—

B.-p. 164° to 166°, 3·1 per cent. of the oil,	d_{15° 0·8650,	α_D − 19° 29′,	$n_{D_{20^\circ}}$	1·47548
,, 166° ,, 168°, 5·6 ,, ,, ,,	d_{15° 0·8645,	α_D − 20° 16′,	$n_{D_{20^\circ}}$	1·47705
,, 168° ,, 170°, 2·6 ,, ,, ,,	d_{15° 0·8632,	α_D − 20° 15′,	$n_{D_{20^\circ}}$	1·47753

All three fractions had a turpentine-like odour, and on oxidation with permanganate in the presence of free sodium hydroxide, yielded sodium nopinate. They consequently contained β-pinene (nopinene), which was thus for the first time detected in large quantity in an essential oil. The fraction of the boiling-point 164° to 166° contained the largest quantity of β-pinene, so that the constants for this fraction possibly approach those of the hydrocarbon most closely. The free nopinic acid obtained from the sodium salt crystallised from benzene in long needles with a silken lustre, of the melting-point 126° to 127·5°.

The principal constituent of the oil is a saturated ketone of the formula $C_{10}H_{16}O$, which is the active variety of pino-camphone, which has been prepared in the inactive form synthetically, by Wallach, by the reduction of nitrosopinene. The characters of the two bodies are as follows:—

i Pinocamphone (Wallach):	*l*-Pinocamphone from hyssop oil:
Boiling-point 211° to 213°.	Boiling-point 211° to 212°.
d_{21° 0·959.	d_{15° 0·9662.
n_D 1·47273.	$n_{D_{20^\circ}}$ 1·47421.
Mol. refr. 44·44.	Mol. refr. 44·40.
Optical rotation 0°.	Optical rotation − 13° 42′.

Oil of Elsholtzia Cristata.

The dry herb, *Elsholtzia cristata*, yields 2 per cent. of an oil which is esteemed in Japan as an antipyretic and diuretic. The oil has a golden-yellow colour and has the following characters:—

Specific gravity	0·970
Optical rotation	− 2·7°
Acid value	0
Ester ,,	1·8
,, ,, (after acetylation)	14·7

The oil contains a ketone, which has been examined by Asahina and Murayama,[1] and which has the following constants:—

Boiling-point at 10 mm.	87° to 88°
,, ,, ,, 760 ,,	210°
Specific gravity	0·9817
Optical rotation	0°
Refractive index	1·4842

[1] Schimmel's *Report*.

It has the formula $C_{10}H_{14}O_2$, and forms a semicarbazone melting at 171° and an oxime melting at 54°.

SPANISH VERBENA OIL.

The so-called Spanish Verbena oil was, until recently, believed to be derived from a verbena species. E. M. Holmes,[1] however, has definitely identified it as a Labiate plant, *Thymus hyemalis*. He has shown that the plant, although called Verbena, does not belong to the Verbena family, but to the Labiate plants, and is nearly allied to the Lemon Thyme cultivated in gardens in this country. In habit, however, it is quite different to Lemon Thyme, having a stiff, rigid, erect growth with woody stems, and small, short, linear leaves, in small axillary tufts along the whole length of the stem, giving it quite a heath-like appearance. Until quite recently much confusion has existed between the different species of thyme found wild on the Continent, the species either hybridising or varying according to the environment, so that there has been considerable difficulty in discriminating between species and varieties. The full description of the plant is as follows:—

Thymus hyemalis J. Lange: An undershrub, ½ to 1 ft. high, flexuosely branched from the base, with erect branchlets, the younger twigs hoary with white hairs, the leaves short and linear with revolute margins, crowded into small tufts in the axils of the leaves, the flowers forming small crowded heads at the tips of the branchlets. The calyx is purplish and two-lipped, the upper lip consisting of two subulate teeth, the lower with three short ovate acuminate teeth, the corolla of a deep rose colour, and nearly twice as long as the calyx.

This species is related to *Thymus hirtus* W. and *Thymus vulgaris* L., and has been placed by Boissier under *Thymus mastichina* L., and by Pourret under *Thymus sparsifolius*, var. *hyemalis*, but by J. Lange it is considered to be a distinct species. Its heath-like habit of growth is, at all events, very distinctive, as well as the small heads of flowers, each of which is nearly stalkless.

The oil is frequently known as Spanish Thyme-Lemon oil. The author and Bennett,[2] have examined a sample, which had the following characters :—

The oil is of a yellow colour, and has an odour recalling that of thyme, but with a strong flavour of lemon, differing in this respect from the oil derived from *Thymus serpyllum* (wild thyme). The latter oil also differs in being laevo-rotatory. The following are the chief characters of the sample examined :—

Specific gravity at 15° C.	0·901
Optical rotation in 100 mm. tube	+ 18° 30′
Aldehydes (principally citral)	20 per cent.
Proportion absorbed by 5 per cent. potash solution	10 ,,
Refractive index at 19° C.	1·4808
,, ,, of first 80 per cent. distilled	1·4779
,, ,, of 20 per cent. residue	1·4980

The portion absorbed by potash solution on separation proved to be somewhat resinous, and gave only a feeble phenol reaction with ferric chloride. A small portion of the oil was fractionated in order to give some idea of its possible constituents. The results were as follows :—

[1] *P. and E.O.R.* (**1912**), 212. [2] *Chemist and Druggist*, 69 (**1906**), 481.

Below 175° C.	Nothing distilled
Between 175° and 180°	10 per cent. was collected
,, 180° ,, 190°	13 ,, ,,
,, 190° ,, 200°	11 ,, ,,
,, 200° ,, 210°	12 ,, ,,
,, 210° ,, 220°	18 ,, ,,
,, 220° ,, 230°	18 ,, ,,
Above 230°	18 ,, residue

Pinene and limonene are probably present. The high boiling-point of the residual fraction, which has a refractive index of about 1·500, points to the presence of a sesquiterpene. The oil is not soluble in 70 per cent. alcohol (10 volumes), but dissolves in 2 volumes of 80 per cent. alcohol.

Other samples of this oil examined had specific gravities up to 0·924 and optical rotations from − 6° to + 10°.

Charabot and Pillet have prepared two samples of Spanish vervain oil, the one distilled from the dried leaves, the other from the flowers, the respective yield being 0·184 per cent. and 0·878 per cent. The oil from the leaves had a rotatory power of − 10° 30′, that from the flowers + 6° 50′. This latter appeared to be more rich in citral than the former, containing 70 per cent. and about 10 per cent. of alcohols.

Oil of Lophanthus Rugosus.

This plant is found wild in North America and in Eastern Asia. The oil has been examined by Vilmorin and Levallois.[1] It is an amber-yellow liquid, soluble in all proportions of 95 per cent., and in 5 volumes of 85 per cent. alcohol. It has the following characters :—

Specific gravity at 4°	0·962 to 0·967
Optical rotation	+ 5° ,, + 6°
Ester value	3·7
,, ,, (after acetylation)	7·4

The principal constituent is methyl-chavicol, and it also contains limonene.

Oil of Lophanthus Anisatus.

The herb, *Lophanthus anisatus*, yields about 0·1 per cent. of essential oil having an odour resembling that of tarragon. It has the following characters :—

Specific gravity	0·948 to 0·964
Refractive index	1·5165
Optical rotation	− 7° 10′
Acid value	2·8
Ester ,,	14·0

The principal constituent is methyl-chavicol.

Oil of Nepeta.

Umney and Bennett[2] have examined the oil distilled in Sicily from the catmint, *Nepeta Cataria*. It has the following characters :—

Specific gravity	0·927
Optical rotation	+ 12°

[1] *Bull. Soc. Chim.* **(1914)**, 342. [2] *Pharm. Jour.*, 75 **(1905)**, 861.

It contains 22 per cent. of menthol of which about 3 per cent. is present in the form of esters. A small amount of a ketone is also present.

The oil of *Nepeta Nepetella* has an odour of peppermint and has the following characters:—

Specific gravity	1·0398 at 20°
Optical rotation	+ 15° 12′
Acid value	45·5
Ester "	245·7
" " (after acetylation)	314·5

It consists largely of the caprylic and valerianic esters of menthol or an allied alcohol.

Oil of Prostanthera Cineolifera.

Baker and Smith[1] have examined the oil distilled from the leaves and stalks of this Labiate plant. The yield was 0·71 per cent. and the oil had an odour of eucalyptol, with the following characters:—

Specific gravity	0·9204
Refractive index	1·4711 at 22°

The oil contains about 3 per cent. of geranyl acetate, pinene, cymene, free geraniol, eucalyptol, traces of phenol (carvacrol ?), and an aldehyde which is probably cuminic aldehyde.

Oil of Bystropogon Mollis.

Bystropogon mollis, known as Argentine mint, yields about 0·4 per cent. of essential oil, which has been examined by Doering.[2] It is a clear liquid of specific gravity 0·918 to 0·920, which on cooling to 12° does not separate any menthol. It boils principally at 210°. The oil contains 2·5 per cent. of furfurol, but no other constituent has been identified.

Oil of Meriandra.

The oil of *Meriandra benghalensis* (*M. dianthera*) has been distilled in the Italian Colony of Eritrea, and is described as "sage-oil". It has been examined by Schimmel & Co.[3] A sample distilled from the herb by these chemists (yield 1·5 per cent.) had the following characters:—

Specific gravity	0·9513
Optical rotation	− 2° 5′
Refractive index	1·4790
Acid value	3·7
Ester "	14·8

On freezing, camphor separates from the oil.

The two oils distilled in Eritrea had the following characters:—

1. $d_{15°}$ 0·9464; α_D − 0° 30′; $n_{D20°}$ 1·47176; acid value 1·0; ester value 11·8; soluble in 2 volumes 70 per cent. alcohol.

2. $d_{15°}$ 0·9526; α_D − 1°; $n_{D20°}$ 1·47548; acid value 5·6; ester value 9·3; soluble in 1·8 volumes 70 per cent. alcohol.

Neither sample possessed its full camphor-content, for in the preparation of No. 1 part of the camphor had separated out during distilla-

[1] *Jour. and Proc. Roy. Soc. N.S.W.*, xlvi. (**1912**), 103.
[2] *Bol. Acad. Nac. Cienc. Cordoha*, 19 (**1913**), 379.
[3] *Bericht*, October, **1911**, 106.

tion in the condensing worm, while from No. 2 it had been partly extracted by freezing out.

Oil of Coleus Amboinicus.

Weehuizen[1] has obtained 25 c.c. of essential oil by distilling 120 kilos of this Labiate plant, using the fresh herb for the purpose. The principal constituent was identified as carvacrol.

Oil of Micromeria.

Micromeria Chamissonis is a plant found wild on the Western Coast of the United States, where it is known as *Yerba Buena*. The oil has been examined by Power and Salway.[2]

The air-dried herb yielded 0·16 per cent. of essential oil having the following characters :—

Specific gravity	0·9244 at 20°
Optical rotation	– 22° 48′

By extraction of the plant with hot alcohol, and steam distilling the extract, an oil was obtained having the following characters :—

Specific gravity	0·9450 at 20°
Optical rotation	– 26° 44′

Oil of Amaracus Dictamnus.

An oil resembling oil of pennyroyal, and probably derived from *Amaracus Dictamnus*, has been examined by Schimmel & Co.[3] The oil has the following characters :—

Specific gravity	0·933 to 0·935
Optical rotation	+ 3° „ + 6°
Acid value	2·3
Ester „	20·9
„ „ (after acetylation)	80

It contains from 66 to 85 per cent. of pulegone, and some menthol and borneol.

Oil of Pycnanthemum.

Pycnanthemum lanceolatum yields an essential oil resembling American pennyroyal oil in odour. It has been examined by Correll[4] who found it to have the following characters :—

Specific gravity	0·918 to 0·936
Optical rotation	– 0·5 „ – 11·1°

It has more recently been examined by E. R. Miller,[5] prior to whose work Correll had identified carvacrol in the oil, and Alden[6] had shown that pulegone was also present. The oil examined by Miller had a specific gravity 0·9017 at 25°, optical rotation – 4·6°, and refractive index 1·4881 at 16°. He showed that in addition to the two bodies above

[1] *Rec. Trav. Chim. Pays-Bas.*, 37 (**1918**), 355.

[2] *Jour. Amer. Chem. Soc.*, 30 (**1908**), 251.

[3] *Bericht*, October, **1906**, 84.

[4] *Pharm. Rev.*, 14 (**1896**), 32.

[5] *Circular* 2, April, **1918**, *Pharm. Expt. Station, University of Wisconsin.*

[6] *Pharm. Review*, 16, 414 (**1898**).

named, the oil contains geraniol, dipentene, one or more terpenes not identified, and a ketone which is probably thujone. The oil from young plants contains as much carvacrol as, but less ketones than the oil from matured plants. The following table gives the character of a number of oils distilled during 1913-1916.

OILS DISTILLED IN 1913.

No. of Sample.	Yield of Oil Per Cent.	Specific Gravity of Oil at 25°.	Optical Rotation in 100 mm. Tube.	Index of Refraction at 18·9°.	Per Cent. of Phenol.	Per Cent. of Ketone.	Acid No.	Ester No.	Saponification No. of Acetylated Oil.
1	0·314	0·9087	- 5·95	1·4962	41·4	3·0	1·49	4·32	166·5
2	0·270	0·8960	- 3·10	1·4915	34·2	—	—	—	—
3	0·276	0·9140	- 4·28	1·4850	21·7	7·0	3·44	17·02	193·7
4	0·442	0·9033	- 3·85	1·4905	31·0	4·0	2·15	11·31	173·8
5	0·376	0·9026	- 3·95	1·4930	38·0	8·0	1·23	5·85	160·4
6	0·431	0·9089	- 2·75	1·4911	32·0	3·0	1·85	7·61	167·1
7	0·426	0·9166	+ 0·25	1·4894	24·0	7·5	3·57	11·00	148·7
8	0·425	0·9239	+ 3·38	1·4893	19·0	40·0	4·13	11·88	118·2
9	0·510	0·9292	+ 0·60	1·4980	38·0	8·0	7·42	15·34	165·8
10	0·432	0·9355	+ 5·20	1·4918	23·0	20·0	30·85	19·20	175·2
11	0·444	0·9284	+ 4·00	1·4905	22·0	20·0	4·85	14·34	142·3
12	0·490	0·9440	+ 4·64	1·4958	29·6	12·0	41·66	27·28	158·8
13	0·522	0·9246	+ 3·70	1·4910	25·0	26·0	10·99	6·09	127·0
14	0·555	0·9290	+ 2·30	1·4910	26·0	12·0	20·95	18·00	171·5
15	0·627	0·9140	+ 12·25	1·4840	14·5	30·0	1·38	6·92	85·8
16	0·720	0·9465	+ 10·05	1·4940	28·0	24·0	22·86	15·46	146·7
17	0·674	0·9239	+ 1·60	1·4902	27·4	12·0	18·60	3·64	184·3
18	0·694	0·9034	- 4·22	1·4888	28·0	5·0	0·54	7·94	180·4
19	0·540	0·9200	- 2·35	1·4995	53·0	5·0	0·60	5·04	193·1
20	0·557	0·9340	+ 11·20	1·4850	20·0	24·4	45·17	33·27	185·2
21	0·402	0·9220	- 0·80	1·4974	42·0	3·0	5·25	17·65	195·8
22	0·555	0·8998	Too highly coloured	1·4831	15·0	3·0	6·01	21·21	190·7
Oil of 1914	—	0·9180	+ 0·05	—	32·0	10·0	1·88	4·75	138·8
1915	0·578	0·9087	+ 8·05	1·4901 at 14·5°	23·0	15·0	0·55	3·80	110·1
1916	0·318	0·9230	- 1·10	1·5025 at 15°	49·0	5·0	0·72	3·75	111·6

VERBENACEÆ.

OIL OF VERBENA.

Most of the commercial so-called verbena oils and verbena extracts are in reality composed of lemon-grass oil, which has almost entirely superseded the older and true vervain oil. The genuine oil closely resembles lemon-grass oil in general characteristics, but is of more delicate odour. It contains a large quantity of citral, which, of course, accounts for its similarity to lemon-grass oil. The plant yielding this oil appears to be *Lippia citriodora* (*Verbena triphylla : Aloysia citriodora*). The exact character of true verbena oil is somewhat obscure on account of the almost universal use of lemon-grass oil under this name.

The so-called Spanish verbena oil is derived from a Labiate plant. The yield of oil from the leaves varies from 0·07 to 0·2 per cent. The oil has the following characters :—

Specific gravity	0·900 to 0·920
Optical rotation	− 10° ,, − 17°
Aldehydes	20 to 40 per cent. (rarely to 50 per cent.)

Australian verbena oil, of uncertain origin, but possibly distilled from the same plant, has a specific gravity about 0·890 and an optical rotation − 6° to − 16°.

Verbena oil is rarely soluble in either 70 or 80 per cent. alcohol.

The principal odorous constituent of the oil is citral. Theulier[1] has identified geraniol, *l*-limonene, a sesquiterpene, and a paraffin hydrocarbon melting at 62·5°. Myrcene has been reported as present, but the oil examined was probably lemon-grass and not true verbena oil. Kerschbaum[2] has made an exhaustive examination of the genuine oil and has isolated therefrom an aldehyde $C_{10}H_{16}O$, which he has named verbenone, and which had the following characters :—

Specific gravity	0·974 at 17°
Boiling-point at 16 mm.	103° to 104°
Optical rotation	+ 66°
Refractive index	1·4995
Melting-point of semicarbazone	208° to 209°

The following results were obtained by Messrs. Roure-Bertrand from oils distilled in Grasse at the time of early flowering :—

	From Leaves.	From Inflorescence.
Percentage yield from the fresh plant	0·195	0·132
a_D	− 14° 16′	− 8° 24′
Esters	3·5 per cent.	3·2 per cent.
Combined alcohols	2·8 ,,	2·5 ,,
Free alcohols	16·5 ,,	13·8 ,,
Citral	35·4 ,,	29·6 ,,

The root yielded 0·014 per cent. and the stems 0·007 per cent. of essence.

Lantana Oils.

Lantana Camara is a tropical plant, common in Java, Farther India, the Philippines and New Caledonia. Bacon[3] has examined the oil distilled from the leaves, and found it to have the following characters :—

Specific gravity at $\frac{30°}{4°}$	0·9132
Optical rotation at 30°	+ 11·5°
Refractive index at 30°	1·4913

A Java oil has been examined, which had a specific gravity 0·952 and optical rotation − 0° 24′.

Kanga[4] has examined samples distilled from the flowers and leaves of plants grown in Farther India, and found them to have the following characters :—

[1] *Bull. Soc. Chim.*, iii. 27 **(1902)**, 1113.
[2] *Bericht*, 33 **(1900)**, 886.
[3] *Philipp. Jour. Sci.*, 4 **(1909)**, 12.
[4] *Arch. der. Pharm.*, 252 **(1914)**, 1.

	From Dried Flowers.	From Fresh Flowers.	From Leaves.
Specific gravity	0·915 at 26°	—	0·921 at 24°
Optical rotation	+ 23·9°	—	+ 1·96°
Refractive index	1·4987 at 26°	1·5031 at 26°	1·4893 at 27°

An oil distilled in the West Indies, from *Lantana odorata* was found to have an odour recalling that of amber and hyssop, and had the following characters :—

Specific gravity	0·9149
Optical rotation	– 1° 36′
Refractive index at 20°	1·4963
Ester value	4·7
,, ,, (after acetylation)	51

CONVOLVULACEÆ.

Oil of Rosewood.

Rosewood oil or rhodium oil as met with in commerce is almost, if not quite, invariably an artificial mixture of several essential oils in which true oil of roses and oil of geranium or Indian geranium are responsible for the rose or rose-geranium odour. The true oil of rhodium is distilled from the wood, either of the stem or the root, of *Convolvulus scoparius* and *Convolvulus floridus*, two species indigenous to the Canary Islands. It is a thick, viscid, pale yellow oil, darkening on keeping. It has an agreeable rose-like odour and sharp aromatic taste. An oil examined by Gladstone, but whose authenticity is uncertain, had a specific gravity 0·906 and an optical rotation – 16°. According to Gladstone, it contains 80 per cent. of a terpene; but the authenticity of the sample examined by this chemist is not guaranteed. It is more probable that a large quantity of sesquiterpene is present. According to Schimmel & Co. the oil solidifies at low temperatures to a mass of acicular crystals melting at 11° to 12°. An oil, also of uncertain origin, examined by these chemists had the following characters :—

Specific gravity	0·951
Optical rotation	+ 1° 30′
Acid value	0
Ester ,, (after acetylation)	151·3

The oil is chiefly used in soap perfumery. The *Bois de rose femelle* is the wood of the so-called Brazilian lign-aloe, and yields the corresponding linaloe oil. It is one of the *Burseraceæ*, and has no connection with the above-described rosewood.

PRIMULACEÆ.

Oil of Cowslip.

The fresh roots of *Primula officinalis*, when strongly bruised, develop more or less rapidly, but always very distinctly, a marked odour. This odour recalls at first that of anise, but that of methyl or amyl salicylate finally predominates.

Further, if the roots thus bruised be treated with ether, a residue is left on evaporation of the ether which, when dissolved in a little water, gives a violet-blue coloration on the addition of dilute ferric chloride.

Since the odour of the essential oil does not pre-exist, it may be inferred that in this root two substances are present, normally separated, which are capable of yielding an essential oil by their reciprocal action.

Two glucosides have been separated from the roots by Goris, Mascré, and Vischniac,[1] which have been termed primeverin and primulaverin, and which are both hydrolysed, yielding the two constituents of the essential oil.

Primaverin has the formula $C_{20}H_{28}O_{13}$ and melts at 206°, and on hydrolysis yields sugars and the methyl ester of β-methoxyresorcylic acid, of the formula.

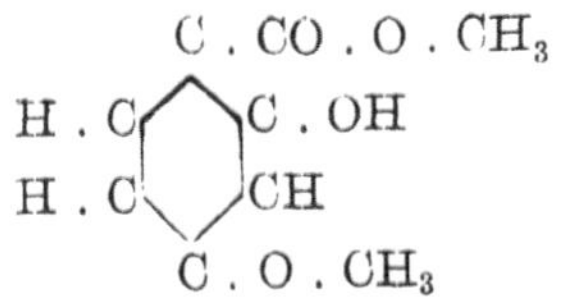

This is the solid constituent of the essential oil, melting at 49°. It has been described by Mutschler[2] as primulla camphor.

Primulaverin, $C_{20}H_{28}O_{13}$, melts at 163°, and on hydrolysis yields the liquid portion of the essential oil, which is the methyl ester of *m*-methoxysalicytic acid, of the formula

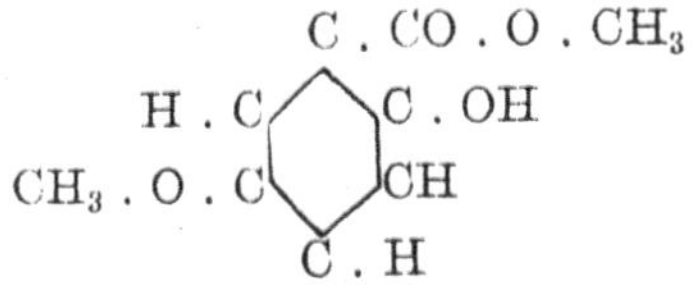

The essential oil from the flowers is practically identical with that from the roots.

RUBIACEÆ.

Oil of Gardenia.

The flowers of *Gardenia florida* and *Gardenia grandiflora* yield, on maceration with petroleum, and distillation of the extract, about 0·07 per cent. of an essential oil having a specific gravity 1·009 and a specific rotation + 1·47°.

According to Parone,[3] the oil contains benzyl acetate, stryroyl acetate, linalol, linalyl acetate, terpeneol and methyl anthraniiate.

Oil of Chione Glabra.

The bark of this tree, which is found in Grenada, yields 1·5 per cent. of essential oil, which has been examined by Dunstan and Henry.[4] It is a liquid, depositing crystals on being cooled to − 20°. The principal constituent is a substance of the formula $C_8H_8O_2$, of specific gravity 0·850, and crystallising at low temperatures.

It forms an oxime melting at 112° and a phenylhydrazone melting at 108°. It appears to be ortho-oxyacetophenone, $C_6H_4 . OH . CO . CH_3$. Traces of indol or a similar nitrogenous compound are also present.

[1] Roure-Bertrand Fils, *Bulletin*, October, **1912**, 3.
[2] *Annalen*, 185 (**1877**), 222.
[3] *Bull. Chim. Farm.*, 41 (**1902**), 489.
[4] *Jour. Chem. Soc.*, 75 (**1899**), 66.

Oil of Nuanua Leaves.

This oil is distilled from the leaves of one of the *Nelitris* species, belonging to the *Rubiaceæ*. The plant is found in Samoa, and the oil has been examined by Schimmel & Co. and found to have the following characters:—

Specific gravity	0·9025
Optical rotation	+ 9° 30′
Refractive index	1·4849
Acid value	2·2
Ester „	7·4

OLEACEÆ.

Oil of Jasmine.

The perfume of the Jasmine flower is principally marketed in the form of enfleurage products or similar preparations, but the oil is of the highest importance, and its examination has led to the preparation of a synthetic jasmine oil of considerable commercial value. The jasmine perfume of commerce is almost entirely prepared from the flowers of *Jasminum grandiflorum*, but as there are many closely allied species the following details[1] of them are of considerable interest.

There are about 100 species and varieties, mostly natives of India, Arabia, China, and tropical regions of the Old World, where this jasmine is found in the wild state. A few occur in Africa and one in South America.

The species commonly found in this country is *Jasminum officinale*, the white jasmine, but it is not known when it was introduced into England. It may be propagated by cuttings or by layering. The flowers only develop on the young shoots, and to secure a good crop the plant should be pruned in the autumn. Three varieties of this species are known, the golden and the silver-edged leaf varieties, and a double-flowered variety.

Jasminum grandiflorum, the Spanish or Catalonian jasmine, is the species generally cultivated in the south of France, where it is grafted on cuttings of the white jasmine. It resembles the latter, but the branches are shorter and thicker, and the flowers much larger and reddish underneath, forming at the extremities of the branches.

The white jasmine cuttings are planted in rows and trenched. Level ground is best, but if rising ground only is available, it is formed into a series of terraces. As soon as the young stem is strong enough shoots of *Jasminum grandiflorum* are grafted upon it. It is then left until the second year, when the long slender branches are trained along light poles, supported horizontally and running the whole length of the rows, the branches being entwined and interlaced between them. At the end of the autumn the plants are banked up with earth to half their height, when the exposed parts die off. The earth is removed in the spring, when the plant begins to grow rapidly, and the flowers appear about July, the flowering period continuing until the middle of October.

Jasminum sambac, var. *trifoliatum*, known in India as "kudda mulla," differs from the other varieties in having solitary flowers, the calyx of which is divided into a large number of segments, and the leaves

[1] *P. and E.O.R.* (1916), 36.

FIG. 31.—Manufacture of flower pomades.

[*Lautier Fils.*

occur in threes at the extremity of the flowering branches and in other parts indifferently. This variety is called the "Tuscan jasmine," as it was first imported into India by the Grand Duke of Tuscany about 1691. The flowers are greatly esteemed in India by the native women, who string them into necklaces.

Two species are found in Madeira. *Jasminum odoratissimum* bears yellow flowers, which retain when dry the natural perfume suggestive of a mixture of jasmine, jonquil, and orange blossom.

Jasminum odoratissimum is also cultivated in Formosa, on account of its fragrant flowers, which are known as shuei flowers. According to Tsuchihashi and Tasaki[1] these yield on extraction with petroleum ether a concrete extract consisting of equal parts of "flower wax" and essential oil. The latter amounted to 0·116 per cent. of the weight of the fresh flowers extracted. It is a reddish-brown oil of specific gravity 0·9309,

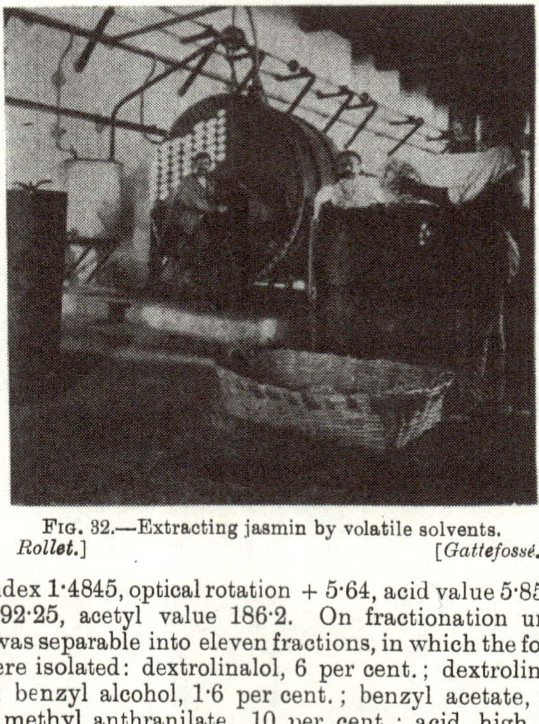

FIG. 32.—Extracting jasmin by volatile solvents.
Rollet.] [*Gattefossé.*

refractive index 1·4845, optical rotation + 5·64, acid value 5·85, saponification value 92·25, acetyl value 186·2. On fractionation under 4 mm. pressure it was separable into eleven fractions, in which the following constituents were isolated: dextrolinalol, 6 per cent.; dextrolinalyl acetate, 6 per cent.; benzyl alcohol, 1·6 per cent.; benzyl acetate, 6 per cent.; indole and methyl anthranilate, 10 per cent.; acid high boiling constituents, sesquiterpene or diterpene alcohol, 57 per cent.

Jasminum azonicum is also a native of Madeira. It bears white flowers almost all the year round, and has long been cultivated in English greenhouses.

Jasminum hirsutum is a native of China and India. It bears large white flowers, and the leaves and stems are hairy, as the botanical name indicates. The young plants are more hairy than older plants. The flowers occur in bunches, sometimes as many as thirty in a bunch.

[1] *J. Chem. Ind. Tokyo* (**1918**), 21, 17.

Jasminum paniculatum is a white-flowered species, used in China for scenting tea.

Jasminum revolutum is also a native of China, and bears yellow flowers with a very fragrant odour.

Jasmin oil prepared from the pomade has the following characters:—

Specific gravity	0·920 to 1·020
Optical rotation	− 1° to + 4° 15′
Esters as benzyl acetate	41 to 73 per cent.

Differences of opinion have been expressed as to the composition of this oil. Verley [1] extracted the pomade with acetone and obtained, after evaporation of this solvent *in vacuo*, 40 grms. of oil, which on distillation *in vacuo* left 40 per cent. of resin and yielded 60 per cent. of essential oil. Verley concluded that the oil was approximately composed of 10 per cent. of linalol and 90 per cent. of phenyl-glycol-methylene-acetal ($C_9H_{10}O_2$), which he termed jasmal. These researches led to the granting of a French patent for the production of this body synthetically, as an artificial jasmin oil. The specification states that 50 grms. of phenol-glycol, 30 grms. of sulphuric acid, diluted with 125 grms. of water, and 100 grms. of formic aldehyde are to be heated on a water-bath. The resulting body when purified boils at 218°, and is said to have a strong jasmin odour. On the other hand, Hesse and Müller,[2] who have examined a very large number of samples of this oil, state that they have most carefully examined the oil, and assert that there is not a trace of this body present, but that the main constituent is benzyl acetate. The following are the figures they give for ten samples of oil prepared by themselves from the pomade:—

No.	Yield per Cent.	Specific Gravity.	Rotation.	Esters—calculated as Linalyl Acetate.	Esters—calculated as Benzyl Acetate.
				Per Cent.	Per Cent.
1	·427	1·015	+ 2° 30′	95·4	73·0
2	·457	1·018	+ 2° 30′	91·5	70·0
3	·429	1·011	+ 2° 30′	95·0	72·9
4	·473	1·009	+ 3° 30′	92·0	70·4
5	·409	1·006	+ 3° 15′	90·3	69·1
6	·395	1·007	+ 3° 10′	90·9	70·7
7	·595	1·014	+ 3° 20′	92·8	71·0
8	·490	1·012	+ 3° 10′	94·1	72·0
9	·526	1·009	+ 3° 30′	95·4	73·0
10	·480	1·015	+ 3° 20′	93·3	71·4

They have made a very careful and exhaustive examination of the oil, and have studied quantitative methods for estimating its components, which led them to the conclusion that the average composition of the oil is as follows:—

[1] *Comptes rendus*, 128 (**1899**), 314.

[2] *Bericht*, 32 (**1899**), 565, 765.

	Per Cent.
Benzyl acetate	65
Linalyl ,,	7·5
Benzyl alcohol	6
Linalol	15·5
Indol	2·5
Jasmone	3
Methyl anthranilate	0·5

Schimmel & Co. could find no jasmal in the oil, and more recently Hesse [1] has found small quantities of indol, methyl anthranilate, and a ketone which he terms jasmone, a sweet-smelling oil of specific gravity 0·945 and boiling at 257° to 258°.

Jasmone forms an oxime melting at 45° and a semicarbazone melting at 201° to 204°. Elze [2] has also isolated geraniol and *para*-cresol from jasmin oil.

There is another compound which can easily be prepared artificially which has a distinct jasmin odour. The styrolene compounds are, as a group, remarkable in that they often possess floral odours. Thus, the *alpha*-substitution products of styrolene $C_6H_5 . CH : CH_2$, such as brom-styrolene $C_6H_5 . CH : CHBr$, have usually a distinct hyacinth-like odour. Styrolene alcohol $C_6H_5 . CH(OH)CH_2OH$ is identical with phenyl-glycol, the basis of the above-mentioned French patent; whilst *secondary* styrolyl acetate $C_6H_5 . CH(O . CO . CH_3)CH_3$ has itself a marked odour of jasmin. This body can be prepared by the action of bromine on boiling ethyl-benzene. The resulting bromide is heated with silver acetate and glacial acetic acid, yielding styrolyl acetate.

Artificial oil of jasmin compounded on the lines of the above-mentioned facts is now a regular article of commerce. It closely resembles the natural perfume, except that it is less delicate, and is well adapted for high-class toilet perfumery.

ERICACEÆ.

Oil of Wintergreen.

This oil is one of those which is so closely imitated by other oils that the commercial article is very seldom the true oil at all. Genuine wintergreen oil is the product of distillation of the leaves of *Gaultheria procumbens;* "commercial" wintergreen oil is almost invariably the product of distillation of the bark of *Betula lenta*, one of the *Betulaceæ;* or an artificial synthetic oil, methyl salicylate. *Gaultheria procumbens* is a plant of North America, known as the tea berry or partridge berry. The leaves are gathered and distilled in primitive apparatus, either ordinary whisky stills, or stills composed of wooden tubs with copper bottoms. The resulting oil, which is yielded to the extent of ·5 to 1 per cent., has a specific gravity 1·180 to 1·193, and is faintly laevo-rotary (under − 1°). It distils between 218° to 221°. Its refractive index is about 1·5350. The bark of *Betula lenta*, the sweet birch, yields about the same quantity of an oil of so exactly similar a nature that it is now sold as oil of wintergreen to the almost entire exclusion of the true *Gaultheria* oil. The oil is the product of decomposition of the glucoside

[1] *Bericht*, 32 (**1899**), 2611; 33 (**1900**), 1585; 34 (**1901**), 291, 2916; 37 (**1904**), 1457.
[2] *Chem. Zeit.*, 34 (**1910**), 912.

gaultherine. The only physical difference between the two oils is that of sweet birch quite inactive optically, both oils being almost pure methyl salicylate $C_6H_4 . OH . CO_2CH_3$. Cahours [1] was the earliest reliable investigator of wintergreen oil, but his work has been shown to be only partially correct. According to him the oil consisted of about 90 per cent. of methyl salicylate and 10 per cent. of a terpene, which he named gaultheriline. Probably he examined an adulterated oil. Trimble and Schroeter [2] stated that both oil of wintergreen and oil of birch contained traces of benzoic acid and ethyl alcohol, together with under 0·5 per cent. of a hydrocarbon (different in the two oils). Power and Kleber,[3] however, have examined a very large number of samples, and give as the result of their work the following summary:—

"*Oil of Gaultheria.*—Contains about 99 per cent. of methyl salicylate, together with a small amount of a paraffin, probably triacontane $C_{30}H_{62}$, an aldehyde or ketone, an apparently secondary alcohol $C_8H_{16}O$, and an ester $C_{14}H_{24}O_2$. To the latter body is possibly due the optical activity of the oil, whereas in the oil of birch it may be assumed that this body has suffered inversion through the process of fermentation by which the latter oil is formed.

"A pure fresh oil of gaultheria should show an optical rotation of not less than $- 0° 25'$ in a tube of 100 mm.

"*Oil of Sweet Birch.*—This oil consists to the extent of about 99·8 per cent. of methyl salicylate, and, in its unrectified state, contains also a very small amount of the above-mentioned paraffin $C_{30}H_{62}$, and the ester $C_{14}H_{24}O_2$, but does not contain the alcohol $C_8H_{16}O$ which is found in gaultheria oil.

"The oil of sweet birch is always optically inactive.

"Both of these oils have a specific gravity ranging usually from 1·180 to 1·187 at 15° C. Both of them, as well as the synthetic methyl salicylate, form a perfectly clear solution with five (5) times their volume of 70 per cent. alcohol at about 20° C., which, in connection with other distinctive characters, is an excellent practical test for their purity.

"Neither the oil of gaultheria nor the oil of sweet birch contains any trace of benzoic acid or its esters, nor does either of them contain any terpene or sesquiterpene."

The above facts easily explain the appearance on the market of artificial methyl salicylate, which is easily prepared by a condensation of methyl alcohol and salicylic acid. It has been objected that the odour of the artificial oil is not so fragrant as that of the natural oil, but so long as the salicylic acid used in its manufacture is pure, no exception can be taken to it. As the synthetic oil is worth less than half the value of the natural product of *Betula lenta*, and mixtures of the two are impossible to detect, it is easy to understand how the producing districts in America started importing the *artificial* oil, which returns to commerce as *natural* oil. The purchase of this oil, so far as its origin is concerned, must therefore be largely a matter of trust in the seller. With the exception of the fact that the true oil of *Gaultheria* is faintly laevo-rotary whilst the oil of *Betula* and the synthetic oil are optically inactive, the limits as regards physical and chemical characters are identical for the three oils. The specific gravity should never vary outside the limits 1·180 to 1·193. The oil should give a clear solution with five times its volume of 70 per cent. alcohol. On saponification

[1] *Annalen*, lii. 331. [2] *Jour. Amer. Chem. Soc.* (**1889**), 398.
[3] *Pharm. Rundschau*, 13 (**1895**), 228.

with alcoholic solution of potash, the oil should give a result equivalent to at least 98 per cent. of methyl salicylate. The salicylic acid resulting from the addition of acid to the saponification product, after a single recrystallisation from alcohol, should melt at 155° to 157°. This will guard against the use of impure salicylic acid in the preparation of the artificial oil.

This oil is sometimes adulterated with petroleum or light camphor oils, both quite easy of detection; also with sassafras oil, which, in spite of its high specific gravity, is easily detected by its lowering the saponification number, and its characteristic odour remaining after the wintergreen odour has been destroyed by saponification. The oil is used to a certain extent in pharmacy, either as a drug or to cover the nauseous taste of other drugs, but more extensively as a flavouring in the manufacture of such confections as the American chewing gums.

Power and Kleber[1] have made an exhaustive examination of wintergreen and birch oils, and consider that they can be discriminated by their odour, as would be indicated from the following results of their work:—

In previous investigations the oils had been saponified by heating with a solution of caustic alkali, and the alkaline solution subsequently distilled, or shaken with a solution of either petroleum benzine or ether. In order, however, that none of the constituents should become decomposed or altered by this treatment, they took advantage of the fact that methyl salicylate combines with cold dilute solutions of the alkalies to form salts of the ester. These are quite unstable, but the potassium compound is readily soluble in water. The oils were, therefore, shaken with successive portions of a cold dilute solution of potassium hydrate, and when the oil had become reduced to about one-third of its original volume an equal volume of ether was added, and the extraction with alkali continued until nothing more was taken up by the latter from the ether solution. After distilling off most of the ether, and the dissipation of the last traces of the latter by heating for a short time on a water-bath, the residues were weighed. The amounts were as follows:—

From 1500 grms. of Gaultheria oil, 15·7 grms. of residue, or 1·05 per cent.
,, 5000 ,, Birch oil, 9·8 ,, ,, 0·196 ,,

The portion of both oils which was soluble in alkali consisted of pure methyl salicylate. The acid separated therefrom had a melting-point of precisely 155·5° C., which is that of pure salicylic acid, and it did not contain a trace of benzoic acid.

The ether residues formed at ordinary temperatures a semi-solid mass, from which, by distillation with steam, a nearly colourless oil was obtained which no longer solidified in the cold. The amount of this was as follows:—

From Gaultheria oil, 6·3 grms. or 0·42 per cent.
,, Birch oil, 2·9 ,, 0·058 ,,

From the residue remaining after distillation with steam there was obtained a substance crystallising in colourless, odourless laminæ, of a pearly lustre, and having an apparently constant melting-point of 65·5° C. The behaviour of this substance towards reagents proved quite conclusively that it is a paraffin, and most probably triacontane $C_{30}H_{62}$. Its analysis gave the following result:—

[1] Schimmel's *Report*, October, **1895**, 50.

C. 85·2 per cent. H. 15·2 per cent.
Calculated for $C_{30}H_{62}$.
C. 85·3 ,, H. 14·7 ,,

The oil, separated as above-mentioned from the paraffin, was readily soluble in 80 per cent. alcohol, and therefore could have contained no hydrocarbon, and especially no sesquiterpene as stated by Trimble. Any benzoic acid contained in the original oil would also have been found in this liquid, as the esters of benzoic acid are very resistant toward a cold solution of potash. Upon saponification of a small portion with a hot alcoholic solution of potash a somewhat oily acid was obtained, but no trace of benzoic acid could be detected.

The oil separated from the paraffin was then subjected to fractional distillation, at first under ordinary pressure and finally in a vacuum. The lower boiling portion formed a crystalline compound with sodium bisulphite, and when isolated in this manner was found to have an odour resembling œnanthol. By oxidation it yielded an acid having the odour of the higher fatty acids. An analysis of its silver salt gave figures corresponding approximately to the formula $C_6H_{10}O_2$.

After the separation of the body which yielded a compound with sodium bisulphite, the oily liquid from the gaultheria was resolved by fractional distillation into essentially two parts, the one distilling at 160° to 165° C. under ordinary pressure, the other at about 135° C. under a pressure of 25 mm. or at 230° to 235° under ordinary pressure. Both of these fractions possessed a highly penetrating and characteristic odour, differing from each other, but which was so intense that a single drop added to 5 c.c. of synthetic methyl salicylate almost completely concealed the odour of the latter.

An analysis of the fraction boiling at 160° to 165° C. gave figures agreeing with the formula $C_8H_{16}O$.

Calculated for $C_8H_{16}O$:—	Found:— 1.	2.
C. 75·0 per cent.	7·49 per cent.	74·9 per cent.
H. 12·5 ,,	12·3 ,,	12·4 ,,

By treatment of this body with benzoyl chloride and a dilute solution of sodium hydrate a benzoic ester was obtained. An attempt to obtain a phenylurethane by treating the substance with phenylisocyanate was not successful. On the other hand, by the reaction with the latter, diphenylurea (melting-point 235° C.) was obtained, due to the elimination of a molecule of water. This indicated that the body in question is a secondary alcohol. The above-mentioned body which combines with sodium bisulphite may thus perhaps be regarded as the corresponding ketone of this alcohol, which could then be readily converted into an acid of the composition $C_6H_{10}O_2$.

The purification of the fraction boiling between 230° and 235° C. was attended with great difficulty, as by distillation a slight elimination of water constantly took place, and the liquid acquired a yellow colour. Several analyses were made of it, without obtaining perfectly constant results. The body was, however, characterised as an ester, and in fact, according to the odour developed after saponification, as the ester of the previously mentioned alcohol $C_8H_{14}O$. The acid obtained from it by saponification was of an oily nature, and had a peculiar unpleasant odour, similar to that obtained from the previously mentioned aldehyde or ketone. The analysis of its silver salt, which is readily decomposed with the loss of acid, also gave a corresponding result. It is therefore probable that this ester is formed from the alcohol $C_8H_{16}O$ and the acid $C_6H_{10}O_2$, and that it possesses the formula $C_{14}H_{24}O_2$, which would require C. 75 per cent., H. 10·7 per cent., whereas the carbon found varied from 75·7 per cent. to 78·4 per cent. and the hydrogen from 8·88 per cent. to 10·40 per cent. The excess of carbon and the deficiency of hydrogen would result from the splitting-off of water.

With regard to the small amount of volatile substance associated with the paraffin in birch oil, it may be said that independently of the aldehyde or ketone, which is also present, it appears to consist chiefly of the above-mentioned ester, and not to contain the body of alcoholic nature which is present in the oil of gaultheria. The presence, therefore, of this intensely odorous alcohol, $C_8H_{16}O$, in the latter oil, imparts to it a specific character, so that by the odour alone it is possible to distinguish it from the oil of birch.

A sesquiterpene alcohol, betulol, $C_{15}H_{24}O$, and its acetic ester have been found in the oil of *Betula lenta*.

Many colour reactions have been suggested for differentiating between the oils of gaultheria and sweet birch and artificial methyl salicylate. Most of these are unsatisfactory, the only one which yields useful results in the author's opinion being that of Umney,[1] which is as follows:—

To five drops of the oil in a test tube add five drops of a 5 per cent. alcoholic solution of vanillin and 1 c.c. of alcohol. Shake well, and add 2 c.c. of concentrated sulphuric acid and mix thoroughly. When this test is applied, pure gaultheria oil produces an intense crimson colour, and pure Betula oil a deep blood-red, while synthetic esters give a yellow colour.

Methyl salicylate has been detected in the following amongst other plants:—

BETULACEÆ.	*Discoverer.*	*Year of Discovery.*
Betula lenta L.	Wm. Procter, jr.	1844
Betula lutea Mich. probably, but not yet determined.		
LAURACEÆ.		
Lindera Benzoïn Meissner	Schimmel & Co.	1885
ROSACEÆ.		
Spiræa ulmaria L.	Schneegans & Gerock	1892
ERYTHROXYLACEÆ.		
Erythroxylon coca Lam.	v. Romburgh	1894
,, *bolivianum* (?)	v. Romburgh	1894
POLYGALACEÆ.		
Polygala senega L.	Langbeck	1881
,, ,, *var. latif.* Torr. and Gray	L. Reuter	1889
,, *baldwinii* Nuttall	J. Maisch	1890
,, *variab.* H. B. K. *albifl.* D. C.	v. Romburgh	1894
,, *javana* D C.	v. Romburgh	1894
,, *oleïfera* Heckel	v. Romburgh	1894
,, *serpyllacea* Weihe (*depressa* Warden).	Bourquelot	1894
,, *calcarea* F. Schultz	Bourquelot	1894
,, *vulgaris* L.	Bourquelot	1894
PYROLACEÆ.		
Hypopitys multiflora Scop. (*Monotropa Hypop.* L.).	Bourquelot	1894
TERNSTRŒMIACEÆ.		
Thea chinensis	Schimmel & Co.	1897

[1] *P. and E.O.R.*, 5 (**1914**), 60.

Ericaceæ.		
Andromeda Leschenaultii	Broughton	—
Gaultheria procumbens L.	Cahours	1843
,, *fragrantissima* Wall	Broughton	1867
(*Gaultheria punctata* Blume	Koehler	1879
,, *Leschenaultii*, D. C.)	Koehler	1879
Gaultheria leucocarpa Blume	Koehler	1879

Numerous other plants are mentioned in Schimmel's *Report*, 1900 (April).

The author has also detected traces of it in some plants of the natural order *Violaceæ*.

Oil of Gaultheria Punctata.

Gaultheria punctata, a plant growing in British India, and also in the Dutch Indies, yields, according to De Vrij, from 0·75 to 1·15 per cent. of essential oil. The oil has the following characters:—

Specific gravity	1·1825 to 1·1877
Optical rotation	0°
Refractive index	1·5348
Methyl salicylate	97·9 to 99 per cent.

It is practically identical with ordinary wintergreen oil.

Oil of Marsh Tea.

Ledum palustre, also known as "wild Rosemary," a plant of this natural order, yields from ½ to 2 per cent. of an essential oil. The plant is one flourishing in marshy places in the northern parts of North America, Asia, and Europe. The leaves and flowering twigs are used for distillation, and the oil obtained is of a reddish-yellow colour and sharp taste and odour, having a specific gravity from 0·930 to 0·960. The oil appears to consist almost entirely of a sesquiterpene alcohol $C_{15}H_{26}O$, which has been termed ledum camphor, and a sesquiterpene possibly identical with that obtained by treating the camphor with dehydrating agents, and which has been named ledene. Ledum camphor forms long white needles, melting at 104° and boiling at 282°, and having a specific rotation of + 8° in alcoholic solution. A ketone of the formula $C_{15}H_{24}O$ is also present in the oil. The oil possesses strongly narcotic properties, and is said to be occasionally illicitly added to beer to render it more intoxicating. This, however, is very doubtful. The oil from the leaves has a specific gravity of ·925 to ·935, and that from the flowering twigs about ·960.

A sample distilled by Schimmel & Co.[1] had the following characters:—

Specific gravity	0·9189
Optical rotation	+ 4° 20′
Refractive index at 20°	1·4845
Acid value	0·9
Ester value	27

No crystals of ledum camphor separated on standing at – 30°.

[1] *Bericht.*, **1919**, 46.

OIL OF RHODODENDRON.

Several species of *Rhododendron* contain essential oils, a distillate (from mixed species) having been examined by Haensel.[1] The yield of oil was 0·123 per cent., which had the following characters :—

Specific gravity	0·862 at 18°
Optical rotation	− 4·33°

The oil contains a small quantity of aldehydes.

Bertrand-Fils[2] have examined a specimen of this oil and found it to have a specific gravity 0·840, and an ester value 63·5.

APOCYNACEÆ.

OIL OF APOCYNUM ANDROSÆMIFOLIUM.

The rhizome of *Apocynum androsæmifolium* yields about 0·016 per cent. of essential oil having the following characters :—

Specific gravity	0·948 at 12°
Optical rotation	+ 0° 50′

The oil distils between 130° and 250°. It contains furfurol, and aceto-vanillone $CH_3O : C_6H_3(OH) . CO . CH_3$, which results from the decomposition of a glucoside, named androsin.

VALERIANACEÆ.

OIL OF VALERIAN.

Oil of valerian is obtained by the distillation of the roots of *Valeriana officinalis*, a plant growing wild, and also cultivated, in Northern and Middle Europe and in Asia. This is the Dutch and Thuringian root usually distilled in Europe, but a much larger yield of a very similar oil is obtained from the Japanese plant *Valeriana officinalis*, var. *angustifolia*.

European roots yield from 0·5 to 1 per cent. of oil, whilst the Japanese roots yield up to 8 per cent.

The characters of the two oils are as follows :—

	European.	Japanese.
Specific gravity	0·920 to 0·960	0·965 to 1·000
Rotation	− 8° ,, − 15°	− 20° ,, − 35°
Refractive index	1·4840 ,, 1·4868	1·4775 ,, 1·4875
Acid value	12 ,, 55	1 ,, 20
Ester ,,	55 ,, 100	90 ,, 135

The oils from the two varieties are similar, and may be used indiscriminately.

Pierlot investigated this oil some years ago, but his results cannot be accepted in the light of our present knowledge of essential oils. Oliviero[3] has carefully examined some European oils which had abnormally low specific gravities—·875 to ·900, and isolated the terpenes pinene, camphene, and limonene; borneol and its formic, acetic, and iso-valerianic esters, terpineol, and probably a sesquiterpene, a sesquiterpene alcohol $C_{15}H_{26}O$, and a crystalline bivalent alcohol of the

[1] *Chem. Zentral.* (**1906**), ii. 1495. 1. [2] *Bulletin*, 4 (**1920**), 1.
[3] *Comptes rendus*, 117 (**1893**), 1096.

formula $C_{10}H_{20}O_2$, melting at 132°. Bertram and Gildemeister[1] have examined the Japanese or "kesso" oil and found in it the terpenes pinene, camphene, and dipentene, terpineol, borneol, bornyl acetate, and bornyl iso-valerianate, a sesquiterpene, a blue oil not identified, and kessyl acetate $C_{14}H_{23}O . CO_2CH_3$. This body is the acetic ester of kessyl alcohol $C_{14}H_{24}O_2$. It is a liquid boiling at 300°. Kessyl alcohol forms rhombic crystals melting at 85°.

Valerian oil is employed to a considerable extent on the continent as a popular remedy for cholera, in the form of cholera drops, and also to a certain extent in soap perfumery.

An oil is also obtained from *Valeriana Celtica*, a native of the Swiss Alps and the Tyrol. Between 1 and 2 per cent. of oil is obtained having an odour recalling that of a mixture of patchouli and chamomiles. Its specific gravity is from 0·960 to 0·970.

Messrs. Schimmel & Co.[2] have distilled a parcel of Mexican root which was probably *Valeriana Mexicana*, known locally as *cuittapatti*. The oil had a disagreeable odour of valerianic acid; its specific gravity was ·949, and it was optically inactive. It appears to consist almost entirely of hydrated valerianic acid.

Oil of Spikenard.

This oil, which is the true spikenard or Indian spikenard, has been described as being obtained from various plants, but there now appears to be no doubt that it is obtained from *Nardostachys Jatamansi*, a plant flourishing in the Alpine Himalayas and district. Kemp[3] distilled 56 lb. of the root and obtained 3 oz. of oil, whilst another parcel of 100 lb. yielded 15 oz. It is a pale yellow oil of specific gravity ·975, and optical rotation about − 20°.

Asahina[4] has examined a Japanese distilled oil, which he found had the following characters :—

Specific gravity	0·9536
Optical rotation	− 11° 30′
Refractive index	1·5710 at 8°
Acid value	0
Ester ,,	45·7
,, ,, (after acetylation)	66·42

He isolated from the oil a sesquiterpene of specific gravity 0·932, boiling at 250° to 254°.

Although its odour may be considered disagreeable by many, it is highly esteemed in the East as a perfume. Its use is practically replaced in Western countries by that of oil of valerian.

COMPOSITÆ.

Chamomile Oil.

There are two varieties of this oil, the Roman chamomile oil, distilled from the flowers of *Anthemis nobilis*, a plant indigenous to the southern and western parts of Europe, and cultivated in Germany, Great Britain, France, and Belgium; and the German chamomile oil distilled from the flowers of *Matricaria chamomilla*.

[1] *Arch. d. Pharm.*, 228 (**1890**), 483.
[2] *Bericht*, April, **1897**, 47.
[3] *Pharmacographia Indica*, ii. 237.
[4] *Jour. Pharm. Japan* (**1907**), 355.

Roman chamomile oil, which is the oil official in the *British Pharmacopœia*, is obtained to the extent of 1 per cent. from the recently dried flowers. It is distilled in England, especially at Mitcham, Sutton, and Long Melford. When first distilled it is of a blue colour, but usually turns to a greenish or greenish-yellow colour on keeping. It has a characteristic flavour and odour, but not of the nature to justify its use in perfumery. Its characters are as follows:—

Specific gravity	0·905 to 0·920
Optical rotation	− 3° ,, + 3°
Refractive index	1·4420 ,, 1·4585
Acid value	1·5 ,, 15
Ester ,,	220 ,, 320

It is chiefly composed of esters of angelic and tiglic acids, two isomeric acids of the formula $C_5H_8O_2$. The following esters have been identified: isobutyl isobutyrate, isobutyl angelate, isoamyl angelate, amyl tiglate, hexyl angelate, hexyl tiglate, and possibly esters of an alcohol anthemol, $C_{10}H_{15}OH$. Traces of the free alcohols and acids, probably due to decomposition of the esters during distillation, also exist. Naudin[1] obtained from the flowers, by extraction with light petroleum, two crystalline substances not included in the above list. One of these he terms anthemene, and considers it to be a hydrocarbon $C_{18}H_{36}$ of the ethylene series. It forms microscopic needles melting at 63°. According to Klobb,[2] however, its formula is probably $C_{30}H_{62}$.

German chamomile oil from *Matricaria chamomilla* has a specific gravity about 0·940. Its colour is intensely blue, often turning greenish-blue on keeping. It is a thick fluid becoming semi-solid on cooling, commencing to thicken at about 14° and congealing at 1°. The oil has the following characters:—

Specific gravity	0·920 to 0·955
Optical rotation	0°
Refractive index	1·3640
Acid value	9 to 50
Ester ,,	5 ,, 35
,, ,, (after acetylation)	115 ,, 160

It is freely soluble in 90 per cent. alcohol. According to older researches, which are not reliable, the oil consists of chamomillol $C_{10}H_{16}O$, a terpene, and trichamomillol $C_{30}H_{48}O_3$ (the blue portion of the oil). As a matter of fact, the only constituents whose identity have been definitely proved are esters of caproic and nonylic acids, with traces of their decomposition products, together with a solid hydrocarbon or mixture of hydrocarbons, which is probably responsible for the solidification of the oil at low temperatures. This body (or mixture) melts at 53° to 54°, and when perfectly pure is quite white. Messrs. Schimmel & Co. consider it to be a hydrocarbon or hydrocarbons of the paraffin series. Possibly it contains the bodies previously isolated from Roman chamomile oil by Naudin (*vide supra*). The exact nature of the blue constituent of the oil is not known, but it is probably identical with that found in other oils, and has been named azulene. It appears to have the formula $C_{15}H_{18}$.

Power and Browning[3] have, in the course of an investigation of the constituents of the flowers of *Matricaria chamomilla*, examined the small amount (23 grms.) of essential oil obtained. The oil distilled

[1] *Bull. Soc. Chim.*, lxi. (**1884**), 483. [2] *Ibid.*, vii. (**1910**), 940.
[3] *Jour. Chem. Soc.* (**1914**), 2280.

over a wide range of temperature, the whole of the distillate being blue in colour, with the exception of a small fraction boiling at 250° to 270° at 15 mm., which was a viscous liquid of green colour. The blue oil contained furfuraldehyde, and deposited, on standing, a small amount of a crystalline substance which was found to be umbelliferone methyl ether. A portion of the essential oil, in ethereal solution, was shaken with aqueous potassium carbonate solution, and a trace of a free fatty acid obtained. No other constituent was identified.

Chamomile oil is often adulterated with cedar and turpentine oils, sometimes with copaiba oil, and, according to some authorities, with milfoil oil. Some of the commercial oil is merely turpentine or lemon oil or a mixture of these oils, distilled over chamomiles. This reprehensible custom of selling low-priced rubbish under unjustifiable names is countenanced by even some reputable firms, who offer in their price lists "chamomile oil" and "chamomile oil with lemon," the latter at about one-fifth of the price of the former.

Oil of Feverfew.

The common feverfew, *Pyrethrum parthenium* (*Matricaria parthenium*), yields a small amount of essential oil of strong, characteristic odour. The green flowering herb yields about ·07 to ·4 per cent. of an oil of specific gravity ·900 to ·960. The dried herb yields rather less oil than the fresh plant. Dessaignes and Chautard [1] investigated this oil and stated that it contained a terpene and an oxygenated body resembling camphor. In all probability this oxygenated body is a mixture of camphor and borneol. Schimmel & Co. obtained ·068 per cent. of oil from the herbs, which had a specific gravity ·960. Even at the ordinary temperature it contained a considerable number of hexagonal crystals, which appear to be borneol. Bornyl esters are also present, but no camphor could be detected.

Oils of Artemisia.

The principal essential oil of this group is that of *Artemisia Absinthium*, oil of wormwood, or oil of absinthe. It is distilled from the herb, a plant indigenous to the mountainous regions of Northern Africa, Southern Europe, and Northern Asia, and cultivated to a considerable extent in North America, especially in the States of Michigan and Indiana, and in Wayne County, New York State.

The oil is of a greenish or greenish-blue colour and of intense odour and taste of the herb itself. It has the following characters:—

Specific gravity	0·900 to 0·955 (rarely as low as 0·885)
Refractive index	1·4660 to 1·4750
Acid value	0 ,, 18
Ester ,,	12 ,, 185
,, ,, (after acetylation)	100 ,, 225

The above figures show that this oil is exceedingly variable in characters.

E. R. Miller [2] has examined six typical samples of Wisconsin oils, and gives the following figures for the ester and acetylation values, which do not indicate so great a variation as the above figures. In the ester determination, the results for saponification in thirty minutes and in sixty minutes are given.

[1] *Jour. prakt. Chem.*, 45 (1848), 45. [2] *Bull. Univ. Wisconsin*, 693.

ESTER DETERMINATIONS.

Sample of Oil.	Gms. of Oil Used.	C.C. of $\frac{N}{2}$ alcohol KOH used.	Length of Time Heated.	C.C. $\frac{N}{2}$ H_2SO_4 Required for 100 c.c. of Solution.	Saponification Number.	Amount of Thujyl Alcohol.	Amount of Ester Expressed as Thujyl Acetate.
1*a*	4·7376	56·7	half hour	7·94	100·4	27·61	35·14
1*b*	5·2484	56·7	one ,,	7·6	99·7	27·42	34·89
2*a*	5·3329	56·7	half ,,	8·45	75·8	20·84	26·53
2*b*	5·4642	56·7	one ,,	8·46	73·78	20·3	25·83
3*a*	5·1840	56·7	half ,,	7·63	100·19	27·55	35·07
3*b*	5·3092	56·7	one ,,	7·64	97·56	26·84	34·16
4*a*	5·2851	56·47	half ,,	7·61	97·58	26·84	34·16
4*b*	4·9747	56·62	one ,,	7·85	97·65	26·85	34·18
5*a*	5·6123	56·81	half ,,	7·5	96·29	26·48	33·70
5*b*	5·6400	56·81	one ,,	7·56	94·32	25·93	33·00
6*a*	5·1156	56·81	half ,,	8·08	89·61	24·64	31·36
6*b*	5·5468	56·81	one ,,	7·6	94·9	26·10	33·21

ACETYLATION VALUES.

Sample of Oil.	Saponification Number.	Corresponding Amount of Alcohol.	Alcohol Combined as Acetate.	Amount of Free Alcohol in Oil = Thujyl Alcohol.	Amount of Acetate present = Thujyl Acetate.
		Per Cent.	Per Cent.	Per Cent.	Per C 9 .
1	157·13	48·97	27·42	21·55	34· 8
2	113·93	34·25	20·30	13·95	28·83
3	134·29	41·07	26·84	14·23	34·16
4	118·44	35·74	26·85	8·89	34·17
5	129·53	39·45	25·93	13·52	33·00
6	125·41	38·06	26·10	11·96	33·12

Roure-Bertrand-Fils [1] have examined two samples distilled by themselves from the wild plants and found them to contain the following:—

	1.	2.
Esters	9·0 per cent.	5·5 per cent.
Combined alcohols	7·0 ,,	4·3 ,,
Free alcohols	71·9 ,,	76·3 ,,
Thujone	8·4 ,,	3·0 ,,

Thujyl alcohol appears to be formed at the expense of the thujone present as the growth of the plant proceeds. Two oils examined by Charabot,[2] which had been distilled at an earlier and a later date from the same plantation, gave the following results:—

	1.	2.
Specific gravity	0·9307	0·9253
Esters	9·7 per cent.	13·1 per cent.
Free thujyl alcohol	7·6 ,,	10·3 ,,
Combined thujyl alcohol . .	9·0 ,,	9·2 ,,
Total ,, ,, . . .	16·6 ,,	19·5 ,,
Thujone	43·1 ,,	35·0 ,,

It contains a considerable proportion of the ketone thujone, varying

[1] *Bulletin*, April (**1906**), 36.

[2] *Comptes rendus*, 130 (**1900**), 923.

from 30 to 45 per cent. Schimmel & Co.[1] have investigated it, and find a very small quantity of phellandrene and cadinene and the merest traces of pinene present. In addition to these bodies, the oil contains free thujyl alcohol and its acetic, isovaleric, and palmitic esters. The blue compound present is probably azulene. A pure sample examined by means of the acetylation and saponification methods gave results equivalent to 14 per cent. of thujyl alcohol and 15 per cent. of thujyl acetate. The oil is frequently adulterated, the favourite adulterant being turpentine. This is best detected by distilling 10 per cent. from the oil. This fraction should dissolve to a clear solution in twice its volume of 80 per cent. alcohol. The presence of even a very small amount of turpentine upsets this test. Wormwood oil has a tonic and stimulating effect on the digestive organs, and is sometimes also used externally. It is a constituent of the well-known liqueur, absinthe.

Paolini and Lomonaco[2] find that the Italian oil contains 10 per cent. of a mixture of *alpha* and *beta*-thujone, which are identified by means of their semicarbazones, and about 48 per cent. of thujyl alcohol, which was separated in the form of its phthalic acid ester. Phellandrene was also isolated as its nitrosite, melting at 105°, and cadinene was separated in the form of its hydrochloride. Thujyl alcohol is present in the free state, and also in the form of acetic, isovalerianic, and palmitic esters. At least two isomeric thujyl alcohols are present including the *delta*-variety.

European wormseed oil (*Oleum cinæ*) is the product of distillation of the unexpanded flower buds of *Artemisia maritima*, although other species probably contribute to the commercial oil. (" American wormseed " oil is the product of one of the *Chenopodiaceæ*, *q.v.*). The plant is indigenous to the Orient and Southern Russia, Levant wormseed (as the buds are termed) being the most highly valued. The principal constituent of wormseed is the well-known drug santonin, the well-known anthelmintic. The greater part, if not the whole, of the oil found in commerce is obtained from the santonin factories, and is said to differ slightly from the oil distilled from the seeds without extracting the santonin; this, however, is very doubtful. The yield of oil is about 2 per cent. It is a thick yellow oil having the following characters :—

Specific gravity	0·915 to 0·940
Optical rotation	− 1° 30′ to − 8°
Refractive index	1·4650 to 1·4690

It is soluble in 2 to 3 volumes of 70 per cent. alcohol.

This oil has been the subject of numerous researches, the results obtained being extremely variable, no doubt due to difference in the species of *Artemisia* used, or to the employment of adulterated oil. Völckel[3] stated that it was a mixture of bodies to which he assigned the names cinæbene and cinæbene camphor, without giving any very definite evidence of their purity. Kraut and Wahlforss[4] found the chief constituent of the oil to be a substance of the formula $C_{10}H_{18}O$. This body is clearly cineol, which is definitely settled as the principal constituent of the oil. Faust and Homeyer[5] assigned to the principal constituent the formula $C_{10}H_{16}O$. Hell and Stürcke,[6] using the oil from *Artemisia vahliana*, one of the species contributing to the Levant wormseed, confirmed the presence of cineol $C_{10}H_{18}O$. The presence of

[1] *Berichte*, April (**1897**), 51.
[2] *Chem. Zentral.* (**1915**), i. 607.
[3] *Annalen*, 38 (**1841**), 110.
[4] *Ibid.*, 128 (**1864**), 293.
[5] *Berichte*, 5 (**1872**), 680.
[6] *Ibid.*, 17 (**1884**), 1970.

cineol was again confirmed by Wallach and Brass,[1] who also obtained a terpene from the oil, which is probably dipentene. Schindelmeiser[2] has identified *i*-α-pinene, terpinene, and *l*-α-terpineol, and Schimmel & Co. have found a terpilenol present. A sesquiterpene of specific gravity 0·917 is present in the higher boiling fractions.

Artemisia annua yields an oil which has been examined by Imada and also by Asahina and Yoshitomi.[3] The oil had the following characters :—

Specific gravity at $\frac{16^\circ}{4^\circ}$	0·908
Optical rotation	− 16·19°
Acid value	1·8
Ester "	34·6
" " (after acetylation)	63·5

A ketone, $C_{10}H_{16}O$, boiling at 182° and of specific gravity at $\frac{14^\circ}{4^\circ}$ of 0·8906 was isolated from the oil, and also laevo-camphor. Eucalyptol is also present. The new ketone, above mentioned, is optically inactive, and yields a semi-carbazone melting at 95° to 96°. It has been provisionally named " artemisia ketone ".

Artemisia Barrelieri yields an essential oil of specific gravity about ·920 which contains thujone. *Artemisia glacialis*, the " mountain wormwood," known commercially as " Genepi des Alpes," yields under ½ per cent. of an oil of specific gravity ·964 to ·970, which solidifies to a buttery consistency at 0°. It has been suggested as a suitable ingredient for making liqueurs to imitate the famous Benedictine and Chartreuse.

Artemisia herba alba, var. *genuina*, a plant of Algerian origin, yields an essential oil known by the Arabs as Oil of *Scheih* or *Chih*. This oil has been examined by Grimal,[4] by Jeancard and Satie,[5] and by Bertrand-Fils.[6] The oil from the variety *densiflora* has also been examined by Schimmel & Co.[7] The former oil contains *l*-camphene, cineol and *l*-camphor.

Artemisia campestris, var. *odoratissima*, is also an Algerian plant, yielding the so-called Oil of Gouft. It has been examined by Jeancard and Satie,[8] and by Bertrand-Fils.[9] The oil contains geraniol, geranyl acetate, *l*-α-pinene, and probably nopinene. Its characters are as follows :—

Constants.	Jeancard and Satie.	Bertrand.
Density	$D^{9.5}_{15}$ 0·8720	D^{15}_{15} 0·8763
Optical rotation	α_D^{10} —15° 20′	α_D^{15} —16° 24′
Acid value	1·12	0·94
Saponification value	14	30·81
Ester value	12·88	29·87
S. V. after acetylation	42	41·07
Total alcohols calculated as $C_{10}H_{18}O$	11·93 per cent.	11·65 per cent.
Combined alcohols " "	3·55 "	8·21 "
Free alcohols " "	8·36 "	3·44 "
Soluble in 96° per cent. alcohol	1/2 vol.	1/2 vol.

The following are the characters recorded by the various observers :—

[1] *Annalen*, 225 (**1884**), 291.
[2] *Apotheker Zeit.*, 22 (**1907**), 876.
[3] *Jour. Pharm. Soc. Japan* (**1917**), 424, i.
[4] *Bull. Soc. Chim.* [3], (**1904**), **31**, 694.
[5] *Ibid.*, 478.
[6] *Bulletin* (**1920**), 4, **1**, 18.
[7] *Bericht*, April, (**1909**), 96; October (**1909**), 19.
[8] *Bull. Soc. Chim.* [3], (**1904**) **31**, 478.
[9] *Bulletin* (**1920**), 4, **1**, 27.

Characters.	Jeancard and Satie.	Grimal.	Schimmel & Co.		Bertrand-Fils.	
			I.	II.	I.	II.
Specific gravity	$D_4^{9.5}$ 0·9540	D_{15}^{15} 0·9456	D_{15}^{15} 0·9192	D_{15}^{15} 0·8994	D_{15}^{15} 0·9432	D_{15}^{15} 0·958
Optical rotation		$[\alpha]_D^{19}$ —15°30′	α_D —5° 20′	α_D +14° 5′	α_D^{17} +2° 6′	—6° 12′
Refractive index		n_D^{20} 1·47274	n_D^{20} 1·45611	n_D^{20} 1·46684	n_D^{17} 1·47120	—
Acid value	8·4	6·46	1·5	4·6	3·5	18·2
Saponification value	66·5	95·69	12·5	39·6	57·4	56·7
Ester value	58·1	89·23	11	35	53·9	38·5
„ after acetylation	129·5	135·38	40·7	163·3	147	173·6
Total alcohols calculated as $C_{10}H_{18}O$	39·44 per cent.	37·13 per cent.	11·54 per cent.	51·17 per cent.	45·44 per cent.	54·8 per cent.
Combined alcohols „ „	16·1 „	24·48 „	3·03 „	9·63 „	14·80 „	10·6 „
Free alcohols „ „	23·34 „	12·65 „	8·51 „	41·54 „	30·64 „	44·2 „
[1] Solubility in 70 per cent. alcohol	2·5 vol.	2 to 2·5 vol.	2·5 vol.	1·8 vol.	1·4 vol.	

[1] All these samples become opalescent on subsequent addition of alcohol.

Artemisia frigida, submitted in the fresh state to distillation with steam, yields about 0·4 per cent. of a greenish oil with a cineol-like odour, and having the following characters :—

Specific gravity	0·927 to 0·933
Optical rotation	− 23° 40′ to − 25° 10′
Acid value	1 to 4
Ester ,,	31 ,, 45
,, ,, (after acetylation)	139 ,, 143

The dried herb gave on distillation a yield of only 0·07 per cent. oil of a darker colour ; d_{22° 0·930 ; acid number, 4·7 ; ester number, 40 ; saponification number, 44·7. From the distillation-water of this second distillation an oil could still be obtained with petroleum ether, which was also darker than the oil from the fresh herb. Rabak [1] gives for the extracted oil d_{22° 0·916 ; acid number 5·3 ; ester number 25 ; saponification number 30·3. The oil has the following composition :—

Free borneol	35·8 per cent.
Borneol combined as heptoic ester	6·8 ,,
Cineol (eucalyptol)	18 to 20 ,,
Fenchone	8 ,, 10 ,,

From the fresh herb of *Artemisia leudoviciana* 0·38 per cent. greenish-yellow oil is obtained, having a strong aromatic odour. Its characters are as follows :—

Specific gravity	0·924 to 0·926
Optical rotation	− 13° to − 17° 20′
Acid value	0 to 4·5
Ester ,,	10 ,, 26
,, ,, (after acetylation)	116

The fresh herb of *Artemisia caudata* yielded 0·24 per cent. yellow oil with a sweetish odour, which points to the presence of methyl chavicol or anethol. Its characters are as follows :—

Specific gravity	0·842 to 0·920
Optical rotation	0° to − 24° 20′
Acid value	0 to 20
Ester ,,	17 ,, 73

Artemisia arborescens yields about 0·6 per cent. of a deep blue essential oil of specific gravity about 0·946, acid value 9·8, and ester value 19·5. According to Jona [2] it contains 13·9 per cent. of alcohols, of which borneol and thujyl alcohol have been identified, and about 13 per cent. of thujone. As constituents of the esters there were found formic, acetic, isovalerianic, pelargonic, palmitic, and stearic acids.

Artemisia vulgaris yields an oil which has the following characters :—

Specific gravity	0·907 to 0·939
Optical rotation	− 8° ,, − 19°
Refractive index	1·4776 to 1·4855 (Indian oil = 1·4620)
Acid value	1 to 2
Ester ,,	16 ,, 103
,, ,, (after acetylation)	50 ,, 206

The oil contains α-thujone.

Artemisia Dracunculus yields the well-known estragon or tarragon oil. This is a pale yellowish-green oil of characteristic odour and taste of the tarragon herb, and having the following characters :—

Specific gravity	0·900 to 0·946
Optical rotation	+ 2° ,, + 10°
Refractive index	1·5028 ,, 1·5138
Acid value	0 ,, 2
Ester ,,	1 ,, 10

[1] *Pharm Review*, 23 (**1905**), 86.

[2] *Ann. Chim. Appl.*, 1 (**1914**), ii. 63.

The oil oxidises on keeping, with the result that the specific gravity may rise to 1·000, or even over this figure.

Tarragon oil contains methyl-chavicol (the "estragol" of Grimaux), ocimene (?), phellandrene, and some unidentified constituents which influence its odour and flavour.

Artemisia lavandulaefolia, a Javan plant, yields an essential oil containing a crystalline compound of the formula $C_{12}H_{14}O_2$, melting at 32° to 33°. It is probably the methyl ester of an unrecognised acid.

Artemisia glutinosa. Bennett[1] has examined the oil distilled from this species in Murcia, Spain. It has the following characters :—

Specific gravity	0·937
Optical rotation	+ 24°
Refractive index	1·4780
Total alcohols as borneol	17·5 per cent.
Esters as bornyl acetate	6·3 per cent.
Phenols	8 per cent.
Aldehydes and ketones	18 per cent.

Terpenes, probably terpinene or dipentene principally, are present. Methyl-chavicol is probably present in the oil.

Oil of Euthamia Caroliniana.

This plant, a member of the Compositæ, is indigenous to Florida and Texas, and occurs in greatest abundance near the coast. It is rarely found on new land or in woods, but frequently in moist sandy fields or lake bottoms. It flowers from September to November, producing an abundance of small lemon-yellow blossoms. A quantity of the fresh herb has been distilled at the Florida station of the Office of Drug and Poisonous Plant Investigations of the Bureau of Plant Industry, the material being gathered in old fields adjacent to the City of Orlando, in Orange County, Florida.[2] The yield of oil from the fresh herb was 0·693 per cent. The oil was pale yellow in colour and possessed a pleasant aromatic odour and an acrid, slightly bitter, taste, with a strong terebinthenous after-taste. The oil has the following characters :—

Specific gravity	0·857
Optical rotation	10° 48′
Refractive index	1·4805

No free acids were present even after storing for fifteen months, and the saponification value was only 6·35. The saponification value after acetylation was 25·3, equivalent to 7·01 per cent. of alcohols. No phenols were present, but a trace of aldehyde was indicated by Schiff's reagent. The oil consisted mainly of dipentene, with a trace of pinene and possibly limonene. On fractionation about 10 per cent. of highly laevo-rotatory fractions was obtained, one with a specific gravity of 0·9056 and optical rotation − 34° 18′, and another of specific gravity 0·9286 and optical rotation − 63° 18′. These fractions were not further investigated. The oil was soluble in 6 volumes of 90 per cent. alcohol and also in 9 volumes of 70 per cent. with slight cloudiness, but did not become clear on adding an excess of 50 volumes.

[1] *P. and E.O.R.* (**1920**), 286. [2] *Jour. Amer. Chem. Soc.* (**1916**), 1398.

OIL OF TANSY.

This oil is distilled from the herb *Tanacetum vulgare*, a native of England, but cultivated in France, Germany, the United States, etc. The yield of oil from the fresh herb is about ·1 to ·2 per cent., and from the dried herb about twice this quantity. The oil has the following characters :—

Specific gravity	0·925 to 0·940
Optical rotation	+ 24° ,, + 35°
Refractive index	1·4570 ,, 1·4600
Acid value	0 ,, 1
Ester ,,	5 ,, 16
,, ,, (after acetylation)	24 ,, 35

English distilled oil is, however, laevo-rotatory, from − 25° − 35°, and occasionally foreign oils vary outside the above limits, the optical rotation being as low as − 3°, and the refractive index rising to 1·4750. It is soluble in 4 volumes of 70 per cent. alcohol. The earliest thorough examination of this oil was made by Bruylants [1] who stated that it contained an aldehyde $C_{10}H_{16}O$, which he called tanacetylhydrine. Semmler examined this body and showed that it was a ketone, and renamed it tanacetone. Wallach has shown, however, that it is identical with that ketone of thuja oil which he had already named thujone. Schimmel & Co. showed that the oil contains traces of borneol and some *laevo*-camphor, a fact previously indicated by Persoz.[2] The English oil, which is strongly laevo-rotatory, also contains much more camphor than the ordinary American oil.

OIL OF TANACETUM BALSAMITA.

Tanacetum balsamita yields about ·07 per cent. of essential oil. The oil has the following characters :—

Specific gravity	0·940 to 0·950
Optical rotation	− 40° ,, − 54°
Ester value	19 ,, 25

On standing in a cold place crystals of a paraffin-like substance were deposited. The oil is not soluble in twice its volume of 80 per cent. alcohol, but is in the same quantity of 90 per cent. alcohol. The oil distilled completely between 207° and 283°, and gave the following fractions :—

207° to 220°	= 10·4 per cent.	240° to 250°	= 12·4 per cent.
220° ,, 230°	= 16·8 ,,	250° ,, 283°	= 28·2 ,,
230° ,, 240°	= 18 ,,	Residue	= 21·2 ,,

OIL OF TANACETUM BOREALE.

Tanacetum boreale yields about 0·12 per cent. of essential oil. It has the following characters :—

Specific gravity	0·960
Refractive index	1·4920
Acid value	30
Ester ,,	40

It is soluble in 2 volumes of 80 per cent. alcohol.

[1] *Berichte*, 11 (1878), 449.   [2] *Comptes rendus*, 13 (1841), 436.

Oils of Achillea.

Achillea millefolium is an aromatic plant known as Millfoil whose leaves yield about 0·2 per cent. of essential oil, of a deep blue colour. It has been examined by several chemists, the results recorded differing according to the locality of origin of the plant. The majority of pure samples have the following characters:—

Specific gravity	0·900 to 0·927
Optical rotation	− 1° „ − 3°
Ester value	20 „ 38
„ „ (after acetylation)	65 „ 84

Sievers[1] has recorded American oils whose specific gravity varied from 0·869 to 0·8935.

The oil has recently been very exhaustively investigated by E. R. Miller.[2] A number of samples distilled from the plant grown near Madison in various stages of development were found to have the following characters:—

No.	Place of Collection.	Condition.	Yield Per Cent.	Specific Gravity at 25°.	Refractive Index at 19·6°.
1	Sunset Point	Dry	0·164	0·902	1·4740
2	Highlands	Fresh	0·212	0·893	1·4718
3	Hammersley Farm	„	0·125	0·9035	1·4885
4	„ „	Dry	0·111	0·908	1·4880
5	„ „	Fresh	0·123	0·9033	1·4855
6	„ „	„	0·111	0·9098	1·4890
7	Piper Farm	—	0·213	0·8966	1·4765
8	Hammersley Farm (Flower Heads only)	Dry	0·85	0·9055	1·4875
9	Hammersley Farm (Stems and Leaves of No. 8)	„	0·067	—	1·4850
10	—	—	—	—	—
11	—	—	—	0·9087	—
12	Med. Herb Garden	Dry	—	—	—
13	—	—	—	—	—
14	Sheboygan Co.	Dry	—	0·8926	—

The results of Miller's investigation may be summarised as follows:—

(1) That a blue oil may be obtained from either the leaves or the flower heads; most oil is obtained from the latter portion of the plant.

(2) That it would not be profitable to distil the lower two-thirds of the plant.

(3) That the blue compound may be obtained from very young plants, though in very small amount.

(4) That drying the plant material apparently produces no change in either the quantity or the quality of the oil.

(5) That the oil contains the following constituents, *l*-α-pinene, *d*-α pinene, *l*-limonene, *l*-borneol, bornyl acetate and other esters of borneol, *l*-camphor, cineol, salicylic acid, aldehydes, formic, acetic, butyric (?) and isovaleric acids, a non-volatile acid or lactone, and a blue constituent of high boiling-point.

Achillea coronopifolia, according to Willdenow, yields an oil of deep blue colour and specific gravity ·924 with a strong odour of tansy.

[1] *Pharm. Rev.*, 25 (**1907**), 215. [2] *Bull. Univ. Wisconsin*, 785.

Achillea moschata yields the so-called *Iva oil*, having the following characters:—

Specific gravity	0·923 to 0·960
Optical rotation	− 12° ,, − 15°
Refractive index	1·4761
Acid value	5 to 21
Ester ,,	18 ,, 44
,, ,, (after acetylation)	91 ,, 115

It contains cineol, valeric aldehyde, *l*-camphor, palmitic acid, and a body $C_{24}H_{40}O_2$, termed ivaol by its discoverer Richenau.[1]

Achillea nobilis also yields an essential oil from the leaves. The oil has been examined by Echtermayer,[2] who found it to have a specific gravity 0·936, optical rotation − 10° 41′, and to contain 18·2 per cent. of esters. It contains borneol, camphene, formic, acetic, and caproic acids in the form of esters, traces of an unidentified phenol, and a hydrocarbon of the formula $(C_{10}H_{16})_n$.

Achillea ageratum yields an oil of specific gravity 0·856. Its composition is unknown.

Oil of Tagetes Patula.

The flowers of this plant yield about 0·1 per cent. of essential oil which has been examined by Schimmel & Co.[3] It has the following characters:—

Specific gravity	0·8856
Optical rotation	− 5° 35′
Refractive index	1·49714
Acid value	2
Ester ,,	18·7
,, ,, (after acetylation)	74·3

Oil of Arnica.

Both the flowers and the root of *Arnica montana* yield essential oils, the former to the extent of about ·1 per cent., the latter up to 1 per cent. The plant is indigenous to Europe, growing upon the Swiss and German mountains, and the more northern plains. It is also met with in Northern Asia and the south-western parts of America. The oils have the following characters:—

	Flower Oil.	Root Oil.
Specific gravity	0·899 to 0·913	0·980 to 1·000[illegible]
Optical rotation	—	+ 1° to − 2°
Refractive index	—	1·5070 ,, 1·5080
Acid value	60 to 130	4 ,, 10
Ester ,,	22 ,, 33	60 ,, 100

The chemistry of this oil has been investigated by Sigel[4] and the following bodies have been identified: phlorol isobutyric ether $C_8H_8(OC_4H_9)_2$, thymohydroquinone dimethyl ether $C_{10}H_{12}(OCH_3)_2$, and phlorol dimethyl ether $C_8H_8(OCH_3)_2$. The flower oil is a blue to bluish-green oil when fresh, turning yellow or brown on keeping. When exposed to cold it becomes solid or semi-solid, owing to the presence of a paraffin hydrocarbon. Esters of lauric and palmitic acid are also present. These oils are only used to a small extent in medicine.

[1] *Annalen*, 155 (**1870**), 148.
[2] *Arch. der Pharm.*, 243 (**1905**), 238.
[3] *Bericht*, October, **1908**, 147.
[4] *Annalen*, 170 (**1873**), 345.

OIL OF CARLINA.

This oil is distilled from the root of the thistle *Carlina acaulis*, indigenous to Central Europe. The yield is about 1·5 to 2 per cent. It is a dark-coloured oil of aromatic taste and odour, having the following characters:—

Specific gravity	1·030 to 1·040
Optical rotation	− 3° ,, − 8°
Refractive index	1·5560 ,, 1·5700
Acid value	2 ,, 3
Ester ,,	4 ,, 5

This oil has often been examined chemically, but a careful inspection of the figures obtained shows that the results are not reliable.

The oil has been, however, carefully investigated by Semmler.[1] It contains about 15 per cent. of a monocyclic sesquiterpene, $C_{15}H_{24}$, boiling at 140° at 20 mm., and of specific gravity 0·8733 and refractive index 1·4920. An oxide, termed "Carlina oxide," $C_{13}H_{10}O$, of specific gravity 1·066 and refractive index 1·5860, forms the principal constituent of the oil. It is a complex furane derivative of the constitution $C_6H_5 . CH : C : CH . (C_4H_3O)$. Semmler has also isolated a sesquiterpene $C_{15}H_{24}$ which he terms carlinene, and a crystalline solid body. Traces of palmitic acid appear to be present.

OIL OF COSTUS ROOT.

The root of *Aplotaxis Lappa*, a plant indigenous to Eastern India, yields about 1 per cent. of an oil with a characteristic odour recalling that of orris or violets. This plant, named as above by Decaisne, is identical with *Aplotaxis auriculata* (De Candolle), *Aucklandia Costus* (Falconer), and *Saussurea lappa* (Clarke). It grows at elevations of 7000 to 12,000 ft. on the north-western Himalayas, being found freely as far east as Sikkim and as far west as Kashmir and Lahore. The oil is of a light yellow colour, having the following characters:—

Specific gravity	0·940 to 0·995 (rarely to 1·010)
Optical rotation	+ 13° to + 25°
Acid value	8 ,, 25
Ester ,,	55 ,, 115
,, ,, (after acetylation)	105 ,, 162

It commences to boil at 275°, and about half distils over below 315°, when decomposition takes place. On keeping, the oil develops an unpleasant odour.

The oil has recently been exhaustively examined by Semmler and Feldstein,[2] who carried out their researches on an oil having the following characters:—

Boiling-point at 11 mm.	from 60° to 215°
Specific gravity at 20°	0·985
Optical rotation	+ 15° 25′
Refractive index	1·5216

From the fraction boiling at 200° to 210° at 11 mm. a lactone of the formula $C_{15}H_{20}O_2$ was obtained, which the authors have named costus lactone. It boils at 205° to 211° at 13 mm., and has a specific gravity

[1] *Chem. Zeit.*, 13 **(1883)**, 1158.

[2] *Berichte*, 47 **(1914)**, 2433, 2687.

1·0891 at 21°, optical rotation + 28°, and refractive index 1·53043. Costic acid, $C_{15}H_{22}O_2$, of specific gravity 1·0501, was also found in the free state. It is an unsaturated bicyclic acid forming a methyl ester boiling at 170° to 175° at 11 mm. In the fraction boiling at 190° to 200° at 11 mm. there was found a lactone of the formula $C_{15}H_{22}O_2$, which has been named dihydro-costus-lactone. A new sesquiterpene alcohol, $C_{15}H_{24}O$, which the authors have named costol, was also isolated. It forms a phthalic acid ester, from which it can be prepared in the pure condition. It was then found to have the following characters:—

Boiling point at 11 mm.	169° to 171°
Specific gravity at 21°	0·983
Optical rotation	+ 13
Refractive index	1·5200

By oxidation with chromic acid it yields an aldehyde. By the action of phosphorus trichloride it yields costyl chloride, which, by the loss of HCl, is converted into the sesquiterpene isocostene, $C_{15}H_{24}$. The authors have also obtained a hydrocarbon of the formula $C_{17}H_{28}$, to which they have assigned the name aplotaxene, belonging to the aliphatic series, and two sesquiterpenes which they have named *a*-costene and *b*-costene, which have the following characters:—

	α-Costene.	β-Costene.
Boiling-point	122° to 126° (12 mm.)	144° to 149° (18 mm.)
Specific gravity	0·9014 at 21°	0·8728 at 22°
Rotation	− 12°	+ 6°
Refractive index	1·49807	1·4905

The approximate composition of the oil is:—

Camphene	about ·04 per cent.
Phellandrene	,, 0·4 ,,
Terpene alcohol	,, 0·2 ,,
α-Costene	,, 6·0 ,,
β-Costene	,, 6·0 ,,
Aplotaxene	,, 20·0 ,,
Costol	,, 7·0 ,,
Dihydrocostuslactone	,, 15·0 ,,
Costus lactone	,, 11·0 ,,
Costic acid	,, 14·0 ,,

Oils of Eupatorium.

Dog fennel oil is distilled from the herb *Eupatorium foeniculaceum*, (*E. capillifolium*), a plant distributed throughout the northern parts of America. The oil is of a golden yellow colour and pepper-like odour, Miller[1] has recently examined a number of samples distilled in Alabama, the yield of oil being from 0·8 to 1·35 per cent. He found eleven samples of the oil to have the following characters (see next page), the rotation varying considerably according to season.

An oil distilled in Florida has been described, having an optical rotation of + 17° 50′, and containing much phellandrene. Miller's more recent work makes the authenticity of this sample doubtful. The oil contains traces of aldehydes and ketones, and, probably, of salicylic acid. The principal constituent of the oil is the dimethyl ether of thymohydroquinone. Phellandrene, borneol, bornyl acetate, linalol and sabinene (?) are present in the oil.

Eupatorium triplinerve, a plant indigenous to tropical America, and

[1] *Bull. University of Wisconsin*, 693 (**1914**), 7.

	When Distilled.	Yield Per Cent.	Specific Gravity, $\frac{25^\circ}{25^\circ}$.	$n_{D_{27^\circ}}$.	a_D for 100 mm. Tube.	Per Cent. OCH_3.	Saponification No.	Saponification No after Acetylating.	Acid No.	Per Cent. of Thymo-hydroquinone Dimethyl Ether.
1	1 May, 1912	0·36	0·93919	1·504479	+ 10·27	18·42	5·6	12·94	0·207	57·63
2	13 ,, ,,	0·47	0·9326	1·502330	+ 7·47	16·87	6·4	13·26	0·206	52·78
3	25 ,, ,,	0·51	0·92378	1·500732	+ 13·82	15·92	5·38	10·61	—	49·81
4	27 June, ,,	0·527	0·91950	1·497808	+ 14·29	14·79	9·45	43·72	—	46·02
5	30 July, ,,	0·567	0·9269	1·49875	+ 10·02	16·20	16·68	56·85	—	50·91
6	8 Sept., 1911	·0960	0·9278	1·499886	– 6·33	—	13·67	—	0·18	52·13
7	13 ,, ,,	0·99	0·9346	1·501014	– 8·20	16·66	12·26	23·35	—	—
8	27 Oct., ,,	1·02	0·9472	1·505866	– 13·99	20·16	—	—	0·25	63·08
9	16 Nov., ,,	1·11	0·9505	1·506143	– 11·77	20·61	10·53	—	0·10	64·49
10	23 ,, ,,	1·03	0·9577	1·506887	– 11·73	21·45	14·43	30·95	—	67·11
11	3 Dec., 1912	0·613	0·96004	1·506422	– 16·70	21·24	10·60	23·83	—	66·46

found wild in other tropical and subtropical countries, yields an essential oil having the following characters :—

Specific gravity	0·966 to 0·981
Optical rotation	+ 3° ,, + 6°
Refractive index	1·5090

According to Semmler,[1] the oil contains a sesquiterpene, and the dimethyl ether of thymohydroquinone. Traces of coumarin are also present.

Oils of Solidago.

Various species of Solidago yield essential oils, of which the best known is the "golden rod oil," which is obtained to the extent of about 1 per cent. from *Solidago odora*, a plant very common in the United States, east of the Rocky Mountains. The oil has the following characters :—

Specific gravity	0·937 to 0·960
Optical rotation	+ 7° ,, + 14°
Refractive index	1·5050 ,, 1·5160
Ester value	6 ,, 10
,, ,, (after acetylation)	18 ,, 25

According to Miller and Moseley,[2] the oil contains about 76 per cent. of methyl-chavicol. Traces of pinene are possibly present, but this is doubtful. Borneol is present in small quantity.

Canadian golden rod oil is obtained from *Solidago canadensis*. It is a yellow oil of aromatic odour, of specific gravity 0·859 and optical rotation – 11° 10′. It consists of about 85 per cent. of terpenes, of which pinene, phellandrene, dipentene, and probably limonene are the constituents. There are also present borneol, bornyl acetate, and cadinene.

Solidago nemoralis is a plant which grows from Quebec to the North-West Territory, south to Florida and west to Texas and Arizona. A sample of the oil obtained from this species was examined by Schimmel & Co.[3] It was bright olive-green in colour, and had a peculiar odour, reminding somewhat of cypress oil. It had a specific gravity 0·8799, optical rotation – 23° 10′, ester number 14·4, ester number after acetylation 38·2, and formed a cloudy solution in about seven or more volumes of 95 per cent. alcohol.

An average of ten samples examined by Miller and Eskew[4] had the following characters :—

Specific gravity $\frac{25°}{25°}$	0·853
Optical rotation	– 16° 17′
Refractive index	1·4739
Saponification value	5·6
,, ,, (after acetylation)	9·4

It is soluble in about 4 volumes of 90 per cent. alcohol, and in about 24 volumes of 70 per cent. alcohol. On fractionation the oil yielded β-pinene and traces of salicylic and acetic acids. The presence of borneol was considered probable, but no definite proof was obtained. Phellandrene was not present, and no reactions were obtained for aldehydes or ketones, phenols or camphor. The chief constituent is pinene, a mixture of the dextro- and laevo-rotatory varieties.

[1] *Berichte*, 41 (**1908**), 509.
[2] Schimmel's *Report*, April, **1906**, 63.
[3] *Ibid.*
[4] *Jour. Amer. Chem. Soc.* (**1914**), 2538.

Solidago rugosa yields 0·5 per cent. of an oil of the following characters :—

Specific gravity $\frac{25°}{25°}$	0·862
Optical rotation	− 12° 8′
Refractive index	1·4800
Saponification value	4·2
,, ,, (after acetylation)	11

It contains pinene and limonene.

Solidago Caroliniana yields an oil which has been examined by Russell.[1] The fresh herb, gathered before flowering, yielded 0·693 per cent. of oil having the following characters :—

Specific gravity at 23°	0·8587
Optical rotation	− 10° 48′
Refractive index	1·4805
Ester value	6·35
,, ,, (after acetylation)	25·3

It consists mainly of dipentene. It also contains pinene, limonene, an aldehyde in traces, and acetic and formic acids in the form of esters.

Oil of Blumea Balsamifera.

Blumea Lasamifera is a plant indigenous to India, and is also found freely in the Malay Archipelago. The essential oil has been examined by Jonas, [2]who found it to have the following characters :—

Specific gravity	0·950
Optical rotation	− 12° 30′
Refractive index	1·4815
Acid value	23·4
Ester ,,	1
,, ,, (after acetylation)	198

He has identified the following compounds in the oil: cineol, limonene, palmitic and myristic acid in traces, 5 per cent. of a phenol not definitely identified, *l*-borneol, *l*-camphor, and a sesquiterpene and a sesquiterpene alcohol. The phenolic body present is probably phloroglucintrimethyl ether.

Oil of Pluchea Foetida.

The fresh herb, *Pluchea foetida*, which is common in the southern parts of the United States, yields 0·025 per cent. of an essential oil, which has been examined by Rabak [3] and found to have the following characters :—

Specific gravity	0·933
Optical rotation	− 10·8°
Refractive index	1·4832
Acid value	4·1
Ester ,,	44
,, ,, (after acetylation)	104

It contains cineol.

[1] *Jour. Amer. Chem. Soc.*, 38 **(1916)**, 1398.
[2] Schimmel's *Bericht*, April, **1909**, 149.
[3] *Mid. Drug. and Ph. Rev.*, 45 **(1911)**, 485.

Oil of Ageratum Conyzoides.

According to Murat, this member of the Compositæ abounds in Annam to an extent which makes it a plague to the cultivators. Distillation takes place at the beginning of February or March; 500 kilos of this plant yielded only 27 grams of essential oil; the yield, a very small one, is therefore 0·0054 per cent.

The oil of *Ageratum conyzoides* from Annam possesses the following properties:—

Specific gravity at 15° C.	1·1090
Optical rotation	− 1° 20′
Acid No.	0·9
Saponification No.	12·1
Ester No.	11·2

Oil of Helichrysum Angustifolium.

Helichrysum angustifolium yields about 0·08 per cent. of essential oil having the following characters:—

Specific gravity	0·892 to 0·920
Optical rotation	+ 4° 25′ ,, − 9° 40′
Refractive index	1·4745 ,, 1·4849
Acid value	up ,, 15
Ester ,,	39 ,, 134
,, ,, (after acetylation)	very variable

Oil of Parthenium.

Parthenium argentatum yields 0·5 per cent. of essential oil which has been examined by Alexander, who considers that a higher yield would be obtained if a suitable distilling apparatus were available. The plant is a native of Mexico, and the oil is of a faint greenish-yellow colour, slightly laevo-rotatory, and of specific gravity 0·886. On fractional distillation it behaved as follows:—

67 c.c. *distilled at* 17 *mm. pressure*	
From 50° to 60°	17 c.c.
,, 60° ,, 80°	20·3 ,,
,, 120° ,, 160°	24·8 ,,
Residue	5·5 ,,

The oil contains *l*-α-pinene, and a sesquiterpene (?) of specific gravity 0·9349, specific rotation − 21° 24′, and refractive index 1·4960 at 16°.

Oil of Fleabane.

This oil, also known as erigeron oil, is distilled from the fresh-flowering herb, *Erigeron canadensis*, a plant widely distributed through North America, and known locally as horseweed, fleabane, butterweed, colt's tail, etc. Only about 0·5 per cent. of oil is obtained, of a light yellow colour and aromatic odour. The oil has the following characters:—

Specific gravity	0·855 to 0·870
Optical rotation	+ 52° ,, + 81°
Acid value	0 ,, 1·5
Ester ,,	34 ,, 109
,, ,, (after acetylation)	67 ,, 108

The known constituents of the oil are the terpene, limonene, terpineol, and traces of citronellal.

Oil of Elecampane.

This oil is distilled from the roots of *Inula helenium*, a plant indigenous to Southern Asia and Siberia, and found to a fair extent in Southern Europe and North America. The oil has the following characters :—

Specific gravity	1·013 to 1·0380
Optical rotation	+ 120 to + 123° 45′
Refractive index	1·5220
Acid value	6 to 8
Ester ,,	160 ,, 180
,, ,, (after acetylation)	186 ,, 199

The oil is semi-solid, owing to the deposition of crystals of alantolactone, the above specific gravities being those taken in the superfused condition. The principal constituent of the oil is alantolactone, with small quantities of alantic acid, alantol, and helenin (which is probably identical with isoalantolactone). Alantic acid has the formula $C_{14}H_{20}\begin{cases}OH\\COOH\end{cases}$, and alantol is isomeric with camphor $C_{10}H_{16}O$.

Oil of Chrysanthemum Japonicum.

The leaves and flowers of this plant yield an essential oil which is known in Japan as kiku oil. It has a specific gravity about 0·885 and boils at 165° to 175°. According to Perrier,[1] the green leaves of the plant yield an oil of specific gravity 0·932 and refractive index 1·4931. It contains a paraffin hydrocarbon, and probably angelic acid in the form of esters.

Oil of Chrysanthemum Cinerariæfolium.

The flowers of this plant, growing largely in Dalmatia and Montenegro, and which, when ground, form the insect powder of commerce, yield about 0·07 per cent. of essential oil. According to Siedler,[2] it contains a paraffin $C_{14}H_{30}$ melting at 54° to 56° and a substance melting at 62°, probably palmitic acid. It also contains a phenol, and probably butyric acid.

Oil of Chrysanthemum Marginatum.

Shinosaki [3] has obtained 0·005 per cent. of a greenish blue essential oil by distilling this plant. The oil has the following characters :—

Specific gravity	0·923 at 15°
Optical rotation	– 46·58°
Refractive index	1·5020 at 20°
Saponification value	16·3
,, ,, (after acetylation)	63·1

[1] *Bull. Soc. Chim.*, iii. 23 (**1900**), 216.
[2] *Ber. d. deutsch. Pharm. Ges.*, 25 (**1915**), 297.
[3] *J. Chem. Ind. Tokyo* (**1919**), 22, 455.

Besides esters and alcohols, a small quantity of aldehyde or ketone, which yields a crystalline compound with sodium bisulphite, is present in the oil.

Oil of Santolina.

The herb *Santolina Chamæcyparissus,* indigenous to Southern Europe, yields about 0·5 per cent. of essential oil, which has been investigated by Francesconi, Scaraffia, and Granata.[1]

The oil has the following characters :—

Specific gravity	0·8732
,, rotation	− 11·74° at 25°

A sample examined by Schimmel & Co. had the following values :—

Specific gravity	0·9065
Refractive index	1·5004
Acid value	6·6
Ester ,,	16·4
,, ,, (after acetylation)	74·2

A terpene is present in the oil, and two isomeric ketones, which have been named α-santolinenone and β-santolinenone, of the formula $C_{10}H_{16}O$, both of which are unsaturated. A saturated ketone of the camphor type of the same formula is also present. α-santolinenone appears to be an optically inactive, racemic ketone, and β-santolinenone is also optically inactive, and probably racemic.

β-Santolinenone yields a liquid oxime and a crystalline compound with hydroxylamine melting at 63° to 64°. This compound is obtained by allowing a solution of hydroxylamine hydrochloride to react with an ether-alcoholic solution of the oil with the addition of sodium bicarbonate.

UMBELLIFERÆ.

Oil of Aniseed.

On account of the great similarity of the oil distilled from *Pimpinella Anisum,* with the star aniseed oil, which constitutes the greater part of the aniseed oil of commerce, this oil will be described, together with the latter, as one of the Magnoliaceæ (*q.v.*).

Oil of Caraway.

This oil is distilled from the seed of *Carum carui,* a plant inhabiting the low-lying lands in many parts of Northern and Central Europe. It is cultivated in England, Germany, Holland, Prussia, Morocco, etc. The Dutch fruit is esteemed very highly, and the oil obtained from it is worth more than that obtained from any other, except English, which, however, is a matter of local sentiment. Schimmel & Co. give the following as the average yields obtained from various fruits :—

[1] *Gazz. Chim. Ital.,* 44 **(1914)**, ii. 150, 354, 46 **(1916)**, ii, 251.

Bavaria, wild	6·5 to 7 per cent.
Finnish	5 to 6 „
Hessian, wild	6 „ 7 „
Moravian, cultivated	4 „
East Frisian	5·5 to 6 „
Russian, wild	3·2 „ 3·6 „
Styrian	6 „
Wurtemburg	5·5 to 6 „
German, cultivated	3·5 „ 5 „
Galician	4·5 „
Dutch, cultivated	4 to 6·5 „
Norwegian, wild	5 „ 6·5 „
East Prussian, cultivated	5 „ 5·5 „
Swedish, wild	4 „ 6·5 „
Tyrolese	6·5 „

Caraway oil has the following characters :—

Specific gravity	0·910 to 0·918 (rarely 0·907)
Optical rotation	+ 70° to + 82° 30′
Refractive index	1·4840 to 1·4890
Carvone	50 to 60 per cent.

As caraway oil consists almost entirely of carvone, of specific gravity 0·964, and limonene of specific gravity 0·850, the amount of carvone present can be very approximately deduced from the specific gravity of the sample.

Various direct methods of determining the carvone have been proposed, based on the preparation of crystalline compounds of this body. The oxime, the sulphuretted hydrogen compound, and the phenylhydrazone have been employed, but only approximate results can be obtained in all these cases. The following are the details recommended in the case of the phenylhydrazone :—

When 5 c.c. phenylhydrazine are added to 5 c.c. of caraway oil the mixture becomes warm owing to chemical combination taking place, and if the action be accelerated by placing the test tube in boiling water for a few minutes a copious crystallisation of carvone phenylhydrazone, $C_{10}H_{14} : N . NH . C_6H_5$, appears, and on cooling the whole solidifies to a crystalline mass. After heating for one hour the reaction is complete, the excess of phenylhydrazine is removed by adding 5 c.c. glacial acetic acid whilst hot, shaking, and diluting with 20 c.c. water. The contents of the test tube are then cooled and filtered through a paper disc by means of a pump, and the crystalline mass washed with water until of a pale yellow colour. By this process not only is the excess of phenylhydrazine removed in aqueous solution as acetate, but nearly all the oily terpene adherent to the crystals is washed away. On crystallising from a definite volume of 95 per cent. alcohol the carvone phenylhydrazone is obtained in long silky pale yellow needles, melting at 106° C., but so difficult to dry without decomposition as to render the determination only approximate.

The sodium sulphite process yields good results. On fractional distillation not more than 25 per cent. should distil below 185° and at least 55 per cent. to 65 per cent. should distil over 200°, a considerable fraction—40 per cent. to 50 per cent.—being obtained from 220° to 230°.

The valuable constituent of this oil is the carvone, and on this account, and as it is easy to separate this body in a fairly pure state, carvone is often used in place of the oil, and is official in the *German Pharmacopœia*. The result is that there is much de-carvonised oil on the market to be disposed of. A portion of the carvone is abstracted by fractional distillation, and the misleading name "rectified caraway oil" often given to the remaining oil, which frequently has a specific gravity of ·890 to ·900 and a correspondingly high rotation. When all the carvone is abstracted the resulting "carvene" as it is called—in reality almost pure limonene, with traces of carvone—is sold as "light" oil of caraway. It is used for perfuming cheap soaps, but it is false economy, as the same value in pure oil or in carvone has a far higher odour value, and is cheaper in the end. "Carvene" is comparable to "citrene," the waste terpenes obtained in preparing terpeneless oil of lemons, and "auranciene," the corresponding terpenes from orange oil. They are all practically useless for perfumery purposes, and are either sold under misleading names or used to adulterate the respective oils from which they have been obtained. There is also an oil known as caraway chaff oil, which appears to be distilled from a mixture of the herb itself, and the chaff obtained on threshing the seeds. "Light" caraway oil examined by the author had a specific gravity ·848 and an optical rotation + 103°. Umney gives a sample of "light" caraway or "chaff" oil as having a specific gravity ·8482 and an optical rotation of – 58·5°. Attempting to determine the question as to when the carvone was formed in the plant, Schimmel & Co. distilled the samples described below, grown by them on their Miltitz fields:—

No. 1 from fresh long-cut, blooming plants with partly ripening fruits.

No. 2 from fresh plants cut at the same time and in the same way, from which, however, the umbels bearing both flowers and ripening fruits had been removed, so that the stalks and the leaves alone were distilled.

No. 3 from plants in a more advanced stage of ripening, and after the time of blooming, but before the full ripening of the fruits.

The following were the physical constants of the oils obtained:—

	Specific Gravity.	Refractive Index.	Optical Rotation.
No. 1	0·882	1·48306	+ 65° 12′
„ 2	about 0·880	1·5083	+ 20° 36′
„ 3	0·9154	1·48825	+ 63° 6′

The odour of oil No. 2 hardly resembled that of caraway seeds, nor did it contain either limonene or carvone, the characteristic constituents of caraway oil, in any perceptible quantity. The small sample was just sufficient for ascertaining the boiling-point; it began to boil at 195° C., the thermometer then rapidly rose to 230° C., and between 230° and 270° about 65 to 70 per cent. distilled over; the residue was resinified.

Sample No. 1 evidently contained much more terpene and less carvone than sample No. 3. Hence it is possible that the latter results from the oxidation of the former during the ripening of the plant. A body of high boiling-point, not yet identified, was detected in small quantity in each of these oils.

The constituents recognised in caraway oil are carvone, *d*-limonene, dihydrocarvone, carveol, and dihydrocarveol, and a base having a narcotic odour which has not been investigated.

Oil of Coriander.

This oil is distilled from the fruit of *Coriandrum sativum*, a native of the Levant and Southern Europe, cultivated in many places all over the world. The average yield from various fruits is given by Schimmel & Co. as follows:—

French	·4 per cent.	Dutch	·6 per cent.
Italian	·5 ,,	Moravian	·8 ,,
Morocco	·2 to ·3 per cent.	East Indian	·15 to ·2 per cent.
Russian	·8 ,, ·1 ,,	Thuringian	·6 ,, ·8 ,,

Thuringia and Russia have, of late, been the most important producers, the crops from these districts sometimes reaching as high a figure as 450 tons. The oil has an intense odour of the seeds, which is much impaired if immature fruits have been used in the distillation. This is explained by the following experiment undertaken by Schimmel & Co. A portion of a crop was gathered in the early summer, whilst the herb was in early flower, and the whole herb distilled. Five weeks later, when half ripe and just starting to seed, another portion was gathered and distilled. In another month, the fully ripe fruit of a third portion was distilled. The result of the examination of the oils obtained is given below in Schimmel & Co.'s own language:—

"1. *Oil from the entire Flowering Plant in the Green State.*—Yield 0·12 per cent. Specific gravity 0·853. Insoluble in 70 per cent. alcohol. Extremely objectionable, bug-like odour. After 2½ months the specific gravity of the oil had increased to 0·856. The optical rotation (which had not been determined immediately after distillation) was then + 1° 2′ in a 100 mm. tube at 18° C. The bug-odour had disappeared almost entirely. It would therefore seem that in the meantime the carrier of the odoriferous principle had become polymerised or otherwise transformed.

"2. *Oil from Green Half-ripe Coriander Herb, with Fruit.*—Yield 0·17 per cent. Specific gravity 0·866. Optical rotation + 7° 10′ at 18°. Soluble in three parts of 70 per cent. alcohol. Odour resembling that of coriander, with a subsidiary odour reminding of bugs. After one month the specific gravity had increased to 0·869.

"3. *Oil from Ripe Coriander Fruit, distilled immediately after harvesting.*—Yield 0·83 per cent. Specific gravity 0·876. Optical rotation + 10° 48′ at 16°. Soluble in three parts of 70 per cent. alcohol. Pure coriander odour."

Pure coriander oil has the following characters:—

Specific gravity	0·870 to 0·885
Optical rotation	+ 7° ,, + 14°
Refractive index	1·4635 ,, 1·4760
Acid value	1 ,, 5
Ester ,,	3 ,, 22

The oil is soluble in 3 volumes of 70 per cent. alcohol.

On fractional distillation from 45 to 55 per cent. is obtained between 190° and 200°, indicating a due proportion of coriandrol.

The oil is sometimes adulterated with orange oil, which causes a lowering of the specific gravity and a considerable rise in the optical rotation. It also interferes with the solubility in 70 per cent. alcohol. Cedar-wood oil and turpentine have also been met with as adulterants.

Semmler[1] isolated the characteristic component of this oil, which he termed coriandrol, an alcohol of the formula $C_{10}H_{17}OH$. Barbier[2] has, however, shown that this is identical with *dextro*-linalol. The only other known constituent of the oil is pinene.

[1] *Berichte*, 24 (**1891**), 206. [2] *Comptes rendus*, 116 (**1893**), 1459.

According to Walbaum, the oil contains about 50 per cent. of linalol, but his results were obtained by a direct acetylation. If the method of Boulez be used, which is necessary in the case of linalol, from 65 to 70 per cent. of linalol is indicated. *Dextro*-α-pinene, *i*-α-pinene, β-pinene, dipentene, cymene, terpinene, geraniol, borneol, acetic esters, and decyl aldehyde are also present in the oil, with, possibly, phellandrene and terpinolene.

Oil of Dill.

Dill Oil is obtained by the distillation of the fruit of *Anethum graveolens* Linnæus (*Peucedanum graveolens* Bentham). It is indigenous to Central and Southern Europe, but is found in many other localities, such as the Caucasus, Persia, and the north-east of Africa and India. Slight botanical differences exist between the European plant and that grown in India, so that the latter has been described as a distinct species (*Anethum Sowa*). A difference of opinion exists as to whether this distinction is justifiable, but the oils obtained from the two plants are certainly not identical. The yield of oil obtained from the seeds is about 3 to 4 per cent.

European dill oil is a pale yellow liquid having the following characters :—

Specific gravity	0·895 to 0·918 (rarely below 0·903)
Optical rotation	+ 70° to + 82°
Refractive index	1·4830 „ 1·4900
Carvone	30 to 60 per cent. (by the sulphite method)

The oil is soluble in from 4 to 9 volumes of 80 per cent. alcohol.

In spite of the difference in odour between this and caraway oils, the composition of the two is almost identical, both consisting nearly entirely of limonene and carvone. Dill oil, however, contains less carvone than caraway oil. Hence the tests given under caraway oil apply here, and stress should be laid on the specific gravity, optical rotation, and fractionation. Not more than 15 per cent. should distil below 185°, and not less than 40 per cent. above 220°. English distilled oils usually have the highest specific gravity, from ·910 to ·916, and are consequently held in the highest esteem. The following are the average fractions obtained from English and German oils :—

	English.	German.
Below 185°	10 per cent.	10 per cent.
185° to 200°	20 „	22 „
200° „ 220°	20 „	22 „
220° „ 230°	48 „	42 „
Above 230°	2 „	4 „

East Indian dill has a much higher specific gravity than European oil, this figure usually varying from ·945 to ·970. Its rotation is from + 40° to + 50°. This high specific gravity is in all probability due to the presence of dill apiol, a body isomeric with ordinary parsley apiol, discovered by Ciamician and Silber. Indian oil contains much less carvone than European oil, as does the Japanese oil, which also possesses a high specific gravity and appears to be nearly identical with Indian oil. A sample of Japanese oil examined by Umney was found to have a specific gravity 0·964 and optical rotation + 50° 30′. The optical rotation of both varies from + 40° to + 50°. Spanish oil is distilled from the herb, with or without the unripe fruits. It has a specific gravity 0·905 to 0·935, optical rotation + 37° to + 60° and refractive index 1·4910 to 1·4970. It does not contain more than 20 per cent. of carvone. Schimmel & Co. have found a large amount of phellandrene in a sample of English, and also in Spanish oil. Apart from this, the only known constituents are limonene and carvone, with traces of a solid body, possibly a paraffin hydrocarbon, and, in Indian oil, dill apiol $C_{12}H_{14}O_4$. Spanish oil contains phellandrene, carvone, limonene, dill apiol, and dill isoapiol. Pure dill oil is used to a considerable extent in pharmacy as a carminative.

African dill oil has recently been examined by J. C. Umney.[1]

The African fruits are slightly larger than the English fruits, and a little paler in colour. Their odour closely resembles the English. They are produced from plants grown from English imported seed, whilst the Indian variety has been ascribed by some to a different species, *Anethum sowa*, which is also the source of the Japanese dill.

The yield to ether of these fruits was 11·6 per cent., comparing very closely therefore with English fruits, which yielded 10·9 per cent. to ether.

The essential oil had the following characters, and contained no fraction heavier than water :—

Specific gravity at 15° C.	0·9117
Optical rotation at 20° C.	+ 77° 3′
Solubility in 80 per cent. alcohol	1 in 6 to 1 in 6½

As a general rule it is found that the yield of essential oil from umbelliferous fruits is higher in temperate climates, as for example from fennel, the Galician and Roumanian varieties of which yield as much as 4 per cent. of oil against 0·75 per cent. from the East Indian fruits, but in this case the difference is not so marked.

If the fruits can be produced in quantity in Cape Colony, they should form a most useful source of supply.

Oil of Fennel.

This oil is distilled from the fruit of several varieties of *Foeniculum vulgare*, which is found all over Europe except in the north and north-east, being especially common on the Mediterranean littoral, and is also found in Asia Minor, Persia, India, and Japan. Two oils are recognised in commerce, the "sweet" and "bitter" oils, the former being more esteemed. The sweet fennel is distilled from *Foeniculum dulce*, but this is probably only a variety of the common wild fennel. The yield of oil obtained is very variable, according to the fruit distilled. In general it averages from 4 to 6 per cent. In the ordinary way, Galicia and Roumania furnish some of the best fennel crops, and from their fruit a large portion of the oil of commerce is distilled. The great variability in the fruits of different districts makes it necessary to fix some limits for the physical characters of the oil, which, however, may be exceeded in individual cases.

Ordinary sweet fennel oil has the following characters :—

Specific gravity	0·964 to 0·976
Optical rotation	+ 6° to + 20° (rarely to + 26°)
Refractive index	1·5280 to 1·5380
Congealing-point	+ 3° ,, + 10°

[1] *P. and E.O.R.* **(1910)**, 290.

The principal constituent of the oil is anethol, and the oil also contains fenchone, α-phellandrene, dipentene, camphene, *d*-pinene, and methyl-chavicol. As anethol is the principal constituent of fennel oil, the congealing-point may be regarded as a fair criterion of the value.

A good oil will contain as much as 60 per cent. of anethol. If necessary the crystalline stearoptene may be separated and examined, but as a rule added solid bodies will alter the other characters of the oil. The oil is soluble in an equal volume of 90 per cent. alcohol. The above tests will guard against the abstraction of anethol, or the addition of the residue of oil from which this body has been abstracted.

One observer gives as the specific gravity and optical rotation of the oil distilled from French bitter fennel, ·910 to ·955 and + 22° to + 48° respectively. Tardy,[1] however, has examined a sample of this oil, and states that its specific gravity is 1·007 at 0°, and its optical rotation + 18° 20′. In the oil he found pinene, dipentene, cymene, phellandrene (?), *d*-fenchone, methyl-chavicol, and anethol, together with anise aldehyde and anisic acid, and a ketone of the constitution

$$C_6H_4\begin{cases}OCH_3\\CH_2 . CO . CH_3\end{cases}$$

Schimmel & Co. have examined a sample of Japanese oil, and give the specific gravity as ·9754, and the optical rotation as + 10° 3′. Umney gives ·9754 and + 15° 5′ for a sample of the same oil. He found 10·2 per cent. of fenchone in the oil.

Foeniculum piperitum yields an oil (*Sicilian fennel oil*) of specific gravity about ·950. Spanish fennel oils, when pure, do not differ materially from other genuine sweet and bitter fennel oils. But much of the commercial Spanish oil is a mixture distilled from the so-called wild, and other fennel species.

The sweet Roman fennel, as it is termed, usually contains much anethol and little or no fenchone. It has the following characters :—

Specific gravity	0·976 to 0·980
Optical rotation	+ 5° to + 16° 30′
Congealing-point	10° to 14·5°

Macedonian fennel yields a very similar oil. Wild bitter fennel, growing in France, Spain, and Algeria, yields an oil of specific gravity 0·905 to 0·925 and an optical rotation up to + 50°. It is practically free from anethol. Indian fennel oil has a specific gravity ·970 and an optical rotation + 21°.

Tardy has examined an Algerian oil of specific gravity 0·991 and optical rotation + 62°. He found present: pinene, phellandrene, fenchone, methyl-chavicol, anethol, a sesquiterpene, and a little thymohydroquinone. Schimmel & Co. find camphene in fennel oil, but are not able to confirm the presence of cymene. Traces of basic compounds were also detected.

Wild bitter fennel oil differs considerably from the above-described oils. Its specific gravity varies from 0·905 to 0·925, and optical rotation + 40° to + 65°. Anethol is present only in traces, but a fair amount of phellandrene is present. Tardy isolated from this oil a crystalline compound melting at 213°, which is possibly di-para-

[1] *Bull. Soc. Chim.*, **xvii**., 660.

methoxystilbene. A crystalline substance melting at 164° to 165° is also present in the oil.

Water fennel oil is obtained from the fruit of *Oenanthe Phellandrium*, the fruit of which yields about 2 per cent. of oil. The oil has the following characters:—

Specific gravity	0·850 to 0·894
Optical rotation	+ 12° ,, + 20°
Refractive index	1·4840 ,, 1·4950

The oil contains pinene, sabinene, and *dextro*-β-phellandrene. Schimmel & Co. have also isolated an alcohol which they termed androl, and which has the following characters:—

Specific gravity	0·858
Optical rotation	− 7° 10′
Refractive index	1·4499
Boiling-point	197° to 198°

It forms a phenyl-urethane melting at 42° to 43°. Its formula is $C_{10}H_{20}O$. They have also isolated an aldehyde, which they named phellandral, and which is, chemically, a tetrahydrocumic aldehyde of the constitution

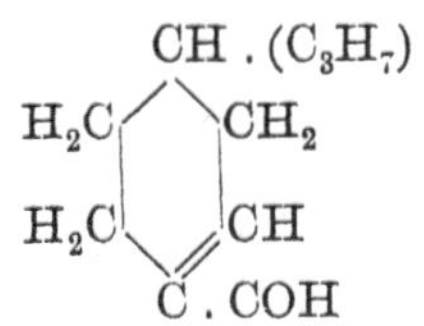

Oil of Cumin.

Cumin Oil or *Cummin Oil* is obtained from the fruit of *Cuminum cyminum*, a native of Upper Egypt and Ethiopia, but cultivated in Arabia, India, China, Malta, Sicily, etc. The bulk of the supply of commerce is obtained from Sicily, Malta, Mogador, and India. The yield is from 2·5 per cent. to 4 per cent. of a light yellow oil having the following characters:—

Specific gravity	0·900 to 0·930
Optical rotation	+ 3° ,, + 8°
Refractive index	1·4940 ,, 1·5070

Oil distilled from Cyprus seeds,[1] however, was found to have the following characters:—

Specific gravity	0·953 to 0·956
Optical rotation	+ 1° 30′ ,, + 1° 39′
Refractive index	1·5100 to 1·5140 at 23°
Aldehydes	47 ,, 52 per cent.

Distillates of high gravity dissolve in 3 volumes of 80 per cent. alcohol, but low gravity oils require 8 to 10 volumes. The oil contains from 25 to 35 per cent. of aldehydes. Wolpian[2] states that he has obtained from cumin oil a fraction boiling at 157° to 158°, which was a terpene which he could not identify with any known terpene. However, it is doubtful whether he obtained it in a state of purity, and the proposed name hydrocuminene appears rather premature, as no derivatives were prepared.

[1] *Bull. Imp. Inst.*, July-September, 1917.
[2] *Pharm. Zeit. für Russland*, 1896, p. 97.

The oil has been exhaustively examined by Schimmel & Co.[1] After repeated fractionation they succeeded in obtaining 11 grms. of a body with boiling-point 158° to 168° and an odour of pinene. Treated with nitrosylchloride a very small quantity of a nitrosochloride was obtained, of which the benzylamine-combination melted at 120° to 123°. The presence of α-pinene could be more readily proved by oxidation into active *d*-pinonic acid. The acid melted at 68° to 69° and was identical with *d*-pinonic acid from Greek oil of turpentine. It follows that the oil contains *i*- and *d*-α-pinene.

The fractions with higher boiling-point contained, besides *p*-cymene, small quantities of terpene hydrocarbons. *p*-Cymene was identified by oxidising with permanganate (hydroxy-*iso*propyl benzoic acid, melting-point 155° to 156°). When the cymene fraction was oxidised it yielded a small proportion of sodium nopinate. The free nopinic acid melted at 125° to 126° and was lævo-rotatory. On oxidation with permanganate it yielded nopinone, of which the semicarbazone melted at 187° to 188°, thus proving the presence of β-pinene. Wolpian, who had isolated considerable proportions of hydrocuminene from the oil, was probably dealing with a sample which was adulterated with turpentine oil, and his hydrocuminene was certainly only a mixture of α-pinene, β-pinene, and *p*-cymene.

Although it was the fraction boiling at 178° which showed the smallest rotation ($\alpha_D + 0° 35'$), it absorbed about 10 per cent. of bromine. With hydrochloric acid it yielded a solid dihydrochloride, melting at 48°, which, when mixed with dipentene dihydrochloride, suffered no reduction of the melting-point. From the same fraction a nitrosochloride was also obtained, but only in very small quantities and only separated out in the solid form upon the addition of water, alcohol, and ether. These facts point to the presence of dipentene.

The presence of terpinene could not be proved. No fraction gave the nitrite reaction, and any considerable quantity of phellandrene is therefore out of the question. The results of the oxidation with permanganate indicate the presence of β-phellandrene, which proves that when small quantities of phellandrene are present, identification by the nitrite reaction is very often a failure.

A small quantity of α-terpineol was isolated, together with a small amount of cuminic alcohol.

The bisulphite compound, after being purified by washing with alcohol and ether, was decomposed by boiling it with soda solution and then distilled by steam. Its constants were as follows: boiling-point 97° to 99° (7 mm.); $d_{15°}$ 0·9731; $\alpha_D + 0° 3'$. When oxidised with chromic acid the aldehyde was converted into cuminic acid, melting-point 113° to 114°. The melting-point of the semicarbazone was 210° to 211°, of the oxime, 55° to 57°. Cuminic aldehyde (cuminol) which has long been known as forming the principal constituent of oil of cumin, was thus isolated in a state of great purity. It caused some surprise, however, to find the first fraction of the aldehyde showing a slight rotation to the right ($\alpha_D + 1° 13'$), but oxidation always gave only cuminic acid. In the course of fractional crystallisation of the semicarbazone they succeeded in obtaining, in addition to cuminic aldehyde semicarbazone, crystallising in flakelets (melting-point 210°

[1] *Report*, October, **1909**, 49.

to 211°), another body, melting-point 200° to 201°, crystallising in thin prisms. This latter semi-carbazone was not found in the other fractions. Further examination failed to confirm the surmise that the two semicarbazones of cuminic aldehyde contained *cis-* and *trans*-modifications. When oxidised with chromic acid, the aldehyde forming the basis of the semicarbazone of the melting-point 200° to 201° was also converted into cuminic acid. The oxime of the aldehyde had a melting-point 72° to 76° (impure). It is therefore probable that, in addition to the cuminic aldehyde, there is present, in small proportions, a hydrogenated cuminic aldehyde.

A Persian cummin seed has recently been found on the London market. This has been examined by Holmes,[1] who reported as follows on the seed :—

"The appearance of a fruit under this name on the London market is of considerable interest, inasmuch as its botanical source has never been identified with certainty, and so far as I am aware there is no authenticated specimen of the Persian plant yielding it in any European Herbarium. This fruit has been in the museum of the Pharmaceutical Society in London for at least thirty years under the name of *Carum nigrum*.

"The botanical and indeed the geographical source of Persian cummin, which is distinguished at Bombay as 'Zeerah Shiah,' i.e. black cummin (in contra-distinction from 'Zeerah Suffed,' i.e. white cummin, the name applied to the cummin fruit, *Cuminum cyminum*, used in Europe), is at present not clearly determined, since there are apparently two or three different, but closely allied, fruits having the cummin flavour, and collected apparently in widely different districts in or about the 30th parallel of N. latitude, and all bearing the name of black cummin.

"The Persian cummin is believed to have come from Mohammerah, near the north-west coast of the Persian Gulf, in the Province of Khuzistan, but of the plant that yields it nothing is known. The fruit, however, agrees in structure with a fruit from Kunawar, in the north-east of the Punjab, which is stated by Dr. Royle (*Illustrations of Himalayan Botany*, p. 229) to be collected there under the name of 'Zeerah Shiah,' or black cummin, and he suggests the name of *Carum nigrum* for the plant, but gives no description of it. He evidently sent the plant, however, to Dr. Lindley, who was at that time professor of botany at the London University, since in the Lindley Herbarium, which is now in the Botanical Museum of Cambridge University, a specimen of a *Carum* received from Dr. Royle, occurs, which is named *Carum gracile* Lindl. This name, with Lindley's description, occurs on page 232 of the same work without any reference to Royle's *Carum nigrum*, so that it is evident that Dr. Royle sent it to Dr. Lindley unnamed, and Lindley's name for the plant must therefore supersede Dr. Royle's *Carum nigrum*."

The yield of oil from the seeds was 2 per cent., and it has been examined by Umney[2] who found it to have the following characters :—

	Persian Cummin Oil.
Yield	2 per cent.
Specific gravity	·911
Optical rotation	+ 7°
Refractive index	1·4980
Aldehydes by absorption with sodium bisulphite	18 per cent.

[1] *P. and E.O.R.* (**1913**), 43. [2] *Ibid.*

Fractionation—

Below 180	2 per cent.
,, 185	24 ,,
,, 190	36 ,,
,, 195	46 ,,
,, 200	50 ,,
,, 205	56 ,,
,, 210	58 ,,
,, 215	63 ,,
,, 220	66 ,,
,, 225	70 ,,
,, 230	76 ,,
,, 235	86 ,,
,, 240	practically all

The odour is distinctly sweeter than the normal cummin oil and a little more reminiscent of caraway, although, from the figures above indicated, it does not seem to contain a higher percentage of cumic aldehyde. Carvone does not appear to be present.

Oil of Lovage.

Oil of lovage is obtained by distilling the roots of *Levisticum officinale* with steam, the yield being under 1 per cent. The oil is a somewhat dark brown viscous liquid having an odour of angelica, and possessing the following characters:—

Specific gravity	1·000 to 1·050
Optical rotation	+ 1° ,, + 7°
Refractive index	1·5390 ,, 1·5520

It is soluble in 3 volumes of 80 per cent. alcohol. Braun[1] has investigated it, but he did not succeed in characterising any single constituent of the oil. Schimmel & Co. have detected terpineol in it, but, so far, the other constituents are unknown. The fresh herb also yields from ·1 to ·2 per cent. of an oil of specific gravity ·905 to ·940 and optical rotation + 15° to + 45°. The fruit yields 1 to 2 per cent. of an oil of specific gravity ·930 to ·940. Lovage oil is chiefly employed in liqueur and cordial making.

Oil of Celery.

All parts of the herb *Apium graveolens*, the common celery, yield essential oils, of which that from the seeds is most valued. Schimmel & Co., however, state that that from the green leaves most exactly reproduces the natural celery flavour. The oil from the seed is obtained to the extent of about 3 per cent., as a liquid of strong celery odour having the following characters:—

Specific gravity	0·860 to 0·895
Optical rotation	+ 60° to + 82° (rarely + 40°)
Refractive index	1·4780 to 1·4860

The constituent present in greatest quantity is dextro-limonene. Ciamician and Silber[2] have recently examined the high boiling fractions of the oil and found therein traces of palmitic acid, guaiacol, and a crystalline phenol of the formula $C_{16}H_{20}O_3$, melting at 66° to 67°. In addition, a sesquiterpene was found, and two bodies, both acids, of a peculiar constitution, sedanolic and sedanonic acids, together with the lactone of the former, sedanolide, which appears to be the chief odor-

[1] *Arch. Pharm.*, 1897, iii. 1. [2] *Berichte*, 30 (**1897**), 492, 1419.

ous constituent of the oil. Sedanonic acid $C_{12}H_{20}O_3$ is a crystalline body melting at 88° to 89°, and is easily converted into its lactone sedanolide $C_{12}H_{18}O_2$. Sedanonic acid $C_{12}H_{18}O_3$ melts at 113°, and possibly occurs as an anhydride in the oil. These two acids are nearly related, and Ciamician and Silber consider that sedanolic acid is *ortho*-oxyamyl-tetrahydrobenzoic acid, and that sedanonic acid is a related ketonic acid. The following formulæ thus express the relations of these bodies :—

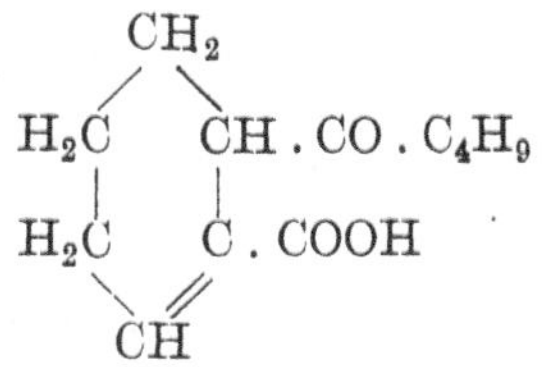

Sedanonic Acid.

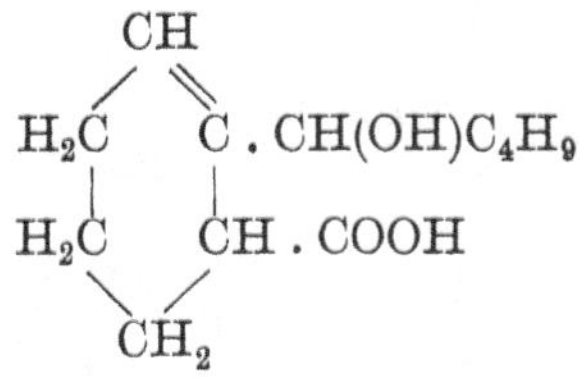

Sedanolic Acid.

Sedanolide, the chief odorous constituent, would thus be tetrahydrobutyl-phthalide of the formula—

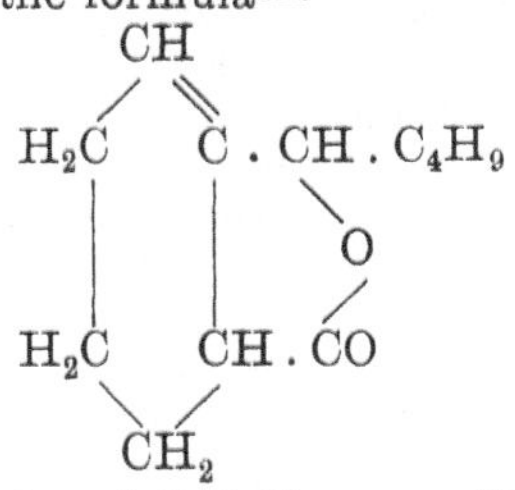

A closely-related body, phthalylisopropylidene, and its reduction products, were prepared artificially and found to have an odour much resembling celery.

Schimmel & Co.[1] have also examined the oil and isolated from it a new sesquiterpene, to which they have given the name selinene. It is probably identical with the sesquiterpene discovered by Ciamician and Silber, but not further investigated by them.

The sesquiterpene fraction was separated into the following two portions :—

First fraction. Boiling-point 120° to 121° (6 mm.); $d_{17.5°}$ 0·9197; a_D + 35° 11′; $n_{D21°}$ 1·49863.

Second fraction. Boiling-point 121° to 122° (6 mm.); $d_{18°}$ 0·9170; a_D + 38° 12′ $n_{D21°}$ 1·4956.

The sesquiterpene, or rather mixture of sesquiterpenes, was a yellow liquid of an agreeable odour, which did not solidify even when cooled considerably.

By passing hydrochloric acid into the ethereal solution of the hydrocarbon a solid chloride was obtained, which, after evaporation of the ether, crystallised out in needles from the reddish-blue residue. The crystallising process of the chloride was materially facilitated by a slight addition of alcohol and the application of a freezing mixture. The product, after being crystallised once, melted at 68° to 70°. The melting-point of a sample which had been repeatedly recrystallised from methyl alcohol was found to be 72° to 74°.

[1] *Report*, April, **1910**, 33.

By splitting off hydrochloric acid with sodium ethylate a hydrocarbon possessing the following properties was recovered from the chloride: Boiling-point 268° to 272° (the bulk 268°); $d_{15°}$ 0·9232; $d_{20°}$ 0·9196; $[a]_D$ + 49° 30′; $n_{D20°}$ 1·50483. The hydrocarbon which was obtained from the dihydrochloride is again converted by means of hydrochloric acid in ethereal solution into the dihydrochloride, melting-point 72° to 74°.

It is therefore possible to obtain the hydrocarbon in the pure state by means of the dihydrochloride, that is to say, in so far as it is at all possible, in view of the possible formation of isomerides, to regenerate a hydrocarbon from a chloride.

Selinene has been investigated by Semmler and Risse[1] who give the following details:—

Selinene has a specific gravity 0·919, optical rotation + 61° 36′, refractive index 1·5092, and boiling-point 128° to 132° at 11 mm. It is shown to be a bicyclic doubly unsaturated hydrocarbon. On reduction it yields tetrahydro-selinene, $C_{15}H_{28}$, of specific gravity 0·8881, rotation + 7°, refractive index 1·48259, and boiling at 126° to 128° at 11 mm. Selinene yields a dihydrochloride, which, when treated with calcium hydroxide, yields the alcohol selinenol $C_{15}H_{20}O$. This body boils at 155° to 163° at 19 mm., and has a specific gravity 0·9627, refractive index 1·50895, and optical rotation + 52° 36′. If the hydrocarbon be regenerated the resulting sesquiterpene has a slightly higher optical rotation, due possibly to isomerisation. The two sesquiterpenes were oxidised with ozone in order to throw light on their constitutions. Natural selinene, to which Semmler assigns the name β-selinene, yields a large amount of an indifferent body, which yields a disemicarbazone melting at 228°. From this the diketone was recovered, and found to have the formula $C_{13}H_{20}O_2$. It boils at 178° to 180° at 11 mm., and has a specific gravity 1·0566, refractive index 1·49994, and optical rotation + 15°. Regenerated selinene, or *ortho-α*-selinene, boils at 128° to 132° at 11 mm., and has a specific gravity 0·9190, refractive index 1·5092, and optical rotation + 61° 36′, yields only a small amount of this diketone, but a large amount of an acid which is termed methyl-selinene diketo-monocarboxylate $C_{15}H_{24}O_4$, boiling at 185° to 190° at 11 mm., and having a specific gravity 1·0635, refractive index 1·47889, and rotation + 4° 24′. These facts are accounted for on the hypothesis that during the elimination of hydrochloric acid from the dihydrochloride a double linkage has been displaced into the ring.

An oil from the herb and seed of the wild celery (*Ache des marais*), distilled in the South of France, possessed the following constants:—

Specific gravity	0·871
Optical rotation	+ 58° 30′
Refractive index	1·4771
Acid number	1·8
Ester ,,	41·5

After rectification with steam (in the course of which process a residue of 7·7 per cent. remained) its properties had undergone considerable change: $d_{15°}$ 0·8541; a_D + 70° 55′; $n_{D20°}$ 1·47489. Soluble in 6 volumes and more of 90 per cent. alcohol with slight turbidity.

An Algerian oil distilled from wild plants had the following characters:—

[1] *Berichte*, 45 (**1912**), 3301.

Specific gravity	0·8467
Optical rotation	+ 69° 18′
Acid number	0
Ester value	14·7
,, ,, (after acetylation)	20·9

The green herb yields ·1 per cent. of an oil of the following characters :—

Specific gravity	0·848 to 0·880
Optical rotation	+ 41° ,, + 60°
Refractive index	1·4780 ,, 1·4810
Ester value	25 ,, 50
,, ,, (after acetylation)	30 ,, 60

Although it has not been thoroughly examined, it is probable that its constituents closely resemble those of the seed oil. The oil is soluble in 10 parts of 90 per cent. alcohol.

Oil of Parsley.

All parts of the common parsley, *Petroselinum sativum*, yield an essential oil, of which that from the green herb itself is most esteemed, although that from the seed is the commoner oil. The yield from the seeds is from 2 to 6 per cent.

The oil has the following characters :—

Specific gravity	1·040 to 1·101
Optical rotation	– 4° 30′ to – 10°
Refractive index	1·512 to 1·525
Acid value	2 ,, 6
Ester ,,	1 ,, 8
,, ,, (after acetylation)	4 ,, 20

It is soluble in 8 volumes of 80 per cent. alcohol, sometimes with turbidity.

Pinene and apiol are present in the oil. Allyl-tetramethoxybenzene and one or more unidentified phenol ethers are also present in small amount, as well as traces of palmitic acid, ketones, and aldehydes. Myristicin has been identified.

If the oil contains more than a certain quantity of the latter body it is semi-solid at ordinary temperatures, whilst in any case crystals are deposited on exposure to cold. German oil is rich in apiol, but French oil contains very little. Apiol is prepared from the oil either as a thick liquid, or in the purer crystalline condition, and is used to some extent in medicine.

The oils from the root[1] and from the whole herb[2] have also been examined and found to have the following characters :—

	Root Oil.	Herb Oil.
Specific gravity	1·049 to 1·012	0·9023 to 1·0157
Optical rotation	+ 1° 24′	+ 1° 16′ ,, + 4° 10′
Refractive index	—	1·5090 ,, 1·5250
Acid value	—	0 ,, 1
Ester ,,	—	5 ,, 14
,, ,, (after acetylation)	—	19 ,, 68

The oil distilled from the herb in flower is similar to the ordinary herb oil, but is slightly laevo-rotatory.

[1] Schimmel's *Bericht*, April, **1894**, 55. [2] *Ibid.*, October, **1895**, 59.

Oil of Ajowan Seed.

This oil is the product of distillation of the fruit of the herb *Ptychotis Ajowan* (*Carum Ajowan*), a plant cultivated in many parts of Egypt, Persia, Afghanistan, and Bengal. The fruit yields from 3 to 4 per cent. of oil. It has the following characters :—

Specific gravity	0·910 to 0·930
Optical rotation	+ 1° „ + 5°
Refractive index	1·4980 „ 1·5005
Phenols	45 to 60 per cent

The characteristic odour of the oil is due to the presence of thymol. In addition, cymene, α-pinene, dipentene, and γ-terpinene are present. The oil itself is used to a certain extent in India for medicinal purposes, but its principal use is for the preparation of thymol. According to Stohmann, carvacrol is also present, but this statement requires confirmation.

Experimental cultivations of ajowan seed have been made in the West Indies, but, so far, not on a commercial scale. Ajowan seeds grown in the Seychelles yield about 9 per cent. of oil containing 38 per cent. of thymol, and seeds grown in Montserrat yield 3 per cent. of oil containing 54 per cent. of thymol.[1]

Oil of Angelica.

All parts of the herb *Angelica officinalis* (*Angelica Archangelica*) yield essential oils, of which the best known is that distilled from the fresh root. The plant is a native of Europe, on mountains or river sides, especially in Lapland, Sweden, Norway, Germany, etc. It is cultivated to a considerable extent in Saxony, but is comparatively rare in Britain. Although the root oil is that mostly used, and held in high esteem, according to Schimmel & Co. the seed oil is preferable to this. Angelica root oil is a light yellow or reddish limpid liquid of strong aromatic odour and pungent taste. The yield from dried roots grown in the Saxon Erzgebirge is from 0·5 to 1 per cent. of oil. The specific gravity varies from ·854 to ·918, that from the dried root usually having a higher specific gravity than that from the fresh root.

The oil has the following characters :—

Specific gravity	0·854 to 0·918
Optical rotation	+ 16° „ + 41°
Refractive index	1·4770 „ 1·4880
Acid value	1 „ 4
Ester „	12 „ 40
„ „ (after acetylation)	50 „ 75

On long standing crystals are deposited, which according to Giordani have the composition $C_{32}H_{62}O_5$, and melt at 74° to 77°. Beilstein[2] showed that the oil contains one or more terpenes. Naudin[3] confirmed this, and Schimmel & Co. have shown that amongst the lower boiling fractions of the oil, dextro-phellandrene exists. Ciamician and Silber[4] showed that the crystalline bodies (of which they separated two) existing in the oil were probably oxyacid anhydrides. Hydrocarbons, probably sesquiterpenes, were also found, which distilled at 240° to 270°,

[1] *Bull. Imp. Inst.*, xvi. **1**, 30.
[2] *Berichte*, 15 (**1882**), 1741.
[3] *Ibid*, 16 (**1883**), 1382.
[4] *Ibid.*, 29 (**1896**), 1811.

and oxypentadecylic acid $C_{15}H_{30}O_3$. They also isolated valerianic and two methyl-ethyl-acetic acids from the oil. The oil obtained from the seeds, which yields about 1 per cent., has a specific gravity ·856 to ·890, and an optical rotation of + 10° to + 13°. It contains phellandrene, methyl-ethyl-acetic and oxymyristic acids.

E. Böcker and A. Hahn have recently isolated a crystalline lactone of the formula $C_{15}H_{16}O_3$ in the last runnings of angelica root oil. From 200 grms. they obtained 10 grms. of a lactone which, when recrystallised from ethyl ether and light petroleum melted at 83°. It boils without decomposition at 250°. The lactone is an unsaturated body; when brominated in a glacial acetic acid solution it yields a dibromide, which, when recrystallised from glacial acetic acid, melts with decomposition at 143° to 145°.

Schimmel & Co.[1] have examined the following samples from various parts of the plant:—

Oil from	$d_{15°}$.	a_D.	$n_{D20°}$.	Acid V.	Ester V.	Solubility in 90 per cent. Alcohol.
Leaves (Sept.)	0·8550	+ 28° 2′	1·47782	0·5	17·6	In 6 vol., with opalescence.
,, (Oct.)	0·8697	+ 22° 8′	1·48044	1·3	22·6	,, 5 ,, ,,
,, ,,	0·8767	+ 20° 11′	1·48316	1·6	21·1	,, 2·5 vol. a. m.
Root . .	0·8733	+ 28° 23′	1·48083	2·5	25·9	,, 3 vol. a. m.
Seed . .	0·8623	+ 12° 12′	1·48681	1·1	18·1	,, 6 vol., with opalescence.

The oil distilled from the whole fresh herb has the following characters:—

Specific gravity	0·855 to 0·890
Optical rotation	+ 8° ,, + 25°
Refractive index	1·4770 ,, 1·4830
Acid value	0 ,, 3
Ester ,,	17 ,, 25
,, ,, (after acetylation)	51

Japanese angelica root, from which an oil is distilled, is from *Angelica anomala* (*Angelica refracta ?*). The oil, which is obtained to the extent of ·1 to ·2 per cent. of the root used, has a specific gravity ·910 to ·915. At 10° it becomes thick, and at 0° it solidifies to a paste. The crystalline mass thus deposited consists mainly of a fatty acid melting at 62° to 63°. The oil distils over between 170° and 310°; the last fractions have a blue-green colour. This oil has a more acrid odour than the Saxon oil, but distinctly recalls musk, although this is to a great extent covered by the other and less pleasant odour of the oil. The chief employment for angelica oil is in liqueur making.

Oil of Carrot.

The seed of the carrot *Daucus carota*, yields from 1 to 1·5 per cent. of oil, which has the following characters:—

Specific gravity	0·870 to 0·945
Optical rotation	− 11° ,, − 40°
Refractive index	1·4820 ,, 1·4920
Acid value	1 ,, 5
Ester ,,	17 ,, 55
,, ,, (after acetylation)	77 ,, 96

[1] *Jour. prakt. Chem.*, ii. 83 (**1911**), 243.

According to Landsberg,[1] the oil contains pinene and an oxygenated body of the formula $C_{10}H_{18}O$. Richter[2] has, more recently, examined the oil and found present isobutyric and palmitic acids in the form of esters, and formic and acetic acids in the free state.

From the laevo-rotatory portions of the saponified oil, which boiled between 150° and 160°, Richter isolated a fraction boiling at 158° to 166° ($d_{14°}$ 0·8633; $\alpha_{D_{15°}}$ − 8·45°) in which he proved the presence of pinene by means of the nitrosochloride and the nitrolbenzylamine. The latter body possessed the same melting-point as a preparation obtained from ordinary *d*-pinene, from which fact Richter concludes that the pinene of oil of carrot seed is also *d*-pinene, an erroneous inference, because, as is well known, *d*- and *l*-pinene yield the same inactive nitrolbenzylamine. Deussen[3] has since shown that pinene is present in the laevo-rotatory form. In a fraction boiling at from 168° to 176° ($d_{14°}$ 0·8624; $\alpha_{D_{15°}}$ − 5·85°) he discovered *l*-limonene, which he identified from the nitrosochloride and the nitrolpiperidine (melting-point 103° to 106°). The portions with a high boiling-point contained a crystalline body $C_{15}H_{26}O_2$, which Richter calls daucol, and which he regards as a divalent sesquiterpene alcohol. This body forms white needle crystals with a silky lustre, melting-point 115° to 116°; $[\alpha]_{D_{14°}}$ − 17·15° 17·47°.

The oil obtained at the period of maturity of the seeds by distillation of the entire plant of the wild carrot of the Puy-de-Dôme, has been examined by Roure-Bertrand Fils.[4]

The oil had a bright green colour.

Its constants are:—

Specific gravity at 15° C.	0·9016
Optical rotation	− 6° 56′
Soluble in 1 volume of 80 per cent. alcohol, but an opalescence is formed on the subsequent addition of alcohol.	
Acid value	2·7
Saponification number	195·4
,, ,, of the acetylated oil	226·8

The green colour of the oil is to be attributed to the fact that the distillation had been carried out in a copper still.

The saponification number indicated above was obtained by boiling for one hour with semi-normal alcoholic potash; after boiling for half an hour a value of only about 169 was found.

Oil of Ammoniacum.

This oil is usually obtained by distilling the gum resin of *Dorema ammoniacum* with water. Persian ammoniacum is derived from this plant, but a certain amount of ammoniacum from other sources is derived from Ferula species. The yield of oil is very small, usually about ·3 per cent. The oil has the following characters:—

Specific gravity	0·885 to 0·894
Optical rotation	+ 1° ,, + 3°
Refractive index	1·4723 ,, 1·4808
Acid value	1 ,, 4
Ester ,,	40·5
,, ,, (after acetylation)	106·4

It distils between 250° and 290°.

[1] *Arch. d. Pharm.*, 228 (**1890**), 85.
[2] *Ibid.*, 247 (**1909**), 391.
[3] *Berichte*, 43 (**1910**), 523.
[4] *Bulletin*, April, **1912**, 30.

A small amount of terpenes distils between 155° and 170°, but the greater part of the oil distils from 250° to 290°.

Semmler, Jonas, and Roenisch[1] have examined this oil, and have identified the following constituents therein: linalyl acetate, citronellyl acetate, a new dihydro-sesquiterpene which has been named ferulene, a new sesquiterpene ketone which has been named doremone, doremyl acetate, cetyl alcohol, and an unidentified terpene boiling between 65° and 80° at 11 mm. Ferulene $C_{15}H_{26}$ is a monocyclic dihydro-sesquiterpene having the following characters:—

Boiling-point	124° to 126° at 7 mm.
Specific gravity at 20°	0·8698
Refractive index	1·48423
Optical rotation	+ 6°

It is probable that this body is not a chemical individual, but is a mixture of about 75 per cent. of a dihydro sesquiterpene, and 25 per cent. of a sesquiterpene not yet separated. Doremone $C_{15}H_{26}O$ is an unsaturated aliphatic sesquiterpene ketone having the following characters:—

Boiling-point	145° to 155° at 12 mm.
Specific gravity at 20°	0·8765
Optical rotation	+ 3° 30′
Refractive index	1·4716

It yields an oxime melting at 88° and a semicarbazone melting at 124°. On reduction with sodium and alcohol, it yields the corresponding alcohol, doremol $C_{15}H_{27}OH$, which occurs in the oil in the form of its acetic ester. Doremol has the following characters:—

Boiling-point	145° to 150° at 12 mm.
Specific gravity	0·8702
Optical rotation	+ 3°
Refractive index	1·4713

Oil of Asafœtida.

The gum resin of certain species of *Ferula*, chiefly, in all probability, *Ferula Narthex*, and *Ferula scorodosma*, both perennial herbs, indigenous to Turkestan, Bokhara, Western Afghanistan, and Cashmir, yields, when distilled with water, about 3 to 20 per cent. of a somewhat foul-smelling oil. The oil has the following characters:—

Specific gravity	0·915 to 0·993
Optical rotation	+ 10° 58′ to − 36°
Refractive index	1·4940 to 1·5260

Semmler[2] has investigated this oil, having separated by fractional distillation under reduced pressure two terpenes, one of which was probably pinene, and a sesquiterpene, which had a lavender-like odour. The remainder of the oil consists chiefly of compounds containing sulphur. According to Brannt, the oil contains allyl sulphide and allyl disulphide, but Semmler denies this. Sulphur compounds of the formulæ $C_7H_{14}S_2$, $C_{10}H_{20}S_2$, $C_8H_{16}S_2$, and $C_{10}H_{18}S_2$ were found, together with an oxygenated body of the formula $C_{10}H_{16}O$, or a multiple of this.

Umney and Bunker[3] have examined the oil from the lump asafœtida, and also from the tears, and arrived at the following conclusions:—

[1] *Inaug. Dissert., Breslau*, **1916**; and *Berichte*, **1917**, 50, 1823.
[2] *Berichte*, 23 **(1890)**, 3530; 24 **(1891)**, 78.
[3] *P. and E.O.R.* **(1910)**, 165.

1. The percentage of volatile oil in the "tears" of asafœtida may equal that in the "mass" variety, or even be slightly higher in good specimens, but in general it is slightly lower.

2. In both varieties the actual percentage of volatile oil present is 12 to 16 per cent.

The "tears" themselves, however, and the tinctures prepared from them are far more pungent in their odour than the "mass" and tinctures from it, despite this approximate equality in the perçentage of oil present. It appeared, therefore, that notwithstanding the greater proportion of oil in the mass the composition of it might be different, and needed further investigation. They found the following amounts of sulphur in the two oils :—

Sulphur in oil from mass = 2·06 per cent.
,, ,, tears = 10·44 ,,

It would appear, therefore, that the oil of the mass differs in composition from the oil of the tears in containing a far lower proportion of sulphur compounds, and is certainly less offensive in smell.

Harrison and Self[1] have examined a number of samples of this oil, distilled by themselves from the drug as found on the London market, with the following results :—

No.	Percentage of Oil.	Sp. Gr. of Oil at 15·5°.	Ref. Index of Oil at 20°.	Rotation of Oil.	Percentage of Sulphur in Oil.
1	11·0	0·976	1·5200	– 10° 16′	28·2
2	16·4	0·975	1·5152	– 9° 0′	21·6
3	9·4	0·917	1·4952	+ 6° 58′	19·3
4	5·8	0·981	1·5207	– 9° 30′	29·2
5	10·1	0·915	1·4942	+ 9° 39′	17·6
6	7·2	0·993	1·5250	– 35° 55′	37·8
7	11·1	0·918	1·4980	+ 0° 50′	20·8
8	15·7	0·920	1·4978	+ 5° 40′	19·7
9	9·6	0·925	1·4993	+ 4° 32′	21·6
10	7·3	0·957	1·5077	– 17° 3′	19·9
11	14·1	0·923	1·4985	—	—
12	16·7	0·927	1·4982	—	—
13	11·9	0·930	1·4997	—	—
14	14·1	0·930	1·4987	—	—
15	17·1	0·929	1·4999	—	—

The sulphur in the oils was determined in the following manner: About 0·5 grm. of the oil was weighed out into a 150 c.c. flask, fitted by a ground neck to a vertical condenser; 5 c.c. of water was then added through the tube, followed by 5 c.c. of HNO_3, and if necessary the flask was gently warmed to start reaction, which then proceeds somewhat vigorously; 3 grms. of KBr in powder was then dropped in through the condenser tube, and the liquid boiled for ten minutes, after

[1] *Pharm. Jour.* **(1912)**, [4], **34**, 205.

which it was cooled, and 5 grms. of NaOH, dissolved in a little water, was added in the same way. The flask was then disconnected and the contents transferred to a platinum dish and evaporated to dryness and ignited. The residue was dissolved in water, hydrochloric acid added, and the liquid evaporated to drive off nitrous and nitric acids, and the sulphate determined in the usual way. A blank determination must, of course, be made with all the materials.

It will be observed that high percentage of sulphur in the oil goes with high specific gravity, high refractive index, and laevo-rotation.

Oil of Galbanum.

Galbanum Oil is the product of distillation of the gum-resin *Peucedanum galbaniflorum* and probably other species indigenous to Persia. From 12 to 24 per cent. of oil is obtained, which has a penetrating odour and sharp taste.

Galbanum oil has the following characters :—

Specific gravity	0·905 to 0·955
Optical rotation	− 10° „ + 20°
Refractive index	1·4840 „ 1·4863

α-Pinene and cadinene have long been recognised as constituents of the oil. Semmler and Jonas[1] have recently examined an oil having the following characters :—

Specific gravity	0·9353 at 25°
Optical rotation	+ 8°
Refractive index	1·4939 at 25°

They detected α-pinene, β-pinene, myrcene, a sesquiterpene which is either cadinene, or is easily converted into cadinene hydrochloride, and a sesquiterpene alcohol, cadinol, of the formula $C_{15}H_{26}O$. Cadinol forms a hydrochloride melting at 117° to 118°.

Oils of Heracleum.

The ordinary heracleum oil is distilled from the fruit of *Heracleum sphondylium*, the cow parsnip. The yield is from 0·5 to 3 per cent. of an oil having the following characters :—

Specific gravity	0·865 to 0·880
Optical rotation	0° „ + 2°
Refractive index	1·4260 „ 1·4330
Acid value	4 „ 16
Ester „	214 „ 276

According to Zincke,[2] it contains octyl alcohol $C_8H_{17}OH$ and its acetic ester $C_{10}HO_2$ and caproic ester $C_{14}H_{28}O_2$. According to Möslinger,[3] ethyl butyrate is also present; and also the lauric acid ester of octyl alcohol, and some hexyl compounds in very small quantities. The oil of *Heracleum giganteum* has been examined by Gutzeit, and found to correspond closely with that just described. Schimmel & Co. give the following interesting details of this oil :—

It is now well known that this oil is obtained from the fruit or the entire umbellate heads of *Heracleum sphondylium* L. A distillate which we have obtained from the dry umbels from which the fruit had been removed is for this reason interesting. The yield of oil amounted

[1] *Berichte*, 47 (**1914**), 2068. [2] *Annalen*, 152 (**1869**), 1.
[3] *Berichte*, 9 (**1876**), 998.

to 0·08 per cent. The odour of the brownish-yellow oil differs distinctly from that of the fruit. d_{15° 0·9273; α_D − 0° 48′; acid number 16·2; ester number 148·6; ester number after acetylation 195·9; soluble in 1·1 volume 80 per cent. alcohol; when more solvent is added opalescence and strong separation of paraffin occur. For comparison we give below the properties of two oils distilled from the fruit only.

1. Yield (compared with dry material) 0·9 per cent.; d_{15° 0·8744; α_D + 0° 38′; acid number 15·9; ester number 215·4; ester number after acetylation 285·3; soluble in 0·8 and more volume 80 per cent. alcohol; not entirely soluble in 70 per cent. alcohol.

2. Yield (calculated on dry material) 1·21 per cent.; d_{15° 0·8798; α_D + 1° 6′; acid number 7·3; ester number 242·4; ester number after acetylation 276·3; soluble in 1 and more volume 80 per cent. alcohol.

The oil from the dry, ripe fruit of *Heracleum giganteum* L. (yield 3·6 per cent.) has the following characters:—

Specific gravity	0·872 to 0·875
Optical rotation	+ 1° to + 1° 30′
Refractive index	1·4230 ,, 1·4255
Acid value	1·5 ,, 4·0
Ester ,,	280 ,, 290
,, ,, (after acetylation)	314 ,, 320

It contains the acetic and butyric esters of hexyl and octyl alcohols, and traces of ethyl butyrate.

The colourless oil had a peculiar odour, similar to that of common heracleum oil.

Oil of Bupleurum Fruticosum.

The oil of this umbelliferous plant, which grows wild abundantly in the island of Sardinia, has been examined by Francesconi and Sanna.[1] It has the following characters:—

Specific gravity	0·831 to 0·869
Rotation	+ 19° ,, + 47°
Refractive index	1·4783 ,, 1·4862
Saponification value	5 ,, 14
,, ,, (after acetylation)	23 ,, 28·5

The yield varied from 0·5 to 4·4 per cent. The oil consisted chiefly of hydrocarbons; it contained only small proportions of alcohols and esters. It was found that the oil-content increases as the plant reaches maturity, and that the flowers contain more oil than the leaves. Plants gathered at high altitudes contain more oil than do those from the plains. The density, optical rotation and index of refraction increase as the plant develops; they attain their maximum during the flowering period, at the end of which they diminish again. The ester-content of the leaves diminishes during the flowering and is greatest in the plants from the mountains. The leaf-oils contain more ester than do the flower-oils; plants grown in a moist atmosphere are deficient in ester. The specific gravity of oils rich in ester is higher, and their rotatory power lower than that of oils which are deficient in ester. The flowers contain a larger proportion of free alcohol than the leaves, and during the flowering period the content of free alcohols increases at the expense of that of the alcohols in combination. The terpene present boils at 167° to

[1] *Gazz. Chim. Ital.*, 41 **(1911)**, i. 395.

169° C.; it has the specific gravity 0·8416 at 14° C.; α_D + 35·70; n_D 1·4862. It polymerises to a white amorphous mass, melting-point 90° C. Dextro β-phellandrene is also present.

Francesconi and Sernagiotto[1] have investigated the alcohol bupleurol, which they isolated recently from the oil. It is best isolated from the high boiling fractions of the oil by means of phthalic anhydride. It has the composition $C_{10}H_{20}O$, and boils at 209° to 210°, and has a specific gravity 0·8490, refractive index 1·4508, and is optically inactive. It has a slight but distinct odour of roses. From its general character it is probably an aliphatic compound. It forms an oily dibromide, and a crystalline urethane, which forms lustrous needles melting at 45°. On oxidation with chromic acid it yields (1) an aldehyde yielding a semi-carbazone melting at 135°; (2) an aldehyde yielding a semi-carbazone melting at 97°; (3) a ketone boiling at 217°; and having a refractive index 1·4419 at 25°; this ketone yields a semi-carbazone melting at 189° to 190°; (4) a red oil which boils at 207°, and was found to be the ester of bupleurol and its corresponding acid, which is formed during the oxidation process. Bupleurol is isomeric with citronellol and with androl. Its formula is probably $CH . (CH_3)_2(CH_2)_3 . C(CH_2) . CH_2 . CH_2 . OH$, which would make it a dihydro derivative of the alcohol nerol. A body, probably an aldehyde, was also isolated at the same time as bupleurol. This body has the formula $C_{10}H_{16}O$, and has the following characters: Specific gravity 0·9264; specific rotation + 14·93°; and refractive index 1·4909.

Oil of Seseli Bocconi.

This oil, distilled from *Seseli Bocconi* (*Crithmum siculum*) to the extent of about 0·3 per cent. has a specific gravity of about 0·855.

According to Francesconi and Sernagiotto,[2] the essential oil contains, in addition to laevo-pinene and *beta*-phellandrene, a di-cyclic aldehyde, another carbonyl compound accompanying the aldehyde, a di-cyclic primary alcohol, an unsaturated secondary alcohol, and formic, acetic, and methyl-butyric acids. The primary alcohol yields, on oxidation with potassium bichromate and sulphuric acid, an aldehyde apparently having the formula $C_{10}H_{16}O$, and being identical with the aldehyde naturally occuring in the oil. It yields a semi-carbazone melting at 148° to 158°. The primary alcohol has the formula $C_{10}H_{18}O$.

Oil of Sea Fennel.

Crithmum maritinum is a plant growing freely on the French coast. According to Borde[3] the stems and leaves on the one hand, and the seeds on the other, yield essential oils, the former to the extent of up to 0·3 per cent., and the latter to 0·8 per cent. The two oils have the following characters:—

	Stems and Leaves.	Seeds.
Specific gravity at 4° .	1·0374 to 1·0492	0·9661 to 0·9809
Optical rotation . .	+ 6° 42′ ,, + 8° 15′	+ 5° 27′ ,, + 8° 9′
Saponification value .	6·4	4 ,, 10

The seed oil contains about 12 to 18 per cent. of *d*-pinene, 40 to

[1] *Att. R. Accad. Lincei* (**1913**), 22, 34. [2] *Ibid.* (**1913**), 116.
[3] *Bull. Soc. Pharm.*, 16, **132**, 395.

45 per cent. of other terpenes, and 40 to 50 per cent. of dill-apiol $C_6H(OCH_3)(O_2CH_2)(CH_2 . CH : CH_2)$.

Delépine[1] has further identified dipentene, cymene, the methyl ether of thymol, and an alcohol with a rose-like odour but not yet identified, as constituents of the oil.

Italian sea fennel oil has been examined by Francesconi and Sernagiotto,[2] who found the oil to have the following characters:—

Specific gravity	0·9816 at 29°
,, rotation	+ 3·18°
Refractive index	1·4978
Saponification number	6·5
,, ,, (after acetylation)	11·24

The principal constituents of the oil were found to be β-phellandrene, dill-apiol, and a terpene, which the authors have named crithmene, and which has the following characters:—

Specific gravity	0·8679 at 12°
Refractive index	1·4806
Boiling-point	178° to 180°

It yields two nitrosochlorides, melting at 101° to 102° and 103° to 104° respectively.

Sardinian sea-fennel oil has been found to contain *p*-cymene.

Oil of Masterwort.

Masterwort Oil is distilled from the root of *Imperatoria ostruthium*, a plant indigenous to the mountain regions of Southern and Central Europe. The root yields about 1 per cent. of oil of characteristic aromatic odour, which, according to Hirzel, contains bodies resembling terpene alcohols. Wagner states that angelic aldehyde is present.

The oil has the following characters:—

Specific gravity	0·876
Optical rotation	+ 66°
Boils at	170° to 190°

Lange[3] has examined an oil distilled from two-year old Tyrolese plants. The yield was 1 per cent.

The oil had the following characters:—

Specific gravity	0·8627 at 20°
Optical rotation	+ 59° 30′
Acid value	0·8
Ester ,,	17·9
,, ,, (after acetylation)	28·34

The esters of isobutyric, isovalerianic, isopropylidene-acetic, formic and acetic acids were present, also free palmitic acid; 95 per cent. of the oil consisted of terpenes, among which were: dipentene, dextro-limonene, α-pinene, and dextro-phellandrene. An alcohol $C_{10}H_{20}O$ and a sesquiterpene yielding a dihydrochloride, melting-point 157° to 157·5°, are also present. Angelic aldehyde, described by Wagner, could not be found.

[1] *Comptes rendus*, 149 (**1909**), 215; 150 (**1910**), 1061.
[2] *Att. R. Accad. d. Lincei, Roma* (5), 22 (**1913**), 231, 321, 382.
[3] *Year Book of Pharmacy* (**1912**), 100.

Oil of Sumbul.

Sumbul Oil or *Musk-root Oil* is obtained from the dried root of *Ferula sumbul*. The yield of oil is only from ·2 to 1·1 per cent. The oil has the following characters:—

Specific gravity	0·932 to 0·964
Optical rotation	$-6° 20'$
Acid value	7
Ester ,,	20 to 90

On standing for a few days after distillation crystals are deposited, which melt at 113° to 114°.

It has a distinct musk-like odour, and in India the root of *Dorema ammoniacum* is often substituted for it and used under the name of sumbul root.

Oils of Parsnip.

Various parts of *Pastinaca sativa* yield essential oils. The seed oil contains the propionic and butyric esters of octyl alcohol. Schimmel & Co.[1] have investigated the oils distilled from the under-mentioned parts of the plant.

They submitted separately to distillation with water vapour, the ripe dried seeds, the umbels, and the roots of the plants.

1. Oil from dried seed of *Pastinaca sativa*.—The bright yellow oil obtained in a yield of 1·47 per cent. had the following properties: $d_{15°}$ 0·8736; a_D $-0° 9'$; $n_{D20°}$ 1·43007; acid number 4·4; ester number 240·6; ester number after acetylation 276; soluble in $2\frac{1}{2}$ and more volumes 80 per cent. alcohol.

2. Oil from dry umbels of *Pastinaca sativa*.—The oil had a dark brown colour, and had a very remote odour of oil of ambrette seeds. It was obtained in a yield of 0·3 per cent.; $d_{15°}$ 1·0168; a_D $-0° 50'$; $n_{D20°}$ 1·50049; acid number 4·2; ester number 62·9; ester number after acetylation 86·2; soluble in 6·5 volumes 80 per cent. alcohol, with separation of paraffin.

3. Oil from the dry roots of *Pastinaca sativa*.—The bright yellowish oil which in odour reminded somewhat of vetiver oil, was obtained in a yield of 0·35 per cent., and had the following properties: $d_{15°}$ 1·0765; a_D $-0° 10'$; $n_{D20°}$ 1·52502; acid number 3·9; ester number 12·6; ester number after acetylation 33·7; not completely soluble in 10 volumes 80 per cent. alcohol; soluble in 0·6 and more volumes 90 per cent. alcohol.

The oil from the wild plant seeds has the following characters:—

Specific gravity	0·869 to 0·885
Optical rotation	0° to $+0° 13'$
Refractive index	1·4280 to 1·4350
Acid value	2 ,, 10
Ester ,,	218 ,, 270
,, ,, (after acetylation)	260 ,, 296

Two samples of parsnip oil examined by Roure-Bertrand Fils had the following characters. Sample No. 1 was a German distilled oil from the seeds only; sample No. 2 was an Italian distilled oil from the stems with flowers and seeds:—

[1] *Bericht* (**1908**), 96.

	No. 1.	No. 2.
Specific gravity at 15° C.	0·8712	0·8970
Optical rotation	± 0°	+ 0° 6′
Solubility in 80 per cent. alcohol	1 volume and over	2 volumes and over
Acid value	9·1	5·6
Saponification number	250·6	228·9
,, ,, (after acetylation)	293·0	251·1

On dilution with 90 per cent. alcohol the Italian sample deposits a large quantity of paraffins. When the oil itself is subjected to refrigeration it thickens a little but does not deposit solid matter.

Oil of parsnip contains a considerable amount of esters of octylic alcohol.

These esters are composed chiefly of normal octyl butyrate.

Oils of Peucedanum.

The root of *Peucedanum officinale* and the fruit of *Peucedanum grande* yield small quantities of oil, having the following characters :—

	P. officinale.	*P. grande.*
Specific gravity	0·902	0·901
Optical rotation	+ 29° 4′	+ 36°

Oil of Eryngium Fœtidum.

The oil distilled from this plant, which is known as *Walang Dœrr* in Java, has the following characters :—

Specific gravity	0·905 at 26°
Optical rotation	+ 0° 42′
Acid value	29·5
Saponification value	65
,, ,, (after acetylation)	327

Oil of Sium Cicutæfolium.

The fresh herb, *Sium cicutæfolium,* yields, according to Rabak,[1] 0·5 per cent. of essential oil, having the following characters :—

Specific gravity	0·8447 at 22°
Optical rotation	+ 63° 40′
Acid value	0
Ester ,,	33

It contains *d*-limonene and an aldehyde, which has not yet been identified.

Oil of Conium Maculatum.

The herb *Conium maculatum* (deprived of flowers) yields about 0·08 per cent. of essential oil, which, according to Haensel, has the following characters :—

Specific gravity	0·950
Acid value	60
Ester ,,	10

The fruit yields 0·018 per cent. of oil of specific gravity 0·895 and saponification value 34.

[1] *Midland Drug and Pharm. Rev.*, 43 **(1909)**, 5.

Oil of Meum Athamanthicum.

The fresh roots of *Meum Athamanthicum* yield about 0·5 per cent. of essential oil which Schimmel & Co. find to have the following characters:—

Specific gravity	0·9807
Optical rotation	− 1° 55′
Refractive index	1·51898
Acid value	10·3
Ester ,,	151·2

Oil of Cnidium Officinale.

The oil from the roots of this plant has been examined by Sakei.[1] He obtained 0·82 per cent. of a bitter oil of specific gravity 1·030 to 1·040. It contains an unsaturated acid $C_{12}H_{19}O_3$ (?), an alcohol of the formula $C_{10}H_{18}O$ and possibly a lactone of the formula $C_{12}H_{18}O_2$.

MYRTACÆ.

Myrtle Oil.

This oil is distilled from the leaves of *Myrtus communis*, probably a native of the south of Europe, although possibly it was introduced here from Western Asia. It flourishes on the Mediterranean littoral, especially in Spain, Italy, and the south of France. The leaves yield from 0·25 to 0·35 per cent. of essential oil. So long ago as 1867 Gladstone examined this oil, but his results are of little value to-day. Jahns has shown that this oil contains dextro-α-pinene, cineol, and a camphor of the formula $C_{10}H_{16}O$. Camphene, and probably a terpene not yet identified at present. Dipentene is present in small quantities. Under the name "myrtol" the fraction of the oil boiling between 160° and 170° has been introduced into commerce. It is not a definite com pound, and probably possesses little value. Von Soden and Elze[2] have found in the higher boiling fractions of this oil an alcohol, which they term myrtenol. It occurs chiefly as its acetic ester.

This alcohol has been examined by Semmler and Bartelt,[3] who obtained it by saponifying the higher boiling fractions of the oil and then fractionating the same. The alcohol has the composition $C_{10}H_{16}O$ and possesses the following characters:—

Boiling-point at 760 mm.	222° to 224°
,, ,, 9 mm.	102·5°
Specific gravity at 20°	0·9763
Optical rotation	+ 45° 45′
Refractive index	1·49668

In addition to myrtenol, small quantities of the alcohols gereniol and nerol are present.

Myrtle oil has the following characters:—

Specific gravity	0·880 to 0·926
Optical rotation	+ 8° ,, + 27°
Refractive index	1·4625 ,, 1·4685
Acid value	0 ,, 2
Ester ,,	13 ,, 135
,, ,, (after acetylation)	30 ,, 190

[1] *Schweiz. Apotheker. Zeit.*, 55 **(1917)**, 226.
[2] *Chem. Zeit.* **(1905)** 1031. [3] *Berichte*, **40 (1907)**, 1363.

The oil varies so much in composition that no solubility values can be usefully assigned to it. Messrs. Schimmel and Co. have given the following figures for oils distilled in different countries:—

	Specific Gravity.	Rotation.	Ester Value.
France	0·890 to 0·904	+ 15° to + 25°	19 to 43
Spain	0·913 „ 0·925	+ 22° „ + 25° 20′	68 „ 86
Corsica	0·883 „ 0·887	+ 22° „ + 27°	13 „ 25
Syria	0·893 „ 0·922	+ 11° „ + 26°	18 „ 31
Asia Minor . . .	0·913	+ 10° 42′	39·4
Cyprus	0·917	+ 8° 11′	20·9
Algeria	0·881 to 0·887	+ 26° to + 27° 30′	17 to 20·6
Dalmatia . . .	0·925	+ 13° 20′	134·8

Oil of Cloves.

The clove tree is a beautiful evergreen, indigenous to the Molucca Islands. It is cultivated in several of the islands to the south of the Moluccas, and in Sumatra, Penang, Malacca, Madagascar, the Seychelles, Bourbon, Mauritius, the West Indies, Zanzibar, and Pemba, our chief supply coming from the two last-named islands. The tree is *Eugenia caryophyllata* (*Caryophyllus aromaticus* Linn.), and most parts of the tree have a very aromatic odour. The spice we know as cloves consists of the dried unexpanded flower buds, which are used in the distillation of the oil. The flower stems are also used, but the oil, although of very similar character, is not of so fine an aroma. As a spice Penang cloves fetch the highest price, but Zanzibar cloves are highly esteemed in commerce; it must be remembered, however, that the majority of "Zanzibar" cloves in reality come from the adjacent island of Pemba.

According to a writer in the *Perfumery and Essential Oil Record*,[1] much vagueness surrounded the source of cloves, the first definite statement being by the Arabian geographer, Ibn Khurdâdbah (A.D. 869 to 885), when he named the spice, together with cocoanuts, sugar, and sandalwood as a product of Java. Marco Polo, the Venetian traveller of the thirteenth century, found cloves in Java, and assumed them to be indigenous to that island. Later navigators showed that both these observers had been misinformed, but it must be borne in mind that the name Java was applied in a general sense by Arabian geographers to the Islands of the Eastern Archipelago. It remained for Niccolo del Conti, another Venetian pioneer trader of the fifteenth century, to clear up the matter. He learned in Java that cloves did not actually grow there, but were brought from the Banda Islands, fifteen days' sail farther east.

At the beginning of the sixteenth century Portugal was the premier maritime power of the world, and while the Spaniards were opening up the new and unknown West, the Portuguese were penetrating to the old but vaguely understood East, and just a few of its world-old mysteries were being revealed to European knowledge. One of these mysteries solved was the origin of certain Oriental spices, notably the

[1] *P. and E.O.R.* (1916), 20.

clove tree, described by Pigafetta, a shipmate of Magellan, as he saw it growing in 1521. For nearly a century the Portuguese had the lion's share of the clove trade. Holland then came to the fore as a naval Power in the East, and the Molucca Islands were wrested from the Portuguese by the Dutch, who adopted drastic measures to secure a monopoly of the spice trade.

In spite of all their efforts their enemies and competitors managed to run many a rich cargo of Eastern products right through to England. In 1609 the East India Company's ship "Consent" arrived in London with 112,000 lb. of cloves, the duty alone on which amounted to £1400, and the impost as much more. The price in bond at that time was from 5s. 6d. to 5s. 9d. per lb. ex bale.

The Dutch, in desperation, endeavoured to eradicate the clove tree from its native islands, and even made annual attacks on all young trees that might have sprung up meanwhile, with a view to confining the growth of this spice to a group of small islands, of which Amboyna is the chief. This practice has only of late years been discontinued (since 1873), and though there is no restriction of cultivation in other islands, the "clove parks" of Amboyna remained the property of the Netherlands Government. It is said that the original Moluccas, or Clove Islands, now produce no cloves at all. By the eighteenth century France had become an important factor in Eastern affairs. The French Governor of Mauritius (or "Isle de France," as it was then styled), De Lozier Bouvet, adopted the suggestion of the financier, Pierre Poivre, to introduce spice cultivation into the island, in spite of the rigorous precautions of the Dutch.

It was not until 1770 that Poivre could revert to his long-cherished scheme regarding these valuable spice trees. The spice plantations thrived, and became famous under the name of "King's Garden," and to-day have developed into the "Royal Botanical Garden," Pamplemousses.[1] Three years later, in 1773, the French transported the clove tree to their Western Colony of Cayenne, where, however, it does not appear to have flourished with any marked success. At the close of the century, an Arab, by name Harameli-bin-Saleh, accompanied a French officer from Zanzibar to Bourbon (where the clove had been introduced about the same time as into Mauritius), and obtained leave to take back a small quantity of clove seeds and plants with him. This was the beginning of the clove cultivation in Zanzibar, Harameli making the first plantation at Mitoni, on the road to Chueni, and the enterprise was most successful. The local Swahili name for the tree is "garafu," presumably a corruption of the French "giroflier". The industry was almost destroyed in Zanzibar by the terrible hurricane of 1872, but in spite of this reverse has in later years recovered its position. Pemba, a small island thirty miles North-West of Zanzibar, is also an important seat of clove cultivation, though the quality of the product is not equal to that of the Zanzibar growth.

The clove also flourishes in the Straits Settlements. Many years ago it was introduced into Penang and cultivated by European planters and wealthy Chinese settlers. Penang cloves have a high reputation for quality, but the Eastern consumption is so great that not a great deal of the spice finds its way on to the European market.

[1] *Mauritius Illustrated* (**Macmillan**), 1916.

The spice usually known as cloves, and which is generally employed for the distillation of the oil, consists of the dried unexpanded flower buds.

The flower stalks, separated from the cloves whilst they are drying, are also dried and exported, chiefly from Zanzibar and Pemba, and yield on distillation 5 to 6 per cent. of oil, as against 14 to 21 per cent. obtained from the buds. The stalks are known in Zanzibar as Vikunia, and in France as "Griffes de Girofle".

The dried fruits are occasionally exported, and are known as mother-cloves. A small amount of essential oil is obtained from them.

Clove buds yield from 14 to 21 per cent. of a heavy essential oil, having the following characters:—

Specific gravity	1·044 to 1·069
Optical rotation	− 0° 20′ „ − 2° 30′
Refractive index	1·5290 „ 1·5355
Eugenol (by absorption)	78 to 98 per cent.

The oil is soluble in 1 to 3 volumes of 70 per cent. alcohol.

The following figures represent samples examined by the Imperial Institute Chemists, the author, and Schimmel & Co., all being of authentic origin:—

	Specific Gravity.	Refractive Index.	Rotation.	Eugenol (by Absorption).
Amboyna	1·0490	1·5305	− 1°	82 per cent.
Seychelles	1·0470	1·5295	− 1° 25′	82 „
„	1·0485	—	− 1° 6′	86 „
„	1·0510	1·5320	− 1° 30′	87 „
„	1·0510	1·5325	− 1° 20′	88 „
Ceylon	1·0456	—	− 2° 30′	87 „
„	1·0477	—	− 2°	85 „
Mauritius	1·0514	—	− 0° 23′	89 „
„	1·0614	—	− 0° 24′	93 „
„	1·0681	—	− 0° 20′	96 „
„	1·0590	1·5345	− 0° 50′	92 „
„	1·0570	1·5338	− 0° 30′	93 „
Zanzibar	1·0550	1·5310	− 1°	92 „
„	1·0525	1·5308	− 1° 30′	91 „
„	1·0530	1·5316	− 1° 20′	94 „

An examination of the oil yielded by Zanzibar cloves from trees of different ages, has been made by the chemists at the Imperial Institute [1] with the following results:—

No		Yield.	Specific Gravity.	Rotation.	Eugenol.
1	8 to 9 years' old	17·8 %	1·056	− 0° 23′	89 %
2	45 „	19·2 %	1·064	− 0° 30′	88 %
3	60 „	18·8 %	1·049	− 0° 29′	84 %
4	60 „ [2]	17·4 %	1·050	− 0° 37′	88 %
5	60 „ [3]	6·5 %	1·056	− 0° 33′	90 %
6	Stems from 1	5·9 %	—	—	—
7	„ 2	6·3 %	1·050	− 0° 42′	89 %

[1] *Bull. Imp. Inst.*, 12 (**1914**), 337. [2] Tree in full flower.
[3] From the mother-cloves (fruit).

The value of oil of cloves depends almost entirely on the percentage of the phenol eugenol (especially when it is required for vanillin manufacture), and as will be seen from the following paragraphs, the method of determination used should always be stated if results are to be comparative :—

Several methods have been suggested for the determination of the eugenol, none of which give absolutely accurate results, but from them useful comparative results may be obtained. Umney suggested the absorption of the eugenol in a Hirschsohn flask, as used for the examination of cassia oil, with a 10 per cent. solution of caustic potash. This strength was found to be too high, and a 5 per cent. or even 3 per cent. solution is now preferred. The unabsorbed portion has a specific gravity of about 0·910, so that this may be allowed for (although rarely done in practice), since the eugenol has a much higher specific gravity. Spurge prefers to saponify the esters first, and to make an allowance for them, but if a 5 per cent. solution be used, this method without any such allowance gives a close approximation to the total amount of eugenol, both free and combined being reckoned as eugenol. The globules of uncombined hydrocarbons have a great tendency to stick round the top of the flask and require some "coaxing" to rise and agglomerate in the neck of the flask. Heat will accelerate and assist this, however. But the source of error lies in the fact that the aqueous solution of potash and potassium eugenate dissolves some of the sesquiterpene, which is thus reckoned as eugenol, and a too high result is obtained. The process proposed by Thom, although more tedious, gives more exact results. This depends on the conversion of the eugenol into benzoyl eugenol. The following are the details, which should be carefully observed in order to secure accurate results :—

About 5 grms. of clove oil are weighed into a beaker of about 150 c.c. capacity, 20 grms. of 15 per cent. sodium hydroxide solution added, and then 6 grms. benzoyl chloride. On stirring, the solid mass of eugenol sodium salt at first formed goes into solution again as it is converted into benzoic ester, with evolution of much heat. In the course of a few minutes the reaction ends, and on cooling a solid crystalline mass of benzoyl eugenol is obtained. To this, 50 c.c. water is added, and the whole warmed on a water-bath until the ester is completely melted to an oil, well stirred, cooled, and the clear supernatant aqueous solution filtered off. The crystalline mass is again washed with two successive quantities of 50 c.c. water, and the resulting impure benzoyl eugenol is recrystallised from alcohol, due allowance being made for its solubility in that medium. 25 c.c. of hot alcohol (90 per cent. by weight strength) are poured through the filter employed in the previous washing operations, in order to dissolve any adherent crystals, into the beaker, and the whole warmed upon the water-bath until complete solution is effected. The solution is then cooled to 17° C., and the crystalline precipitate thrown upon a small weighed filter paper, filtered into a 25 c.c. cylinder, and washed with 90 per cent. alcohol until the filtrate exactly measures 25 c.c. The filter and crystals are then removed to a weighing bottle, dried at 100° C. until constant, and then weighed. From the total weight the weights of the filter paper and of the weighing bottle are deducted, from which the benzoyl eugenol is calculated. To the latter weight 0·55 grm. are added, being the weight of pure benzoyl eugenol dissolved by 25 c.c. 90 per cent. alcohol at 17° C. as determined by experiment.

This final quantity gives the amount of benzoyl eugenol, from which the amount of eugenol is easily calculated, eugenol having the formula $C_{10}H_{12}O_2$, and benzoyl eugenol $C_{17}H_{16}O_3$, so that $\frac{164x}{268y} \times 100 =$ the percentage of eugenol, if x equals the weight of benzoyl eugenol obtained, and y the weight of oil used in the estimation. Under these circumstances the eugenol-content should not fall below 75 per cent., or if estimated by absorption with potash not below 82 per cent., usually from 85 to 90 per cent.

Thom has recognised the inaccuracy of the foregoing, and now recommends the following process. It consists in heating 5 grms. of the oil in a water-bath with 20 c.c. of a 15 per cent. soda solution for thirty minutes. After allowing the hydrocarbons to separate, the eugenol soda solution is run off, and the hydrocarbons washed with dilute soda solution twice, the washings being added to the original soda solution. The reaction is now effected at water-bath temperature with 6 grms. of benzoyl chloride. The whole is allowed to cool, and the crystalline mass transferred to a beaker with 55 c.c. of water. It is heated in order to melt the crystals, and well agitated with the water to wash the benzoyl eugenol. This washing is repeated twice. The crystalline mass is then transferred to a beaker with 25 c.c. of 90 per cent. alcohol, and warmed till complete solution takes place. The solution is allowed to stand till the bulk of the crystals have separated out, and is cooled to 17° and filtered through a paper 9 cm. in diameter, previously dried and tared. The filtrate measures about 20 c.c. and the crystals are washed with more alcohol until it measures 25 c.c. The paper and crystals are then dried in a weighing-glass and weighed, the temperature of drying being not more than 101° C. The solubility allowance for 25 c.c. of alcohol is 0·55 grm. The total eugenol is calculated from the formula

$$P = 4100 \frac{a + 0{\cdot}55}{67\, b}$$

where P is the percentage, a the weight of benzoyl eugenol obtained, and b is the weight of clove oil used.

Verley and Bolsing propose the following method. It depends on the fact that acetic and other anhydrides react with phenols in excess of pyridine. Eugenol reacts readily, forming eugenyl acetate and acetic acid, the latter combining with pyridine to form pyridine acetate. This compound reacts towards indicators such as phenolphthalein in the same way as acetic acid and therefore a titration is possible. Verley and Bolsing use from 1 to 2 grms. of the oil, which is placed in a 200 c.c. flask, and 25 c.c. of a mixture of acetic anhydride (15 parts) and pyridine (100 parts). The mixture is heated for thirty minutes on a water-bath, the liquid cooled, and 25 c.c. of water added. The mixture is well shaken and titrated with normal potash, using phenolphthalein as indicator. A blank experiment is carried out without the eugenol, and the difference between the titration figures in c.c. of normal alkali, multiplied by 0·582, gives the amount of eugenol in the sample taken.

As the boiling-points of eugenol and caryophyllene are both high, no appreciable portion of the oil should distil below 245°, or even a few degrees higher. These tests will ensure the presence of a due proportion of eugenol, a point very necessary to take into account, because of the frequent abstraction of part of the eugenol, which is used in the manufacture of vanillin, or sold as pure eugenol. Absence of turpentine of petroleum is further ensured by the ready solubility of the oil in 90 per cent. alcohol.

The principal constituent of oil of cloves is the phenol eugenol, together with about 3 per cent. of aceteugenol (or, according to Spurge,[1] from 7 to 17 per cent.—which is unlikely). The sesquiterpene caryophyllene contributes the principal "neutral" constituent of the oil. Traces of esters of eugenol, methyl alcohol, furfurol, vanillin, salicylic acid, methyl benzoate, amyl-methyl ketone, and heptyl-methyl ketone have also been detected. The sesquiterpene caryophyllene exists in both the α- and the β-modifications. Valeric aldehyde is probably present in traces.

A series of newly-discovered constituents, present in traces only, has also been described by H. Masson.[2] These bodies are as follows: methyl-amyl-carbinol, methyl-heptyl-carbinol, benzyl alcohol, and dimethyl-furfurol.

A body termed caryophyllin, $C_{10}H_{16}O_4$, has been stated to be present.[3]

The oil from clove stems does not differ materially in characters from that of ordinary cloves. The specific gravity, however, may fall as low as 1·040. The eugenol-content rarely falls below 85 per cent. The stem oil has been found to contain traces of naphthalene, a sesquiterpene alcohol, $C_{15}H_{26}O$,[4] and a crystalline body $(C_{21}H_{30}O)_5$ [5] (?), melting at 146°.

The oil from mother-cloves has been examined by Schimmel & Co., who found it to have the following characters:—

Specific gravity	1·0983
Optical rotation	− 3° 11′
Refractive index	1·54332
Phenols	88 per cent.

The phenols consisted of a mixture of eugenol and a solid phenol melting at 44° to 45·5°, and boiling (at 754 mm.) at 309° to 310°.

Clove leaves yield about 4 to 5 per cent. of oil having the following characters:—

Specific gravity	1·0320 to 1·067
Optical rotation	− 0° 30′ to − 2°
Eugenol	75 to 93 per cent.
Refractive index	1·532 to 1·539

Oil of Eugenia Smithii.

Dawkins[6] has obtained 0·44 per cent. of oil from the leaves of this tree. It has the following characters:—

	(1)	(2)
Specific gravity	0·866	0·863
Optical rotation	+ 35°	+ 34·6°
Refractive index	1·4701	1·4675

It contains 80-90 per cent. of *d*-α-pinene, 4 per cent. of esters, and 3·7 per cent. of alcohols.

[1] *Pharm. Jour.*, 70 (**1903**), 701, 757.
[2] *Comptes rendus*, 149 (**1909**), 630, 795.
[3] *Monatshefte* (**1905**), 379.
[4] *Berichte*, 45 (**1912**), 1392.
[5] *Ibid.*, 42 (**1909**), 380, 680.
[6] *Proc. Roy. Soc. Victoria, Chem. Abstracts*, (**1917**), ii. 2715.

OILS OF EUCALYPTUS.

The oils distilled from the leaves of the various species of this enormous genus of the *Myrtaceæ* are in many cases so dissimilar that the generic term, eucalyptus oil, is meaningless unless the species is quoted. Several species, however, yield oils which are very similar to each other, and on this basis the *British Pharmacopœia* describes as eucalyptus oil the product obtained from *Eucalyptus globulus* and other species, giving definite characters of the oil, the two chief of which are the presence of a due proportion of cineol (eucalyptol) and the absence of appreciable quantities of the terpene phellandrene. The great variability in the oil from different trees will make it convenient to deal with this oil, firstly with reference to the cineol value, and afterwards with especial reference to the characters of the oils from those trees which in many cases are in no way similar to "*Eucalyptus oil, British Pharmacopœia*". It is commonly believed that whatever medicinal value this oil has—and it has been, and still is, used to a very large extent as a prophylactic against influenza—is due to the presence of cineol, or, as it was previously called, eucalyptol, so that whether this be, in fact, the valuable constituent or not, in commerce the value of the oil is made to depend on its cineol-content.

The quantitative determination of the cineol present is a matter of importance. The earlier attempts in this direction gave distinctly too low results. The chief method adopted was a series of fractionations, retaining all those fractions distilling between 173° and 190°, cooling them to - 15°, filtering off the still liquid portion, retaining the crystals of cineol, refractionating the liquid portions and again freezing, subsequently weighing the crystallised cineol. An average loss of about 10 per cent., in the author's opinion, takes place in this process, assuming the oil to contain 50 to 60 per cent. of cineol. Scammell's process above-mentioned has been more successfully applied to the quantitative determination of cineol, and yields the most satisfactory results so far, although an error of several per cent. is scarcely avoidable.

There is some diversity of opinion on this matter, some chemists maintaining that a very close result is obtained, others that very large errors occur. In the author's opinion, in the hands of a careful worker, results accurate to within 3 per cent. can be obtained, so long as the cineol-content is not below 40 per cent. If the amount of cineol is lower than this, the method is inapplicable, and the oil must previously be fractionated.

To a known weight of oil from 1 to 1½ times its weight of phosphoric acid of specific gravity 1·75 should be added, drop by drop, the oil being kept cold and continually stirred. The crystalline magma formed is pressed between filter paper, after as much as possible has drained off; and when the adherent terpenes and phosphoric acid have been removed as far as possible, the crystals are decomposed by hot water in a graduated tube. On cooling, the cineol is measured, and from its specific gravity (·930) the weight is easily calculated. The separated cineol should readily crystallise on cooling to - 3°, otherwise it must be regarded as impure and the process repeated. Oils rich in cineol yield a correspondingly high fraction distilling between 170° and 190°.

If the oil be first diluted with petroleum ether, before treatment with phosphoric acid, the results are rather more concordant in the hands of different analysts.

Schimmel & Co. recommend the absorption of the cineol by a 40 to 50 per cent. solution of resorcin in water, and reading the unabsorbed portion in the neck of a Hirschsohn flask.

Ten c.c. of the oil containing cineol are mixed in a cassia flask of 100 c.c. capacity with so much 50 per cent. resorcinol solution that the flask is filled for about four-fifths. The mixture is shaken thoroughly for five minutes, and the oil portions which have not entered into reaction are brought into the neck of the flask by adding resorcinol solution, and their volume determined. By subtracting the volume from 10 the cineol-content of the oil is obtained, which is then expressed in per cent. by volume by multiplication with 10.

Messrs. Schimmel & Co. have since modified the method by recommending the oil to be first fractionated and the portion boiling between 170° to 190° C. to be treated with the resorcin solution. In some cases this gives results fairly concordant with those of the phosphoric acid method usually adopted, as the following results will show:—

	Cineol by Direct Absorption with Resorcin Solution.	Cineol by Absorption with Resorcin after Fractionation.	Cineol by Phosphoric Acid Process.
Cajuput Oil, normal . .	62 per cent.	54 per cent.	52 per cent.
,, ,, abnormal .	53 ,,	17 ,,	14 ,,
Eucalyptus Oil I. . .	82 ,,	68 ,,	65 ,,

C. T. Bennett[1] has, however, shown that this method is quite unreliable, only yielding accurate results in certain isolated cases.

In other cases, however, the results obtained are obviously too high. An oil which yielded 95 per cent. distilling between 170° to 190° C., all of which was absorbed by resorcin solution, had the following characters:—

Specific gravity	0·924
Optical rotation	Nil
Cineol by phosphoric acid method	78 per cent.
,, by resorcin method	95 ,,

Other samples gave results as under:—

	Cineol by Direct Absorption.	Cineol by Absorption after Fractionation.	Cineol by Phosphoric Acid Process.
Eucalyptus Oil A	100 per cent.	89 per cent.	70 per cent.
,, ,, B	Crystallised	97 ,,	75 ,,
,, ,, C	,,	92 ,,	73 ,,

It is evident that this modified method is not absolutely accurate since other constituents, besides cineol, are undoubtedly included in the portion boiling between 170° and 190° C. A further objection is the separation of the solid crystalline double compound with oils rich in cineol.

[1] *P. and E.O.R.* (**1912**), 269.

The method has been criticised by H. G. Harding,[1] who states that a pure white oil from *Eucalyptus dives*, which contained no cineol, showed an absorption of 32 per cent. by the resorcinol test. Rectifying the oil and applying the test to the portion distilling between 170° and 190° does not entirely remove the difficulty, as the results are always slightly high, owing to retention of other constituents by the resorcinol solution. His experiments show that the method is more reliable when the percentage of cineol does not exceed 40 to 50 per cent., oils containing a higher percentage requiring to be diluted. Ordinary turpentine oil must not be used for dilution, as it is likely to produce serious errors. By employing the fraction of turpentine distilling between 156° and 160° for dilution the error is minimised, and the following method is recommended: 100 c.c. of the oil to be tested is distilled, the portion distilling between 170° and 190° collected, and this is diluted to 100 c.c. with the turpentine fraction. If a trial shows the percentage of cineol to be above 70 per cent., the cineol fraction is further diluted with the turpentine so that the percentage is not over 50. The temperature is noted, and 6 to 10 c.c. is shaken with warm 55 per cent. resorcinol solution. After five minutes' shaking more solution is added so as to bring the oil into the graduated neck. It is then cooled and the volume read.

A method for the determination of cineol has recently been proposed by Dodge.[2]

The process is based on the fact that the terpenes of the essential oils to be examined are readily oxidised at 0° C. by a 5 to 6 per cent. solution of potassium permanganate, whilst the eucalyptol remains unattacked. Ten c.c. of essential oils are gradually added to 400 or to 100 c.c. of permanganate solution, according to whether the essential oil is more or less rich in terpenes. When the reaction is finished, the solution is allowed to stand in the cold for twelve to eighteen hours, with occasional agitation; sulphurous acid or a mixture of sodium sulphite and hydrochloric acid is added and the oily portion of the residue is brought up into the neck of the flask, from which it is removed by means of a slender pipette. This oil is washed with a little alkali, then transferred to a graduated tube where its volume is determined which indicates the percentage of eucalyptol.

This method cannot be relied on, as the action of potassium permanganate is such as to react with some and not with other bodies present in various types of eucalyptus oil.

The whole question of cineol determination in eucalyptus oil has recently been carefully studied by Turner and Holmes,[3] who consider that all methods so far published are either inaccurate or at best only approximate. They suggest the following process:—

The determination of cineol in cineol-bearing oils by means of arsenic acid is carried out as follows:—

Deliver from a pipette 10 c.c. of the oil into a glass dish (preferably a round-bottom one) of 50 c.c. capacity, which is imbedded in finely cracked ice. Add 10 c.c. of concentrated arsenic acid (containing about 85 per cent. arsenic acid), and stir until precipitation is complete. When the mixture ceases to congeal further, allow to stand ten minutes in the ice. *At this point* if the mixture forms a hard mass, indicating

[1] *Analyst* **(1914)**, 475. [2] *Jour. Ind. and Eng. Chem.*, **4**, 592.
[3] *P. and E.O.R.* **(1915)**, 21.

an oil rich in cineol, 5 c.c. of purified petroleum ether should be added, and the mass mixed well. Transfer immediately to a hardened filter paper by means of a pliable horn spatula, spread evenly over the surface of the paper, and lay a second hardened filter paper over the top. Outside of the hardened filters place several thicknesses of absorbent or filter paper, and transfer the whole to an ordinary letter-press, bringing to bear all the pressure possible for about one minute. Change the outside absorbent papers and press again, repeating the operation, if necessary, until the cineol arsenate is apparently dry and separates readily when touched with a spatula. The pressing is *not* complete when a hard mass remains which is broken up with difficulty. The method usually requires two changes of filter paper, pressing each time for about two minutes. If left too long in the press the compound may decompose. Now transfer completely the compound by means of the horn spatula to a glass funnel inserted into a 100 c.c. cassia flask with neck measuring 10 c.c., graduated in $\frac{1}{10}$ c.c. Wash the precipitate into the flask with a stream of hot water from a wash bottle, assisting the disintegration with a glass rod. Place the flask in boiling water and rotate until the compound is thoroughly broken up; add enough water to cause the cineol to rise into the neck of the flask, cool to room temperature, and read off the volume; on multiplying the latter by 10 the percentage of cineol in the oil is obtained.

In judging whether or not petroleum ether should be added the following rule should be observed: Add enough petroleum ether to soften the cineol arsenate, so as to obtain a plastic mass; the quantity necessary never exceeds 5 c.c., and decreases with oils containing less than 80 per cent. of cineol. The object of adding petroleum ether is merely to soften the hard mass and to aid in the separation of non-cineol constituents of the oil; a large excess of petroleum ether will decompose the compound.

The above method is applicable directly to all oils containing above 50 per cent. of cineol; in oils containing lower proportions of cineol the precipitate is not solid enough to permit convenient handling; and if the cineol-content drops below 25 per cent. the separation of cineol arsenate is not quantitative. It was found that the addition of an equal volume of eucalyptol to such oils (i.e. mixing 5 c.c. of the oil with 5 c.c. of eucalyptol) successfully overcomes this difficulty; it then only becomes necessary to subtract from the volume of cineol, as observed in the neck of the flask, 5 c.c., and to multiply the difference by 10, in order to obtain the percentage of cineol in the oil.

In order to test the reliability of the method the authors prepared, to begin with, various mixtures of cineol with turpentine oil, and ascertained their cineol-content in the manner above described, with the following results:—

1. 50 volumes per cent. cineol + 50 volumes per cent. turpentine oil; found 49·5 per cent.; 50 per cent. cineol.

2. 60 volumes per cent. cineol + 40 volumes per cent. turpentine oil; found 59 per cent.; 60 per cent. cineol.

3. 75 volumes per cent. cineol + 25 volumes per cent. turpentine oil; found 74 per cent.; 75 per cent. cineol.

The authors found that the application of the method to mixtures containing less than 50 per cent. cineol is not practicable.

The latest method suggested for the determination of eucalyptol is that of Cocking.[1] It is based on the formation of a compound of ortho-cresol and eucalyptol, which was first obtained by Bellucci and Grassi.[2] This compound melts, according to Cocking, at 55·2°, and he attempts to determine eucalyptol by noting the freezing-point of a mixture of 3 grams of the oil and 2·1 grams of ortho-cresol. His experiments were carried out with a mixture of pure eucalyptol and terebene, and with a mixture of pure eucalyptol and oil of copaiba. But as eucalyptus oil contains various other bodies, and further, since ortho-cresol freezes at about 30° so that a varying mixture of the compound (cresineol), melting at 55·2°, and ortho-cresol melting at 30°, is under consideration it is obvious that such an empirical method cannot be relied on. Bennett and Salamon[3] have examined this method, and consider that it is only useful as a rapid method in a works for grading eucalyptus oils all derived from the same variety of Eucalyptus.

Eucalyptus oil is produced principally in Australia and Tasmania, which practically supply the world's requirements. Small amounts are distilled elsewhere, as in Algeria, the South of France, Italy, Portugal, the Transvaal, Mexico, etc., etc., but commercially they are of little importance.

The systematic examination of the Eucalyptus oils of Australasia is almost entirely due to R. T. Baker and H. G. Smith, of the Technological Museum, Sydney, N.S.W., and the greater part of the details here recorded is acknowledged to their numerous valuable publications.

From the purely commercial point of view the following notes[4] will indicate the direction in which distillation of the oil for the markets is tending:—

Economically the industry is becoming of some importance to Australia. Most of the old species which used to do duty for oil production have been replaced by others economically better in every way, and the oil of *Eucalyptus globulus* only remains largely as a name, its pride of place being taken by the products of other species. *Eucalyptus dumosa*, *Eucalyptus oleosa*, etc., have also fallen back as oil-producing plants. The species now mostly worked for pharmaceutical oils in New South Wales and Victoria are *Eucalyptus polybractea*, the "Silver-leaf Mallee" of Victoria and the "Blue Mallee" of New South Wales; *Eucalyptus Smithii*, which is a good yielding species and perhaps the best of all eucalyptus oils for pharmaceutical purposes; and a recent species, *Eucalyptus Australiana*, which produces an excellent oil, and no doubt will be much heard of in the future. This oil gave 79·5 per cent. eucalyptol by the resorcinol method. It does not contain objectionable volatile aldehydes except in traces, nor does it contain aromadendral. In *Research on the Eucalypts* this species is referred to under *Eucalyptus amygdalina*, and it was there shown that if fractionally separated during the original distillation the first-hour oil was richer in eucalyptol. This discovery remained unused for over twelve years, but is now being employed commercially in the Nerrigundah and Yourie districts of New South Wales, where the phellandrene has been replaced by eucalyptol.

There are a few compounds which have been described as occurring in eucalyptus oils, which are, so far as is known, not found in other oils.

[1] *P. and E.O.R.*, **1920**, 282.
[2] *Gazz. Chim. Ital.*, 43, II., 712.
[3] *P. and E.O.R.*, **1920**, 302.
[4] Private communication from Mr. H. G. Smith to the author.

Eudesmol, which has been also named eucalyptus camphor, was first isolated from the oil of *Eucalyptus piperita* [1] by H. G. Smith, who found it to be a crystalline body melting at 79° to 80°, and having the formula $C_{10}H_{16}O$, and being, in all probability, an oxide. Semmler and Mayer,[2] however, consider it to be a bicyclic sesquiterpene alcohol, and this has since been confirmed by Semmler and Tobias,[3] who give the following as its characters :—

Melting-point	78°
Boiling-point	156° at 10 mm.
Specific gravity	0·9884 at 20°
,, rotation	+ 31° 21′ in chloroform solution
Refractive index	1·5160 at 20°
Formula	$C_{15}H_{26}O$

The same chemists have also made an examination of the sesquiterpene alcohol isolated from oil of *Eucalyptus globulus* by Schimmel & Co.,[4] which they have named globulol. This body has the following characters :—

Melting-point	88·5°
Boiling-point	283° at 755 mm.
Specific rotation	– 35° 29′ in chloroform solution
Formula	$C_{15}H_{26}O$

Smith [5] has also isolated a new aldehyde from certain eucalyptus oils which he has named aromadendral. It has been suggested by Schimmel & Co. that this was identical with cuminic aldehyde, which had previously been found in eucalyptus oil by themselves. An examination of Smith's work on the subject, however, leaves no room for such an assumption, and it is clear that aromadendral is not identical with cuminic aldehyde.

Baker and Smith have recorded the following details (see pages 342-44), of 118 species of eucalyptus oils, arranged in groups according to their dominating constituent.

Aromadendral was extracted fron the higher-boiling portion of the oil of *Eucalyptus rostrata* by the aid of sodium bisulphite in the usual manner. When purified it had an odour reminiscent of cumin aldehyde, was almost colourless, being slightly yellow in appearance, and was highly laevo-rotatory. The specific gravity at 20° C. = 0·9534; specific rotation $[\alpha]_D$ – 73·94°; refractive index at 20° = 1·5066; the oxime melted at 84° to 85°, and the phenylhydrazone at 105° to 106° C. By arranging these results in tabulated form with those of cuminaldehyde the differences appear distinctly marked :—

	Cuminaldehyde.	Aromadendral.
d 15° C.	0·9818	0·9569
Optical Rotation	Nil	$[\alpha]_D$ – 73·94°
Melting-point Oxime	58° to 59°	84° to 85°
,, Phenylhydrazone	126° ,, 127°	105° ,, 106°
Refractive Index at 20°	—	1·5066

[1] *Jour. of Proc. Roy. Soc., N.S.W.*, 33 **(1899)**, 86.
[2] *Berichte*, 45 **(1912)**, 1390.
[3] *Ibid.*, 46 **(1913)**, 2026.
[4] *Report*, April, **1904**, 51.
[5] *Jour. of Proc. Roy. Soc., N.S.W.*, 34 **(1900)**, 1.

Eucalyptol-pinene Oils.

No.	Species.	Refractive Index n_d 16° C.	Specific Gravity 15°/15° C.	Specific Refractive Energy $\frac{n_d - 1}{d}$.	Solubility in Alcohol (·8722 at 15·5° C.) 1 Volume Requires	10 Times Solubility × Refractive Index.
1	*E. Smithii*	1·4706	·9238	·5094	1·05	15·44
	Do., oil of "suckers"	1·4707	·9151	·5144	1·15	16·91
	Do., com. crude 5·03	1·4689	·9172	·5112	1·05	15·42
2	*E. Bridgesiana*	1·4723	·9327	·5064	1·05	15·46
3	,, *Risdoni*	1·4733	·9373	·5049	1·05	15·47
4	,, *pulverulenta*	1·4688	·9280	·5049	1·1	16·15
5	,, *dealbata*	1·4705	·9268	·5077	1·1	16·17
6	,, *stricta*	1·4711	·9254	·5090	1·1	16·18
7	,, *polyanthema*	1·4736	·9346	·5067	1·1	16·21
8	,, *oleosa*[1]	1·4746	·9319	·5093	1·1	16·22
9	,, *cordata*	1·4695	·9265	·5067	1·15	16·89
10	,, *cinerea*	1·4706	·9198	·5116	1·15	16·91
11	,, *populifolia*	1·4709	·9246	·5093	1·15	16·91
12	,, *Cambagei*[1]	1·4720	·9243	·5106	1·15	16·92
13	,, *sideroxylon*	1·4725	·9219	·5125	1·15	16·93
14	,, *pendula*	1·4732	·9337	·5068	1·15	16·94
15	,, *bicolor*	1·4734	·9266	·5109	1·15	16·94
16	,, *Maideni*	1·4736	·9253	·5117	1·15	16·94
17	,, *cneorifolia*[1]	1·4747	·9194	·5163	1·15	16·96
	Do., F. & Co., crude	1·4774	·9375	·5092	1·1	16·25
18	*E. maculosa*	1·4741	·9278	·5109	1·17	17·24
19	,, *Morrisi*	1·4693	·9191	·5106	1·2	17·63
20	,, *squamosa*[1]	1·4692	·9202	·5099	1·2	17·63
21	,, *globulus*	1·4720	·9243	·5106	1·2	17·66
	Do., Platypus bd. 4 years	1·4697	·9153	·5131	1·15	16·90
	Do., do., 8 years old	1·4738	·9392	·5044	1·15	16·95
22	*E. polybractea*[1]	1·4736	·9197	·5149	1·2	17·68
	Do. commer. dist. 6·04	1·4686	·9286	·5046	1·0	14·68
	Do., do., crude dist. 7·05	1·4692	·9282	·5055	1·05	15·42
	Do., same oil rectified	1·4676	·9254	·5053	1·05	15·41
23	*E. hemilampra*	1·4735	·9310	·5086	1·2	17·68
24	,, *longifolia*	1·4738	·9249	·5122	1·2	17·68
25	,, *intertexta*	1·4748	·9323	·5092	1·2	17·69
26	,, *Behriana*	1·4765	·9272	·5139	1·2	17·72
27	,, *Stuartiana*	1·4709	·9194	·5122	1·25	18·38
28	,, *eugenioides*	1·4747	·9220	·5148	1·25	18·43
29	,, *amygdalina*	1·4760	·9104	·5228	1·25	18·45
30	,, *punctata*[1]	1·4774	·9297	·5135	1·25	18·46
31	,, *Rossii*	1·4741	·9202	·5152	1·35	19·90
32	,, *resinifera*	1·4755	·9194	·5172	1·35	19·91
33	,, *Seeana*	1·4706	·9146	·5145	1·37	20·14
34	,, *camphora*	1·4733	·9072	·5217	1·4	20·62
35	,, *rostrata* var. *borealis*	1·4747	·9251	·5131	1·4	20·64
36	,, *viminalis* var. (*a*)	1·4711	·9169	·5138	1·45	21·33
37	,, *goniocalyx*	1·4746	·9097	·5218	1·8	26·54
38	,, *ovalifolia* v. *lanceolata*	1·4711	·9119	·5166	2·0	29·42
39	,, *salmonophloia*[1]	1·4738	·9069	·5225	3·5	51·58
40	,, *quadrangulata*	1·4692	·9075	·5170	4·0	58·76
41	,, *Bosistoana*	1·4732	·9175	·5158	5·0	73·66
42	,, *melliodora*	1·4706	·9041	·5205	6·0	88·23
43	,, *redunca*	1·4720	·9092	·5191	6·0	88·32
44	,, *conica*	1·4733	·9259	·5112	6·0	88·39
45	,, *propinqua*[1]	1·4788	·9035	·5299	8·0	118·30
46	,, *odorata*[1] (Faulding)	1·4775	·9150	·5218	Insoluble 10 volumes 70 % alcohol.	Soluble 1 volume 80 % alcohol.
47	,, *occidentalis*[1]	1·4774	·9128	·5230		
48	,, *dumosa*[1]	1·4760	·9130	·5213		
49	,, *microcorys*	1·4747	·9059	·5240		
50	,, *gracilis*[1]	1·4771	·9107	·5239		
51	,, *paludosa*	1·4773	·9095	·5248		

[1] Denotes the presence of a small amount of aromadendral in the oil.

Pinene Oils; Phellandrene Absent.

No.	Species.	Refractive Index n_d 16° C.	Specific Gravity $\frac{15}{15}$° C.	Specific Refractive Energy $\frac{n_d - 1}{d}$.	
52	*E. botryoides*	1·4787	·8802	·5439	None soluble in less than 7 volumes of 80 per cent. alcohol.
53	,, *calophylla*	1·4788	·8751	·5471	
54	,, *dextropinea*	1·4741	·8806	·5383	
55	,, *diversicolor*	1·4747	·9134	·5197	
56	,, *lævopinea*	1·4769	·8964	·5320	
57	,, *saligna*	1·4760	·8940	·5324	
58	,, *Wilkinsoniana*	1·4774	·9016	·5295	

Pinene-sesquiterpene Oils; Phellandrene Absent.

No.	Species.	Refractive Index n_d 16° C.	Specific Gravity $\frac{15}{15}$° C.	Specific Refractive Energy $\frac{n_d - 1}{d}$.	
59	*E. affinis* [1]	1·4921	·9270	·5301	None soluble in less than 1 volume 80 per cent. alcohol; the majority insoluble in 10 volumes 80 per cent. alcohol.
60	,, *apiculata*	1·4934	·9123	·5408	
61	,, *Baeuerleni*	1·4841	·8890	·5445	
62	,, *corymbosa* [1]	1·4895	·8867	·5520	
63	,, *eximia*	1·4889	·8999	·5433	
64	,, *intermedia* [1]	1·4878	·8838	·5519	
65	,, *lactea*	1·4898	·8794	·5570	
66	,, *maculata*	1·4861	·9035	·5380	
67	,, *nova-anglica*	1·4900	·9062	·5407	
68	,, *paniculata*	1·4801	·9096	·5278	
69	,, *patentinervis*	1·4948	·8784	·5633	
70	,, *rubida*	1·5011	·9089	·5513	
71	,, *tesselaris*	1·4881	·8962	·5446	
72	,, *trachyphloia* [1]	1·4901	·8915	·5497	

Oils in which Aromadendral is a Pronounced Constituent; Phellandrene is Absent.

No.	Species.	Refractive Index n_d 16° C.	Specific Gravity $\frac{15}{15}$° C.	Specific Refractive Energy $\frac{n_d - 1}{d}$.	
73	*E. albens*	1·4836	·9188	·5263	With the exception of No. 74, all are soluble in either 1 or 2 volumes of 80 per cent. alcohol.
74	,, *hemiphloia*	1·4910	·9084	·5405	
75	,, *marginata*	1·4946	·9112	·5428	
76	,, *punctata* var. *didyma*	1·4868	·9068	·5368	
77	,, *rostrata*	1·4896	·9018	·5429	
78	,, *salubris*	1·4841	·9013	·5358	
79	,, *tereticornis*	1·4934	·9308	·5301	
80	,, *viridis*	1·4828	·9027	·5348	
81	,, *Woollsiana*	1·4895	·8998	·5440	

Phellandrene Oils Containing Piperitone.

No.	Species.	Refractive Index n_d 16° C.	Specific Gravity $\frac{15}{15}$° C.	Specific Refractive Energy $\frac{n_d - 1}{d}$.	
82	*E. coriacea*	1·4902	·9120	·5375	Mostly insoluble in 10 volumes 80 per cent. alcohol; none more soluble than with 1 volume 80 per cent. alcohol.
83	,, *delegatensis*	1·4881	·8645	·5646	
84	,, *dives*	1·4894	·8883	·5509	
85	,, *fraxinoides*	1·4908	·8762	·5601	
86	,, *Luehmanniana*	1·4937	·8846	·5581	
87	,, *obliqua* [1]	1·4934	·8944	·5528	
88	,, *oreades*	1·4945	·8935	·5534	
89	,, *piperita*	1·4838	·9221	·5247	
90	,, *radiata*	1·4863	·8814	·5517	
91	,, *Sieberiana*	1·4886	·8947	·5461	
92	,, *vitrea*	1·4882	·8967	·5384	

[1] Denotes the presence of a small amount of aromadendral in the oil.

PHELLANDRENE OILS IN WHICH THE SESQUITERPENE IS A PRONOUNCED CONSTITUENT.

No.	Species.	Refractive Index n_d16° C.	Specific Gravity $\frac{15^\circ}{15^\circ}$ C.	Specific Refractive Energy $\frac{n_d - 1}{d}$.	
93	*E. acmenoides*	1·5056	·9266	·5466	No. 102 is the only oil more soluble than with 1 volume 80 per cent. alcohol; a large number were insoluble in 10 volumes 80 per cent. alcohol.
94	,, *angophoroides*	1·4881	·9207	·5301	
95	,, *capitellata*	1.4828	·9176	·5261	
96	,, *crebra*	1·4844	·8989	·5388	
97	,, *Dawsoni*	1·5144	·9528	·5399	
98	,, *fastigata*	1·4873	·8948	·5446	
99	,, *Fletcheri*[1]	1·4881	·8882	·5495	
100	,, *gomphocephala.*	1·4815	·8752	·5501	
101	,, *hæmastoma*	1·5013	·9196	·5451	
102	,, *macrorrhyncha*	1·4802	·9166	·5239	
103	,, *melanophloia*	1·4950	·8959	·5526	
104	,, *microtheca*	1·4895	·8866	·5521	
105	,, *nigra*	1·4871	·8838	·5511	
106	,, *ovalifolia*	1·4921	·8911	·5522	
107	,, *Planchoniana*	1·4878	·9166	·5322	
108	,, *pilularis*	1·4961	·8924	·5559	
109	,, *robusta*	1·4801	·8899	·5395	
110	,, *siderophloia*	1·5000	·9081	·5506	
111	,, *sideroxylon* v. *pallens*	1·4884	·9167	·5328	
112	,, *stellulata*	1·4902	·8766	·5589	
113	,, *viminalis*	1·4855	·9088	·5342	
114	,, *virgata*	1·5015	·9352	·5363	

OILS NOT CLASSIFIED; CONTAINING GERANIOL AND ITS ACETIC ACID ESTER, CITRAL, CITRONELLAL, ETC.

No.	Species.	Refractive Index n_d16° C.	Specific Gravity $\frac{15^\circ}{15^\circ}$ C.	Specific Refractive Energy $\frac{n_d - 1}{d}$.	
115	*E. citriodora*	1·4651	·8887	·5233	Soluble in 1·5 vols.
	Do. Mr. Ingham, Qld.	1·4678	·8829	·5298	70°/₀ alco. at 16° C.
116	*E. Macarthuri*	1·4793	·9271	·5172	Soluble in 1·3 volumes 70°/₀ alcohol.
	Do. cont. 64·73°/₀ ester	1·4763	·9252	·5148	
	Do. cont. 68·8 °/₀ ester	1·4768	·9287	·5134	
117	*E. Staigeriana*	1·4871	·8708	·5594	Insoluble in 6 volumes 80°/₀ alcohol.
118	,, *aggregata*	1·5062	·9701	·5128	Insoluble in 10 vols. 80°/₀ alcohol.

H. G. Smith[2] has also isolated a sesquiterpene, which is present in many eucalyptus oils in the portions boiling at above 255°, and which, in the oil of *Eucalyptus hæmastoma*, represents as much as 50 per cent. of that fraction. It does not yield crystallised derivatives, but, on the other hand, when dissolved in glacial acetic acid, it gives characteristic colour-reactions with acids and also with bromine. When submitted to the action of bromine vapour the oil acquires first a crimson-red colour, then violet, and finally indigo-blue. The sesquiterpene, prepared as pure as possible by fractional distillation, boils at atmospheric pressure at 260° to 265°, and has at 19° the specific gravity 0·9249. Smith proposes for the sesquiterpene the name "aromadendrene".

This sesquiterpene, according to Schorger,[3] may also be present in the turpentine oil from *Pinus Lambertiana*.

[1] Denotes the presence of a small amount of aromadendral in the oil.
[2] *Jour. of Proc. Roy. Soc.,N.S.W.*, 35 (1901), 124.
[3] *U.S. Dept. Agric. Forest Service, Bull.* 119.

A ketone has also been isolated from the oil of *Eucalyptus dives* which Smith has termed piperitone. This body, of the formula $C_{10}H_{18}O$, has the following characters:—

Specific gravity	0·9393 at 17°
Boiling-point	224° to 225°

On reduction with sodium it yields an alcohol, $C_{10}H_{18}O$, melting at 155° to 156°.

Finally, Robinson and Smith [1] have separated a phenol from the oil of *Eucalyptus linearis*, which they have named tasmanol.

The phenol was removed from the crude oil in the usual manner by shaking with aqueous sodium hydrate, washing the aqueous solution with ether to remove adhering oil, acidifying and extracting with ether. The residue, which contained a small amount of acetic and butyric acids, was washed with dilute sodium carbonate, extracted with ether, the ether removed and the phenol distilled. It boiled at 268° to 273° C. (uncor.) and at 175° under 25 mm. pressure. It was optically inactive, the specific gravity at 23° was 1·077, and the refractive index at 22° was 1·5269. Besides being soluble in the alkalis the phenol is soluble in ammonia, partly soluble also in sodium carbonate but not in bicarbonate. It also dissolves slightly in boiling water. The reaction with ferric chloride in alcoholic solution is characteristic, the deep red colour which is first formed remaining persistent for days, after the alcohol has evaporated. The odour reminds one somewhat of carvacrol under certain conditions. It contains one methoxy group and appears to have two phenolic groups in the para position to each other.

The following are the principal varieties of eucalyptus oil that have been described:—

Oil of Eucalyptus Acaciæformis.

This tree is the narrow-leaved peppermint gum of New South Wales. The oil has the following characters:—

Specific gravity	0·8864
Optical rotation	+ 35·7°
Refractive index	1·4713 at 20°

It contains *d*-pinene, aromadendrene, and geranyl acetate (?).

Oil of Eucalyptus Acervula.

This eucalyptus is a Tasmanian tree, and is also found in South Australia. The oil has the following characters:—

Specific gravity	0·8956
Optical rotation	− 1·1°
Refractive index	1·4756 at 18°
Saponification value	32·8
Eucalyptol	21 per cent.
Yield	0·212 ,,

It contains *d*-pinene, phellandrene, eucalyptol, geraniol, geranyl acetate, a liquid paraffin hydrocarbon, and a solid (paraffin?) stearoptene melting at 55° to 56°.

[1] *Jour. of Proc. Roy. Soc., N.S.W.*, 48 (**1914**), 518.

Oil of Eucalyptus Albens.

This tree is found principally in South Australia. Its oil has the following characters:—

Specific gravity	0·9044
Optical rotation	− 6·5°
Refractive index	1·4780 at 20°
Saponification value	8·5
Eucalyptol	30 per cent.
Yield	0·1 ,,

The oil contains eucalyptol, pinene, aromadendral, and esters.

Oil of Eucalyptus Acmenoides.

This tree is the white mahogany of Queensland and New South Wales. The oil contains phellandrene, aromadendrene, pinene (?), and eucalyptol. It has the following characters:—

Specific gravity	0·925
Refractive index	1·5065
Saponification value	5·7
Yield	0·09 per cent.

Oil of Eucalyptus Aggregata.

This tree is found in New South Wales, the oil having the following characters:—

Specific gravity	0·956
Optical rotation	+ 27·13°
Refractive index	1·5062 at 16°
Ester value	112·2
Yield	0·04 per cent.

It contains *d*-*a*-pinene, and the amyl ester of eudesmic acid.

Oil of Eucalyptus Amygdalina.

In commerce the oil known as *Eucalyptus amygdalina* is understood to be an oil containing much phellandrene and very little eucalyptol. It is an oil having a specific gravity about 0·855 to 0·885 and an optical rotation of from − 25° to − 80°. This oil of commerce is probably the distillate from various species, especially from *Eucalyptus dives*. But as the oil is one of great importance in commerce, where it is used to a considerable extent for the purpose of "floating" mineral ores in separation processes, it is important that the exact nature of the true *amygdalina* oil should be understood. The matter has been finally cleared up by Baker and Smith,[1] who give the following account of the tree and its oil:—

"*Historical.*—This eucalyptus was described by Labillardier in his *Plants of New Holland* (1806).

"*Remarks.*—This name, *Eucalyptus amygdalina*, has almost since the beginning of Eucalyptology been associated with the Tasmanian flora, for it was from Van Diemen's Land that Labillardier obtained his specimens. It is, however, with much reluctance that we announce that we have failed to find this tree on the mainland of Australia.

"The tree passing as *Eucalyptus amygdalina* in the Eastern States[2]

[1] *Eucalypts of Tasmania* (1912), p. 62. [2] *Vide* under *Eucalyptus Australiana*.

is of medium size with a 'Peppermint' bark, opposite, sessile, cordate-lanceolate 'sucker' leaves, medium-sized lanceolate normal leaves, small hemispherical fruits with a truncate or depressed rim, and the leaves yielding a particular oil.

"As far as our researches go, trees having these characteristics do not appear to occur in Tasmania, where the name was originally ascribed to this species. Thus the trees going under the name of *Eucalyptus amygdalina* in Tasmania and *Eucalyptus amygdalina* in Victoria and New South Wales are not the same.

"As, however, the name has become so interwoven with the eucalyptus literature of Victoria and New South Wales in connection with the economics of the tree found there, it is now almost next to impossible to supersede the name of the tree of the Eastern States, or at least without adding to the already long nomenclature of the genus.

"That the latter is not Labillardier's tree we are firmly convinced, and on the following grounds:—

"1. The plate of *Eucalyptus amygdalina* of Labillardier in his *Plants of New Holland* depicts the Tasmanian Eucalypt, and certainly not the mainland one.

"2. The 'sucker' leaves of the former are petiolate, alternate, and the latter opposite, sessile, cordate, obtuse.

"3. The normal leaves of the former are smaller and narrower.

"4. The fruits of the Tasmanian tree are practically identical in shape with those of *Eucalyptus dives* Sch., and not hemispherical like those of the mainland *Eucalyptus amygdalina.*

"5. The oil of the Tasmanian tree closely approaches in chemical composition that of *Eucalyptus dives* Sch.

"Labillardier's name for the Tasmanian tree must, of course, stand, as that has priority, and to us it appears too late in the day to alter the specific appellation of the other; but if a systematic distinction is necessary in future, we would suggest that it might be known, to botanists at least, as *Eucalyptus amygdalina*, var. *Australiana*. This form is faithfully figured by Mueller in his *Eucalyptographia*, and is also illustrated by us in our work *Eucalypts and Their Essential Oils* (p. 168). The varieties recorded by Bentham, we find, belong to *Eucalyptus Risdoni* rather than *Eucalyptus amygdalina*.

"*Essential Oil.*—Material of this species—known as 'Black Peppermint'—was received from various localities in Tasmania, collected at various times of the year, in order that definite conclusions might be formed as to the specific differences between the oil of *Eucalyptus amygdalina* of Tasmania and that of the tree known in New South Wales and Victoria as *Eucalyptus amygdalina*. Although the general character of the oil of the Tasmanian tree places it in the '*amygdalina* group' of these oils, yet it differs considerably from that of the New South Wales form, and more closely approaches in constituents and physical properties the oil of *Eucalyptus dives*, with the exception that the Tasmanian *Eucalyptus amygdalina* contains a little more eucalyptol than does that of *Eucalyptus dives*. Commercially the oil could be utilised for purposes similar to those to which that of *Eucalyptus dives* is put, but it would not pay to submit it to fractional distillation in order to separate the eucalyptol portion, as is now often carried out with the oil of the New South Wales form, any more than it would pay to do so with the oil of *Eucalyptus dives*. The yield of oil of the New South

Wales form of *Eucalyptus amygdalina* is almost twice that obtained with the Tasmanian trees, and even *Eucalyptus dives* appears to yield a greater quantity of oil than does the *Eucalyptus amygdalina* of Tasmania. The oil of the Tasmanian *Eucalyptus amygdalina* differs from that of the New South Wales form in that it contains much less eucalyptol, has a very high laevo-rotation, contains more phellandrene, and is much less soluble in alcohol. The odour of the oil is also much less aromatic."

The oil has the following characters :—

Specific gravity	0·8668 to 0·8848
Optical rotation	− 59·1° „ − 75·1°
Refractive index	1·4761 to 1·4790 at 18°
Eucalyptol	12 „ 24 per cent.
Yield	1·62 „ 2·04

The chief constituents of the oil are phellandrene, eucalyptol, piperitone, and a sesquiterpene.

OIL OF EUCALYPTUS ANDREWSI.

This oil is obtained from a tree growing in the north-east of New South Wales. It has the following characters :—

Specific gravity	0·8646
Optical rotation	− 41·5°
Refractive index	1·4854
Ester number	4·3
Yield	1·27 per cent.

It contains *l*-phellandrene, piperitone, and a sesquiterpene.

OIL OF EUCALYPTUS ANGOPHOROIDES.

This tree is found in New South Wales. The oil has the following characters :—

Specific gravity	0·9049
Optical rotation	− 12·7°
Refractive index	1·4881 at 16°
Yield	0·185 per cent.

It contains phellandrene, pinene, eucalyptol and a sesquiterpene.

OIL OF EUCALYPTUS APICULATA.

Eucalyptus apiculata is a New South Wales tree, the leaves of which yield 0·65 per cent. of essential oil containing pinene, piperitone, and a small amount of eucalyptol, and having the following characters :—

Specific gravity	0·9056 to 0·9112
Optical rotation	− 8·6° „ − 9·3°
Refractive index	1·4934 at 16°
Saponification value	8·7 to 10·1

OIL OF EUCALYPTUS AUSTRALIANA.

The tree described up till 1915 under the name *Eucalyptus amygdalina*, growing on the Australian mainland, was referred to under that oil. The importance of the oil sold under that name in commerce warrants an exhaustive examination of the scientific nomenclature of the trees bearing the name. This has now been undertaken by Baker and Smith,[1] who have now re-named the mainland tree *Eucalyptus Australiana.*

When these scientists studied the Eucalypts of Tasmania, Labillardier's species of *Eucalyptus amygdalina* necessarily received much attention, as it was from that island he obtained his material, which, of course, stands as the authenticated species. The name, however, has long since been also given to a tree on the mainland, and it was only when investigating the species for the above research, that differences were detected in the two trees, and these have since been followed up. In fact, the differences were so marked that even then they raised the continental form to varietal rank under the name of *Eucalpytus amygdalina,* var. *Australiana,* being loath to introduce another species name to the already long list of eucalypts. They were prepared to let it stand at that, but the technology of this eucalyptus has since come so much to the front in the commercial world, that they considered it best in the interest of applied, as well as pure science, to give it specific rank, for which they propose the name *Eucalyptus Australiana* for the mainland tree.

The probability of this oil becoming one of the most important producers of eucalyptus oil induced the authors to investigate it very fully. They were able to show that, by fractional separation at stated times during the primary distillation, an oil richer in eucalyptol could be obtained, particularly if the portion which came over during the first hour was separated. The eucalyptol was thus more easily distilled than the alcoholic bodies and other oil constituents in the leaf. By taking advantage of this peculiarity, and working the trees growing at Nerrigundah, Yourie, and neighbouring districts of New South Wales in the same way, it has been possible to produce a product of a fairly constant character, high in eucalyptol-content, and answering to the requirements demanded for a first-class pharmaceutical eucalyptus oil. The results have been so satisfactory that the species is, at the above localities, now worked in this way, the first-hour oil being sold for pharmaceutical purposes, the remainder being employed for mineral separation and for other industrial uses.

The remarkably high yield of oil given by this species enables this mode of working to be profitably undertaken, and the amount obtained during the first hour is almost as great as that from many other cineol-producing species when these are distilled right out. The second-hour oil thus becomes practically a subsidiary product. Little advantage appears to be derived from distilling the leaves for a longer period than two hours, as the amount of oil thus obtained is but small. The first-hour oil, when properly prepared, is almost water-white, which fact appears to be due to the presence of the phenol, tasmanol—common to this class—containing a methoxyl group, a constitution which does not permit the formation of a quinone, as is the case with the oils of the other large class of cineol-producing eucalypts. The amount of volatile aldehydes in the crude oil is but small, so that altogether no rectification of the first-hour oil is needed before placing it on the market.

The crude oil has the following characters :—

Specific gravity	0·9157
Optical rotation	+ 2·8°
Refractive index	1·4644 at 20°
Eucalyptol	70 per cent.

[1] *Jour. Proc. Roy. Soc., N.S.W.,* xlix., 514.

The oil contains eucalyptol, piperitone, esters, and some free alcohols, as well as the phenol tasmanol.

The following analyses are of the first-hour distillates, which are being put on the market for medicinal purposes:—

Specific Gravity at 15° C.	Rotation α_D.	Solubility in 70 per cent. Alcohol.	Refractive Index at 20° C.	Cineol.
		Vols.		
0·9188	+1·7°	1·1	1·4614	Over 70 per cent.
0·9188	+0·3°	1·1	1·4621	,, ,,
0·9190	- +0·0°	1·1	1·4622	,, ,,
0·9190	+1·3°	1·1	1·4633	75 per cent. phosphoric acid method.
0·9200	+1·5°	1·05	1·4627	Over 70 per cent.
0·9193	+0·9°	1·1	1·4628	84 per cent. resorcinol method.
0·9186	+0·5°	1·05	1·4624	Over 70 per cent.
0·9186	-0·5°	1·1	1·4620	69·5 per cent. phosphoric acid method.
0·9195	+0·4°	1·1	1·4622	Over 70 per cent.
0·9198	+0·3°	1·1	1·4617	,, ,,
0·9191	+1·3°	1·1	1·4633	73 per cent. phosphoric acid method.
0·9193	+1·5°	1·15	1·4631	Over 70 per cent.
0·9199	+0·1°	1·1	1·4622	,, ,,
0·9202	-1·2°	1·15	1·4636	71 per cent. phosphoric acid method.
0·9211	+1·4°	1·1	1·4628	74 ,, ,, ,, ,,
0·9179	+1·6°	1·15	1·4632	Over 70 per cent.
0·9195	+1·4°	1·1	1·4631	,, ,,
0·9196	+1·2°	1·1	1·4625	81 per cent. resorcinol method. 77 per cent. by phosphoric acid method.

The second-hour's distillate and the third-hour's distillate are quite similar to each other. They contain about 30 per cent. of eucalyptol, together with the less desirable constituents of the oil.

OIL OF EUCALYPTUS BÆUERLENI.

This tree is the brown gum of New South Wales. Its oil has the following characters:—

Specific gravity	0·8895
Optical rotation	+ 4·55°
Refractive index	1·4841 at 16°
Saponification value	10
Yield	0·33 per cent.

It contains pinene, eucalyptol, eudesmol, and aromadendrene.

OIL OF EUCALYPTUS BAILEYANA.

This is one of the numerous trees known locally as "stringy bark," and is confined chiefly to the north of New South Wales and the adjacent southern portion of Queensland. The leaves yield about 1 per cent. of oil, of a melissa-like odour, which, however, is partially covered by the cineol present. Its specific gravity is variable, usually about 0·940. It appears to contain up to 25 or 30 per cent. of cineol, some citral, and phellandrene.

Oil of Eucalyptus Behriana.

This tree is one of the Mallees of New South Wales, South Australia, and Victoria. It yields an oil having the following characters :—

Specific gravity	0·9237
Optical rotation	+ 3·7°
Refractive index	1·4709 at 20°
Eucalyptol	48 per cent.
Yield	0·61 ,,

It contains eucalyptol, pinene, and phellandrene.

Oil of Eucalyptus Bicolor.

Eucalyptus bicolor is known as the Bastard Box. The oil has the following characters :—

Specific gravity	0·9259
Optical rotation	+ 1·2°
Refractive index	1·4734 at 16°
Saponification value	3·6
Yield	0·52 per cent.

It contains eucalyptol and a sesquiterpene.

Oil of Eucalyptus Bosistoana.

This tree is also known as a Bastard Box in Victoria. The oil has the following characters :—

Specific gravity	0·9078
Optical rotation	+ 9° 25′
Saponification value	3·2
Refractive index	1·4676 at 20°
Eucalyptol	38 per cent.
Yield	0·97 ,,

It contains eucalyptol, pinene, and aromadendrene.

Oil of Eucalyptus Botryoides.

This tree is the Bastard Mahogany of Victoria and Queensland. The oil contains *d*-*α*-pinene, aromadendrene, and traces of eucalyptol. It has the following characters :—

Specific gravity	0·877
Refractive index	1·4787 at 16°
Saponification value	15 to 21
Optical rotation	+ 23·9°
Yield	0·1 per cent.

Oil of Eucalyptus Bridgesiana.

This is the Gippsland woolly-butt of Victoria. Its oil has the following characters :—

Specific gravity	0·912 to 0·925
Optical rotation	+ 1·8° ,, + 3·3°
Refractive index	1·4716 to 1·4729 at 20°
Eucalyptol	73 ,, 78 per cent.
Yield	0·57 ,, 0·75 ,,

Oil of Eucalyptus Calophylla.

This tree, the Western Australian Bloodwood, yields an essential oil with the following values :—

Specific gravity	0·8756
Optical rotation	+ 22·9°
Refractive index	1·4788 at 16°
Ester value	10·5
Yield	0·25 per cent.

It contains *d*-pinene, aromadendrene, traces of eucalyptol, and acetic esters. Cymene is probably present.

Oil of Eucalyptus Cambagei.

This tree is another of the so-called Bastard Boxes of New South Wales and Victoria. The oil has the following characters:—

Specific gravity	0·919 to 0·921
Optical rotation	− 0·6° ,, + 5·6°
Refractive index	1·4720 at 16°
Yield	0·83 per cent.

It contains eucalyptol, pinene, aromadendral, and eudesmol.

Oil of Eucalyptus Campanulata.

This tree is the Bastard Stringy Bark of New South Wales. The oil contains eucalyptol, piperitone, and eudesmol, with a large amount of phellandrene, and has the following characters:—

Specific gravity	0·8804
Optical rotation	− 25·8°
Refractive index	1·4856 at 18°
Saponification value	7·6
Yield	0·85 per cent.

Oil of Eucalyptus Camphora.

This tree is the Swamp Gum of Victoria and New South Wales. Its oil has the following characters:—

Specific gravity	0·9071
Optical rotation	+ 1·3°
Refractive index	1·4733 at 16°
Saponification value	4·4
Yield	1·34 per cent.

The oil contains 50 per cent. of eucalyptol, pinene, and eudesmol.

Oil of Eucalyptus Capitellata.

This oil has the following characters:—

Specific gravity	0·9175
Optical rotation	+ 4·4°
Saponification value	4·2
Refractive index	1·4772 at 20°
Eucalyptol	28 per cent.
Yield	0·1 ,,

It contains eucalyptol, pinene, phellandrene, and a sesquiterpene.

Oil of Eucalyptus Carnea.

This oil has only been partially examined. It consists chiefly of *d*-α-pinene with traces of eucalyptol.

Oil of Eucalyptus Cinerea.

This tree is the Argyle Apple of New South Wales. The oil has the following characters :—

Specific gravity	0·911 to 0·9225
Optical rotation	+ 2·4° ,, + 4·25°
Refractive index	1·4706 at 16°
Eucalyptol	59 per cent.
Yield	1 to 1·3 ,,

It contains eucalyptol, pinene, and esters.

Oil of Eucalyptus Citriodora.

This tree, the "citron-scented" eucalyptus, is found along the coast of Queensland and as far south as Port Jackson. It has been described as a variety of the New South Wales "spotted gum," *Eucalyptus maculata* (which itself possesses very little odour), but Baker and Smith consider it to be a separate species. This variety yields a typical scented eucalyptus oil, having the following characters :—

Specific gravity	0·864 to 0·905
Optical rotation	− 1° ,, + 2°
Refractive index	1·4540 ,, 1·4678

It contains a very large quantity of citronellal, one specimen, according to Schimmel & Co., having been found to contain as much as 95 per cent. It also contains geraniol and traces of pinene. This oil is of a most pleasant odour, and is admirably suited for soap perfumery, possessing a far more delicate odour than ordinary citronella oil, which it somewhat resembles in its general characteristics.

Oil of Eucalyptus Cneorifolia.

This eucalypt, one of the Mallees, is chiefly found in Kangaroo Island, South Australia. It is a comparatively small plant, seldom reaching more than 12 to 15 ft. in height. It is known locally as the "narrow-leaved eucalyptus". The plant was originally looked upon as a variety of *Eucalyptus oleosa*, but it is now understood to be a separate species. The oil came into commerce some few years ago under the name of *oleosa*, but is quite different from the oil which was originally known under the same name. The oil is very rich in cineol and has the following characters :—

Specific gravity	0·899 to 0·929
Optical rotation	− 3·5° ,, − 12·5°
Refractive index	1·4747 at 16°

It has a secondary odour recalling that of cumin. It contains from 50 to 70 per cent. of eucalyptol and a small quantity of aromadendral and a sesquiterpene. This oil is much valued as a medicinal oil, and usually conforms admirably to pharmacopœial requirements.

Oil of Eucalyptus Coccifera.

This oil, a Tasmanian eucalypt, has the following characters :—

Specific gravity	0·881
Optical rotation	− 35·8°
Saponification value	4·9
Refractive index	1·4831 at 24°

It contains phellandrene piperitone, eudesmol, and traces of eucalyptol.

Oil of Eucalyptus Conica.

This oil has the following characters:—

Specific gravity	0·9057
Optical rotation	+ 5·2°
Refractive index	1·4733 at 16°
Saponification value	6·26
Yield	0·59 per cent.

It contains pinene, traces of aromadendral, and 40 per cent. of eucalyptol.

Oil of Eucalyptus Cordata.

This tree is a Tasmanian eucalypt, the oil having the following characters:—

Specific gravity	0·9138
Optical rotation	+ 9·3°
Saponification value	14·8
Refractive index	1·4695 at 16°
Eucalyptol	62 per cent.
Yield	2·32 ,,

Oil of Eucalyptus Coriaceæ.

This tree is the white gum of most of the Australian Colonies. Its oil has the following characters:—

Specific gravity	0·8947
Optical rotation	− 29·3°
Saponification value	4·6
Refractive index	1·4846 at 20°
Eucalyptol	5 per cent.
Yield	0·45 ,,

It contains phellandrene, piperitone, eucalyptol, and a sesquiterpene.

Oil of Eucalyptus Corymbosa.

This tree is the New South Wales and Queensland Bloodwood. Its oil has the following characters:—

Specific gravity	0·881 to 0·887
Optical rotation	− 8·5°
Refractive index	1·4895 at 16°
Yield	0·06 per cent.

It contains pinene, eucalyptol, and aromadendral.

Oil of Eucalyptus Cosmophylla.

This tree is a South Australian eucalypt, whose oil has the following characters:—

Specific gravity	0·9108
Optical rotation	− 3·2°
Saponification value	5·6
Refractive index	1·4659 at 20°
Eucalyptol	53 per cent.
Yield	0·62 ,,

It contains pinene, eucalyptol, aromadendral, and sesquiterpenes.

Oil of Eucalyptus Crebra.

This tree is the narrow-leaved iron bark of several of the Australasian Colonies, and is found in India. The oil has the following characters:—

Specific gravity	0·8986
Optical rotation	− 3° to − 10·9°
Refractive index	1·4670 to 1·4844 at 16°

It contains phellandrene, eucalyptol, and aromadendrene.

Oil of Eucalyptus Dawsonii.

This tree is the New South Wales Staley gum, and its oil contains aromadendrene as its principal constituent. It has the following characters:—

Specific gravity	0·9418
Refractive index	1·5144 at 16°
Saponification value	13·3
Yield	0·172 per cent.

It contains phellandrene and eucalyptol, as well as aromadendrene.

Oil of Eucalyptus Dealbata.

This tree is found in New South Wales and Queensland. The oil has the following characters:—

Specific gravity	0·9261
Optical rotation	+ 3·7°
Refractive index	1·4705 at 16°
Yield	0·86 per cent.

It contains eucalyptol, pinene, and aromadendrene.

Oil of Eucalyptus Delegatensis.

This tree is the gum-topped stringy bark of Tasmania, New South Wales, and Victoria. The oil has the following characters:—

Specific gravity	0·8664
Optical rotation	− 48·4°
Refractive index	1·4828 at 24°
Yield	1·3 to 1·9 per cent.

It contains phellandrene, piperitone, and a sesquiterpene.

Oil of Eucalyptus Dextropinea.

This tree is one of the stringy barks of New South Wales. The oil has the following characters:—

Specific gravity	0·874 to 0·883
Optical rotation	+ 24·2°
Refractive index	1·4688 at 21°
Saponification value	22·1
Yield	0·83 per cent.

The oil contains a considerable amount of *d*-α-pinene, a small amount of eucalyptol, and a little geranyl acetate.

Oil of Eucalyptus Diversicolor.

This tree is a Western Australian Karri. The oil has the following characters :—

Specific gravity	0·9145
Optical rotation	+ 30·1°
Refractive index	1·4747 at 16°
Ester value	53·2
Yield	0·83 per cent.

It contains *d*-pinene, eucalyptol, and acetic esters.

Oil of Eucalyptus Dives.

This tree is found in Victoria and New South Wales, and is known as the broad-leaved peppermint. It is the tree which was originally known as *Eucalyptus amygdalina*, and its oil has the following characters :—

Specific gravity	0·882 to 0·889
Optical rotation	− 55° to − 70°
Refractive index	1·4894 at 16°

Its principal constituent is phellandrene, but it also contains piperitone.

Oil of Eucalyptus Dumosa.

This plant is one of the Mallees (dwarf eucalypts forming the "scrub" of parts of Victoria, New South Wales, and South Australia). The oil is of a good quality from a medicinal point of view, having the following characters :—

Specific gravity	0·900 to 0·916
Optical rotation	+ 2° ,, + 6°
Refractive index	1·4760 at 16°
Yield	1 per cent.

It contains pinene, eucalyptol, and aromadendral.

Oil of Eucalyptus Eugenoides.

This tree is a white stringy bark of Victoria, Queensland, and New South Wales. The oil has the following characters :—

Specific gravity	0·912 to 0·913
Optical rotation	+ 3·4° ,, + 4·7°
Refractive index	1·4747 at 16°
Saponification value	6·9
Yield	0·7 to 0·8 per cent.

The oil contains pinene and eucalyptol.

Oil of Eucalyptus Eximia.

This tree is the white or yellow Bloodwood of New South Wales. The oil has the following characters :—

Specific gravity	0·900
Optical rotation	+ 27°
Refractive index	1·4889 at 16°
Yield	0·462 per cent.

It contains pinene and aromadendrene.

Oil of Eucalyptus Fastigata.

This tree is the Cut tail of New South Wales. Its oil has the following values :—

Specific gravity	0·892 to 0·900
Refractive index	1·4873 at 16°
Acid value	4·9
Ester ,,	7·5
Yield	0·11 per cent.

It contains *d*-pinene, phellandrene, eudesmol, and eucalyptol.

Oil of Eucalyptus Fasciculosa.

This tree is a Tasmanian eucalypt. Its oil contains pinene, eucalyptol, and esters, and has the following characters :—

Specific gravity	0·904
Optical rotation	+ 6·3°
Saponification value	22·1
Refractive index	1·4789 at 20°
Eucalyptol	15 per cent.

Oil of Eucalyptus Fletcheri.

This tree is a New South Wales "Box" tree. It yields an essential oil having the following characters :—

Specific gravity	0·880 to 0·895
Optical rotation	− 12·8°
Refractive index	1·4881 at 16°
Yield	0·294 per cent.

It contains about 5 per cent. of eucalyptol, pinene, and aromadendral.

Oil of Eucalyptus Fraxinoides.

The New South Wales "white ash" yields an essential oil having the following characters :—

Specific gravity	0·869
Optical rotation	− 28·5°
Refractive index	1·4908 at 16°
Yield	1 per cent.

It contains pinene, eudesmol, and about 5 per cent. of eucalyptol, and probably some citral.

Oil of Eucalyptus Globulus.

This tree, from which a typical eucalyptus oil is obtained, is the Tasmanian "blue gum". (Blue gum, however, is a name applied to a number of other eucalypts in New South Wales, Victoria, and Western Australia.) This plant was discovered by Labillardier in Tasmania in 1792, and introduced into Europe in 1856 by Ramel. It is a native of Tasmania and Victoria, although found to a certain extent in other colonies. The fresh leaves yield from ·8 to 1·5 per cent. of oil of a light yellow colour. The fact that this oil was regarded in Europe as the typical eucalypt yielding a medicinal oil, together with the occurrence of such numerous and closely similar species in Australia, account for the fact that eucalyptus leaves from various trees were indiscriminately mixed and distilled, the resulting product being marked oil of *Eucalyptus globulus*. Hence the difference observed by various chemists in the properties of this oil. The pure globulus oil, which is now largely replaced by other eucalyptus oils, has the following characters :—

Specific gravity	0·910 to 0·930
Optical rotation	+ 0° 30′ to + 12°
Refractive index	1·4600 to 1·4700 at 20°
Yield	0·8 to 1·5 per cent.

The oil contains from 50 to 70 per cent. of eucalyptol, *d*-α-pinene, an unidentified terpene, valeric, butyric and caproic aldehydes, iso-amyl alcohol, a pineocarvol ester, and a sesquiterpene alcohol which has been named globulol. This is a crystalline compound melting at 88·5°. Eudesmol and a sesquiterpene are also present.

Oil of Eucalyptus Gomphocephala.

This tree is a Western Australian Tuart tree. Its oil has the following characters:—

Specific gravity	0·876
Refractive index	1·4815
Ester value	25·7

It contains phellandrene.

Oil of Eucalyptus Goniocalyx.

This tree is known in Victoria as the "spotted gum" or "white gum," but in New South Wales it is usually spoken of as "blue gum". The oil is bright yellow, of suffocating odour, and has the following values:—

Specific gravity	0·911 to 0·9125
Optical rotation	+ 4·3° to + 7·1°
Refractive index	1·4746 at 16°
Yield	0·96 per cent.

It contains 50 to 60 per cent. of eucalyptol, α-pinene, eudesmol, and valerianic and acetic esters.

Oil of Eucalyptus Gracilis.

This tree inhabits the same districts as *Eucalyptus dumosa*, and is also found in Queensland. Its oil contains a fair proportion of cineol, and its cultivation has attracted some attention in Victoria. It has the following characters:—

Specific gravity	0·910
Optical rotation	+ 1·35°
Refractive index	1·4771 at 16°
Yield	0·9 per cent.

It contains eucalyptol, pinene, and aromadendral.

Oil of Eucalyptus Gunnii.

This tree is the Cider gum of Tasmania. Its oil has the following values:—

Specific gravity	0·9014
Optical rotation	+ 1·5°
Refractive index	1·4752 at 18°
Saponification value	6·7
Yield	0·39 per cent.

It contains about 40 per cent. of eucalyptol, *d*-pinene, *l*-phellandrene, esters, and a sesquiterpene.

OIL OF EUCALYPTUS HÆMASTOMA.

This tree is the Scribbly gum of most of the colonies. Its oil contains much phellandrene, eucalyptol, and aromadendrene, and has the following characters :—

Specific gravity	0·9195
Refractive index	1·5013 at 16°
Yield	0·28 per cent.

OIL OF EUCALYPTUS HEMILAMPRA.

This tree is found in New South Wales. The oil has the following values :—

Specific gravity	0·928
Optical rotation	+ 6·9°
Refractive index	1·4735 at 16°
Yield	0·38 per cent.

It contains eucalyptol, pinene, aromadendrene, and an aldehyde.

OIL OF EUCALYPTUS HEMIPHLOIA.

This tree is found in South Australia. The oil contains eucalyptol, pinene, and aromadendral. Its characters are as follows :—

Specific gravity	0·9117
Optical rotation	− 6·8°
Refractive index	1·4854 at 20°
Eucalyptol	20 per cent.
Yield	0·55 ,,

OIL OF EUCALYPTUS INTERTEXTA.

This tree is known in New South Wales as the Candle Bark. Its oil has the following characters :—

Specific gravity	0·9076
Optical rotation	+ 10·2°
Refractive index	1·4748 at 16°
Yield	0·2 to 0·64 per cent.

It contains eucalyptol, *d*-pinene, and aromadendrene.

OIL OF EUCALYPTUS LACTEA.

This tree is known in New South Wales as "Spotted Gum". The oil contains pinene and eucalyptol, and has the following characters :—

Specific gravity	0·875 to 0·883
Optical rotation	− 0·6° ,, − 1·2°
Refractive index	1·4898 at 16°

OIL OF EUCALYPTUS LÆVOPINEA.

The oil from this "Silver-top Stringy Bark" has the following characters :—

Specific gravity	0·875 to 0·888
Optical rotation	− 30·7° ,, − 33·3°
Refractive index	1·4697 at 19°
Yield	0·66 per cent.

The oil contains much laevo-pinene and about 5 per cent. of eucalyptol.

Oil of Eucalyptus Largiflorens.

This tree is a Tasmanian eucalypt. Its oil contains eucalyptol, pinene, and esters, and has the following characters:—

Specific gravity	0·9155
Optical rotation	+ 5·5°
Refractive index	1·4676
Eucalyptol	50 per cent.
Yield	0·67 ,,

Oil of Eucalyptus Longifolia.

This oil contains eucalyptol, pinene, and a sesquiterpene, and has the following characters:—

Specific gravity	0·9226
Optical rotation	+ 2·8°
Refractive index	1·4738 at 16°
Yield	0·535 per cent.

Oil of Eucalyptus Loxophleba.

This tree is the Western Australian York Gum. The oil contains phellandrene and eucalyptol, and has the following values:—

Specific gravity	0·8828
Optical rotation	+ 5°

Oil of Eucalyptus Luehmanniana.

This oil has the following characters:—

Specific gravity	0·879
Optical rotation	− 31·3°
Refractive index	1·4937 at 16°
Yield	0·29 per cent.

It contains phellandrene, eucalyptol, piperitone, and probably citral.

Oil of Eucalyptus Macarthuri.

This is one of the most interesting of the eucalyptus oils as it differs materially from nearly every other one. It has been investigated fully by Smith.[1]

The crude oil of *Eucalyptus macarthuri*, obtained by steam distillation from fresh material of leaves and terminal branchlets, is reddish in colour owing to the presence of a small amount of free acid in the original oil. In appearance, odour, etc., it resembles more than anything else the crude oil of *Darwinia fascicularis*, but the higher boiling portion consists largely of eudesmol, the stearoptene of eucalyptus oil, which constituent is absent in *Darwinia* oil. Although containing this stearoptene no crystals were obtained when the crude oil was placed in a freezing mixture, eudesmol being so exceedingly soluble in the oil.

The oil had the following characters:—

Specific gravity	0·9245
Optical rotation	+ 3·6°

[1] *Jour. Proc. Roy. Soc., N.S.W.*, xxxiv. 142.

It was found to contain about 10 per cent. of free geraniol and rather over 60 per cent. of geranyl acetate. This oil should be of considerable perfume value.

Oil of Eucalyptus Macrorhyncha.

This tree is a red Stringy Bark of New South Wales. The oil has the following characters:—

Specific gravity	0·929
Refractive index	1·4746 at 20°
Eucalyptol	30 per cent.
Yield	0·27 „

It contains eucalyptol, phellandrene, pinene, eudesmol, and a sesquiterpene.

Oil of Eucalyptus Maculata.

This tree is a Spotted Gum of New South Wales and Queensland. Its oil has the following characters:—

Specific gravity	0·896 to 0·920
Optical rotation	+ 5·2° „ + 8·2°
Refractive index	1·4861 at 16°
Yield	0·23 per cent.

Oil of Eucalyptus Maculosa.

This tree is also known as Spotted Gum. The oil has the following values:—

Specific gravity	0·9158
Optical rotation	+ 3·4°
Refractive index	1·4741 at 16°
Eucalyptol	50 per cent.
Yield	1·06 „

The oil contains eucalyptol, *d*-pinene, and aromadendrene.

Oil of Eucalyptus Marginata.

This tree is a Western Australian Jarrah tree. The oil contains pinene, cymene, eucalyptol, aromadendral, and a sesquiterpene. It has the following characters:—

Specific gravity	0·9117
Optical rotation	− 8·5°
Refractive index	1·4946 at 16°
Yield	0·2 per cent.

l of Eucalyptus Megacarpa.

The oil of eucalyptus megacarpa[1] contains pinene, limonene, dipentene, about 30 per cent. of eucalyptol, and up to 10 per cent. of terpinene. The latter terpene was identified in the fraction 170° to 190°, by the formation of its nitrosite melting at 155°.

Oil of Eucalyptus Microcorys.

This is the sallow wood of New South Wales or "turpentine tree" of Queensland. It yields an oil having the following characters:—

[1] H. G. Smith, *J. Proc. Roy. Soc., N.S. Wales* (**1918**), 52, 529.

Specific gravity	0·895
Optical rotation	+ 18·4°
Refractive index	1·4747 at 16°
Yield	0·56 to 2 per cent.

Cineol, pinene, and aromadendrene appear to be the only well-ascertained constituents, but according to some the oil, when properly prepared, forms an excellent perfume.

Oil of Eucalyptus Microtheca.

This is a common tree in Australia, the characters of its essential oil being as follows:—

Specific gravity	0·8855
Optical rotation	− 27·2°
Refractive index	1·4839 at 20°
Eucalyptol	5 per cent.
Yield	0·23 ,,

The oil contains phellandrene, pinene, cymene, eucalyptol, and a sesquiterpene.

Oil of Eucalyptus Morrisii.

This oil has the following characters:—

Specific gravity	0·910 to 0·916
Optical rotation	+ 3·8° ,, + 6·3°
Refractive index	1·4693 at 10°
Yield	1·6 to 1·7 per cent.

It contains about 60 per cent. of eucalyptol and some *d*-pinene.

Oil of Eucalyptus Muelleri.

The oil from this tree has the following characters:—

Specific gravity	0·9096
Optical rotation	+ 10·4°
Refractive index	1·4629 at 24°
Yield	1·28 per cent.

The oil contains 60 per cent. of eucalyptol, pinene, and geranyl acetate.

Oil of Eucalyptus Nigra.

This tree is a Black Stringy Bark of New South Wales. Its oil has the following values:—

Specific gravity	0·874
Optical rotation	− 34·8°
Refractive index	1·4871 at 16°
Yield	0·041 per cent.

It contains phellandrene and about 5 per cent. of eucalyptol.

Oil of Eucalyptus Nova Angelica.

This tree is a Black Peppermint of New South Wales. Its oil contains pinene, phellandrene, eucalyptol, and aromadendrene. It has the following characters:—

Specific gravity	0·907
Optical rotation	+ 14·5°
Refractive index	1·4900 at 16°
Yield	0·5 per cent.

Oil of Eucalyptus Obliqua.

This tree is a "Stringy Bark" of the western part of Australia, but is found in the Eastern Colonies as well. Its oil contains phellandrene, aromadendral, and traces of eucalyptol, and has the following characters :—

Specific gravity	0·883 to 0·890
Optical rotation	− 24·2° „ − 28·8°
Refractive index	1·4934 at 16°
Yield	0·66 to 0·77 per cent.

Oil of Eucalyptus Occidentalis.

This oil has the following characters :—

Specific gravity	0·9135
Optical rotation	+ 9°
Refractive index	1·4774 at 16°
Yield	0·95 per cent.

It contains about 35 per cent. of eucalyptol, pinene, aromadendral, and a sesquiterpene.

Oil of Eucalyptus Odorata.

This is a South Australian Box tree, whose oil has the following characters :—

Specific gravity	0·899 to 0·922
Optical rotation	− 1·1° „ + 2·3°
Refractive index	1·4639 at 20°
Eucalyptol	up to 86 per cent.
Yield	1·87 „

It contains eucalyptol, cuminic aldehyde, pinene, and phellandrene.

Oil of Eucalyptus Oleosa.

This tree forms one of the Mallee scrub group, which are all fairly rich in cineol. It is well known in Victoria and South Australia and is also found in the interior extra-tropical parts of Western Australia. The leaves yield about 1·3 per cent. of an oil, having the following characters :—

Specific gravity	0·925
Optical rotation	− 5° to + 5°
Refractive index	1·4746 at 16°
Yield	1·0 to 1·3 per cent.

It is very rich in eucalyptol, and contains pinene, aromadendral, and a sesquiterpene.

According to Baron Müller, it is the best-known solvent for fossil resins, and "is unique for many technological applications". This oil is not identical with much of the so-called *Oleosa* oil of commerce, which is in reality obtained from *Eucalyptus cneorifolia*.

Oil of Eucalyptus Oreades.

This oil has the following characters:—

Specific gravity	0·887
Optical rotation	− 22·8°
Refractive index	1·4945 at 20°
Yield	1·16 per cent.

It contains phellandrene, piperitone, and eudesmol.

Oil of Eucalyptus Ovafolia.

This oil has the following characters:—

Specific gravity	0·9058
Optical rotation	− 9°
Refractive index	1·4921 at 16°
Yield	0·27 per cent.

It contains phellandrene, pinene, eucalyptol, and a sesquiterpene.

Oil of Eucalyptus Paludosa.

This tree occurs in New South Wales, Victoria, and Tasmania. The oil has the following characters:—

Specific gravity	0·902 to 0·9056
Optical rotation	+ 9·9° ,, + 14·2°
Refractive index	1·4773 at 16°
Yield	0·243 per cent.

It contains pinene, eucalyptol, a sesquiterpene, and a valerianic ester.

Oil of Eucalyptus Paniculata.

This tree is a "White Iron Bark" found in most of the colonies. Its oil has the following characters:—

Specific gravity	0·901
Optical rotation	+ 7·8°
Refractive index	1·4801 at 16°
Yield	0·1 per cent.

It contains eucalyptol, aromadendrene, aromadendral, and an alcohol not yet identified.

Oil of Eucalyptus Patentinervis.

This tree is known in New South Wales as a Bastard Mahogany. The following are the characters of the oil:—

Specific gravity	0·8735
Optical rotation	− 15·7°
Refractive index	1·4948 at 16°

It contains limonene, a sesquiterpene, and citral.

Oil of Eucalyptus Pendula.

This oil has the following characters:—

Specific gravity	0·9155
Optical rotation	+ 5·3°
Refractive index	1·4732 at 16°
Yield	0·84 per cent.

It contains about 50 per cent. of eucalyptol, pinene, and aromadendrene.

OIL OF EUCALYPTUS PERRINIANA.

This oil has the following characters :—

Specific gravity	0·912
Optical rotation	+ 8·9°
Refractive index	1·4651 at 15°
Eucalyptol	68 per cent.

It contains eucalyptol, pinene, and a sesquiterpene.

OIL OF EUCALYPTUS PHLEBOPHYLLA.

This tree is known as the Weeping Gum in Tasmania. It yields an essential oil having the following characters :—

Specific gravity	0·8766 to 0·8925
Optical rotation	– 22·4° ,, – 32·5°
Refractive index	1·4702 at 15°

It contains *l*-α-pinene, about 10 per cent. of eucalyptol, phellandrene, eudesmol, and a sesquiterpene.

OIL OF EUCALYPTUS PILULARIS.

This oil has the following characters :—

Specific gravity	0·885 to 0·903
Optical rotation	– 4·3° ,, + 11·1°
Refractive index	1·4961 at 16°
Yield	0·07 to 0·18 per cent.

It contains *l*-phellandrene, a sesquiterpene, eucalyptol, pinene, and an alcohol not yet identified.

OIL OF EUCALYPTUS PIPERITA.

This tree is found in New South Wales, Victoria, and Queensland, and is known as a "Peppermint Tree". The oil contains eudesmol, pinene, eucalyptol, phellandrene, and a sesquiterpene. Probably piperitone is also present. The oil has the following characters :—

Specific gravity	0·911
Optical rotation	– 2·7°
Refractive index	1·4838 at 16°

OIL OF EUCALYPTUS PLANCHONIANA.

This oil has the following characters :—

Specific gravity	0·904 to 0·915
Refractive index	1·4878 at 16°
Yield	0·01 to 0·06 per cent.

It contains phellandrene and a sesquiterpene.

OIL OF EUCALYPTUS PLATYPUS.

This oil, according to Earl,[1] has the following characters :—

Specific gravity	0·9045
Optical rotation	+ 9·1°
Refractive index	1·4675
Yield	1 per cent.

It contains about 60 per cent. of eucalyptol, 20 to 25 per cent. of pinene, and some phellandrene and aromadendrene.

[1] *Proc. Roy. Soc. Victoria*, *Chem. Abstracts*, **1917**, ii. 1715.

Oil of Eucalyptus Polyanthema.

This oil has the following characters:—

Specific gravity	0·9281
Optical rotation	+ 5·2°
Refractive index	1·4736 at 16°
Yield	0·825

It contains pinene and eucalyptol.

Oil of Eucalyptus Polybractea.

This is one of the Mallee eucalyptus trees found in New South Wales. Its oil has the following characters:—

Specific gravity	0·9143 to 0·930
Optical rotation	0° ,, − 2°
Refractive index	1·4592 to 1·4736 at 16°
Yield	1·35 per cent.

It contains up to 90 per cent. or over of eucalyptol, with small amounts of pinene and aromadendral.

Oil of Eucalyptus Populifolia.

This oil has the following characters:—

Specific gravity	0·920 to 0·923
Optical rotation	+ 0·41° ,, + 1·2°
Refractive index	1·4709 at 16°
Eucalyptol	about 60 per cent.
Yield	0·66 to 0·86 per cent.

It contains eucalyptol, pinene, and a sesquiterpene.

Oil of Eucalyptus Propinqua.

This oil has the following characters:—

Specific gravity	0·899
Optical rotation	+ 4·5°
Refractive index	1·4788 at 16°
Yield	0·235 per cent.

It contains eucalyptol, pinene, and aromadendral.

Oil of Eucalyptus Pulverulenta.

This oil has the following characters:—

Specific gravity	0·9236
Optical rotation	+ 2·1°
Refractive index	1·4686 at 16°
Yield	2·2 per cent.

It contains eucalyptol and pinene.

Oil of Eucalyptus Punctata.

This tree is found in Queensland and Victoria. Its essential oil has the following characters:—

Specific gravity	0·913 to 0·9297
Optical rotation	− 2·3° to + 4°
Refractive index	1·4774 at 16°
Eucalyptol	46 to 65 per cent.
Yield	0·63 ,, 1·19 per cent.

It contains eucalyptol, aromadendral, and cuminic aldehyde.

Oil of Eucalyptus Quadrangulata.

The characters of this oil are as follows :—

Specific gravity	0·9064
Optical rotation	+ 10·6°
Refractive index	1·4692 at 16°
Yield	0·69 per cent.

It contains *d*-pinene and eucalyptol.

Oil of Eucalyptus Radiata.

This tree is known as "White-top Peppermint" in New South Wales. The oil has the following values :—

Specific gravity	0·869 to 0·875
Optical rotation	− 64·8° „ − 77·8°
Refractive index	1·4863 at 16°
Yield	1·22 to 1·88 per cent.

It contains pinene, phellandrene, eucalyptol, and piperitone.

Oil of Eucalyptus Redunca.

This is a white gum found in Western Australia. The oil has the following characters :—

Specific gravity	0·9097
Optical rotation	+ 12·1°
Refractive index	1·4720 at 16°
Yield	1·2 per cent.

It contains *d*-pinene, eucalyptol (about 40 per cent.), and aromadendrene.

Oil of Eucalyptus Regnans.

This tree is a Tasmanian "Swamp Gum". Its oil has the following characters :—

Specific gravity	0·880 to 0·888
Optical rotation	− 28·4° „ − 31·1°
Refractive index	1·4882 to 1·4901 at 20°
Yield	0·8 per cent.

It contains phellandrene, eudesmol, geranyl acetate, piperitone, traces of eucalyptol, and a sesquiterpene.

Oil of Eucalyptus Resinifera.

This oil has the following values :—

Specific gravity	0·9098
Optical rotation	+ 2·2°
Refractive index	1·4755 at 16°
Yield	0·414 per cent.

It contains eucalyptol.

Oil of Eucalyptus Risdoni.

This oil has the following characters :—

Specific gravity	0·9045 to 0·916
Optical rotation	− 0·3° to − 14·6°
Refractive index	1·4733 at 16°
Yield	1·35 per cent.

It contains eucalyptol, phellandrene, and probably piperitone and amyl acetate.

Oil of Eucalyptus Rosii.

This oil has the following characters:—

Specific gravity	0·9168 to 0·9215
Optical rotation	+ 7·2°
Refractive index	1·4741 at 16°
Yield	0·72 per cent.

It contains pinene and eucalyptol.

Oil of Eucalyptus Rostrata.

This oil has the following characters:—

Specific gravity	0·895 to 0·9065
Optical rotation	− 11·7° to − 15·3°
Refractive index	1·4896 at 16°
Yield	0·1 to 0·4 per cent.

It contains cymene, aromadendral, eucalyptol, phellandrene, and pinene.

Oil of Eucalyptus Rubida.

This oil has the following characters:—

Specific gravity	0·9067
Optical rotation	+ 3·2°
Refractive index	1·5011 at 16°
Yield	0·008 per cent.

It contains pinene, a sesquiterpene, and a little eucalyptol.

Oil of Eucalyptus Rudderi.

This oil has the following values:—

Specific gravity	0·942
Optical rotation	− 8·5°
Refractive index	1·4898 at 20°
Yield	0·3 per cent.

It contains pinene, aromadendral, and about 5 per cent. of eucalyptol.

Oil of Eucalyptus Saligna.

This tree is found in New South Wales. The oil has the following characters:—

Specific gravity	0·886 to 0·894
Optical rotation	+ 32·9° ,, + 35·1°
Yield	0·12 to 0·255 per cent.

It contains *d*-pinene, eucalyptol, eudesmic acid, amyl ester, a free alcohol, and a valerianic ester.

Oil of Eucalyptus Salmonophloia.

This oil has the following characters:—

Specific gravity	0·9076
Optical rotation	+ 6·3°
Refractive index	1·4738
Yield	1·44 per cent.

It contains eucalyptol, pinene, and aromadendrene.

Oil of Eucalyptus Salubris.

This oil has the following values :—

Specific gravity	0·902
Optical rotation	− 5·8°
Refractive index	1·4841 at 16°
Yield	1·4 per cent.

It contains *d*-pinene, cymene, eucalyptol (10 per cent.), geranyl acetate, and aromadendral.

Oil of Eucalyptus Santalifolia.

This tree is a Tasmanian eucalypt. Its oil contains pinene, limonene, eucalyptol, and sesquiterpenes; it has the following characters :—

Specific gravity	0·884
Optical rotation	− 37·7°
Refractive index	1·4736 at 20°
Eucalyptol	12 per cent.
Yield	0·41 ,,

Oil of Eucalyptus Siderophloia.

This tree is a "Red Iron-bark" of Queensland. Its oil has the following characters :—

Specific gravity	0·9067
Optical rotation	+ 14·4°
Refractive index	1·5000 at 16°
Yield	0·056 per cent.

It contains phellandrene, pinene, and eucalyptol.

Oil of Eucalyptus Sideroxylon.

This tree is the "Red Flowering Iron-bark" of New South Wales and Victoria. Its oil has the following characters :—

Specific gravity	0·919 to 0·923
Optical rotation	+ 1·35° ,, + 3·1°
Refractive index	1·4725 at 16°

It contains about 60 per cent. of eucalyptol, pinene, a sesquiterpene, and an aldehyde.

Oil of Eucalyptus Smithii.

This oil is of very great importance, since it is perhaps the best eucalyptol-pinene oil obtainable from any species. H. G. Smith gives the following information in regard to this tree (which has been named after him) :—

The time is rapidly approaching when it will be considered advisable, and found to be profitable, to cultivate or re-afforest with the best species for the required kinds of essential oils obtainable from the members of this great genus. The time seems opportune, therefore, to direct attention to the advantages which *Eucalyptus Smithii* offers for cultivation.

The growth of *Eucalyptus Smithii*, under natural conditions of soil and climate, is considerable, and is more rapid with the so-called "suckers," or new growths, which spring from the stumps of the felled mature trees than with plants grown from seed, although with the seedlings the increase in height usually averages, for the first year or two, from 6 ins. to a foot per month.

Eucalyptus Smithii is a well-defined species and has a somewhat extensive range. The oil obtained by steam distillation is one of the richest in cineol-content of all eucalyptus species so far examined. The yield of oil is also good, as much as from 20 to 26 lb. of oil being obtainable commercially from the leaves and terminal branchlets which can be packed in a 400 gallon tank. The oil consists very largely of cineol, and the terpene is dextro-rotatory pinene. In commercial samples the amount of high boiling constituents is often very small indeed, the original distillation not being carried to the end, and 96 per cent. or more of crude commercially distilled oil may come over below 190° C.

In addition to pinene and eucalyptol, the oil contains a phenol, traces of aldehydes, *d*-eudesmol, butyl-butyrate (?), and a sesquiterpene. A small quantity of a solid paraffin melting at 64° is also present.

The great importance of this oil from the commercial point of view rendered an examination of samples obtained under different conditions desirable, and the following samples have been examined by H. G. Smith :—

(*a*) Leaves from lopped trees, seven months' growth ; collected May, 1913. (*b*) Leaves from lopped trees, fifteen months' growth ; collected May, 1913. (*c*) Leaves from seedlings, twelve months' growth ; collected June, 1914. (*d*) Leaves from seedlings two and a half years' old ; collected July, 1914. (*e*) Leaves from cultivated tree at Marrickville ; collected June, 1915. (*f*) Leaves from general material, partly young ; collected January, 1915. (*g*) Leaves from general material collected three weeks later than (*f*). (*h*) Leaves from old trees ; collected March, 1913.

The constants, etc., given by the crude oils from the above material were as follows :—

	Specific Gravity at 15° C.	Rotation. α_D.	Refractive Index.	Solubility in 70 % Alcohol.	Saponification Number.	Eucalyptol per Cent.
(*a*)	0·9098	+ 7·6°	1·4636 at 20°	Required. 1·6 vols.	4·8	67·4
(*b*)	0·9157	+ 6·5°	1·4635 at 20°	1·2 „	5·6	74·2
(*c*)	0·9116	+ 9·2°	1·4650 at 19°	2·1 „	1·3	61·5
(*d*)	0·9139	− 7·6°	1·4634 at 18°	1·4 „	4·1	69·0
(*e*)	0·9198	+ 4·7°	1·4672 at 16°	1·2 „	2·7	75·0
(*f*)	0·9156	+ 5·3°	1·4571 at 26°	1·1 „	3·3	80·7
(*g*)	0·9154	+ 5·1°	1·4574 at 25°	1·1 „	3·1	79·0
(*h*)	0·9210	+ 4·2°	1·4604 at 22°	1·1 „	1·3	85·2

The eucalyptol was determined by the resorcinol method.

FIG. 33.—*Eucalyptus Smithii*, R.T.B.

Photograph taken in 1911. Tree was felled in 1906; top branches first removed for oil distillation in 1908. Photograph shows three years' growth. Material has been collected from this tree for oil distillation four times in seven years. Hill Top, New South Wales. *Journ. Roy. Soc. N.S. Wales*, vol. xlix., Plate XIII.

Oil of Eucalyptus Squamosa.

This oil has the following characters:—

Specific gravity	0·9182
Optical rotation	0°
Refractive index	1·4692 at 16°
Yield	0·65 per cent.

It contains about 60 per cent. of eucalyptol, pinene, and aromadendral.

Oil of Eucalyptus Staigeriana.

This tree is the Lemon-scented Iron-bark of Queensland, the essential oil being of particular interest, as it differs entirely from ordinary eucalyptus oils. The oil has the following characters:—

Specific gravity	0·8708 to 0·880 (0·901 ?)
Optical rotation	− 43·1°
Refractive index	1·4871 at 16°
Yield	2·5 per cent.

The oil has been carefully investigated by Baker and Smith. In the course of fractional distillation only a few drops came over below 175°, but between that temperature and 193° C. 65 per cent. distilled; this fraction had specific gravity at 23/15° C. = 0·851, and refractive index at the same temperature, 1·4798. Between 193° and 265° C., 23 per cent. distilled, the specific gravity of this = 0·8828 at 23° C., and refractive index 1·4846 at the same temperature. Between 265° and 300° C. only 2 per cent. distilled; this had a refractive index 1·5033, and was quite acid. On again rectifying the first fraction (65 per cent.) nothing came over below 175° C.; but 40 per cent. of the fraction distilled between 175° to 177° C.; the rotation of this in 100 mm. tube was − 66·2° ($[\alpha]_D$ − 78·18°), the specific gravity = 0·8468 at 22/15° C., and the refractive index = 1·4793. Between 177° and 179° C. 23 per cent. distilled; the rotation was − 64·8°, the specific gravity = 0·8533° at 22/15° C., and the refractive index = 1·482. Between 179° and 184° C. 17 per cent. distilled; the rotation was − 60·9°, the specific gravity = 0·8534, and the refractive index = 1·4828.

The crystallised tetrabromide was formed in the usual way with the portion distilling 175° to 177° C., and these crystals, when purified to constant melting-point with acetic ether, melted at 104° to 105° C. From the physical properties of the several fractions it may be assumed that nearly the whole of the terpene present was laevo-limonene, and that the oil contained about 60 per cent. of that terpene.

An aldehyde determination in the usual way, with 30 per cent. sodium bisulphite solution, gave 16 per cent. of aldehydes. The preparation of the non-aldehydic oil was carried out quantitatively to act as a check on the other determinations. Thirty grms. of oil gave 25·4 grms. of non-aldehydic constituents equal to 15·33 per cent. of aldehydes.

The rotation of the non-aldehydic oil was − 46·8°, the specific gravity = 0·8658 at 22/15° C., and the refractive index = 1·4831. It was very aromatic.

The ester determination was carried out on the non-aldehydic oil. The saponification number was 28·3, equal to 9·9 per cent. of ester calculated as geranyl acetate, or 8·32 per cent. on the original oil. An ester determination on the original oil gave 8·6 per cent.

A portion of the non-aldehydic oil was acetylated in the usual way, and this gave a saponification number 78·47, which represents 15·14 per cent. of free alcohol, or 12·72 per cent. of geraniol in the original oil.

The oil of *Eucalyptus Staigeriana* may be stated to have approximately the following constitution :—

Limonene	60·00 per cent.
Geraniol	12·72 „
Geranyl acetate	8·32 „
Citral	16·00 „
Undetermined	2·96 „
	100·00 „

Oil of Eucalyptus Stellulata.

This oil has the following characters :—

Specific gravity	0·871
Optical rotation	− 26·1°
Refractive index	1·4902 at 16°
Yield	0·3 per cent.

It contains phellandrene, a sesquiterpene, and traces of eucalyptol.

Oil of Eucalyptus Stricta.

This oil has the following characters :—

Specific gravity	0·9246
Optical rotation	− 3·4°
Refractive index	1·4711 at 16°
Yield	0·5 per cent.

It contains pinene, eucalyptol, and eudesmol.

Oil of Eucalyptus Stuartiana.

This oil has the following characters :—

Specific gravity	0·916
Optical rotation	+ 4·7°
Refractive index	1·4790 at 16°
Yield	0·4 per cent.

It contains about 50 per cent. of eucalyptol, pinene, and some esters.

Oil of Eucalyptus Tæniola.

This oil has the following characters :—

Specific gravity	0·8864
Optical rotation	− 27·6°
Refractive index	1·4872 at 17°
Yield	0·66 per cent.

It contains 7 per cent. of eucalyptol, phellandrene, eudesmol, piperitone, and a sesquiterpene.

Oil of Eucalyptus Tereticornis.

This oil has the following characters :—

Specific gravity	0·9218
Optical rotation	− 9·3°
Refractive index	1·4934 at 16°
Yield	0·48 per cent.

It contains pinene, eucalyptol (about 5 per cent.), aromadendral, and a sesquiterpene.

An Indian oil, examined by Ghose,[1] had the following characters :—

Specific gravity	0·903
Optical rotation	− 22·3°
Refractive index	1·4905

Oil of Eucalyptus Tessellaris.

This oil has the following values :—

Specific gravity	0·896
Optical rotation	+ 8·8°
Refractive index	1·4881 at 16°
Yield	0·16 per cent.

It contains pinene, cymene, eucalyptol, and aromadendrene.

Oil of Eucalyptus Trachyphloia.

This oil has the following characters :—

Specific gravity	0·893
Optical rotation	+ 9·8°
Refractive index	1·4901 at 16°
Yield	0·2 per cent.

It contains pinene, aromadendrene, aromadendral, and eucalyptol.

Oil of Eucalyptus Umbra.

This oil has the following characters :—

Specific gravity	0·8963
Optical rotation	+ 37·4°
Yield	0·16 per cent.

It contains *d*-pinene, eucalyptol, and some esters.

Oil of Eucalyptus Unialata.

This tree is a Tasmanian eucalypt. Its oil contains about 60 per cent. of eucalyptol, *d*-pinene, and a sesquiterpene; and has the following characters :—

Specific gravity	0·918
Optical rotation	+ 3·1°
Refractive index	1·4690 at 18°
Yield	0·9 per cent.

Oil of Eucalyptus Urnigera.

This oil has the following characters :—

Specific gravity	0·9088
Optical rotation	+ 10·6°
Refractive index	1·4638 at 23°
Yield	1·13 per cent.

It contains about 60 per cent. of eucalyptol, *d*-pinene, and geranyl acetate.

[1] *P. and E.O.R.* **(1919)**, 145.

Oil of Eucalyptus Vernicosa.

This oil has the following characters:—

Specific gravity	0·9038
Optical rotation	+ 11·3°
Refractive index	1·4651 at 18°
Yield	0·8 per cent.

It contains about 60 per cent. of eucalyptol, and pinene.

Oil of Eucalyptus Viminalis.

This tree is found in most of the Australasian Colonies. The oil has the following characters:—

Specific gravity	0·904 to 0·916
Optical rotation	+ 3·6° „ + 4·2°
Refractive index	1·4799 to 1·4840 at 20°
Eucalyptol	35 per cent.
Yield	0·35 to 0·74 per cent.

It contains eucalyptol, phellandrene, aromadendrene, and *d*-pinene.

Oil of Eucalyptus Virgata.

This oil has the following characters:—

Specific gravity	0·888 to 0·915
Optical rotation	− 20·9° „ − 35·8°
Refractive index	1·5015 at 16°

It contains about 20 per cent. of eucalyptol, and *l*-phellandrene.

Oil of Eucalyptus Viridis.

This oil has the following characters:—

Specific gravity	0·9006
Optical rotation	− 8°
Refractive index	1·4828 at 16°
Yield	1·06 per cent.

It contains eucalyptol, *l*-pinene, and aromadendral.

Oil of Eucalyptus Vitrea.

This oil has the following characters:—

Specific gravity	0·886
Optical rotation	+ 29·8°
Refractive index	1·4828 at 16°
Yield	1·48 per cent.

It contains phellandrene, eucalyptol, a sesquiterpene, piperitone, and citral (?).

Oil of Eucalyptus Wilkinsoniana.

This oil has the following values:—

Specific gravity	0·8944
Optical rotation	− 23·9°
Refractive index	1·4774 at 16°
Yield	0·975 per cent.

It contains *l*-pinene and eucalyptol.

Oil of Eucalyptus Woollsiana.

This oil has the following characters:—

Specific gravity	0·889 to 0·905
Optical rotation	− 9·5° „ − 16°
Refractive index	1·4839 to 1·4880 at 20°
Yield	0·5 per cent.

It contains pinene, aromadendral, and eucalyptol.

OILS OF THE ANGOPHORAS.

So far only six species of the genus *Angophora* have been described. They are all Australian plants, and are closely allied to the Eucalypts, and believed by Baker and Smith to be the prototypes of the latter; these scientists have investigated the plants, and Smith[1] has examined the essential oils.

The following table gives the yields of oil distilled from material cut as would be done for commercial oil distillation :—

Angophora Bakeri	0·31 per cent.
,, *melanoxylon*	0·19 ,,
,, *sp.* undescribed	0·13 ,,
,, *lanceolata* (Sydney)	0·013 ,,
,, ,, (Warialda)	0·005 ,,
,, *intermedia*	traces
,, *subvelutina*	none
,, *cordifolia*	,,

The oil from *Angophora Bakeri* has the following characters :—

Specific gravity	0·8719
Optical rotation	+ 35·6°
Refractive index	1·4660 at 22°

The crude oil was light lemon-coloured, had an indistinct odour at first, but distinctly a secondary aromatic one. The chief constituents were dextro-rotatory pinene, geraniol, geranyl-acetate, geranyl-valerianate, a small amount of a sesquiterpene, together with a little volatile aldehyde, and most probably a minute quantity of amyl-acetate. No other terpene than pinene was detected and cineol was quite absent.

The oil from *Angophora melanoxylon* has the following characters :—

Specific gravity	0·8809
Optical rotation	+ 24·9°
Refractive index	1·4678 at 21·5°

The constituents of this oil are identical with those of *Angophora Bakeri*, the geranyl valerianate, and the sesquiterpene being present in rather larger amount.

Angophora lanceolata yields an oil having the following characters :—

Specific gravity	0·927
Refractive index	1·4946 at 20°
Saponification value	24·2

The constituents are similar to those of the other species just described.

The approximate composition of the oils from *Angophora Bakeri* and *Angophora melanoxylon* are as follows :—

	Angophora Bakeri. Per Cent.	*Angophora Melanoxylon.* Per Cent.
Dextro-rotatory pinene	78·0	72·0
Free geraniol	4·5	5·1
Geranyl acetate	6·7	6·9
Geranyl valerianate	6·1	9·6
Water, volatile aldehydes, and low-boiling esters	1·0	1·0
Sesquiterpene and undetermined	3·7	5·4
	100·0	100·0

[1] *Jour. and Proc. Roy. Soc., N.S.W.*, xlvii. (**1913**), 106.

Oils of Leptospermum.

The lemon-scented *Leptospermum Liversedgei* is a New South Wales shrub, from 6 to 12 ft. in height, which yields about 0·23 per cent. of essential oil, which has been examined by Baker and Smith.[1] The oil has the following characters:—

Specific gravity	0·8895
Optical rotation	+ 9·2°
Refractive index	1·4903 at 10°

The principal constituents in the oil are: (1) the aldehyde citral; (2) an alcohol, which is probably geraniol; (3) an acetic acid ester, which is probably geranyl acetate; (4) the terpene pinene which was dextrorotatory; and (5) a sesquiterpene, which is probably the constituent which gives the laevo-rotation to the higher boiling portion. Limonene could not be detected by any method, and was, therefore, absent; nor was phellandrene present. The whole of the aldehyde appears to be citral, as proof of the presence of any other aldehyde could not be obtained, and two determinations by Flatau and Labbe's method failed to give any indication for citronellal. The secondary odour of the oil, from which the aldehydes had been removed, strongly resembled that of geraniol. The oils of the Leptospermums, with this exception, do not appear to have been investigated chemically, so that the occurrence of citral in the oil of *Leptospermum Liversedgei* is of some scientific interest.

The approximate composition of the oil is as follows:—

Citral	35·0 per cent.
Geranyl acetate	5·35 ,,
Geraniol	9·74 ,,
d-pinene	25·0 ,,
Sesquiterpene and undetermined	24·91 ,,
	100·00 per cent.

According to Challinor, Cheel, and Penfold, the leaves of *Leptospermum flavescens*, var. *citratum*, yield about 1·8 per cent. of oil having the following characters:—

Specific gravity	0·884
Optical rotation	+ 3·6°
Refractive index at 20°	1·4730

It contains about 90 per cent. of aldehydes, which consist of citral and citronellal in approximately equal amount. A small quantity of an alcohol (possibly geraniol or citronellol) was found, and a trace of a phenol. Probably aromadendrene is present.

Oil of Darwinia Fascicularis.

This oil has been investigated by Baker and Smith.[2]

Darwinia fascicularis is a virgate shrub, varying in size in different localities, about 5 ft. being the maximum height in the coast form. The mountain variety rarely exceeds 2 ft. The leaves are slender, numerous, and crowded, and vary in length from four to eight lines. The flowers although small are attractive, being sometimes all white, or pink and white.

[1] *Jour. and Proc. Roy. Soc., N.S.W.*, xxxix. (1905), 124.
[2] *Ibid.*, xxxiii. (1899), 163.

It is entirely an Australian plant, being principally found in the neighbourhood of Port Jackson. It occurs on sandy soil mostly, and covers many hundred acres of ground between Botany, La Perouse, and the coast. It has been found north of Manly and also on the Blue Mountains at King's Tableland, Wentworth, and Lawson.

The yield of oil obtained was from 0·318 to 0·456 per cent., having a specific gravity 0·9154, and being slightly dextro-rotatory.

It contains geraniol, to the extent of 57 to 65 per cent. in the form of acetate, and about 13 per cent. in the free state.

Oil of Darwinia Taxifolia.

This oil has the following characters:—

Specific gravity	0·8734 at 21°
Optical rotation	− 6·5°
Yield	0·313 per cent.

This oil contains only about 5 to 6 per cent. of esters and 8 per cent. of free alcohol. It is probable that the alcohol present is linalol and not geraniol.

Oil of Darwinia Grandiflora.

Baker and Smith [1] have described the essential oil distilled from the leaves of *Darwinia Grandiflora*, which was obtained to the extent of 0·12 per cent. This oil had the following characters:—

Specific gravity	0·915
Optical rotation	+ 23·1°
Refractive index	1·4773
Ester value	100·4

The oil contains *d*-pinene, geranyl acetate, and probably geranyl butyrate.

Cajuput Oil.

This oil is distilled from the leaves of several species of *Melaleuca*, myrtaceous shrubs growing abundantly in the Indian Archipelago, the Malay Peninsula, and other places. Most of the oil of commerce is yielded by *Melaleuca minor*. A great portion of the oil is distilled on the islands of Bouru and Banda, whence it is brought to Macassar and the neighbourhood for shipment. The bulk of the oil is exported in wine bottles, although shipment in drums, to save the excessive freight, has been resorted to. The oil has a powerful camphoraceous odour, due largely to its high cineol-content. It is usually of a green colour, generally due to contamination with copper, but can be obtained white by re-distillation. The oil has the following characters:—

Specific gravity	0·917 to 0·930
Optical rotation	− 1° ,, − 4°
Refractive index	1·4650 ,, 1·471

Normal oil contains up to 65 per cent. of cineol. In addition to this body, to which the oil owes its value, terpineol and terpinyl acetate have been found, and, according to Voiry,[2] butyric, valeric, and benzoic aldehydes and probably laevo-pinene. The low specific gravity of many

[1] *Jour. and Proc. Roy. Soc., N.S.W.*, l. **(1916)**, 181.
[2] *Jour. Pharm. Chim.* **(1888)**, 149.

commercial samples is accounted for by adulteration with turpentine or petroleum. The oil is not used to any extent internally, but is employed externally as a stimulant and anti-spasmodic. Eucalyptus oil with a high cineol-content is well able to replace it for most purposes. Genuine Cajuput oil should contain at least 50 per cent. of eucalyptol, as determined by the phosphoric acid process.

Much of the obscurity surrounding the origin of Cajuput oil has been cleared up by the systematic investigation of the species by Baker and Smith,[1] who state that there had been imported into Europe from the East, about the beginning of the seventeenth century, an oil under the name of "Cajoepoeli" (according to Linnæus' spelling), but under a later spelling "Cajuput". At that time, and long after its introduction, the botanical origin was ascribed to Linnæus' species (*Melaleuca leucadendron*), as the specimens forwarded to Linnæus were reputed to be taken from trees from which the oil was obtained, and he evidently described it under that impression, as shown by his original specimen now in the possession of the London Linnean Society and labelled by him "Cajoepoeli" and afterwards by Smith as *Melaleuca leucadendron, vera*.

This reputed origin of the oil, however, was shown later by Roxburgh to be an error, being the true source of "Cajuput" a Melaleuca which he named *Melaleuca cajuputi*, but this was found later to be identical with *Melaleuca minor*, described earlier in 1813 by Smith in *Rees Cyclopædia*, Vol. XXIII, and so quite a distinct tree from that to which Linnæus had given the above name. However, many European systematists in the last century regarded the two as one, but the early Indian botanists being very emphatic over the matter, always kept them distinct, and Baker and Smith's investigations support the latter botanists.

Schimmel & Co.[2] have examined the oils obtained from the narrow-leaved and the broad-leaved varieties of *Melaleuca leucadendron* grown in Amani. They find that eucalyptol is entirely absent from the broad-leaved variety, and only present to a small extent, as compared with true cajuput oil, in the narrow-leaved variety. The oil from the broad-leaved variety contains about 78 per cent. of methyl-eugenol. The two oils had the following characters :—

	Broad Leaved.	Narrow Leaved.
Specific gravity	1·0019	0·9056
Optical rotation	– 3° 44′	– 0° 22′
Refractive index	1·5250	1·4794
Acid value	0·4	0·7
Ester value	7·6	11·8
,, ,, (after acetylation)	15·2	84·8

Since the original description by Linnæus was published, several species have been described which had the general facies of his tree but differed in some important characters.

The following *Melaleuca* oils have been described in addition to that from *Melaleuca minor* :—

[1] *Jour. and Proc. Roy. Soc., N.S.W.*, xl. **(1906)**, 60; xli. **(1907)**, 196; xliv. **(1910)**, 592; xlv. **(1911)**, 365; xlvii. **(1913)**, 193.
[2] *Report*, October, **1914**, 53.

Oil of Melaleuca Maideni.

This tree is found in New South Wales and Queensland. The oil has the following characters:—

Specific gravity	0·9199 to 0·9234
Optical rotation	– 0·7° „ – 4·2°
Refractive index	1·4744 to 1·4800 at 22°
Eucalyptol	26 to 39 per cent.

The oil contains, in addition to eucalyptol, *l*-pinene, *l*-limonene, an alcohol, a sesquiterpene alcohol, a sesquiterpene (?), and benzaldehyde (?).

Oil of Melaleuca Smithii.

This is the common broad-leaved *Melaleuca* of certain districts of New South Wales. Three samples examined by Baker and Smith had the following characters:—

Locality.	Yield per Cent.	Specific Gravity at 15°.	Eucalyptol per Cent.	Rotation α_D.	Refractive Index at 22 .	Saponification Number.
Rose Bay, Sydney 18/12/1911	0·607	0·8815	About 1	+ 11·8°	1·4812	3·1
Terrigal, N.S.W. 21/12/1911	0·923	0·9003	less than 2	+ 6·7°	1·4824	3·3
Gosford, N.S.W. 20/9/1899	0·495	0·8976	about 5	+ 5·8°	1·4806	6·5

In addition to eucalyptol, the oil contains benzaldehyde (?) *l*-pinene *l*-limonene, dipentene, and a sesquiterpene alcohol, which has been termed melaleucol. It is suggested that this alcohol may be related to the alcohol nerolidol, in the same way as geraniol is to linalol. The identity of this body, however, is hardly established by experimental evidence.

Oil of Melaleuca Thymifolia.

This oil has the following characters:—

Specific gravity	0·9134
Optical rotation	+ 2·1°
Refractive index	1·4665 at 23°
Yield	2·28 per cent.

It contains about 53 per cent. of eucalyptol.

Oil of Melaleuca Linarifolia.

This oil has the following characters:—

Specific gravity	0·9129
Optical rotation	+ 2·5°
Refractive index	1·4741 at 22°
Yield	1·214 per cent.

It contains much terpenes and but little eucalyptol.

Oil of Melaleuca Uncinata.

This tree yields 1·246 per cent. of essential oil which has the following characters:—

Specific gravity	0·9211 at 21°
Optical rotation	+ 7·5°
Refractive index	1·4760 at 21°

The oil is rich in eucalyptol, and also contains *d*-*a*-pinene, a sesquiterpene, and a crystalline body, which may be an alcohol, melting at 72·5°. In this the name uncineol and the formula $C_{10}H_{18}O$ is suggested

Oil of Melaleuca Nodosa.

This oil from the Melaleuca has the following characters :—

Specific gravity	0·8984
Optical rotation	+ 11·6°
Refractive index	1·4689 at 18°
Yield	0·664 per cent.

The oil contains much *d*-pinene, eucalyptol, and a sesquiterpene.

Oil of Melaleuca Trichostachya.

This oil has the following characters :—

Specific gravity	0·914 to 0·915
Optical rotation	+ 2·3° „ + 3·1°
Refractive index	1·4636 „ 1·4655
Yield	1·25 to 2·58 per cent.

The oil contains a very large amount of eucalyptol, terpenes, traces of a phenol, and terpinyl acetate.

Oil of Melaleuca Bracteata.

The oil from the *Melaleuca* differs from those from all other members of the genus. The yield of oil is about 1 per cent., having the following characters :—

Specific gravity	1·0320 to 1·0360
Optical rotation	− 1·4° „ − 3·1°
Refractive index	1·5325 „ 1·5350

The principal constituent of the oil is methyl-eugenol, with some eugenol, free cinnamic acid, cinnamic aldehyde, cinnamic alcohol (?), phellandrene, and cinnamyl-(?)-cinnamate.

Oil of Melaleuca Genistifolia.

This oil has the following characters :—

Specific gravity	0·8807
Optical rotation	+ 32·7°
Refractive index	1·4702 at 22°
Yield	0·526 per cent.

About 2 per cent. of eucalyptol is present, with 80 to 90 per cent. of pinene, a small amount of a sesquiterpene, and 5 to 6 per cent. of terpinyl acetate.

Oil of Melaleuca Gibbosa.

This oil has the following characters :—

Specific gravity	0·9138
Optical rotation	+ 4·5°
Refractive index	1·4703 at 20°
Eucalyptol	61·5 per cent.
Yield	0·16 „

The oil contains eucalyptol, pinene, and a sesquiterpene.

Oil of Melaleuca Pauciflora.

This oil is quite different from ordinary *Melaleuca* oils. It contains no pinene, the terpenes present consisting of limonene and (probably) dipentene, with a little eucalyptol, a sesquiterpene, terpineol, and terpinyl acetate. The characters of the oil are as follows:—

Specific gravity	0·9302
Optical rotation	+ 3·3°
Refractive index	1·4921 at 24
Yield	0·3 per cent.

Oil of Melaleuca—(Niaouli).

Oil of "Niaouli" is generally stated to be obtained from *Melaleuca viridiflora*, from New Caledonia. The name was given to the plant yielding the oil by Brongniart and Gris, but as it is quite different from the true *Melaleuca viridiflora* of Solander, the plant requires a new specific name, and should not be described as *viridiflora*. The oil has the following characters:—

Specific gravity	0·908 to 0·929
Optical rotation	− 1° ,, + 1°
Refractive index	1·4723

It contains *d*-α-pinene, *l*-limonene, terpineol, eucalyptol, valeric and benzoic aldehydes, and terpinyl valerianate (?).

Oil of Bay.

This oil is the product of distillation of the leaves of *Pimenta acris* (*Myrcia acris*) and probably other closely allied species, plants flourishing in the West Indies. The yield of oil obtained is from 2 to 3 per cent., the oil coming over in two portions, firstly, a light oil which distils over rapidly, then a heavy oil distilling over more slowly. This, of course, means nothing more than that there are constituents of bay oil which have widely different boiling-points. The fact that the *United States Pharmacopœia* of 1880 gave the specific gravity of bay oil as about 1·040, caused the heavier fractions to appear in commerce under that name. Pure normal distillates, however, have a specific gravity of ·965 to 1·010, and oils with a higher specific gravity must be regarded as suspicious, being possibly adulterated with oil of cloves or oil of pimento. Occasionally distillates with a specific gravity of 1·020 or thereabouts are found with phenol values of 65 to 75 per cent., but these are rare.

The erratic manner in which this oil is distilled in the West Indies results in a large majority of the oils reaching this country being merely fractions of Bay oil. Umney[1] quotes the following (see opposite page) as representing commercial samples which he has observed. Those of very low specific gravity, however, may be actually adulterated.

Reference to the figures recorded shows that the percentage of phenols in oils having a specific gravity between 0·936 and 0·938 may vary as much as from 32 to 48 per cent. The strong probability is that the reason of the variation lies in the changes which the principal terpene, myrcene, undergoes on keeping, and especially on exposure.

Efforts are being made to improve the Bay Leaf industry in the West Indies, with a view to a more constant quality of oil being imported.

[1] *P. and E.O.R.* **(1916)**, 295.

Some interesting details of the experimental work being undertaken is given by Sir Francis Watts, D.Sc., F.I.C., Commissioner of Agriculture

Phenol Percentage.	Specific Gravity.	Phenol Percentage.	Specific Gravity.
20	·864	48	·948
24	·858	50	·985
26	·875	50	·951
26	·879	50	·967
28	·895	51	·953
28	·896	52	·924 Dominica
30	·896	52	·977
30	·898 Montserrat	52	·944
32	·936	52	·952
33	·945 Dominica	52	·954
34	·908	53	·969
34	·941	54	·945
36	·936	54	·949
36	·937	54	·950 Montserrat
36	·943	55	·946 Dominica
37	·932	55	·966
40	·941	55	·966
40	·948	56	·978
41	·929	56	·955
42	·948	56	·956
43	·931	58	·971
44	·935	60	1·004
45	·925 Dominica	60	·965 U.S.A.
45	·926	60	·966
45	·958	60	·974
45	·933	60	·984 Martinique
47	·937	62	·960
48	·937	64	·977
48	·938	76	1·011

for the West Indies. He states that, in connection with the island of Montserrat, "the results obtained from the experimental plot, 1 acre in area, established in 1908, have given further proof that at the current price obtained for bay oil the cultivation of this tree is a matter worth the attention of persons having lands that are at present unused, particularly on the higher lands of the island. Details of the planting and cultivation up to the end of 1912 in connection with this plot are given in the last annual report on the Botanic Station, and, generally, the results obtained in 1913 are a decided improvement on those of the two previous years.

"Briefly, the plot was planted in 1908, and reaping of the leaves from which the oil is distilled was commenced in January, 1911. Since that time monthly reapings have been made. The yield of fresh leaves, and the actual amount of oil obtained from them for the three years ending December, 1913, is as follows:—

	Yield of Leaves, lb.	Actual Yield of Oil.
1911	1368	6,189 c.c. = 221 oz. = 14 lb. approx.
1912	1940	8,615 ,, = 307 ,, = 19 ,, ,,
1913	2510	12,764 ,, = 456 ,, = 28½ ,, ,,

"Not only has there been an increase in the total amount of leaves reaped from the plot, but also an increase in the proportion of oil obtained from the leaves, which in 1912, on the average, was 16·7 oz., and in 1913 18·3 oz. per 100 lb. of leaves as reaped.

Date of Distillation.	Amount of Fresh Leaves.	Yield of Oil, c.c.	Ounces per 100 lb. Fresh Leaves.	Specific Gravity.	Phenol-content, per Cent.
January 3rd	165	764	16·5	·9410	50
January 31st	162	773	16	·9496	54
March 3rd	162	732	16	·9806	70
April 1st	185	931	18	·9443	58
May 1st	153	783	18	·9620	63
May 6th	180	909	18	·9515	60
June 12th	160	905	20	·9288	47
June 26th	176	1115	22·7	·9714	66
July 16th	188	1149	21·8	·9520	60
August 22nd	160	814	18	·9421	53
September 26th	174	846	17·4	·9219	45
October 31st	195	801	14·7	·9274	51
November 28th	163	830	18·1	·9242	48
December 15th	104	639	21·7	·9491	62

"The particular object of making reapings monthly was to decide if at any season of the year the leaves yielded a higher percentage of oil or an oil of better quality; the data collected from the reapings made in 1913 are shown in the preceding table.

"In considering these results it is noticed that there are abrupt differences in certain months, e.g. the high specific gravity and phenol-content in March, the low phenol-content on 12 June, and the striking fall in the proportion of oil obtained in October, that need explanation; but, on the whole, the results are in accord with those obtained in the previous year, when June and July gave better returns than any two contiguous months of the year.

"Experiments both here and in Antigua have shown that specific gravity and phenol-content are considerably influenced by the kind of leaves collected for distillation, and the larger the amount of immature growth included in the reapings the less satisfactory these two factors prove to be. This is again clearly demonstrated in two tests made for the purpose, the first on 15th December, 1913, when only mature leaves were distilled. The specific gravity and phenol-content in this case were as shown in the table, and in the test for comparison made on 16th January, 1914, young growth preponderated, and the specific gravity of the oil was 0·8856 and the phenol-content only 33 per cent. In view of this it is premature to say whether the rise in yield and the better quality of the oil obtained in the middle of the year during the last two years are due solely to the absence of young growth at the time. For the first six months of 1914 the average yield per 100 lb. of leaves has risen to 20 oz. It has been a rule-of-thumb practice to dry the leaves for three days before distillation, but no scientific examination of this point has been carried out, and it is not known whether the drying increases or diminishes the oil yield."

The distillations were repeated in 1917 with the following results:—

Date Distilled, 1917.	Weight of Green Leaves distilled, lb.	Oz. of Oil per 100 lb. Leaves.	Specific Gravity.	Phenol-content per Cent.
January 12th	159	16·7	·9382	50
January 19th	159	17·6	·9414	56
January 26th	161	17·9	·9480	55
February 2nd	157	17·7	·9440	56
February 9th	162	19·9	·9513	56
February 16th	177½	16·7	·9504	57
February 23rd	171	17·8	·9450	50
March 14th	177	16·1	·9462	55
March 22nd	162	16·6	·9542	57
March 27th	162	18·7	·9480	59
April 11th	155	19·8	·9486	60
April 18th	145½	18·8	·9470	53
May 9th	150	20·0	·9530	57
May 14th	150	18·2	·9453	56
May 17th	144	18·4	·9354	55
June 13th	150	16·1	·9467	56
June 21st	141	16·1	·9413	54
July 11th	154	17·8	·9332	53
July 18th	158	17·6	·9433	55
July 26th	158	15·1	·9285	52
August 3rd	170	18·4	·9493	59
August 9th	162	17·1	·9480	54
August 16th	148	18·7	·9470	55
August 23rd	158	15.6	·9390	52
August 30th	141	18·1	·9520	55
September 6th	164	18·1	·9312	50
September 14th	147	18·2	·9420	55
September 20th	138	17·9	·9437	56
September 27th	153	18·0	·9340	50
October 4th	159	18·6	·9330	50
October 12th	156	16·8	·9390	48
October 18th	135	20·1	·9513	60
October 25th	145	20·0	·9563	63
November 2nd	148	19·0	·9550	63
November 8th	168	17·8	·9490	58
November 15th	148	18·5	·9483	60
November 23rd	144	19·0	·9392	52
November 29th	148	19·4	·9512	60
December 6th	145	17·6	·9446	54
December 13th	154	15·5	·9345	55
December 20th	153	15·5	·9347	54

A sample of the oil from each distillation was forwarded to and examined at the Government Laboratory at Antigua. It is seen that the average yield of oil per 100 lb. of the leaves reaped is 17·8 oz., compared with an average of 17·3 oz. on 40 distillations in the previous year. The interesting fact is also brought out by the data, that there has been considerable improvement in the quality of the oil obtained, as indicated by the specific gravity and phenol-content, and a comparison with the averages obtained in three previous years shows the following results:—

Season.	Specific Gravity.	Phenol-content per Cent.
Average 1914-1916	·9351	50·9
1917	·9441	55·0

This improvement in the quality of the oil cannot be accounted for, as the methods of distillation were strictly comparable with those followed in the previous year.

As indicating the characters of the various fractions, and showing how different the quality of the oil may be according to the time accupied in the distillation, 150 lb. of fresh leaves were distilled and the fractions coming over during each of nine hours examined. The results were as follows :—

Oil Obtained in c.c.		Phenol-content.		Specific Gravity.	
1st hour . .	519	1st hour . .	23	1st hour . .	0·8669
2nd ,, . .	126	2nd ,, . .	78	2nd ,, . .	0·9241
3rd ,, . .	98	3rd ,, . .	89	3rd ,, . .	1·0259
4th ,, . .	96	4th ,, . .	93	4th ,, . .	1·0381
5th ,, . .	69	5th ,, . .	95	5th ,, . .	1·0409
6th ,, . .	58	6th ,, . .	96	6th ,, . .	1·0432
7½th ,, . .	53	7½th ,, . .	99	7½th ,, . .	1·0434
9th ,, . .	38	9th ,, . .	96	9th ,, . .	1·0436

According to Fishlock[1] there exists, in addition to the recognised type of *Pimenta acris*, which produces the bay oil of commerce, in various parts of the West Indies a variety closely similar in appearance, but yielding an oil of strong lemon scent and of little commercial value. In Porto Rico this is known as "lemoncillo". This lemon-scented variety is common in Dominica and St. Lucia. In Tortola it is the kind commonly met with, the true type being rare, whereas this condition is reversed in St. Jan. In Montserrat it is reported to be absent.

Fishlock examined the experimental tree plot at the Government Station at Tortola and reported thereon as follows :—

"The trees in the plot were planted in November, 1914, by Mr. C. A. Gomes; they had been raised from seed carefully hand-picked from a single tree having excellent characters growing in the station. This tree came from St. Jan in 1902 with some others, and was planted by me at the station. Only two of the trees then planted have survived; one is a tree having good characters and undoubtedly of the true variety; the other is unmistakably of the false or lemoncillo type. The latter tree flowers earlier than the true variety, and, consequently, I do not think there has been cross fertilisation.

"Upon careful examination of the plants in the plot referred to we found that, of sixty-two living plants, twenty-two, or 35·5 per cent., were of the false or lemoncillo type. This appears to indicate that the original plant, the one planted in 1902, is a hybrid, and that its progeny are splitting up into two distinct types—roughly, 65 per cent. coming rue and 35 per cent. reverting to the lemoncillo type.

"These observations may help to throw some light upon a subject which has already received some amount of attention. It is, of course,

[1] *P. and E.O.R.* (1916), 303.

of the highest importance that in the production of a high-class bay oil there should be no admixture of leaves of the lemoncillo type."

It is suggested that the distillation of the oil should be carried out at a central distillery under competent control, so that a uniform product can be assured.

A sample of bay oil from Fiji, examined at the Imperial Institute in 1909, differed considerably from ordinary commercial bay oil. The oil was pale brown, and had the characteristic odour of bay oil, with, in addition, a distinct anise-like odour. On examination it furnished the following constants: Specific gravity at 15° C., 0·961; optical rotation in 100 mm. tube at 20° C., − 1° 58′. The oil was soluble in its own volume or more of 90 per cent. alcohol, and contained only 23 per cent. of phenols instead of the 60 per cent. found in good bay oil, and was unusually rich in methyl ethers. In view of recent investigations recorded in the *Kew Bulletin*,[1] it seems probable that the Fiji sample consisted of oil derived from the leaves of the form of *Pimenta acris* known as "Bois d'Inde Anise".

Bay oil has the following characters (subject to the above remarks on the irregularity of the distillation):—

Specific gravity	0·965 to 0·985
Optical rotation	− 0° 30′ to − 3°
Refractive index	1·5100 to 1·5200
Phenols	55 to 68 per cent.

Our knowledge of the chemistry of this oil is chiefly due to Mittmann,[2] who first thoroughly investigated the oil, and to Power and Kleber,[3] who cleared up several points left in an obscure state by the first-named chemist. According to Mittmann, pinene and probably dipentene are present, together with the principal constituent eugenol, and its methyl ether. He also believed a diterpene to be present, but this was probably formed by polymerisation during the distillations he effected. Power and Kleber do not agree that either pinene or dipentene are present, but state that the only terpene is laevo-phellandrene. The oxygenated compounds which these chemists state they have identified are eugenol, methyl-eugenol, chavicol, methyl-chavicol, and citral. In addition to this they found a hydrocarbon of the formula $C_{10}H_{16}$ with the abnormally low specific gravity ·802. This, and other considerations exclude the probability of this hydrocarbon being an ordinary terpene, and the chemists above named term it myrcene, and it is now recognised as an olefenic terpene.

The oil is sometimes adulterated with oils of cloves and pimento, which cause an increase in the specific gravity. Turpentine is also added at times. This alters the optical rotation and specific gravity, and if the oil be distilled in a small fractionating flask the bulk of the turpentine is obtained in the early fractions. Assuming that pinene is not present at all in oil of bay, turpentine may be detected by adding to, say, the first 10 per cent. distilled, an equal volume of amyl nitrite and two volumes of glacial acetic acid. The mixture is kept cold in ice and salt, and a mixture of equal parts of glacial acetic acid and hydrochloric acid is added drop by drop so long as a blue colour is developed. If pinene is present a white crystalline precipitate of pinene nitrosochloride $C_{10}H_{16}NOCl$ is formed. Ten per cent. of turpentine can be

[1] *P. and E.O.R.*, October, 1918, p. 278.
[2] *Arch. Pharm.*, xxvii., **529**.
[3] *Pharm. rendus*, *New York*, 13 (**1895**), 60.

detected with certainty in this way. The oil should, when freshly prepared, also be easily soluble in 90 per cent. alcohol, and when kept for some time should be soluble with a slight turbidity. This oil is largely employed in perfumery, especially in the preparation of the favourite toilet article, bay rum.

Oil of Pimento.

Pimento Oil or *Allspice Oil* is distilled from the fruit of *Pimenta officinalis* Linn. (*Eugenia pimenta* De Candolle). The plant, a handsome evergreen, is indigenous to the West Indies, being found especially on calcareous soil near the coast on the islands of Cuba, Hayti, Trinidad, Domingo, Antigua, the Leeward and Windward Islands, and in abundance in Jamaica. It also occurs in Mexico, Costa Rica, and Venezuela. The berries are gathered in an unripe condition, as when ripe they are filled with a sweet pulp lacking the characteristic aroma of the green berry. These unripe berries, dried in the sun, form the pimento of commerce, and Jamaica alone exports, chiefly from the port of Kingston, about £50,000 worth per annum. The fruit, which is about the size of a small pea or large peppercorn, has an aromatic odour somewhat recalling cinnamon, nutmeg, and pepper, with a predominating odour of cloves. This explains the origin of the name allspice. It yields from 3 to 4½ per cent. of oil on distillation, of a dark reddish colour and of the same aromatic odour as the fruit.

The oil contains the phenol-eugenol with small quantities of cineol, laevo-phellandrene, caryophyllene, eugenol-methyl ether, and palmitic acid. The eugenol is, of course, responsible for the clove-like odour of the oil, and the other bodies, together with one or more as yet unknown constituents, modify the characteristic odour of the oil. A certain amount of resin, due no doubt to the unstable nature of some of the constituents, is usually present. The actual amount of eugenol in the oil is approximately estimated by absorption with 5 per cent. caustic soda solution, as in the case of oil of cloves.

The oil has the following characters:—

Specific gravity	1·024 to 1·056
Optical rotation	$-0° 40'$ to $-5°$
Refractive index	1·5250 to 1·5350
Phenols	65 to 80 per cent.

It is soluble in 2 volumes of 70 per cent. alcohol, with at most slight turbidity.

Pimento leaves yield an oil of similar composition, having the following characters:—

Specific gravity	1·026 to 1·030
Optical rotation	$-4°$ to $-5° 30'$
Phenols	65 to 70 per cent.

Oil of Amomis Jamaicensis.

The wild pimento of Jamaica, as it is called, is *Amomis jamaicensis*, a tree growing to a height of 40 feet, found on the limestone in the woods of Upper Clarendon and in the Dry Harbour mountains. The tree is a new species only named in 1919 by the Government botanists, and experiments are being undertaken with a view to developing the distillation of its essential oil. In an experimental distillation by the

Government Department, 100 lbs. of the leaves yielded 202 c.c. of a pale yellow mobile oil. This has been examined in the laboratory of the Imperial Institute and found to have the following characters:—

Specific gravity	0·8895
Optical rotation	− 6°
Ester value	4·2
„ „ (after acetylation)	129·4

It is soluble in 2·5 volumes of 70 per cent. alcohol, becoming cloudy with 6 volumes.

The principal constituent appears to be linalol, so that the oil has no resemblance to ordinary pimento oil.

Cheken-leaf Oil.

Cheken-leaf Oil is the product of distillation of the leaves of *Myrtus cheken*, a plant indigenous to Chili. The yield is about 1 per cent. of a pale yellowish-green oil, much resembling ordinary myrtle oil. Its odour recalls that of myrtle, eucalyptus, and sage oils. According to Weiss,[1] its specific gravity is about ·880, and its optical rotation about + 20°; it consists of pinene (75 per cent.), cineol (15 per cent.), and substances not further examined.

Oil of Backhousia.

Backhousia citriodora is a small tree indigenous to Queensland. The leaves yield about 0·7 per cent. of essential oil which resembles lemon-grass oil closely in odour. The oil has the following characters:—

Specific gravity	0·895 to 0·900
Optical rotation	0° to − 0° 30′
Refractive index	1·4860 to 1·4890
Citral value	93 to 96 per cent.

The oil probably contains small quantities of a sesquiterpene.

Oil of Mespilodaphne.

Mespilodaphne pretiosa is a South American tree, known as "Prisprioca". It is a native of Rio Janeiro and Minas Geraes. The branches yield 0·5 per cent. of an essential oil having an odour of cinnamon. It has the following characters:—

Specific gravity	0·8912
Optical rotation	+ 7° 25′
Refractive index	1·4690
Esters	4·65 per cent.
Total alcohols (as linalol)	51·8 „

The wood oil separates on distillation into a light and a heavy oil. The light oil has a specific gravity 0·953, optical rotation + 8° 48′, and refractive index 1·501. It contains 35 per cent. of esters calculated as linalyl acetate, and 66·6 per cent. of total alcohols calculated as linalol. The heavy oil has a specific gravity 1·055, optical rotation + 3° 8′, and refractive index 1·5450. It consists chiefly of linalyl or geranyl benzoate.

[1] *Arch. der Pharm.*, 226 (1888), 666.

OIL OF RAVENSARA AROMATICA.

The leaves of *Ravensara aromatica*, a large laurel tree growing on the high plateaus of Madagascar, yield a large amount of essential oil, which has been examined by Ferrand and Bonnafores.[1] The oil consists principally of a terpene, with a small amount of an oxygenated body not identified. The terpene (?) boils at 171° to 172° and has a specific gravity 0·881 and refractive index 1·4616.

OIL OF AGONIS FLEXUOSA.

Crozier[2] has obtained about 1 per cent. of essential oil from this shrub, which closely resembles eucalyptus oil, containing 62 to 72 per cent. of eucalyptol, cymene, pinene, traces of alcohols, esters and phenols, but no aldehydes.

ROSACEÆ.

OIL OF ROSES.

Although at the present day oil or otto of rose is distilled almost entirely in Bulgaria and France, only a very limited amount being produced elsewhere, there can be no doubt that the art of distilling roses originated in Persia, and was practised on a large scale at a very early date. The earliest record of this fact was discovered by Prof. Flückiger in an old work in the National Library in Paris. This document states that between the years 810 and 817 of the Christian era, during the reign of Kaliph Mamoun, the province of Faristan was required to pay an annual tribute of 30,000 bottles of rose water to the Treasury of Bagdad. It is also stated by Istakhri that a considerable quantity of rose water was produced throughout Faristan, and was from thence sent to China, India, Yemen, Egypt, Andalusia, and Magreb (in the Barbary States, especially to Morocco). The most important factories were at Dschur (now called Firuzabad), situate between Shiraz and the coast.

The art of distilling roses was probably introduced into Western countries by the Arabs.

The essential oil of the rose was described by Rossi in the latter half of the sixteenth century, but was not discovered until 1612 in Persia, and then only by accident. Sawyer gives the following historical details of the discovery of otto of rose:—

The date of the discovery by the Persians of this valuable essence has been ascertained by Langlès to be 1612. He was thoroughly acquainted with the languages and literature of the East, and in his excellent little book[3] he shows that previous to this date (1612) no mention of otto of rose is made in any work written in Oriental language, and that the circumstances attending the discovery are for the first time described in a work written in Persian by Mohammed Achem, entitled *Tarykh montekheb lubâb*, which is a history of the Great Moguls from the year 1525 to 1667. In this work mention of the essence is twice made.

[1] *Bull. Sci. pharmacol.*, 20 (1913), 403.
[2] *P. and E.O. Record*, 9 (1918), 58.
[3] *Recherches sur la découverte de l'Essence de Roses*, Paris, 1804.

The first time, it occurs in the chapter entitled: "Marriage of the Princess Nour-Djihan with the habitant of the paradise Djihan-guyr; inventions and discoveries of the Queen of the World". The second mention is made in the chapter entitled: "History of the seventh year of the reign of Djihan-guyr, and Feast of the New Year". According to the translation, it appears that at the commencement of this fête, the mother of Nour-Djihan having presented to the prince the "essence of rose water" which she had prepared, the prince thereon attached to this discovery his illustrious name, and called the perfume A'ther Djihanguyry (Perfume of Djihanguyr); at the same time presenting to the princess a necklace of the value of 30,000 rupees.

Another version of the circumstances connected with this discovery is recounted by Manucci, a Venetian physician who resided about forty years in India. He bases it on the *Annals of the Mongolian Empire*, an important work translated by Catron. According to this description the fête was of a very costly and magnificent character, luxurious amusements of every sort being provided, the princess even carrying her extravagance and prodigality to the extent of causing rose water to flow through a canal constructed in the flower gardens. It then happened that whilst the emperor and the princess were walking along the bank of this canal they noticed that an oily stratum had concentrated on the rose water and was floating on its surface; this was carefully collected and recognised by the entire Court as the most delicate of perfumes.

The distillation of otto of rose to-day, however, is essentially a European industry.

The Bulgarian rose plantations are situated in the Balkans, and from the slopes of these mountains the greater portion of the otto of rose of commerce is obtained. The most important centre of the industry is Kezanlik. The area under cultivation is included between the 24th and 26th degrees of longitude east, and the 42nd and 43rd degrees of latitude north. The geological formation is chiefly syenite, the decomposition of which has produced a very fertile soil. The roses flourish best in sandy sun-exposed slopes, with a south or south-eastern aspect. The most important districts are (1) the department of Stara-Zagora, including the cantons of Kezanlik, Nova-Zagora, and Stara-Zagora; (2) the department of Pazardjik, chiefly the canton of Pechtera; (3) the department of Philippopolis, including the cantons of Karlovo, Tchirpan, Novo-Selo, and Brezovo.

The rose country occupies especially the two valleys of the Toundja (canton of Kézanlik) and of the Strema (canton of Karlova) both bounded on the lower side by a spur of the Balkans called Sredna-Gora. The south side of this mountain supports numerous rose farms, but their produce is inferior to that obtained from the south side of the Balkans proper, not only in consequence of less favourable climatic and geological conditions, but because the two indispensable factors of the work of distillation, namely, the water, and the wood for fuel, are much less easy to obtain there by the distillers.

The rose cultivated in this district is *Rosa damascena*, a red rose, but the hedges of the plantations are often made up of *Rosa alba*. The latter are often distilled with the red roses, but as the resulting oil contains more stearoptene than that of the *Rosa damascena*, it is rarely distilled alone.

The flowers are gathered before they commence to open, in the early morning just before sunrise, and the picking continues till ten or eleven o'clock, or on cloudy days during the whole day, as it is important that the roses should not be picked when the sun's heat is on them. A garden of an acre when well planted will yield about 100 lb. of flowers every day for three weeks. The roses are carried to the distilleries, which are very numerous, and distilled as quickly as possible. However, if the flowers come in too quickly, they are sometimes left for twenty-four hours before being distilled, the resulting oil being thereby impaired in value. The distilleries are primitive buildings, sometimes little better than wooden sheds. On one side the stills are kept in rows. These are copper alembics about 3 to 5 ft. in height, resting on a furnace built of bricks. The average content of each still is about 20 gallons, the charge usually being 10 kilos of flowers and 75 litres of water.

The condenser is a straight or worm tube passing through a vat of water, into which cold water continually runs. A brisk fire is kept up for an hour to an hour and a half, and when 10 litres of liquid are obtained the fire is drawn. At times 15 litres are distilled over, but the result is an otto containing a larger amount of stearoptene. The still is then opened and the spent petals, or rather flowers, for the green

FIG. 84.—A, copper alembic; B, condenser; C, receiver.

plants are seldom separated, are thrown away, and the residual hot water is returned to the still with cold water to make up the 75 litres, with a fresh charge of flowers. This operation is repeated until, as a rule, 40 litres of rose water have been collected. These 40 litres are now distilled, and the first 5 litres are collected in a long-necked flask. The residual 35 litres are used for distilling fresh flowers. The 5 litres distilling over are cloudy, and the oil drops gradually rise and collect in the neck of the flask. When it has all risen, it is removed by a small tin funnel with a tiny orifice for the water to be drawn off. The yield is variable, a warm humid spring with intervals of strong sunshine being more favourable than an uninterrupted hot dry season. The average yield in Bulgaria is about 1 kilo of otto from 3000 kilos of rose leaves.

Adulteration of Bulgarian otto of rose is so common as to render it a matter of the highest importance to purchase this expensive product from dealers of repute, as the adulteration practised has been raised to a fine art. With every advance in our knowledge of the chemistry of otto of rose, the unscrupulous advisers of the Bulgarian sophisticators have prepared cheap artificial products similar in characters to the natural constituents, so that it is necessary to judge the quality of an otto of rose with great care, since the physical characters are closely simulated by the artificial adulterants.

Formerly palmarosa oil was almost the only adulterant used. To-day geraniol, citronellol, and similar bodies are used to the exclusion of the crude palmarosa oil. It is surprising that an industry of so much importance should have been allowed to go on for so many years fettered by two such great evils; firstly, the enormous adulteration practised, secondly, the defective methods of preparation, resulting in the spoiling of the fine odour of much of the otto. Although climate and soil have so great an influence on the yield and quality of the oil, there can be no doubt that there are localities where rose cultivation and distillation could be carried on in a scientific as well as an honest way.

Formerly the dealers were responsible for the greater part of the adulteration, but now the small growers come in for their share of the illicit profits, and not only do they add the adulterant to the distillate, but frequently sprinkle the leaves with geraniol before distillation, believing that they so obtain a better "blend".

The value of a pure oil depends on the quality of the flowers used and on the care taken in distillation. It also depends on the amount of solid stearoptene present, for this portion of the oil is quite inodorous.

The value of otto in the locality of its distillation is gauged almost entirely by its melting-point. Consequently spermaceti or even paraffin wax has been added to counteract the reducing influence of the liquid geranium oil. Recently, too, guaiacum wood oil (so-called), an oil of fine odour which solidifies with a fine crystalline structure at ordinary temperatures, and consists of bodies much resembling those in oil of roses, has been used as an adulterant.

In this connection it may be mentioned that no standard for the quantity of stearoptene present in natural rose oil can be fixed, as climatic conditions appear to influence the ratio of the solid to the liquid portion very greatly. From 10 to 20 per cent. occurs in Bulgarian oil, whereas that obtained in the south of France, in England, and in Germany, has been stated to contain from 25 to 65 per cent. of stearoptene. In the case of so high a content as the latter figure it is certain that the oil was not prepared in the normal way, probably being the result of the distillation of rose water. Further, steam distilleries are being erected, and so far as one can judge at the present moment, an otto of different properties is obtained when distilled by steam instead of in the usual manner.

In Germany, Schimmel & Co. have laid out plantations near Klein-Miltitz, and although the natural resources of climate and soil are not so favourable here as in Bulgaria or Southern France, the scientific and careful way in which the oil is obtained to a certain extent counterbalances this, and a fairly fine otto is obtained. The amount, however, is only very small. In Algeria a small amount of otto is produced, but the yield is less than in Bulgaria. In Persia and Cashmere, however, it is said that nearly 1 lb. of otto is obtained from 1500 to 2000 lb. of petals.

Anatolia, Hungary, Russia, and Spain also produce a small amount of otto of rose, but apart from Bulgaria, France is the only important rose-growing country.

The roses which are under cultivation in France for perfumery purposes exclusively are high in price, because the May rose bushes (Centifolia or Damascena) are of fairly weak growth; they do not grow more

FIG. 35.—Bulgarian rose distillery. [Gattefossé

than 60 cm. in height, and only bear a few small roses of little weight.

This is also a reason why these roses are specially used for "enfleurage," hot or cold, for the manufacture of pomades, oils, and rose extracts.

But there are on the Azur coast, and in many other districts in France, important rose gardens, which have been planted with a view to the sale of cut flowers for florists.

In the Lyon district (in Touraine) the florists usually sell the plant; in the South, they export during all the winter the prematurely opened blossoms. Thanks to rational irrigation, to a system of removable glass houses covering hundreds of acres, and to the application of special skill in pruning and manuring, these roses realise up to 1·50 fr. each in the winter, and it is clear that the entire profit from these rose gardens is cleared before the spring flowering commences. At that time all the rose bushes flower at the same time. The flowers are no longer of any value, and are left at the disposal of pickers for a small consideration.

Up to 1910 all these roses were sold at 0·10 fr. or 0·20 fr. per kilo; from that time onwards, the red variety, "Brunner," admitted to be the best, increased in value, and fetched 30, 40, and even 60 c. per kilo. The pink varieties, especially Van Houtte and Paul Nabonnand, the yellow varieties, Safranos, and the white varieties, Druski, remained at lower figures. These flowers are obtainable in millions of kilos, and it is estimated that 5000 to 10,000 kilos of roses are required to produce one kilo of essence. French otto has a clear and penetrating perfume, similar to that of the freshly-cut flower.

In France only steam stills are used, to the exclusion of apparatus used with open fires, which in burning the flower somewhat imparts to the essential oil a slightly burnt odour; the French ottos obtained with steam do not offer this drawback.

The "Rose de l'Hay," a comparatively new rose, is a very strong grower. In less than two years a plant becomes a thick bush, full of leaves, and more than 2 metres high. It is very prolific in flowers, producing in the month of May more than 2 kilos of roses, and continuing to flower for at least five months.

The otto distilled from the "Rose de l'Hay" is of a greenish-yellow tint, but becomes yellow on keeping, and has a relatively low melting-point, but an excellent perfume.

It will be very easy to propagate this variety all over France, and to obtain, in addition to the crop of the South, an important quantity of otto.

The characters of otto of rose vary considerably according to the locality in which the trees are grown, and also in one locality from season to season according to climatic and other conditions. All these factors must be taken into consideration in forming an opinion on otto of rose, and it must be remembered that no published figures can be accepted as final or exhaustive. The following may be taken as being typical, however, of the best ottos produced in various localities:—

BULGARIAN OTTO OF ROSE.

Specific gravity $\frac{30^\circ}{15{\cdot}5^\circ}$	0·849 to 0·858
Optical rotation	− 1° 30′ to − 4°
Refractive index at 25°	1·4580 to 1·4650
Melting-point	19° ,, 22°
Total alcohols	68 to 78 per cent.
Citronellol	28 ,, 34 ,,
Acid value	0·1 to 0·3
Ester ,,	0·7 ,, 1·2
Stearoptene content	15 to 20 per cent.

French Otto of Rose.

In France numerous different roses are distilled, each yielding its own characteristic otto, so that the resulting products will vary according to the percentage of each given rose in the distilling material.

In 1904 Jeancard and Satie [1] published a paper dealing with the analysis of otto of rose. They distilled a sample from the sepals only of *Rosa centifolia*, and found it to have the following characters:—

Congealing-point	8°
Stearoptene	51·13 per cent.
Specific gravity	0·8704 at 15°
Rotation	− 41°
Melting-point of stearoptene	14°
Acid value	6·12
Ester ,,	16·28
Total alcohols	13·99 per cent.
Citronellol	13·56 ,,

This sample, as will be observed, contained practically no geraniol, the alcoholic constituents consisting entirely of citronellol.

An oil distilled from the whole flowers, on the other hand, had the following characters:—

Specific gravity	0·879 at 15°
Rotation	− 3°
Alcohols	88·5 per cent.
Citronellol	22·4 ,,
Congealing-point	25·5°
Stearoptene	32·2 per cent.

A similar oil, distilled without cohobation of the distillation waters, and so distilled that 1 kilogram of roses gave 1 litre of rose water, had the following quite different characters:—

Congealing-point	25·9°
Stearoptene	58·88 per cent.
Acid value	2·24
Ester ,,	14·70
Total alcohols	32 per cent.
Citronellol	15·10 per cent.

Dupont and Guerlain [2] have found samples of French otto of rose to have specific gravities varying between 0·8225 and 0·8407 at 30°, optical rotations from − 6° 45′ to − 8° 3′, and stearoptene-content from 26 to 35 per cent.

An authentic otto distilled from tea-roses only was found to contain 72 to 74 per cent. of stearoptene.

Two analysis of otto from the rose d'hai have been published by Schimmel & Co. The figures are as follows:—

	1.	2.
Specific gravity at 30°	0·8706	0·8653
Rotation	− 1° 26′	− 3° 20′
Refractive index at 25°	1·46486	1·46159
Melting-point	13°	18·6°
Acid value	1·8	1·8
Ester ,,	6·3	5·6
Total alcohols	91 per cent.	91·8 per cent.

As a matter of fact, the total alcohols in this particular type of otto consist almost entirely of geraniol, and are nearly free from citronellol.

[1] *Bull. Soc. Chem. de Paris* (**1904**), 934.

[2] *Comptes rendus*, 123 (**1896**), 700.

The Nabonnand roses are to-day members of the "Roses des jardins", or mixed garden roses, used for distillation purposes. These roses, distilled by themselves, yield an otto having a specific gravity 0·8537 at 30°, rotation − 2·2°, melting-point 19°, and ester number 42. As judged by the usual standards set up by limited experience, this otto would be condemned by non-experts on account of its extraordinarily high ester-content.

FIG. 36.—Plantation of red roses (Brunner) on the Côte d'Azur. [*Gattefossé.*

The Van Houtte roses, too, have a quite different character from any other, in so far as they yield an otto with the very high optical rotation − 11°, and the Brunner rose yields an otto which melts at 28°, rarely below this.

The writer has examined authentic otto distilled from the Rose d'Hai during 1916, and found it to melt at 8° and to have a specific gravity of about 0·880. Its alcohol value was 90 per cent., consisting almost entirely of geraniol.

Mixed roses, in the fair average proportion in which they are distilled in practice in the South of France, will give analytical values approximating to the following, but of course their limits are exceeded if certain roses are used in greater proportion :—

FIG. 37.—Plantation of white roses (Drusky) on the Riviera. [Gattefossé.

Specific gravity	0·859 to 0·874
Rotation	− 2° ,, − 4°
Saponification value	15 ,, 35
Melting-point	19° ,, 28°
Total alcohols	70 to 78 per cent.
Geraniol	35 ,, 42 ,,
Citronellol	35 ,, 45 ,,

German Otto of Rose.

	1. German Oil, Miltitz, 1900.	2. German Oil, Miltitz, 1900.
	at 40°	at 40°
Specific gravity	0·8372	0·8304
	at 30°	at 33°
,, ,,	0·8438	0·8360
	at 30°	at 33°
Rotatory power	− 0° 44′	− 0° 52′
Congealing-point	+ 29·0°	+ 30·0°
Stearoptene-content	28·50 per cent.	39·97 per cent.
Acid number	2·3	0
Saponification number	4·5	10·4
,, ,, after acetylisation	188·7	171·14
,, ,, ,, formylation	59·21	47·9
Calculated from these figures—		
Total alcohols $C_{10}H_{18}O$	60·44 per cent.	54·00 per cent.
Citronellol in the formylation product	16·49 ,,	13·34 ,,

Hungarian Otto of Rose.

Specific gravity at 30°	0·8471
Refractive index at 20°	1·4661

Spanish Otto of Rose.

Samples distilled in Chinchilla have been examined and show the following characters :—

Specific gravity	0·825 to 0·845
Optical rotation	− 1° ,, − 3°
Refractive index at 25°	1·4520 ,, 1·4568
Melting-point	25° ,, 29°
Acid value	3 ,, 5
Ester ,,	9 ,, 12
Total alcohols	about 45 per cent.

Russian Otto of Rose.

A sample distilled from flowers grown in the Imperial Appanages of Kachetie (Caucasus) which had been cultivated from Bulgarian cuttings, and has been examined by Roure-Bertrand Fils,[1] had the following characters :—

Specific gravity at 30°	0·8368
Optical rotation	−4° 16′
Solidifying-point	23°
Liquefying point	25°
Stearoptene	33·5 per cent.
Acid value	5·7
Ester ,,	4·7
,, as geranyl acetate	1·66 per cent.
,, value of acetylated oil	157·2
Total alcohols	49·6 per cent.
Citronellol	34 ,,
Geraniol	15·6 ,,

That is, the total alcohols contained 68·5 per cent. of citronellol and 31·5 per cent. of geraniol.

An otto of rose distilled in the Caucasus has also been examined very fully by the same firm, who found it to have the following characters :—

[1] *Report*, October, **1902**, 23.

I. Physical Characters of the Complete Otto.

Colour of the otto	Pale yellow
Specific gravity at 15° C.	0·8618
Optical rotation, a_D at 27° C.	− 2° 28′
Solidifying-point	+ 19° C.
Temperature at which the otto becomes completely liquid	+ 23·5° C.

II. Solid Portion, Insoluble in 75 per Cent. Alcohol.

Stearoptene	23·1 per cent.
Elæoptene (by difference)	76·9 ,,
Melting-point of the stearoptene	+ 31·5° C.
Coefficient of saponification of the stearoptene	6·5

III. Elæoptene.

Specific gravity at 15° C.	0·8809
Optical rotation, a_D at 15° C.	− 2° 26′
,, ,, of the acetylated oil	− 1° 16′
Decrease in rotation	1° 10′
Solubility in 70 per cent. alcohol	1 vol., then cloudy
Solidifying-point	Becomes cloudy but does not solidify at − 15° C.
Acid value	1·4
Ester ,,	10·3
Saponification value	11·7
,, ,, of the acetylated oil	259·5
,, ,, ,, formylated oil	121·3
Percentage of esters, as citronellyl acetate	3·63 per cent.
Combined alcohols } as citronellol, $C_{10}H_{20}O$	2·9 ,,
Free ,, } as citronellol, $C_{10}H_{20}O$	86·8 ,,
Total ,, } as citronellol, $C_{10}H_{20}O$	89·7 ,,
Percentage of citronellol (by formylation)	35·8 ,,
,, ,, geraniol (by difference)	53·9 ,,
Composition of the alcoholic mixture: citronellol	39·9 ,,
Composition of the alcoholic mixture: geraniol	60·1 ,,

IV. Results Calculated to the Complete Otto.

Stearoptene	23·1 per cent.
Free acids (as acetic acid)	0·37 ,,
Esters (as citronellyl acetate)	2·79 ,,
Citronellol	27·53 ,,
Geraniol and phenylethyl alcohol	41·44 ,,
Terpenes, etc. (by difference)	4·77 ,,

Spartan Otto of Rose.

For some few years past there has been a commercial utilisation of the rose flowers growing naturally in the valley of Lake Burdur in the interior of Asia Minor.

Lake Burdur is situated between the town of that name and Isbarta. This latter is the ancient Sparta of Pisidia, which was a flourishing town in the days of Alexander the Great. The whole of this district is inhabited by an essentially agricultural population, who own land of magnificient fertility. The climate is mild, and the valley, protected by the mountains against cold and frost, is most favourably situated for the culture of roses.

Most of the peasants own stills for the distillation of their flowers, while there are other cultivators in a rather better position who buy up the roses of those whose production is very small and who possess no distillation apparatus of their own.

The quantity of essence produced during 1912 was 300 kilos, and in 1913 to 400 kilos. Up to 1912 all this was purchased by the agents of Bulgarian or Constantinople houses, who bought up all that was offered, selling it again to their fellow-countrymen who exported it again under their own seals, principally as Bulgarian otto of rose. So it happened that quite an important production was entirely unknown to Western consumers.

During the last few years, however, upon the initiative of certain of the greater distillers, a combine was formed to bulk the entire production of Sparta and Burdur, and sell it direct to Europe instead of passing it through Bulgaria. This was, of course, an advantage, not only to the growers, but also to the European consumers, in that it did away with intermediate expenses and profits.

The product is then sent down to Smyrna, the principal port of Asia Minor, by the railway that connects Burdur with the coast, to the exporting agents.

The otto is of extreme delicacy of perfume and of high melting-point, as the following analysis shows :—[1]

Specific gravity $\frac{30}{15}$	0·850
Optical rotation	− 3° 30′
Refractive index at 25°	1·4605
Melting-point	24° to 25° C.

PERSIAN OTTO OF ROSE.

Specific gravity at 30°	0·830
Optical rotation	− 9°
Melting-point	21 to 22°

It is obvious from the above figures that the locality of production must be considered in forming an opinion as to the purity of otto of rose.

Adulteration with palmarosa oil is revealed by the odour apart from the obvious alteration in the physical characters of the sample. Alcohol is frequently added in small quantities as it causes the stearoptene to become less soluble, and so raises the apparent melting-point.

Alcohol can be detected by distilling a few c.c. of the sample with water, and testing the distillate by the usual iodoform reaction. If the sample be washed with warm water in a separator, and the refractive index of the washed otto be examined, it will be found to be higher than that of the original otto. If this excess be more than 0·001 it is almost certainly due to alcohol.

The author, a few years ago, observed a new adulterant in Bulgarian otto of rose, which caused the optical rotation to be abnormally high. This has since been identified by Schimmel & Co.[2] as gurjun balsam oil. This is detected by mixing 5 drops of nitric acid with 5 c.c. of glacial acetic acid, and then shaking with 5 drops of the otto in question. In the presence of gurjun oil a reddish-violet colour develops within a minute or two.

The total alcohols are determined by acetylation in the usual manner, and the citronellol by formylation. The formylation process is carried out as follows :—

[1] *P. and E.O.R.* **(1913)**, 338. [2] *Report*, April, **1912**, 108.

Ten c.c. of the oil are heated for one hour on a sand-bath in an acetylation flask with twice its volume of formic acid (specific gravity, 1·226). It is advisable to add a few pieces of pumice to assist the boiling. The contents of the flask are cooled and 100 c.c. of water added, and the whole transferred to a separator. The acid layer is run off, and the oil is washed with water until the washings are neutral. It is then dried by means of anhydrous sodium sulphate and filtered. The formylated oil is then saponified with alcoholic potash in the usual manner, free acids being first neutralised, and the percentage of citronellol in the original oil calculated from the following formula :—

$$\text{Total citronellol} = \frac{0{\cdot}156 \times x \times 100}{w - (0{\cdot}028x)},$$

where x is the number of c.c.'s of normal alcoholic potash absorbed and w the weight of formylated oil.

It was found by experiment that practically the same results were obtained in the case of geranium oils, whether the oil was first saponified or not. It is, therefore, sufficiently accurate to formylate the oil without first decomposing the esters, since the formic acid apparently acts on the alcohols whether combined or not.

Jeancard and Satie [1] determine the stearoptene as follows :—

Ten grams of the otto are weighed into a beaker and 50 c.c. of acetone are added. The liquid is cooled to − 10° or thereabouts, without shaking, and then filtered through a funnel kept in a refrigerating mixture. The stearoptene is washed several times with cold acetone, and dried on the weighed filter paper *in vacuo* over sulphuric acid.

Burgess finds that for the estimation of the stearoptene 85 per cent. alcohol gives the most accurate results. He finds that 5 grams of the otto (in these experiments he also used 5 grams of a standard solution of stearoptene in geranium oil, and in citronellol—see table), in 25 c.c. of the solvent—or, if a weaker alcohol be used, 50 c.c.—is the most suitable quantity for the determination. Solution is effected by means of heat, and the liquid is then cooled in ice for five or six hours. The stearoptene is then filtered off on paper over a filter pump and washed, once if 85 per cent. alcohol be used, twice if a weaker spirit be employed ; the paper with the separared stearoptene is then dried *in vacuo* over sulphuric acid for twelve hours. It is not necessary to tare the filter paper, since the stearoptene can be removed in one cake from the paper, and weighed on a watch glass.

The appended table (p. 403) shows the results obtained with the various solvents on otto of rose, and on standard solutions of stearoptene in geraniol, and in citronellol. With acetone, he found in some cases that a further separation of stearoptene took place, so that he has recorded the results in the form of the two weighings of the stearoptene where necessary.

By addition of alcohol and freezing the stearoptene can be almost entirely separated on a filter pump and washed with alcohol and finally dried. It should then melt between 32° and 37°, usually at 33° to 34°. Spermaceti, and most samples of paraffin wax, will cause this to be rather higher, as will also "guaiacum wood" oil if present in more than very small quantities. Salol and antipyrin have been found as solid adulterants.

[1] *Bull. Soc. Chim.*, Paris (**1904**), 934.

STEAROPTENE DETERMINATIONS. BURGESS'S TABLE.

Solvent.	Otto of Rose.		Geranium Oil. 5 per Cent.		Geranium Oil. 10 per Cent.		Geranium Oil. 15 per Cent.		Citronellol. 5 per Cent.		Citronellol. 10 per Cent.		Citronellol. 15 per Cent.	
	I.	II.	I.	II.	I.	II.	I.	II.	I.	II.	I.	II.	I.	II.
	Per Cent.		Per Cent.		Per Cent.		Per Cent.		Per Cent.		Per Cent.		Per Cent.	
75 per Cent. Alcohol . .	[1] 16·4	[1] 14·2	[1] 5·26	[1] 5·28	[1] 10·64	[1] 10·54	[1] 16·22	[1] 15·64	[1] 4·50	[1] 4·46	[1] 9·84	[1] 9·78	[1] 13·94	[1] 14·[illegible]6
85 ,, ,, . .	—	—	4·52	4·42	9·56	9·76	15·26	15·14	4·34	4·30	9·10	9·10	13·18	13·78
90 ,, ,, . .	15·04	14·74	4·26	4·16	—		13 72	13·76	3·68	3·74	8·92	8·96	13·00	13·72
94 ,, ,, . .	14·4	14·1	3·84	3·88	8·58	8·56	13·40	13·44	3·20	3·16	8·24	8·50	13·00	12·66
Acetone	13·8	14·0	3·68	3·84	4·82 1·88 —— 6·70	4·66 2·62 —— 7·28	13·38	13·44	3·12	3·08	4·62 2·14 —— 6·76	4·62 2·00 —— 6·62	12·16	12·16
Acetic Ether	8·46 4·63 —— 13·09		—		—		—		—		—		—	

[1] In all these cases the precipitate and filter paper were distinctly oily, showing imperfect solution of the eleoptene.

Spermaceti or stearin will be easily detected, whilst paraffin wax or the solid constituent of guaiacum wood oil (which is of an alcoholic nature) are not so easy of detection. As the stearoptene of pure rose oil consists of hydrocarbons, no fatty acids will be obtained by saponification, whereas spermaceti and stearin will yield on saponification and decomposition of the resulting salts with hydrochloric acid, palmitic and stearic acids respectively, which may be identified by their melting-points (palmitic acid from 55° to 62°, and stearic acid from 62° to 68°). Paraffin will be indicated by the more granular structure of its crystals than that of the normal stearoptene, whilst the solid portion of guaiacum wood oil consists chiefly of a solid alcohol and will give an acetylation value which the pure stearoptene will not.

The following tables are given by Schimmel :—

	1.	2.	3.	4.	5.	6.
	German Oil, Miltitz, 1900.	German Oil, Miltitz, 1900.	Bulgarian Oil (Kazanlik).	Bulgarian Oil (Kazanlik).	Oil from Philippople from our Expert. Distilled by Himself.	Purchased.
Specific gravity .	at 40° 0·8372	at 40° 0·8304	at 30° 0·8566	—	—	—
,, ,, .	at 30° 0·8438	at 33° 0·8360	at 25° 0·8598	at 25° 0·8594	at 25° 0·8634	at 25° 0·8614
Rotatory Power .	at 30° − 0° 44′	at 33° − 0° 52′	at 30° − 2° 20′	at 25° − 2° 43′	at 25° − 2° 58′	at 25° − 2° 36′
Congealing-point .	+ 29·0°	+ 30·0°	+ 20·0°	+ 20·4°	+ 19·6°	+ 20·4°
Stearoptene-content . . .	28·50 %	39·97 %	18·62 %	18·83 %	18·49 %	20·55 %
Acid Number .	2·3	0	0	2·3	0	0
Sap. Number .	4·5	10·4	11·03	7·0	11·1	10·1
Sap. Number after Acetylisation .	188·7	171·14	216·0	204·1	214·7	211·0
Sap. Number after Formylation .	59·21	47·9	92·9	96·8	104·0	100·6
Calculated from these figures :— Total Alcohols $C_{10}H_{18}O$ . .	60·44 %	54·00 %	70·88 %	66·27 %	70·37 %	68·93 %
Citronellol in the Formylation Product . . .	16·49 %	13·34 %	25·88 %	26·96 %	28·97 %	28·03 %

Values obtained direct in analysing oils free from stearoptene :—

Spec. Gravity at 15°	0·8857	0·8886	—	0·8862	0·8774	—
Rotatory Power 20°	− 0° 56′	− 1° 2′	—	− 2° 49′	− 2° 38′	—
Sap. Number after Acetylisation .	271·55	261·64	—	255·8	254·8	—
Sap. Number after Formylation .	74·70	81·22	—	125·28	127·4	—
Total Alcohol .	93·78 %	90·80 %	—	87·04 %	86·63 %	...
Citronellol . .	20·81 %	22·63 %	—	34·90 %	35·50 %	—

Of the chemistry of oil of roses, although fairly well understood now, much controversial matter has been written. Eckart [1] states that a small

[1] *Inaugural Dissertation*, Breslau, 1891.

quantity of ethyl alcohol exists in the oil. If this is so, which Poleck denies, it is in all probability the result of fermentation taking place in the leaves, by keeping them too long before distillation. Markovnikoff and Reformatsky[1] originally stated that the liquid portion of rose oil consisted chiefly of an alcohol of the formula $C_{10}H_{20}O$, which they termed roseol. Poleck and Eckart claimed that this alcohol had the formula $C_{10}H_{18}O$, and was identical with geraniol. They, however, proposed the name rhodinol (which was applied to an alcohol $C_{10}H_{20}O$ from geranium oil by Barbier and Bouveault) for it. Bertram and Gildemeister[2] then showed that the alcohols from rose oil were in reality a mixture of two bodies, of which they agreed that geraniol $C_{10}H_{18}O$ was the chief constituent. Tiemann and Schmidt explained these differences, proving that the alcohols consisted of about 75 per cent. of geraniol $C_{10}H_{18}O$ and 25 per cent. of an alcohol $C_{10}H_{20}O$, which was identical with that obtained by reducing citronellic aldehyde. Hence they naturally insisted on the name citronellol, whilst Eckart claimed to have isolated an alcohol of the formula $C_{10}H_{20}O$ which he named rhodinol. For many years rhodinol was regarded as an impure citronellol, but the two bodies are now known to be separate individuals. Much confusion has unnecessarily been caused by various chemists giving names to alcohols they had isolated from different oils in an impure state, and refusing to alter these names after it had been conclusively proved that they were, when pure, identical with already known bodies. Lemonol, réuniol, and roseol are all either geraniol or citronellol, or mixtures of them. The relationships of rhodinol and citronellol are discussed in Volume II. Traces of esters of these two alcohols, and of free acids resulting from their decomposition also occur. The stearoptene was regarded by Markovnikoff as consisting of a single paraffin hydrocarbon of the formula $C_{16}H_{34}$, melting at 37°, but Schimmel & Co. have separated two hydrocarbons melting at 22° and 41° respectively from the stearoptene. This has been confirmed by Dupont and Guerlain. These chemists suggest that the ester, which appears to be found in larger quantity in French oils than in any other, is an important odoriferous constituent, and is destroyed by repeated distillation with water. They separated the liquid portion of the oil and found it had an optical rotation of − 10° 30′, which decreased to − 7° 55′ after hydrolysis. Charabot and Chiris invariably find rose water acid, which tends to confirm this hypothesis.

Phenyl ethyl alcohol has recently been shown to be a normal constituent of otto of rose, but it appears to be mostly, if not entirely lost or destroyed by the Bulgarian method of distillation. Nerol is also present in small amount as well as, probably, farnesol, a sesquiterpene alcohol $C_{15}H_{26}O$, present in oil of cassie flowers. Eugenol, linalol, and nonylic aldehyde are also present in traces.

Oil of Almonds.

Oil of bitter almonds is obtained from the kernels (seeds) of the ordinary bitter almond, *Prunus Amygdalus* (*Amygdalus communis*), a tree which is cultivated in Europe, Asia, Northern Africa, and California. The kernels of the peach and apricot yield essential oils which are practically identical with that from the almond, and many samples of "almond oil" of commerce are no doubt derived from peach and

[1] *Jour. Prakt. Chem.*, II, 48 **(1893)**, 293. [2] *Ibid.*, II, 49 **(1894)**, 185.

apricot kernels. No distinction will be made between these three oils, as they differ in no sensible degree, except for the fact of their actual origin.

Bitter almond oil does not exist as such in the almond kernel, but results from the hydrolysis of a glucoside, amygdalin, under the influence of the naturally existing ferment emulsin. The fatty oil is obtained from the kernels by expression, and the press-cake crushed and digested with water, when the following reaction ensues :—

$$\underset{\text{Amygdalin.}}{C_{20}H_{27}NO_{11}} + 2H_2O = \underset{\text{Benzaldehyde.}}{C_7H_6O} + \underset{\text{Dextrose.}}{2C_6H_{12}O_6} + HCN.$$

Amygdalin, taking up two molecules of water, yields benzaldehyde dextrose, and hydrocyanic acid. Amygdalin is a crystalline body, without any smell of the bitter almond, and does not yield the oil except under the influence of a hydrolytic agent, such as the natural ferment emulsin, or by boiling with dilute acids. The action of the ferment is destroyed by heat or by warm alcohol. Hence if dried and powdered b.tter almonds are shaken with boiling water and distilled, no oil is obtained. After the fixed oil has been expressed the press cakes are ground up and soaked for about twenty-four hours in twice their weight of water, to which a quantity of salt is usually added. The whole is then subjected to distillation. Some trouble, however, is experienced during the process, as the large quantity of albuminoids present causes excessive frothing. To remedy this, the press-cakes are coarsely powdered and at once immersed in boiling water to coagulate the albuminoids and dissolve the amygdalin. The emulsin is, of course, rendered inactive, so that on cooling, a quantity of emulsin of the fresh cake in cold water is added to the previously treated mass. This is allowed to stand, until the emulsin will have converted the whole of the amygdalin into essential oil. The mixture is now distilled. As hydrocyanic acid is a very deadly substance, it is necessary to use great care that none of the vapour is allowed to escape into the air. The distillation is effected by direct steam in most cases. The oil of almonds so obtained contains a considerable amount of hydrocyanic acid, the remainder being principally benzaldehyde, $C_6H_5 . COH$. The absolutely natural oil is a regular commercial article, but much is deprived of its hydrocyanic acid before being sold, and is then listed at a correspondingly high price as "Oil of Almonds (S.A.P.)" (i.e. *Sine Acid. Prussic.*). The following methods are used to deprive the oil of this poisonous constituent :—

The oil is mixed with its own volume of water, and the mixture left on a water-bath with red oxide of mercury, slaked lime, and ferrous chloride, out of contact with the air. After all the prussic acid has been decomposed, the oil is rectified and contains no trace of prussic acid. This process was suggested by Redwood. Liebig used oxide of mercury only. Mackay prefers agitating at intervals for forty-eight hours with a mixture of lime and liquor potassæ. To detect any traces of prussic acid left in the oil, a little of it is dissolved in alcohol, and a few drops of solutions of ferrous sulphate and ferric chloride are added. Slight excess of caustic soda solution is then added, and the precipitate is dissolved by the addition of dilute hydrochloric acid, when a blue colour or precipitate (due to the formation of Prussian blue) appears if any hydrocyanic acid is present. Benzaldehyde is very readily oxidised by the oxygen of the atmosphere to benzoic acid, so that the oil almost invariably contains traces of this body, and if kept in bottles that are not quite full the oxidation becomes more rapid, and crystals of benzoic acid are deposited. This is more especially the case when the oil has been freed from prussic acid. According to Schimmel & Co., 10 per cent. of alcohol added to the oil retards this oxidation.

There appears to be little doubt that the action above indicated for the decomposition of amygdalin does not take place as simply as indicated. Recent researches by Rosenthaler[1] and others indicate that the enzyme is not a single body but a mixture of ferments, and that the system amygdalin-emulsin gives rise to the following series of reactions, which end up with the oil containing benzaldehyde, hydrocyanic acid, and benzaldehyde-cyanhydrin:—

1. Amygdalin, under the influence of amygdalase (one of the ferments), yields mandelonitrile glucoside and glucose.
2. Mandelonitrile glucoside, under the action of prunase (another ferment), yields *dextro*-benzaldehyde-cyanhydrin and glucose.
3. *Dextro*-benzaldehyde-cyanhydrin is resolved by *d*-oxynitrilase into benzaldehyde and hydrocyanic acid.
4. Benzaldehyde and hydrocyanic acid unite again, under the influence of *d*-nitrolese, to form *dextro*-benzaldehyde-cyanhydrin, and also unite, naturally, to form the inactive cyanhydrin.
5. *d*-oxynitrilase effects the resolution of the inactive cyanhydrin, giving rise to laevo-benzaldehyde-cyanhydrin.

Bitter almond oil is a highly refractive liquid of powerful odour having the following characters:—

Specific gravity	1·045 to 1·070
Optical rotation	0° to + 0° 10′
Refractive index	1·5320 to 1·5450

A higher specific gravity indicates a very high amount of hydrocyanic acid, or an abnormal amount of benzaldehyde-cyanhydrin.

The oil is more soluble in water than almost any other essential oil, 1 part dissolving in 300 parts of water at ordinary temperature.

Bitter almond oil, deprived of its prussic acid, has the following characters:—

Boiling-point	179°
Specific gravity	1·049 to 1·055
Optical rotation	0° to + 10′
Refractive index	1·5420 to 1·5460

The oil should not contain more than 1 per cent. of benzoic acid as determined by titration with decinormal alkali: higher percentages indicate that the oil is old or has been badly stored, and its odour and perfume value are proportionally lower as the benzoic acid increases.

The constituents of the oil are, as above indicated, benzaldehyde, to which the almond odour is due, hydrocyanic acid, and benzaldehyde-cyanhydrin, $C_6H_5CH(OH)CN$, which, however, is easily decomposed into its components by distillation with water, so that an oil which has been deprived of its hydrocyanic acid directly after distillation may be practically free from benzaldehyde-cyanhydrin.

The principal adulterant of this oil is synthetic benzaldehyde which is almost impossible to detect, unless it contains, as it very frequently does, traces of chlorine, due to impurities formed in the course of manufacture.

[1] *Arch. der Pharm.*, 251 (**1913**), 85.

Another adulterant, of a much grosser character, is oil of mirbane. This is the cheap almond oil substitute so largely used for perfuming common toilet soaps. Chemically it is nitrobenzene $C_6H_5NO_2$, more or less mixed with impurities, of which the most common is nitrotoluene, which sometimes itself forms the greater part of cheap nitrobenzene. Indeed, nitrotoluene in any great quantity may be regarded as an adulterant of nitrobenzene. The latter, when pure, is a yellowish liquid of specific gravity at 0° of 1·200, boiling at about 206°, and solidifying at + 2° to + 3°. It has a coarse almond-like odour, and is poisonous when taken internally, and irritating to the skin when used externally. The cheapness of artificial benzaldehyde should discourage its use in even the very cheap toilet soaps. Nitrotoluene $C_6H_4(CH_3)NO_2$ exists in three isomeric modifications, and nitroxylene $(B_6H_3)(CH_3)_2NO_2$ in more still. These bodies are found to a considerable extent in the cheaper qualities of nitrobenzene. Consequently it is important that commercial samples should have physical characters in approximate agreement with those above quoted.

To detect the presence of this objectionable substitute in oil of almonds a little of the oil is warmed with iron filings and acetic acid. The nitrobenzene is reduced to aniline $C_6H_5NH_2$, which is distilled off and collected. To the distillate a few drops of solution of ordinary chloride of lime is added. If aniline be present the liquid yields the characteristic violet colour. Pure benzaldehyde combines with sodium bisulphite to form a crystalline compound without the characteristic almond odour. Samples adulterated with nitrobenzene, when shaken with excess of bisulphite of sodium solution, so that the benzaldehyde is entirely combined, then have the characteristic coarse nitrobenzene odour.

The determinations of importance in the examination of this oil are (1) the determination of chlorine, or at all events its detection qualitatively, if present; (2) the determination of benzaldehyde, and (3) the determination of hydrocyanic acid.

The Determination of Chlorine.—Many methods have been suggested for the detection and determination of chlorine, but most of them are merely qualitative. Of these the following are useful :—

(*a*) By simple combustion. This method depends on the fact that on combustion a portion of the chlorine present in the organic compounds existing as impurities, forms hydrochloric acid, which can be detected by means of silver nitrate.

A piece of filtering paper about 2 inches by 1 inch and rolled in the shape of a spill, is saturated with the oil, the excess of oil is thrown off and the paper is placed in a small porcelain dish which in its turn reposes in a larger dish, measuring about 8 inches in diameter. The paper is ignited, and a beaker of a capacity of about 2 litres, moistened inside with distilled water, is quickly placed over the paper. The sizes of the dishes used must be such that the rim of the larger dish must stand out well all round from the beaker. After the flame is extinguished the beaker should be left in position for about one minute, after which the products of combustion which have been deposited against the moist sides of the beaker are rinsed out with a little distilled water, and filtered. The filterate, acidulated with a drop of nitric acid, must remain clear when silver nitrate solution is added.

Care must be taken that the filter paper is free from chlorine. This should be determined by a blank experiment. This process has been strongly recommended by Schimmel & Co., but there is considerable doubt whether it will detect minute quantities of chlorine.

(*b*) The sodium test is distinctly more delicate, and is carried out as follows :—

A piece of pure metallic sodium about half the size of a pea is dropped into 0·5 c.c. of the oil in a dry test tube, and heated until all chemical action has ceased. The test tube and contents are immersed whilst still hot in 10 c.c. of distilled water in a porcelain dish. The solution is filtered, acidulated with nitric acid, and silver nitrate solution added. Any turbidity or opalescence indicates the presence of chlorine compounds.

A blank test should be performed to ensure the absence of chlorine in the materials employed.

(*c*) The lime combustion method gives good results as a qualitative test, although it will often fail to detect very minute traces, say below 0·08 per cent. It is carried out as follows :—

A mixture of 1 c.c. of benzaldehyde and 5 grams of lime are placed in the bottom of a platinum crucible about 3 ins. in depth, a layer of lime placed on the top and the crucible covered and gently heated; a considerable escape of benzaldehyde takes place, so that it is obvious that the results can have no quantitative value.

The temperature is then gradually increased until no further vapours are driven off. The mixture is then dissolved in dilute nitric acid, filtered and tested in the usual manner with silver nitrate.

(*d*) The method of Carius for the determination of chlorine in organic compounds is, of course, absolutely quantitative, but is very tedious. Salomon [1] has recommended the following method, and appended are the results obtained compared with those yielded by the Carius' method and the obviously inadequate lime method.

About 1 gram of benzaldehyde is heated in a retort with 40 c.c. of concentrated sulphuric acid, the fumes collected in a solution of silver nitrate, and the heating continued until no further precipitate insoluble in hot dilute nitric acid is obtained in the silver nitrate solution. This takes about three hours.

The solution of silver nitrate is acidified with dilute nitric acid, boiled so as to decompose any silver sulphite that might have been formed, and the precipitate filtered, washed, etc.

A slight modification consists in using a mixture of 40 c.c. concentrated sulphuric acid and 5 c.c. concentrated nitric acid, the nitric acid being added last to the mixture of sulphuric acid and benzaldehyde.

The hydrochloric acid was evolved more rapidly, and slightly higher results were obtained in certain cases, due probably to better oxidation.

In using this method the heating must be very gradual, and the mixture should not be allowed to froth until the end of the operation.

The Determination of Benzaldehyde.—It is doubtful whether there is any accurate method for the determination of benzaldehyde in essential oil of almonds, in the presence of hydrocyanic acid, except in so far as the actual free aldehyde is concerned. The most modern investigation of this question is that of F. D. Dodge.[2]

The results obtained by the use of the methods for chlorine determina-

[1] *P. and E.O.R.* (**1917**), 41.
[2] *Eighth International Congress of Applied Chemistry*, Washington (**1912**), xvii. 15.

tion described on p. 389 are as follows, and certain of the samples were checked by means of a Carius determination:—

No.	Lime Method Chlorine, per Cent.	Sulphuric Acid Method, per Cent.	Nitro-sulphuric Acid Method, per Cent.	Carius Method, per Cent.
1	·19	1·83	2·14	2·2
2	Not weighable	0·13	0·13	0·15
3	·20	1·8	1·9	2·0
4	·19	1·7	1·7	1·9
5	nil	trace	trace	trace
6	·17	—	1·63	—
7	·28	—	2·66	—
8	·17	—	1·42	—

The United States Pharmacopœia have adopted the method devised by Sadtler, but this is not very satisfactory. It is based upon the formation of the phenylhydrazone. The benzaldehyde is added to an excess of phenylhydrazine in alcoholic solution, and the phenylhydrazone is filtered off. The uncombined phenylhydrazine is then neutralised with semi-normal hydrochloric acid, and the amount of acid required is deducted from the quantity required to neutralise the same quantity of phenylhydrazine in a blank experiment. Unfortunately, the end point with methyl orange is not sharp, and the method cannot be considered a satisfactory one. The low minimum of benzaldehyde required for benzaldehyde and bitter almond oil (85 per cent.) is practically an admission that low results are obtained by this process. The method devised by Ripper gave better values, although these, too, are only approximately accurate. Ripper shakes up the aldehyde with a bisulphite solution of known strength and titrates back the excess of bisulphite with iodine solution at a low temperature. According to Dodge, the most suitable manner of carrying out this method is as follows: 0·15 gram aldehyde or bitter almond oil is weighed in a flask containing exactly 25 c.c. of a, say, $\frac{1}{5}$ n. bisulphite solution, and brought to solution by gentle shaking. The flask is then closed, left standing in ice water for one and a half to two hours, after which the ice-cold solution is titrated with deci-normal iodine solution, starch being used as an indicator. A blank test is carried out in the same way, the benzaldehyde-content being calculated from the difference (1 c.c. deci-normal iodine solution = 0·0053 gram benzaldehyde). The reaction is completed as soon as the blue colour persists for a few seconds. By this method Dodge, in the case of pure benzaldehydes, found from 94·8 to 99·0 per cent.

Determination with hydroxylamine (Walther's method as modified by A. H. Bennett) offered no advantages as compared with the iodometric process, inasmuch as in the case of pure benzaldehydes only 93·5 and 95·1 per cent. were found. On the other hand, the estimation with phenylhydrazine recommended by Denis and Dunbar gives useful, although at times rather too high, values (95·1 to 100·9 per cent.), but in this method the unstable character of the reagent is a disadvantage. In this process 10 c.c. of a solution of the oil of 5 per cent. strength are mixed with 10 to 15 c.c. of a fresh 10 per cent. solution of phenylhydrazine, the liquid shaken, the flask tightly closed, and allowed to stand in the dark for twelve hours. 200 c.c. of water are now added, and the liquid filtered through a Gooch crucible, and the precipitate washed with cold water, then with 10 c.c. of 10 per cent. alcohol, dried for three hours *in vacuo* over sulphuric acid at 70° to 80°, and weighed. The weight multiplied by 5·408 gives the number of grams of benzaldehyde in 100 c.c. of the solution.

Kleber's method, based on the use of phenylhydrazine is also accurate for the determination of the free benzaldehyde. Slightly modified by Schimmel & Co. this method is as follows:—

To 2 c.c. of a solution of the oil (about 5 per cent. strength), 10 c.c. of a fresh alcoholic solution of phenylhydrazine are added and allowed to stand for an hour in a stoppered flask holding about 50 c.c. Twenty c.c. of deci-normal hydrochloric acid are then added and the flask gently shaken. Ten c.c. of benzene are then added and the mixture is shaken well and poured into a separator. The acid layer separates clear, and is measured and filtered, and 20 c.c. of the filtrate is titrated with deci-normal potash, using di-ethyl-orange as indicator. From this the titration figure for the full volume of the acid liquid is calculated. A blank experiment is carried out without the almond oil, and each c.c. of deci-normal potash in excess of that used in the estimation = 0·0106 gram of benzaldehyde.

Dodge has proposed another method for the quantitative estimation of benzaldehyde. This is based upon Cannizzaro's reaction:—

$$2C_6H_5CHO + KOH = C_6H_5COOK + C_6H_5CH_2OH.$$

He made tests with alcoholic potash solution, which showed that the reaction does not proceed quantitatively when the solution is of less than twice normal strength. Good results were obtained with a 2·5 normal solution. Ten c.c. is allowed to stand for twenty-four hours at ordinary temperature with from 1 to 2 grams of benzaldehyde, after which the unattacked alkali is titrated back with semi-normal hydrochloric acid. A blank experiment is carried out without the almond oil. The difference shows how much of the alkali has entered into reaction, expressed in c.c. of semi-normal potash. By multiplying the number of c.c. by the molecular weight of benzaldehyde (106) and dividing by the quantity of oil employed in the test, the aldehyde-content is obtained.

In the case of pure artificial products, Dodge found this figure to be from 99·03 to 100·3 per cent. He lays stress upon the fact that in measuring the 2·5 normal potash solution it is necessary to proceed with great care, because so small a difference as 0·05 c.c. represents an error of not less than 1 to 2·5 per cent. benzaldehyde.

In the case of natural bitter almond oil the method gave no useful results, presumably because the saponification of the benzaldehyde-cyanohydrin into mandelic acid was incomplete.

It is apparent that in natural bitter almond oil all these methods only estimate the free benzaldehyde, and not that which is present in combination with hydrocyanic acid. This is shown by the following results (see table on p. 412).

The Determination of Hydrocyanic Acid.—Hydrocyanic acid may be approximately estimated by dissolving 1 gram of oil in 5 c.c. of alcohol, and adding 50 c.c. of water. Then add ammonio-silver nitrate solution and shake well. Acidify slightly with nitric acid, and collect, wash, and dry the silver cyanide precipitated. Ignite and weigh the silver, four parts of which correspond to practically one of hydrocyanic acid.

	Sample No. 1.	Sample No. 2.	Sample No. 3.
HCN	3·6 per cent.	2·8 per cent.	2·4 per cent.
Equivalent quantity of cyanohydrin . . .	17·6 „	—	—
Calc. quantity of free benzaldehyde	82·4 „	86·2 per cent.	88·2 per cent
Estim. by Ripper's method .	81·3 to 82·1 per cent.	80 per cent.; 81·4 per cent.	
Estim. with hydroxylamine .	78·9 per cent.	—	—
Estim. with phenylhydrazine (Denis and Dunbar's method)	83·2 „	83·6 per cent.; 87·2 per cent.	85·2 per cent.

For an exhaustive examination of the various processes proposed for the determination of hydrocyanic acid, the reader is referred to a series of papers by Runne in the *Apotheker Zeitung*.[1]

Oil of Cherry Laurel.

Cherry Laurel Oil is distilled from the leaves of *Prunus laurocerasus*, an evergreen shrub, a native of South-eastern Europe and Asia Minor. Most of the oil is distilled in the south of Switzerland and in Italy. It is obtained also in England and Germany, together with the official cherry laurel water, by distilling the leaves in the ordinary way. It has the following characters:—

Specific gravity	1·050 to 1·066
Optical rotation	+ 0° 12′ „ − 0° 46′
Refractive index	1·5400 „ 1·5440

It contains benzaldehyde, prussic acid, benzaldehyde-cyanhydrin, and possibly traces of benzyl alcohol, and, indeed, differs very little from oil of bitter almonds.

Oil of Wild Cherry Bark.

The bark of *Prunus virginiana* yields about ·2 per cent. of essential oil of specific gravity about 1·050, and optically inactive. It closely resembles oil of bitter almonds.

Oil of Meadow Sweet.

This oil is obtained by the distillation of the flowers of *Spiraea Ulmaria*, which yield about 0·25 per cent. of oil.

It contains salicylic aldehyde, $C_6H_4(OH)(COH)$, a terpene, methyl saliglate, vanillin, and hetrotropin.

CALYCANTHACEÆ.

Oil of Calycanthus.

Scalione[2] has examined the essential oil of *Butneria occidentalis*, or *Calycanthus occidentalis*, which he obtained to the extent of 0·27 per cent. of the leaves and twigs. The oil had the following characters:—

[1] *Apotheker Zeitung*, 24 (**1900**), 288, 297, 306, 314, 325, 333, 344, 356.
[2] *Jour. Ind. Eng. Chem.*, 1916, 729.

Specific gravity at 25°	0·9295
Optical rotation	+ 7° 28′
Refractive index	1·4713
Saponification value	54·3

On fractionation the oil gave the following results. The oil was fractionated with a 3-bulb Le Bel fractionating column:—

Fraction No.	Fractionation Temperatures °C.	Per cent. Distilled over.	Specific Gravity 15° C.	Specific Rotation at 15° C.	Index of Refraction 15° C.
1	154-170	8·51	0·8810	+ 16·42	1·4590
2	170-180	54·83	0·9119	+ 5·85	1·4585
3	180-190	9·47	0·9215	+ 6·56	1·4650
4	190-200	2·48	0·9290	+ 9·52	1·4665
5	200-220	8·75	0·9635	+ 36·42	1·4817
6	220-over	10·22	0·9365	+ 18·85	1·4793
7		5·01	0·9171		1·4884

Fractions 5, 6, and 7 were obtained under reduced pressure at 150°-180°. The boiling-points of these fractions were redetermined at atmospheric pressure.

The oil was found to contain about 8 per cent. of pinene, 60 per cent. of eucalyptol, 9 per cent. of borneol, 19 per cent. of linalyl acetate, with a little camphor, methyl salicylate, and sesquiterpene alcohol.

From the dry twigs of *Calycanthus floridus* Miller, Taylor, and Eskew[1] obtained from 0·25 to 0·53 per cent. of a pale yellow essential oil having a camphoraceous odour. Its characters are as follows:—

Specific gravity at $\frac{25°}{25°}$	0·9136 to 0·9209
Optical rotation	+ 2·84° ,, + 6·6°
Refractive index at 26°	1·4675 ,, 1·4753
Saponification value	12·5 ,, 16·6
,, ,, (after acetylation)	65·7 ,, 75·1

Eucalyptol is present in the oil, as well as α-pinene, salicylic acid (?), borneol and, possibly, linalol. Bornyl esters of formic and acetic acids are also present.

RUTACEÆ.

Lemon Oil.

This oil, one of the most important of all the essential oils from the commercial point of view, is obtained by various processes from the fresh peel of the ordinary lemon, *Citrus limonum* (*Citrus medica* subsp. *Limonum*). The tree is a native of India, and was introduced into Southern Europe, probably first into Spain, and thence to the South of France and Italy. There are numerous varieties and hybrids, but the ordinary lemon forming the fruit of commerce is the Genoese lemon. This is cultivated on the Mediterranean littoral, especially between Nice and Genoa, in Calabria, Sicily, Spain, and Portugal, and on nearly the whole coast of Liguria. Many of the British Colonies are now cultivating this fruit, Australia for many years past having filled its own requirements for lemons; the trees are also extensively grown in New South Wales, Florida, California, and Jamaica, and will, probably, shortly produce the oil in considerable quantities. The oil is obtained in the South of Europe between the months of November and March, or April, and the processes used are the following: (1) The spugna or sponge process, by which

[1] *Jour. Amer. Chem. Soc.*, 36 (**1914**), 2182.

the oil is removed from the burst oil cells of the peel by a sponge. (2) The *écuelle* method, by which the lemons are rolled about in vessels covered with spikes and the oil collected in a receptacle in the handle of the vessel. (3) Machine processes. (4) Distillation. This method gives a very inferior oil, and is often applied to the peels from which all the oil obtainable by other processes has been obtained. (5) Expression in bags in ordinary presses, now rarely, if ever employed.

The sponge method is the typical process in use in the chief lemon oil districts—Sicily and Calabria; the *écuelle* method being employed chiefly in the South of France and North of Italy, and to a certain extent, of course, in the south.

The sponge method of obtaining lemon oil has been described by many manufacturers. Probably the description by Mr. Arthur Barrett [1] is as accurate as any. It is as follows: "The principle on which the extraction of the essence is carried out may be illustrated in this way. If you hold a piece of lemon peel up to the light, and turn it inside out, a fine shower of mist will be seen to be forcibly ejected. This is not all oil, but a mixture of oil and water. Most people are unpleasantly acquainted with this phenomenon, though many may not have actually seen it, for in peeling a lemon or orange with the fingers a little of the oil is often ejected into the eye, causing a considerable amount of pain. By turning the lemon peel inside out almost the whole of the essence is removed from the peel, for each little globule of oil appears to be surrounded by water, and the liquid which remains adherent to the peel consists principally of water. As it is impossible to turn every piece of peel actually inside out, the following method is adopted: One man takes a lemon in his hand, and with three rapid strokes with a large knife cuts off nearly all the peel in three slices. The central portion which is left consists of most of the pulp with a little of the peel—top and bottom. This is simply pressed for making lemon juice. The slices pass to a second workman, who sits on a low chair with a common quality bath sponge, worth about sixpence, in one hand. With the other he presses the slice of peel against the sponge, pressing the edges of the peel only with his fingers, the object being to press the convex piece of lemon peel as nearly flat as possible. The amount of pressure used is very slight, and at first sight it seems incredible that the oil globules can have been broken; but, if you try the experiment of turning this exhausted peel inside out, nothing more can be extracted. The sponge is periodically squeezed. One man working in this way can extract $1\frac{1}{2}$ lb. (English) essence of lemon per day. To ensure the cells being fully charged with moisture it is usual to allow the lemons to stand in water for a short time; and I myself propose washing the lemons in a stream of running water."

The Scorzetta process is a variation of the ordinary sponge method.

The fruits are cut into two halves, the pulp is removed with a kind of spoon, and all the parts of the peel are then pressed against a sponge by constantly turning it in the palm of the hand. This method has the advantage that the peels remain unbroken, and retain their original appearance. After they have been deprived of the oil, they are salted, and then exported under the name "Salato". The pulp also remains unimpaired, so that it yields its full supply of lemon juice. After having passed twice through the pressers it is used for feeding cattle.

[1] *Year Book of Pharmacy* (**1892**), 505.

The *écuelle* process, which is employed in the North of Italy and the South of France, depends on the preliminary breaking of the oil globules by means of spikes or knives, and varies in detail, according to the quantity to be operated upon. A simple *écuelle* is a saucer-shaped vessel, about 8 to 10 ins. in diameter, usually made of copper, tinned inside, and covered inside with short spikes, about $\frac{1}{4}$ to $\frac{1}{3}$ in. long. In the bottom of the saucer is a hole about $\frac{1}{2}$ to $\frac{1}{3}$ in. in diameter, leading to a hollow tube about 4 or 5 ins. long. A fruit is placed in the vessel, and, by means of a rapid rotary motion, the oil cells are broken by the spikes. The escaping oil flows into the hollow tube, which is periodically emptied.

Various purely mechanical appliances have been devised for obviating the necessity of hand labour, but most of these have proved of little value.

An apparatus, known as the thermopneumatic extractor, has been

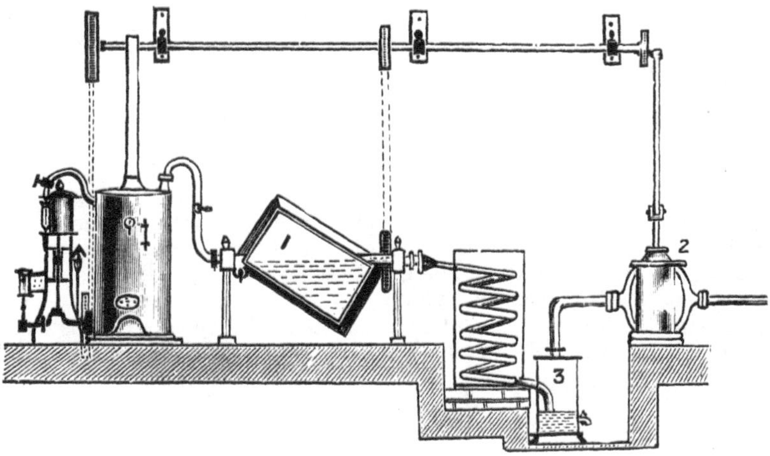

FIG. 38.

invented by Montfalcone for the extraction of all varieties of *citrus* oils. The inventor claims that a far higher yield of oil of equal quality is obtained by the use of his apparatus, which is described as follows:—

1 is a hollow cylinder of sheet iron, the interior surface of which is studded with a large number of metallic knife-points. This cylinder revolves on two axles, attached to it at diagonally opposite points. Its capacity is such that, when charged, it is capable of holding 1200 or 1500 lemons. The cylinder having been properly charged, say with the above-mentioned number of lemons, together with a small quantity of water, it is set in motion by aid of the shafting and pulleys, driven by the small steam engine. The fruits are thereby made to come in contact with the metallic lancets, and their whole surface gradually becomes punctured sufficiently to cut or rupture the cells containing the essential oil, which escapes. The cylinder, 1, has double walls, the space between them being intended for the circulation of a current of steam, which is admitted to heat the water and to facilitate the extraction of the oil, while, at the same time, the vapours are rarefied or aspirated by the vacuum-pump, 2. The axles of the cylinder on both sides are hollow: that on the side look-

ing towards the engine is embedded in such a manner that steam from the boiler may be admitted at will, either into the double walls of the cylinder or into the interior of the cylinder itself; while the hollow passage in the other axle communicates with a condensing worm, the outlet of which descends into a cylindrical vessel, 3, intended to receive the condensed products, consisting of water and essential oil. When the apparatus is first set in motion, the pump begins to produce a vacuum during the first revolutions of the cylinder. Steam being now cautiously admitted into the double walls of the cylinder, the water is raised to boiling at a comparatively low temperature, and the vapours charged with the essential oil rapidly pass over into the receptacle. With a two-horse-power engine and a boiler designed for a five-horse-power engine, five such machines may be driven at once, and the operation, including the time required for charging and emptying, is completed in three-quarters of an hour.

The United States Department of Agriculture have recently devised a machine for the peeling of *citrus* fruits and the expression therefrom of the essential oil. Although only in the experimental stage, it promises to be very successful.

Lemon oil vari s to a considerable extent, according to the period of the harvest, to the season, and to the climatic conditions, so that it can only be properly judged by the analyst when all these features are taken into account.

The greater part of the oil of commerce is produced in the district of Messina, on the north-east of the island of Sicily, and the adjacent districts along the coast to the west; a considerable amount is produced in the Palermo province, and all along the east coast of the island; and a large amount also in Calabria, which is chiefly exported from Reggio. An approximate idea of the relative production of these districts is seen from the export returns during 1905, which were as follows:—

From Messina	742,567 kilos.
,, Reggio	95,319 ,,
,, Palermo	26,041 ,,
,, Catania	4,317 ,,

Apart from the alteration in rotatory power suggested above as coinciding with the time of the crop, although possibly having no direct relation to the quality of the oil, it is found that the oil obtained from round about Barcelona (near the northern coast) and from the plains of Syracuse (on the south-eastern coast) possess the highest average rotations of all the oils, this figure varying from + 64° to + 65°. The following figures are found for the various districts:—

From + 57° to + 71° . .	Messina and neighbourhood, Nizza di Sicilia.
,, + 61° ,, + 63° . .	Acireale, S. Teresa di Riva, Scaletta, S. Lucia. Patti, S. Agata, S. Stefano.
,, + 63° ,, + 64° . .	Catania, Giarre, Giardini, Acireale, Lentini.
,, + 64° ,, + 65° . .	Barcelona, Syracuse.

Palermo oils are usually about + 57° to + 61°, and the Calabrian oils + 59° to + 62° 30′. In the author's experience these limits are of very small value and cannot be taken as in the least degree conclusive. Cool rainy weather appears to have some effect on the optical rotation, but all one can say conclusively is that the average rotation observed during one season's crop is sometimes 2° or 3° higher or lower than that of the preceding year. The Palermo oil and the Messina oil differ somewhat in their odour, etc., so that a trained nose can discriminate between them.

The table on opposite page, covering analysis by Schimmel & Co., gives an outline of the differences between the oils of different districts during a given season (1908).

Lemon oil contains a considerable number of constituents, of which the principal is the aldehyde citral, which is present to the extent of from 4 to 5·5 per cent., rarely up to 6 per cent. Burgess and Page[1] have isolated a hydrocarbon, which is interesting as being quite different from the terpenes so generally found in the citrus oils. It has the following characters :—

Origin.	Pressed in	$d_{15°}$.	$\alpha_{D_{20°}}$.	α_D of the First 10 per Cent.	Residue on Evaporation.	Citral
Palermo :	January	0·8578	+ 60° 2′	+ 57° 42′	2·6 per Cent.	5·2 per Cent.
	February	0·8590	+ 57° 15′	+ 55° 10′	3·0 ,,	5·4 ,,
	March	0·8577	+ 59° 44′	+ 56° 40′	2·6 ,,	6·4 ,,
S. Agata :	January	0·8580	+ 59° 10′	+ 56° 8′	2·8 ,,	6·3 ,,
	February	0·8583	+ 58° 40′	+ 55° 50′	3·6 ,,	5·5 ,,
	March	0·8575	+ 61° 36′	+ 58° 12′	2·6 ,,	5·5 ,,
Barcelona :	January	0·8570	+ 62° 4′	+ 58° 30′	2·2 ,,	5·9 ,,
	February	0·8580	+ 60° 40′	+ 57° 50′	2·4 ,,	5·7 ,,
	March	0·8575	+ 59° 0′	+ 54° 24′	2·8 ,,	5·2 ,,
Messina :	January	0·8583	+ 57° 53′	+ 52° 24′	3·0 ,,	6·5 ,,
	February	0·8586	+ 58° 26′	+ 53° 38′	3·0 ,,	6·0 ,,
	March	0·8591	+ 56° 50′	+ 53° 30′	3·2 ,,	5·7 ,,
S. Teresa :	January	0·8600	+ 59° 18′	+ 58° 0′	3·2 ,,	7·1 ,,
	February	0·8607	+ 57° 20′	+ 54° 50′	3·0 ,,	7·2 ,,
	March	0·8610	+ 57° 24′	+ 54° 36′	2·8 ,,	6·6 ,,
Taormina :	January	0·8569	+ 62° 20′	+ 60° 40′	2·4 ,,	5·8 ,,
	February	0·8572	+ 62° 16′	+ 59° 26′	2·7 ,,	6·0 ,,
	March	0·8580	+ 60° 52′	+ 59° 10′	2·4 ,,	5·3 ,,
Giardini :	January	0·8575	+ 61° 0′	+ 58° 4′	2·8 ,,	5·7 ,,
	February	0·8589	+ 59° 20′	+ 56° 40′	2·8 ,,	6·0 ,,
	March	0·8585	+ 59° 56′	+ 56° 48′	3·0 ,,	5·3 ,,
Mascali :	January	0·8569	+ 61° 44′	+ 58° 40′	2·4 ,,	4·7 ,,
	February	0·8574	+ 62° 40′	+ 59° 18′	2·8 ,,	4·4 ,,
	March	0·8578	+ 60° 14′	+ 56° 4′	2·7 ,,	4·3 ,,
Acireale :	January	0·8588	+ 60° 12′	+ 57° 24′	2·6 ,,	6·5 ,,
	February	0·8583	+ 59° 38′	+ 56° 20′	2·2 ,,	5·6 ,,
	March	0·8590	+ 58° 22′	+ 55° 0′	2·8 ,,	5·9 ,,
Acicatena :	January	0·8580	+ 60° 6′	+ 56° 28′	3·2 ,,	4·9 ,,
	February	0·8586	+ 60° 14′	+ 58° 0′	3·5 ,,	5·1 ,,
	March	0·8580	+ 59° 12′	+ 56° 24′	3·0 ,,	4·9 ,,
Catania :	January	0·8576	+ 61° 52′	+ 59° 20′	2·8 ,,	5·0 ,,
	February	0·8580	+ 58° 16′	+ 55° 26′	2·6 ,,	5·4 ,,
	March	0·8589	+ 59° 50′	+ 56° 50′	2·8 ,,	5·1 ,,
Syracuse :	January	0·8580	+ 59° 28′	+ 57° 24′	2·9 ,,	6·7 ,,
	February	0·8583	+ 60° 35′	+ 59° 22′	2·8 ,,	7·0 ,,
	March	0·8575	+ 62° 6′	+ 60° 32′	2·6 ,,	5·3 ,,

Specific gravity	0·7257
Optical rotation	0°
Refractive index	1·4066 at 15°
Boiling-point	023° to 124°

[1] *Jour. Chem. Soc.*, 85 (**1904**), 1328.

Analysis agreed with the formula C_8H_{16} or C_8H_{18}. It is probably octylene. Pinene appears to be present in traces, as well as camphene, phellandrene, and γ-terpinene. The bulk of the oil, however, consists of *d*-limonene which is present to the extent of about 90 per cent. Methylheptenone, octyl aldehyde, nonyl aldehyde, citronellal, terpineol, linalyl acetate, geranyl acetate, bisabolene, cadinene and methyl anthranilate have all been found present in traces.

Citraptene, or "lemon camphor," a solid substance, which appears to be dimethoxy-coumarin, is present to the extent of about 2 per cent. in the oil. This is a non-volatile substance dissolved out from the peel by the expressed essential oil.

Citraptene, which appears to be identical with limettin, has been carefully examined by E. Schmidt.[1]

The early work of Brissenot, Blanchet, and Sell and Mulder is referred to, as is also the analysis by Berthelot. This brings us up to the first of the recent workers, Crismer, who gave 143° to 144° as the melting-point, and $C_{10}H_{10}O_4$ as the formula of the stearoptene of lemon oil. Schmidt isolated a crystalline body melting at 146° to 147° and having a blue fluorescence in alcohol solution, and he states that this body is the citraptene of Crismer, and the limettin of Tilden and Beck. Theulier had already stated that a crystalline body melting at 145° and an amorphous one melting at 76° were the chief constituents of the solid mass obtained on the distillation of lemon oil; of these citraptene is the principal.

Schmidt gives 146° to 147° as the melting-point of this body, for which the name citraptene is retained. Three analyses gave the following results:—

Carbon.	*Hydrogen.*	*Calculated for* $C_{11}H_{10}O_4$.	
63·92	5·04	C	63·98
63·83	5·01	H	4·85
63·82	4·75	O	31·17

This formula is in agreement with that assigned to limettin by Tilden and Beck.

Citraptene contains two methoxyl groups; when treated by Zeisel's method to remove the methyl groups, the demethylated citraptene, $C_9H_6O_4$, yields phloroglucin and acetic acid on fusion with alkali.

Citraptene has the general character of an acid anhydride, according to Schmidt, its corresponding acid being unstable and rapidly converted into a lactone.

Citraptene appears to be a di-methoxy-coumarin of the formula—

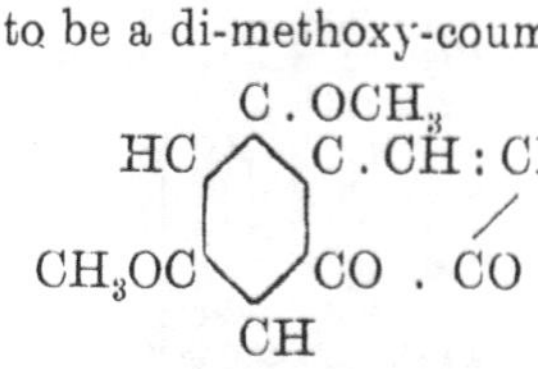

Lemon oil has the following characters:—

Specific gravity	0·854 to 0·862	(usually 0·857 to 0·860).
Optical rotation	+ 54° ,, + 66°	(,, + 57° ,, + 62°).
Refractive index	1·4745 ,, 1·4760	(rarely below 1·4750).
Fixed residue (at 100°)	2 to 6 per cent.	(,, over 4 per cent.).
Citral	4 ,, 6 ,,	(usually 4·2 to 5·5 per cent.).

[1] *Archiv. der Pharm.* **(1904)**, 288.

During the latter part of June and the month of July in some seasons, a number of samples of undoubted authenticity have been examined and found to have an abnormally low optical rotation sometimes as low as + 51°. In some cases the odour of the lemon oil is quite abnormal, and gives one the impression that it had been contaminated with neroli oil. The reason of this appears to be that the spring lemons ("bianchetti"), which are rarely used for pressing for oil, as the oil is inferior and the yield small, are in such seasons being pressed in large quantities; and where this neroli odour is found, it is probable that some quite immature fruits have been used for pressing.

If pure lemon oil be carefully distilled from a three-bulbed fractionating flask, and the first 10 per cent. collected, its optical rotation will rarely differ from that of the original oil by more than 5°, or at most 6°, the rotation of the fraction being lower than that of the oil. The size of the flask and the rate of distillation cause considerable variation in this figure, so that unless a marked difference is found, such as would result from adulteration with an appreciable quantity of turpentine, too much stress must not be laid upon the result of a test which, after all, is empirical in its nature.

Adulteration, which was until recently very frequent, is still common. Turpentine was the regular adulterant, with, at times, the poorer quality distilled oil of lemons. But adulteration with turpentine is now so easily detected that the sophistication is frequently carried out in a more scientific manner. Mixtures with the proper specific gravity and optical rotation can easily be made up from turpentine and orange oil—the poorer qualities of the latter of course being used—and such mixtures are often used to adulterate the oil. But the most formidable adulterant from the analyst's point of view is one that has only come into vogue during the last few years, *viz.* the terpenes obtained in manufacturing the "terpeneless" or concentrated oils of lemon and oil of orange, the latter being sometimes added to turpentine to raise the optical rotation.

The terpenes are sometimes used alone, sometimes together with a little citral obtained from lemon-grass oil. It is a common custom to export oil of lemons with a guaranteed citral-content, and to sell it upon that basis. In judging of the value of such a basis for the market value of the oil, the following points should be noted :—

1. The value of the oil depends on its percentage of all the oxygenated constituents, which are soluble in weak alcohol.

2. The terpenes are practically odourless and insoluble, therefore valueless for the general purposes for which lemon oil is employed.

3. The percentage of terpeneless oil obtained by careful fractionation is an indication of the value of the oil.

Some samples of lemon oil produced in Spain have recently been examined, and, as instancing the differences between oils manufactured in different districts, have the following characters which would be considered impossible in the case of pure Sicilian oils :—

Specific gravity	0·853 to 0·862
Optical rotation	+ 63° 32′ „ + 65° 30′
Refractive index	1·4737 „ 1·4738
Citral	3·1 to 3·2 per cent.

The optical rotation of the first 10 per cent. distilled was found to be about 3° higher than that of the original oil.

The most important method of analysis applicable to lemon oil is the determination of the citral-content. A great many methods have been suggested for this, most of which yield only approximate results.

From a lengthy experience in these processes, the author has no hesitation in saying that the most accurate method for the estimation of citral in lemon oil is that devised by J. Walther,[1] with a slight modification suggested by A. H. Bennett.[2]

This process depends on the formation of an oxime with hydroxylamine in alcoholic solution, and determination of the amount of hydroxylamine used in the process, according to the equation :—

$$C_9H_{15}CHO + H_2NOH = C_9H_{15} : NOH + H_2O.$$

The process is carried out as follows: A mixture of 20 c.c. of lemon oil, 20 c.c. of 90 per cent. alcohol, 20 c.c. of a 5 per cent. solution of hydroxylamine hydrochloride in 80 per cent. alcohol, and 8 c.c. of normal alcoholic potash solution is gently boiled for thirty minutes under a reflux condenser. A similar mixture, without the lemon oil, is treated in exactly the same manner. After cooling, the condensers are washed down with water to restore any traces of hydroxylamine to their respective flasks, and the contents of the flasks diluted with about 200 c.c. of water. A few drops of phenolphthalein are added to the flasks, and the contents first neutralised with semi-normal alkali. The hydroxylamine is now titrated with semi-normal sulphuric acid solution, using methyl orange as indicator. A few drops of the indicator should be added to the flask, so that a general indication of the approach of the end reaction is given, but the end reaction itself should be observed by the familiar process of "spotting" on a white tile, on drops of the indicator. Each cubic centimetre of half-normal acid used in the blank experiment, over that used in the assay experiment, is equivalent to 0·076 grm. of citral.

A process depending on the use of cyanacetic acid was proposed by the author[3] some years ago, which depends on the condensation of cyanacetic acid with citral in alkaline solution, according to the following equation :—

$$C_9H_{15}CHO + CH_2CN . COOH = C_9H_{15}CH . C\begin{cases} CN \\ COOH \end{cases} + H_2O.$$

The procedure is as follows: 200 c.c. of the oil are concentrated to 25 c.c. in a Wurtz flask under a pressure which should not exceed 15 mm. The 175 c.c. of terpenes contain a little citral, but the quantity is so small that it may be neglected. Ten c.c. of the residue, of which the specific gravity is known, are shaken with a solution of 5 grms. cyanacetic acid and 5 grms. ordinary KOH in 30 c.c. water in a small absorption flask with a long graduated neck, and the non-absorbed residue read off as usual. (The specific gravity of citral is so near that of the concentrated oil used that it may be taken as identical.)

This process does not give satisfactory results, and is only quoted here so that the author may definitely withdraw it as a process associated with his name.

E. M. Chace[4] has elaborated a method for the determination of citral depending on the fact that fuchsine-sulphurous acid develops a red coloration with aldehydes, the depth of which depends on the amount

[1] *Pharm. Zentral.*, 40 (1899), 621; 41 (1900), 585.
[2] *Analyst*, 34 (1909), 14.
[3] *Chemist and Druggist*, 56 (1900), 376.
[4] *Jour. Amer. Chem. Soc.*, 28, 1472.

of aldehyde present. The determination is carried out by comparison with solutions of citral of known strength.

The solution of magenta-sulphurous acid is prepared as follows: 0·5 grm. of magenta is dissolved in 100 c.c. of water and a solution containing 16 grms. of sulphur dioxide is then added. The liquid is allowed to stand until it is decolorised and its volume is made up to a litre with water. Such a solution cannot be kept longer than two or three days.

Secondly, alcohol of 95 per cent. strength must be prepared absolutely free from aldehyde. For this purpose the alcohol is kept for several days in contact with a caustic alkali, it is then distilled and the distillate is boiled under a reflux condenser for a few hours with meta-phenylene-diamine hydrochloride (25 grms. of the hydrochloride per litre). The alcohol is then separated by distillation and made up to the desired concentration.

Lastly, a 0·1 per cent. solution of citral in 50 per cent. alcohol is prepared.

These solutions are made up at a temperature of 15° C., and all the operations are carried out at this temperature. It is particularly recommended that a higher temperature be avoided.

Two grms. of each of the essential oils to be examined are diluted to 100 c.c. with the alcohol free from aldehyde. Four c.c. of each of the solutions are introduced into cylinders of identical dimensions, 20 c.c of alcohol free from aldehyde are added, then 20 c.c. of magenta-sulphurous acid, and the volume is made up to 50 c.c. with alcohol. After thoroughly mixing, the solutions are placed in a water-bath at 15° C., and compared in regard to their colours with the necessary standards. An error of 1 per cent., however, may be made in this determination.

R. S. Hiltner[1] has devised a process depending on the yellow colour developed when citral reacts with *meta*-phenylene-diamine. The process is best carried out as follows:—

A standard solution of citral is made containing 0·001 grm. per c.c.

A 1 per cent. solution of meta-phenylene-diamine hydrochloride in dilute alcohol is prepared, well shaken with bone black and filtered.

About 1·5 grms. of oil of lemon are weighed into a 50 c.c. flask, and made up to the mark with 95 per cent. alcohol. From 2 to 4 c.c. of this solution are transferred to a Nessler tube, 10 c.c. of the reagent added, and the liquid made up to 25 c.c. with alcohol. A similar amount of the standard solution is treated in the same way, and the colours developed accurately matched.

The amount of citral is easily calculated from the amount of the standard solution used.

L. D. Little[2] has devised the following method:—

Reagent.—Dissolve ·200 grm. of diaminophenol hydrochloride (commercially known as Amidol) in 100 c.c. of 65 per cent. alcohol, preferably distilled over KOH. The use of aldehyde-free alcohol does not seem to make any difference in the results, as acetaldehyde has no apparent effect upon the reagent. Recently distilled absolute alcohol is sufficiently exact, as the reagent remains clear and colourless for several hours.

Standard Citral Solution.—A solution of pure citral in 50 per cent. alcohol, containing ·001 grm. per c.c.

[1] *Jour. Ind. Eng. Chem.*, 1 **(1909)**, 798.
[2] *Jour. Amer. Pharm. Assoc.* **(1914)**, 3, 553.

Solution of Oil of Lemon.—Weigh from 1 to 2 grms. of the oil and dilute to 50 c.c. with 50 per cent. alcohol.

Place 2 c.c. of the standard citral solution in a 250 mm. colorimeter tube (using preferably a Schreiner colorimeter), add 20 c.c. of 65 per cent. alcohol and 15 c.c. of diaminophenol reagent, and make up to 50 c.c. with 65 per cent. alcohol. Place 2 c.c. of the oil solution in the other tube with 15 c.c. of the reagent and make up to 50 c.c. with 65 per cent. alcohol, mixing the contents of both tubes thoroughly and allowing them to remain for five to ten minutes at rcom temperature, when the maximum colour is reached. The reading and calculation are made at once, or a reading can be made at the expiration of fifteen or twenty minutes in duplicate. The calculation of percentage of citral-content in the extract is made by placing the standard citral tube at the 30 mm. mark and adjusting the tube containing the extract under examination so that the two small disks of colour as observed through the two immersion tubes are similar in tint.

C. Kleber's[1] method, which gives very fair results, depends on the use of phenyl-hydrazine.

Two c.c. of oil, accurately weighed, are mixed with 10 c.c. of 2 per cent. alcoholic phenyl-hydrazine solution, and allowed to stand one hour in a glass-stoppered bottle of 50 c.c.; then 20 c.c. of deci-normal hydrochloric acid and 10 c.c. of benzene are added, and the whole shaken thoroughly. The mixture is transferred to a separator, and the lower layer of 30 c.c. run through a small filter. Twenty c.c. of filtrate are titrated, after the addition of ten drops of ethyl orange 1 : 2000, with deci-normal caustic potash to distinct yellow coloration. By a similar blank experiment without oil the acid equivalent of the hydrazine solution is determined. From the difference between the two experiments the citral-content is calculated, each c.c. of deci-normal caustic potash corresponding to 0·0152 grm. citral. A number of analyses made in this way with mixtures of definite quantities of citral with terpenes were found to be well in accordance with the calculated results.

An absorption method depending upon the formation of a soluble sulphonic acid with sodium sulphite using phenolphthalein as an indicator, gives fairly good results :—

$$C_9H_{15}CHO + 2Na_2SO_3 + 2H_2O$$
$$= C_9H_{17}(NaSO_3)_2COH + 2NaOH.$$

Prof. Tiemann, in his elaborate work on the aldehydes of lemongrass oils has, among other important reactions, shown that if citral be agitated with a neutral solution of sodium sulphite, a sodium salt of the sulphonic acid results, as shown by the above formula. By taking advantage of the sodium hydrate formed by this reaction, phenolphthalein is used as an indicator to show when all the citral has been combined; and the fact that the above sodium salt is quite soluble, has made the estimation of citral fairly accurate and simple. The procedure is as follows: A saturated solution of sodium sulphite is prepared. To 50 c.c. of such solution in a Hirschsohn flask, 25 c.c. of the oil are added, and two drops of an alcoholic solution of phenolphthalein. The whole is then heated on a water-bath to nearly boiling-point, constantly shaking. A deep red colour appears almost at once, which shows that the action has commenced. A few drops of sulphurous acid are then cautiously added, and this is continued until no further colour is pro-

[1] Schimmel's *Bericht*, April, 1912, 64.

duced after a further addition of SO_2. The oil is then measured by adding sufficient sulphite solution to drive the oil into the neck of the flask. The obvious advantage of this method is that the end of the reaction may be ascertained to a certainty, while the above bisulphite method depends on the continual shaking for a period of not less than one hour.

The only objection to this process is that the reading of the meniscus in the neck of the flask is a little difficult owing to the presence of flocculent matter.

Numerous other methods for the determination of citral in lemon oil have been proposed, but the foregoing may be taken to cover the most accurate, and to embrace all that are likely to be of any practical value.

E. M. Chace[1] has published an elaborate study on the detection of traces of pinene in lemon oil, deducing from the presence of even a very

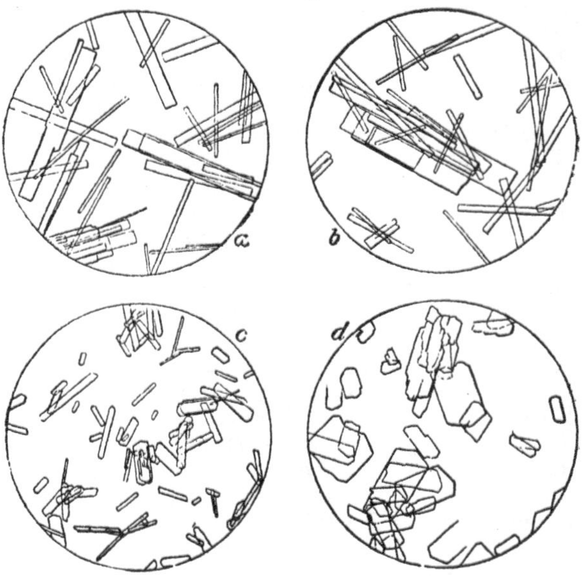

FIG. 39.—Photomicrograph of crystals from lemon oil (× 100).
a, *b*, Limonene nitroso-chloride crystals from lemon oil; *c*, Limonene and pinene nitroso-chloride crystals from a lemon oil mixed with 5 per cent. of turpentine; *d* Pinene nitroso-chloride crystals from turpentine.

minute quantity in the oil that turpentine is present. As pinene appears to be a natural constituent of the oil and adulteration with turpentine would not be practised except on a scale that would render it remunerative, there appears to be no doubt that Chace's deductions are inaccurate, and the presence of traces of pinene cannot be held to indicate adulteration of lemon oil.

Chace points out that the nitrosochlorides of pinene and of limonene crystallise in different forms, and makes use of this property to detect pinene. For this purpose he distils, in a Ladenburg flask, 5 c.c. from 50 c.c. oil, and from this he prepares the nitrosochloride with the use of ethyl nitrite according to Wallach's method. The mixture is left for fifteen minutes in a freezing mixture; the crystals which separate off are then collected, washed with 50 c.c. 95 per cent. alcohol, the mother-

[1] *Jour. Amer. Chem. Soc.*, 30 (**1908**), 1475.

liquor again placed in the freezing mixture for fifteen to twenty minutes, and the crystals which are thus still formed are added to those first obtained. The crystals, well washed with alcohol, are, after drying, dissolved in as little chloroform as possible, and a little hot methyl alcohol is then added so that on cooling crystallisation takes place. Finally, a little more methyl alcohol is added, and the crystals are then filtered off. For the examination under the microscope the crystals are best embedded in oil. In this manner Chace claimed to be able to detect an admixture of even 2 per cent. turpentine oil, whilst as little as 0·5 per cent. turpentine oil could be detected when a three-bulb Glinsky fractionating head was employed.

Böcker[1] has elaborated the following method for the determination of the hydrocarbons present in the concentrated or so-called terpeneless and sesquiterpeneless oils of lemon of commerce :—

Citral is first estimated in 10 c.c. of oil by the sulphite method. When the oil which has not entered into reaction is less than 6 c.c., the estimation is repeated with another 5 or 10 c.c. The oil left over from these tests is bulked, and 5 c.c. of this citral-free oil is placed in a 600 to 700 c.c. separator, into which, immediately previously, 500 c.c. of alcohol of precisely 51 per cent. by volume, cooled down to from 0 to $-$ 2° C., has been introduced. The separator is closed with a cork and the contents are repeatedly shaken, when the aromatic bodies of the oil are dissolved by the alcohol, while the hydrocarbons are left behind almost quantitatively. The separator is then placed in a cooling-bath at 0° C., cork downwards, for six to ten hours. It is then taken out of the freezing mixture, carefully turned back to its proper position and placed in a stand. When the alcoholic solution has cleared to a point at which only a slight film remains (which in certain conditions may take up to two days) it is drawn off to about 10 c.c., any oil-drops which may still adhere to the sides of the funnel being rinsed down with ice-cold 51 per cent. alcohol, so that all the oil which has remained undissolved is brought together. When the mixture has become perfectly clear, the oil, being freed as far as possible from the last traces of alcoholic solution, is transferred to a measuring tube calibrated to $\frac{1}{10}$ c.c., and the vessel rinsed out with a little more ice-cold 51 per cent. alcohol. It is advisable to use a separating funnel of which the lower part ends in a narrow graduated tube. As soon as the oil is completely cleared (when the froth is very persistent a few drops of dilute acetic acid should be added), the volume is read off and calculated to the original oil.

The method may be completed by estimating not only the quantity but also the character of the hydrocarbons. Thus, from 100 to 200 c.c. of oil is fractionally distilled *in vacuo*. The distillation is only continued up to the point where the separate fractions of about 10 c.c. each still show dextro-rotation. These fractions are put together as the "terpene-containing portion" of the oil, the rest constituting the "sesquiterpene-containing portion". Each of these portions is then freed from citral by means of Na_2SO_3, and in both cases the residue is treated with 100 times its quantity of ice-cold 51 per cent. alcohol, as described above, a large glass flask being, if necessary, substituted for the separator. The alcohol solution is separated by means of a siphon from the portion which has remained undissolved. The oils which have been separated are estimated quantitatively, and from the values thus obtained the per-

[1] *Jour. Prakt. Chem.* **(1914)**, 89, 199.

centage of terpenes and sesquiterpenes in the original oil is calculated. For purposes of further identification the rotation is estimated, and, if necessary, the characteristic derivatives are prepared. The terpenes, of which the principal constituent is *d*-limonene, are characterised by a pronounced dextro-rotation; they yield a liberal proportion of limonene tetrabromide, whereas the sesquiterpenes, which consist chiefly of bisabolene, are laevo-rotatory and may be identified from the bisabolene trihydrochloride.

Oil of Orange.

There are two varieties of ordinary orange oil in commerce, almost identical in composition and properties, but with odours and flavours somewhat different. These are the sweet and bitter orange oils, obtained in the same manner as oil of lemon, from the peel of *Citrus aurantium* subsp. *sinensis* and *Citrus aurantium* subsp. *amara* respectively. With regard to the origin of the trees, Prof. Sadebeck (*Plant Cultivation in the German Colonies, and their Products*) says: "The orange tree is probably indigenous to South-eastern Asia, and thence has spread to the Sunda Islands, further India, Persia, Arabia, Syria, North Africa, and Southern Europe. Towards the end of the ninth century of the Christian era, it appeared in Arabia, and since A.D. 1002 in Sicily." An oil is prepared by distillation, but it is much inferior in every respect to the expressed oil. Considerable parcels of West Indian oil have been placed on the London market, the examination of which shows this to be equal if not superior to Sicilian oil.

According to Umney [1] the chief centres of cultivation in the Mediterranean littoral are Hyères, Cannes, Grasse, Nice, Mentone, Bordighera, San Remo, Savone, and Nervi, near Genoa. In Algeria there are many districts where orange trees flourish. Perhaps the most beautiful plantations of Europe are the orange groves of Milis in Sardinia, those of Soller in the island of Majorca, the gardens of Sorrente, near Naples, the Messina plantations near the foot of Mount Etna, and those of Reggio in Calabria.

The sweet orange is largely cultivated in the West Indies and in the Southern States of North America. In Florida it grows wild in great abundance and is also exclusively cultivated for its fruit.

The bitter or Seville orange is known as *Citrus bigaradia* (Duhamel), *Citrus aurantium* var. *amara* (Linnæus), and *Citrus vulgaris* (Risso), of which some twelve varieties are known. The sweet or Portugal orange is known as *Citrus aurantium* (Risso), of which no fewer than nineteen varieties are known.

Orange trees require plenty of air and light to enable them to resist the attacks of parasitic insects and various diseases to which they are subject. Pruning is carried out in May or June, after the flowering period. The inner branches are carefully removed, also dead wood and decayed growth.

The flowers, which expand towards the end of April, furnish on distillation the sweet oil known as neroli oil, and orange flower water as a by-product. Preference is given to the products of the bitter orange, which differs from the sweet orange in its larger and more numerous flowers, the leaves having larger winged petioles and the

[1] *P. and E.O. R.* (**1916**), 246.

fruits having a rougher rind, covered with oil vessels, and containing an acid and very bitter juice.

A tree twenty to thirty years old yields annually from 15 to 20 kilos of flowers. In Grasse alone 200,000 to 300,000 kilos are produced annually. A hundred kilos yield at least 40 kilos of orange flower water and 100 grms. of oil.

At an age of twenty to twenty-five years the trees produce the maximum yield of fruits, usually from 600 to 1000 oranges per annum. In Spain and Algeria trees exist which produce 2000 to 3000 fruits annually. The peel is used for liqueurs and for the manufacture of marmalade, as well as for the production of the essential oil.

There are three distinct types of essential oil obtained from the orange, namely, the flower oil, known as neroli oil, the oil from the leaves and young shoots known as petitgrain oil, and the peel oil, known simply as "orange" oil.

Sweet orange oil, prepared in a similar manner to lemon oil, has the following characters :—

Specific gravity	0·848 to 0·853
Optical rotation	+ 94° ,, + 99°
Refractive index	1·4728 ,, 1·4745

On distilling the first 10 per cent. in a Ladenburg flask, this fraction should have an optical rotation, should be equal to or at most about 1° lower than that of the original oil. The aldehyde-content, calculated as decylic aldehyde, determined by the phenylhydrazine method described under lemon oil, varies from 1·4 to 2·5 per cent.

On evaporation on a water-bath, sweet orange oil yields from 1·5 to 4 per cent. of fixed residue, which has an acid value of 11 to 30 and an ester value of 115 to 170. Spanish orange oil, however, which is expressed from a dark-skinned orange, yields from 7 to 9 per cent. of fixed residue.

West Indian orange oil occasionally has a refractive index as low as 1·4700, but this is an exceptionally low value.

According to Schimmel & Co., Jamaican sweet orange oil contains from 2·3 to 3·2 per cent. of aldehydes calculated as decylic aldehyde from a phenylhydrazine determination, whilst the bitter oil contains only 0·75 to 1·5 per cent.

The oil consists principally of dextro-limonene which is present to the extent of about 90 per cent. or over. Decylic aldehyde is one of the principal odorous constituents of the oil; linalol, *l*-terpineol, *d*-terpineol, nonyl alcohol, methyl anthranilate and traces of caprylic acid esters are also present.

Bitter orange oil only differs very slightly from sweet orange oil, and the composition of the two oils is practically identical. The physical characters of the oils vary to a trifling extent, those of bitter orange oil being as follows :—

Specific gravity	0·850 to 0·856
Optical rotation	+ 90° ,, + 95°
Refractive index	1·4720 ,, 1·4748
Non-volatile residue	2·5 to 4·5 per cent.

The optical rotation of the first 10 per cent. distilled should be higher than, or not more than 1° lower than the rotation of the oil itself.

An orange oil prepared in an abnormal method from Manila oranges has been examined by Gibbs and Agcaoili,[1] according to whom the

[1] *Philipp. Jour. Sci.*, vii., A. 6.

orange commonly found in the Manila market is a green, loose-skinned orange of the mandarin orange type. From the peel the essential oil was obtained by rubbing under water to disintegrate the oil cells, straining the liquid through muslin and extracting the squeezed pulp with petroleum ether. This was then shaken with the watery liquid and separated. This was repeated three times. The petroleum ether was evaporated at a pressure of 12 to 15 mm., at a temperature below 50°, until all the solvent was expelled. The yield and characters of the oil in three cases are as follows:—

No.	Oil from 100 Peels.	Specific Gravity.	Rotation.	Refractive Index.	Ester Number.
1	—	0·8416	+ 58·9°	1·4560	13
2	19·33 grs.	0·8620	+ 83·4°	1·470	7
3	14·0 grs.	0·8481	+ 79°	1·468	10

OIL OF TANGERINE ORANGE.

The Tangerine oil of commerce, also known as oil of Mandarin orange, on account of the custom of presenting the orange in question to the Mandarins in Japan, is obtained by expression from the peel of *Citrus nobilis* (*Citrus madurensis*). The tree is cultivated in Italy, Malta, the Azores, and several other places.

The peel of the fruit yields an essential oil with a slight blue-violet fluorescence, due to the presence of a small quantity of methyl anthranilate. The oil has the following characters:—

Specific gravity	0·854 to 0·859
Optical rotation	+ 67° ,, + 75°
Refractive index	1·4748 ,, 1·4778
Non-volatile residue	2 to 4 per cent.

The optical rotation of the first 10 per cent. distilled is about 2° to 3° higher than that of the original oil.

A Spanish sample of Tangerine oil has been found[1] to have the following characters:—

Specific gravity	0·8665
Refractive index	1·4790
Non-volatile residue	8 per cent.

A number of Mandarin orange oils prepared in Japan have been examined,[2] but as they appear usually to be prepared by distillation they do not always agree in physical characters with Italian oils, nor are they, as a rule, nearly so fragrant. These samples had the following characters:—

No.	Specific Gravity.	Optical Rotation.	Optical Rotation of 1st 10 per Cent. Distilled.
1	0·8489	+ 92° 35′	+ 91° 32
2	0·8528	+ 73° 36′	+ 66° 10′
3	0·8541	+ 68° 5′	+ 60° 48′
4	0·8483	+ 92° 20′	—
5	0·8478	+ 90° 56′	—
6	0·849	+ 90° 35′	—
7	0·848	+ 94°	—

[1] Schimmel's *Bericht*, October, **1911**, 39.

[2] Schimmel's *Report*, October, **1914**, 21; *P. and E.O.R.*, 5 **(1914)**, 72.

The principal constituent of this oil is *d*-limonene. Dipentene is also present. Methyl anthranilate and methyl methyl-anthranilate are also present, and traces of an unknown aldehyde.

The determination of the anthranilic acid esters is of importance in the essential oils of this group which contain it. The only method of any value is that of Hesse and Zeitschel,[1] which is as follows:—

About 25 grams of the oil are dissolved in two to three times this quantity of dry ether and the solution cooled to 0° in a freezing mixture, and a mixture of 1 volume of sulphuric acid in 5 volumes of ether added drop by drop until no further precipitation takes place. The precipitate is collected in a filter and washed with dry ether. The compound of the ester with sulphuric acid is then dissolved in water, and titrated with semi-normal alkali, using methyl orange as indicator. If S grams of oil be used, and A c.c. of semi-normal alkali be required for the titration, then the percentage of methyl anthranilate is

$$\frac{A \times 3 \cdot 775}{S}.$$

Oil of Neroli.

This oil is distilled from the fresh flowers of the bitter and the sweet orange. The former, which is the more valuable, is known as "Oil of Neroli, Bigarade" (which is ordinary neroli oil), the latter as "Oil of Neroli, Portugal". The greater part of the neroli oil of commerce is distilled in the south of France, one of the chief centres being Vallauris. The yield of oil varies greatly according to the time of the season when the flowers are collected. The gathering of the flowers commences about the last week in April and continues till the last week in May, or possibly a little later. The early gathered flowers yield about half a gram of oil per kilo, whilst those gathered late in May yield double this quantity. The petals are separated from the sepals and covered with water in the still, and the contents heated by means of coils of superheated steam. The oil coming over is separated, and the water is sold as orange flower water.

There are numerous orange gardens in Spain, Algeria, Sicily, Greece, and Turkey in Asia, but only a small amount of oil is distilled in these countries. A small quantity of oil, which is equal in perfume value to the oil distilled in France, is prepared in Tunis and Algeria, the principal Tunisian centre being the district of Nabeul, where about 60,000 kilos of flowers are dealt with annually. The distillation is in the hands of Arabs, many of whom use small stills which take only 3 kilos of flowers.

The neroli oil industry is also being well developed in Malaga.

From the port of Malaga northward along the River Guadalhorce right up to Alora, a distance of 50 kilometres, the flowers are grown in great profusion. On some days as much as 6500 kilograms of flowers are dealt with at the distillery established at Alora. Needless to say, however great the arrival of blossom it must be dealt with immediately.

In the photograph reproduced [2] one of the series of stills used at Alora for this work is illustrated. It takes a charge of from 400 to 450 kilos of flowers.

[1] *Berichte*, 34 **(1901)**, 296. [2] From *P. and E.O.R.* **(1913)**, 137.

FIG. 40.—Gathering orange flowers at Bar-sur-Coup, near Grasse. [*Rowe Bertrand Fils.*

Lautier Fils.] FIG. 41.—Spreading the orange flowers prior to distillation. [*Parfumerie Moderne.*

French neroli oil usually has the following characters :—

	Bigarade Oil.	Portugal Oil.
Specific Gravity . . .	0·870 to 0·881	0·865 to 0·870
Optical Rotation . .	+ 1° 30′ ,, + 9° 8′	+ 25° ,, + 45°
Refractive Index . .	1·4675 ,, 1·4738	1·4730 ,, 1·4750
Ester No.	30 ,, 70	about 15

FIG. 42.—Neroli oil still at Alora.

FIG. 43.—Orange garden at Nabeul (Tunisia). [Vial.

These limits, however, are from time to time found to be too narrow, some Bigarade oils being slightly laevo-rotatory, and, occasionally, the ester value being lower than 30. But most oils fall within the above limits.

The following tables, compiled by M. Jean Gras of Cannes, show the characters of oils distilled during two seasons :—

1901 NEROLI OILS.

Distilled on	Specific Gravity at 15°.	Optical Rotation at 22°.	Saponification No.	Solubility in 75 per Cent. Alcohol at 24°.	Yield of Oil per 1000 kilos.
May 23 . .	0·875	+ 5° 2′	39	in two volumes	0·900
,, 24 . .	0·874	+ 4° 28′	40	,, ,,	0·870
,, 25 . .	0·875	+ 5° 25′	41	,, ,,	0·980
,, 26 . .	0·875	+ 5° 10′	37	,, ,,	1·020
,, 27 . .	0·875	+ 5° 45′	34	,, ,,	1·150
,, 28 . .	0·875	+ 5° 46′	37	,, ,,	1·160
,, 29 . .	0·874	+ 6° 15′	31	,, ,,	1·200
,, 30 . .	0·874	+ 4° 55′	33	,, ,,	1·120
,, 31 . .	0·875	+ 5°	34	,, ,,	1·150
June 1 . .	0·874	+ 6°	32	,, ,,	1·020
,, 2 . .	0·873	+ 4° 40′	36	,, ,,	1·210
,, 3 . .	0·875	+ 7°	39	,, ,,	1·400
,, 4 . .	0·876	+ 6° 46′	34	,, ,,	1·500

1904 NEROLI OILS.

Distilled on	Yield per Cent.	Specific Gravity at 15° C.	Optical Rotation l=100 mm. t=20°.	Saponification No.	Solubility in Alcohol of 80 per Cent. at 22°.
May 11	0·65	0·874	+1° 31′ (May 11–12)	43	1 vol. in 1 vol. alcohol, insoluble when more alcohol was added (May 11–12)
,, 12	0·80	0·874			
,, 13	0·89	0·874	+1° 45′	45	1 vol. in 1·5 vol. alcohol, insoluble when more alcohol was added
,, 14	0·81	0·875	+2° 8′	46	insoluble
,, 15	0·83	0·874	+1° 32′	45	do.
,, 16	0·78	0·874	+1° 33′ (May 16–17)	47	do.
,, 17	0·87	0·874			
,, 18	0·77	0·875	+2° 8′	53	do.
,, 19	0·77	0·875	+1° 45′	56	do.
,, 20	0·77	0·875	+2° 20′	45	do.
,, 21	0·87	0·875	+2° 21′	47	1 vol. in 1 vol. alcohol, insoluble when more alcohol was added
,, 22	0·90	0·876	+2° 21′	46	1 vol. in 1·2 vol. alcohol, insoluble when more alcohol was added
,, 23	1·00	0·875	+2° 30′	43	do.
,, 24	1·03	0·874	+2° 44′	45	1 vol. in 1 vol. alcohol, insoluble when more alcohol was added
,, 25	1·06	0·874	+3° 5′	46	do.
,, 26	1·07	0·874	+3° 8′	44	do.
,, 27	1·07	0·874	+3° 30′	45	do.
,, 28	1·02	0·874	+3° 12′	47	do.
,, 29	1·11	0·874	+3° 31′	48	do.
,, 30	1·15	0·874	+3° 48′	50	1 vol. in 1 vol. and in every proportion
,, 31	0·91	0·874	+2° 51′	48	do.
June 2	0·80	0·875	+4° 20′	49	do.
Total Quantity	0·97	0·875	+2° 50′	47	1 vol. in 1 vol. alcohol, insoluble when more alcohol was added

Roure-Bertrand Fils[1] state that certain constituents of neroli oil are destroyed by steam distillation in the ordinary method. They state that slight modification in the distillation process produces distinct alterations in the physical characters of the oil, especially the optical rotation. They have therefore carried out an exhaustive examination of neroli oil made by an extraction process, with subsequent distillation.

FIG. 44.—Arabian still in Tunisia (native distillation of neroli oil). [*Viel.*

For this purpose they treated in May and at the end of October about 300 kilos of flowers.

The flowers were extracted by three washings with petroleum spirit. The three solutions were united, dried, and filtered, then the solvent was distilled off, care being taken to expel the last traces *in vacuo*. The residue from this distillation is a brown, viscous mass composed of vegetable wax, the colouring matter and the essential oil of the flower.

This brown residue was then subjected to distillation in a current of steam. In a suitable rectifying apparatus the essential oil alone is carried over, free from any waxy matters. The water condensed in the course of this operation is exhausted with ether, after saturation with common salt, and the evaporation of the ether leaves a residue which is termed the *soluble essential oil*. The portion which floats on the surface of the condensed water and which is isolated by simple decantation is termed the *direct essential oil*; the name of *total essential oil* is given to that which is obtained by mixing the two former oils in proportion to their yields.

The characters of the essential oils so obtained are set out in the following tables:—

(*a*) *Direct Essential Oils.*

	May.	Autumn.
Density at 15° C.	0·8883	0·8866
Index of Refraction n_d^{15}	1·474	1·469
Rotatory Power	– 0° 40′	– 4° 28′
Solubility in 80 per Cent. Alcohol	1 vol. and over	1 vol. and over
Acid Value	0·7	0·7
Coefficient of Saponification	71·4	96·6
Total Esters (as Linalyl Acetate)	25·0 per cent.	33·8 per cent.
Coefficient of Saponification of the Acetylated Oil	172·2	185·9
Total Alcohol (as Linalol)	54·4 per cent.	59·3 per cent.
Methyl Anthranilate	2·43 ,,	1·57 ,,

[1] *Bulletin*, April, **1910**, 43.

(*b*) *Soluble Essential Oils.*

	May.	Autumn.
Density at 15° C.	0·9434	0·9124
Index of Refraction n_d^{15}	1·498	1·483
Rotatory Power (l = 100 mm.)	– 1° 46′	– 3° 20′
Solubility in 80 per Cent. Alcohol	1 vol. and over	1 vol. and over
Acid Value	0·7	1·05
Coefficient of Saponification	61·6	58·8
Total Esters (as Linalyl Acetate)	21·5 per cent.	20·6 per cent.
Coefficient of Saponification of the Acetylated Oil	164·7	173·7
Total Alcohols (as Linalol)	51·6 per cent.	54·8 per cent.
Methyl Anthranilate	13·2 ,,	12·4 ,,

(*c*) *Total Essential Oils.*

	May.	Autumn.
Density at 15° C.	0·8899	0·8887
Index of Refraction n_d^{15}	1·478	1·476
Rotatory Power (l = 100 mm.)	– 0° 48′	– 4° 6′
Solubility in 80 per Cent. Alcohol	1 vol. and over	1 vol. and over
Acid Value	0·7	1·0
Coefficient of Saponification	70·2	95·8
Total Esters (as Linalyl Acetate)	24·6 per cent.	33·4 per cent.
Coefficient of Saponification of the Acetylated Oil	163·1	179·5
Total Alcohol (as Linalol)	51·0 per cent.	57·0 per cent.
Methyl Anthranilate	3·53 ,,	2·74 ,,

Spanish neroli oil differs in character from the French distillate. The figures for two typical samples examined by Umney[1] are as follows:—

SPANISH NEROLI OIL.

(Distilled at Alora.)

Specific gravity	·860
Optical rotation	+ 25°
Esters	10·9 per cent.

Solubility 1 in 1 vol. of 90 per cent. alcohol. Not soluble in 80 per cent. alcohol.

SPANISH NEROLI OIL.

(Distilled at Malaga.)

Specific gravity	·861
Optical rotation	+ 24°
Esters	8·2 per cent.

Solubility 1 in 1 vol. of 90 per cent. alcohol. Not soluble in 80 per cent. alcohol.

Spanish neroli oils examined by Schimmel gave the following results:—

[1] *P. and E.O.R.* (**1913**), 137.

	Bigarade (Bitter).	"Portugal" (Sweet).
Specific gravity	0·871	0·857
Optical rotation	+ 10° 54′	+ 42° 47′
Acid value	1·37	1·6
Ester value	37·67	6·86
Methyl anthranilate	0·5 per cent.	0·3 per cent.

Spanish oils have a specific gravity up to 0·885 and an optical rotation varying from + 9° to + 29°.

Various other oils distilled in different parts of the world have been examined with the following results:—

Sicilian Oil.

This oil has the following characters:—

Specific gravity	0·860 to 0·924
Optical rotation	+ 2° 54′ ,, 56° 30′
Refractive index	1·4680 ,, 1·4740
Ester value	6 ,, 127

Algerian Oil.

Specific gravity	0·872 to 0·877
Optical rotation	+ 5° 42′ ,, + 6° 6′
Ester value	72 ,, 91

Syrian Oil.

Specific gravity	0·8758
Optical rotation	+ 1° 6′
Ester value	51·5

Umney and Bennett have examined a Chinese neroli oil derived from *Citrus triptera*. It had a specific gravity 0·850 and optical rotation + 35°. It contained 4·8 per cent. of esters calculated as linalyl acetate, and 21 per cent. of free alcohols as linalol. It contained much limonene, camphene, linalyl acetate, methyl-anthranilate and a paraffin.

Methyl anthranilate, determined as described under tangerine orange oil, is present to the extent of 0·2 to 1·5 per cent. in ordinary neroli oil, or 10·5 to 16 per cent. in the oil extracted from the distillation water.

Neroli oil contains *l*-*α*-pinene, *l*-camphene, limonene, dipentene, decyl aldehyde, phenyl-ethyl-alcohol, nerol, linalol, *d*-terpineol, nerolidol, geraniol, acetic esters of the various alcohols present, indol, traces of benzoic, palmitic, and phenyl-acetic acids in the form of esters, linalyl acetate, and one or more paraffin hydrocarbons. None of these bodies, however, are responsible for the characteristic fluorescence of the oil. Ernst and Hugo Erdmann [1] have examined the oil, and found that the portion which distils about 115° at a pressure of 10 mm. leaves on saponification a crystalline acid of melting-point about 140°. This acid was completely identified as anthranilic acid (*o*-amidobenzoic acid), and it occurs in the oil as its methyl ester. The pure body, methyl-*o*-amidobenzoate, is an oil solidifying at low temperatures in crystals, melting at 24·5° and boiling at 127° at 11 mm. Its specific gravity is 1·163 and 26°. It is easily soluble in dilute mineral acids (on account of its basic character), alcohol or ether, and yields a crystalline hydrochloride. It is strongly fluorescent. Schimmel & Co. have found in the oil obtained by

[1] *Berichte* (**1899**), 1213.

the extraction of the fresh flowers with petroleum, and extracting the petroleum extract with alcohol, traces of phenyl aceto-nitrile and a crystalline body melting at 158°, a ketone with a jasmine-like odour, possibly identical with jasmone, and a high boiling sesquiterpene alcohol (nerolidol ?), with traces of a base apparently belonging to the nicotine series.

OIL OF PETIT-GRAIN.

The original oil of petit-grain was obtained by the distillation of the small unripe orange berry, which, according to Pereira, rapidly underwent decomposition. The petit-grain oil of to-day, however, is distilled from the leaves and young shoots of the bitter orange.

The best petit-grain oil is distilled in the south of France, a certain amount also in Algeria and Spain, and a large amount of less perfume value is distilled in Paraguay, which form the bulk of the oil of commerce.

The distillation in Paraguay is carried on in small factories in the immediate vicinity of orange forests, as the total absence of all means of communication renders it impossible to centralise the production. Every group of factories is managed and superintended by one employee. The woods have always the character of timber forests of the most diverse varieties, among which is found a sort of undergrowth of smaller dimensions, with patches almost completely covered with orange trees, which the inhabitants call *manchones*. Such *manchones* are scattered about the forests at certain distances.

In specially favourable situations the *bigaradier* predominates, in others the latter is found mixed with the sweet orange and with another species called *assessù*, which is a variety of the two first-named. This species is used preferably for the manufacture of petit-grain oil.

Although the orange forests are very numerous in Paraguay, they afford but rarely the preliminary conditions necessary for normal working, as they are mostly far removed from inhabited centres. The leaves are gathered all the year round, but chiefly between October and April, during which period the bulk of the oil is produced. For this purpose the trees are cut down at a height of about 2 ins. from the ground ; the leaves and young fruit are distilled together, and the wood of the trees is used as fuel for the boiler.

During the past few years a number of samples of Paraguay petit-grain oil have been found which had an optical rotation up to + 8° or + 9°. This appears to be due to the distillation material containing small, immature fruits, the essential oil of which contains much dextro-limonene. Normal Paraguay petit-grain oil has the following characters :—

Specific gravity	0·885 to 0·900
Optical rotation	− 3° to + 11° (usually under + 5°)
Refractive index	1·4590 to 1·4655
Acid value	1 „ 3
Ester „	105 „ 166

The oil is soluble in 1 to 2 volumes of 80 per cent. alcohol, but in the case of oils with a high dextro-rotation, the presence of limonene may cause a slight turbidity in the solution. Subject to the same consideration, the oil is soluble in 3 to 5 volumes of 70 per cent. alcohol.

French petit-grain oil has the following characters :—

Specific gravity	0·888 to 0·897
Optical rotation	− 1° to − 8°
Esters	50 to 70 per cent.

Schimmel & Co.[1] give the following as the characters of petit-grain oils of various origins :—

Origin.	Specific Gravity.	Rotation.	Esters.
Calabria	0·8746	+ 18°	26·3 per cent.
Spain	0·8849	+ 8° 25′	28·6 ,,
Jamaica	0·8846	− 6° 30′	28·7 ,,
West Indies	0·8531	+ 43° 36′	2·1 ,,
South America	0·887	+ 2°	36·5 ,,
Syria	0·8857	− 3° 24′	27·1′ ,,

It is obvious that the W. Indian sample is not a normal petit-grain oil at all, but has been distilled chiefly from immature orange fruits.

The twigs and leaves of the lemon, lime, and mandarin trees also yield similar oils, which have been termed "petit-grain oils". The lemon product has been termed "oil of petit-grain citronnier".

Roure-Bertrand Fils[1] have prepared four different petit-grain oils from the branches of the bitter orange, the sweet orange, the lemon, and the mandarin-tree, and have found these oils to possess the constants given below :—

	Sweet Orange.	Bitter Orange.	Lemon.	Mandarin.
$d_{15°}$	0·8930	0·8584	0·8806	1·0643
α_D	− 4° 20′	+ 53° 52′	+ 30° 44′	+ 2° 20′
Saponification Value	161·4	12·5	43·8	265·0
Ester Value after Acetylation	203·0	47·4	130·6	—
Solubility	In 3·5 vols. 70 per cent. Alcohol.	In 0·5 vols. 90 per cent. Alcohol.	In 1 vol. 85 per cent. Alcohol.	In 1 vol. 90 per cent. Alcohol.

The oil distilled from the corresponding parts of the lemon tree has an odour resembling that of petit-grain oil, with a secondary odour of lemon. It has the following characters :—

Specific gravity	0·870 to 0·895
Optical rotation	+ 14° ,, + 35°
Ester value	14 ,, 46
Aldehydes	20 to 30 per cent.

It contains camphene (?), limonene, geraniol, linalol (?) and citral.

Jeancard and Satie give the following figures for the "petit-grain" oil of lemon :—

	Specific Gravity.	Rotation.	Esters.
Lemon Petit-grain	0·8768	+ 13° 20′	12·25 per cent.

Mandarin petit-grain oil has the following characters :—

Specific gravity	0·990 to 1·064
Optical rotation	+ 3° ,, + 8°
Saponification value	150 ,, 190

The oil appears to vary in character to a considerable extent according to the proportion of twigs and leaves present in the distillation material. This oil may contain 65 per cent. of methyl-anthranilate.

[1] *Bulletin*, October, **1910**, 42.

Lime "petit-grain" oil has the following characters:—

Specific gravity	0·875 to 0·878
Optical rotation	+ 35° ,, + 40°
Ester value	23 ,, 28

An oil distilled from the unripe orange fruits, analogous to the old-fashioned petit-grain oil, is manufactured and offered under the name of "oil of orange peas". Its specific gravity is about ·852 to ·854 and its optical rotation + 75°. It is intermediate in composition between orange oil and neroli oil. It contains methyl-anthranilate and also traces of a pyrrol compound.

Umney and Bennett have examined a Buenos Ayres oil resembling ordinary South American petit-grain oil, but having more free alcohols and less esters than those oils. Pinene, dipentene, furfural, geraniol, linalol, and geranyl acetate were found to be present. Its specific gravity was 0·887, optical rotation + 2°, esters 36·5 per cent., and free alcohols as geraniol 38·4 per cent.

Terpinyl acetate has been used recently as an adulterant of petit-grain oil, and in 1912 a serious dispute took place as to the purity of a sample suspected of containing this adulterant, several chemists holding antagonistic opinions on the subject. The sample was eventually referred to the author, and terpinyl acetate was found to be present, the following method of investigation being adopted.

The oil had the following characters:—

Specific gravity at 15·5°	0·8965
Optical rotation	+ 1° 10′
Refractive index at 20°	1·4650
Ester value (as linalyl acetate)	40·5 per cent.

The peculiar odour of the oil suggested the presence of an added ester, the indication being that of terpinyl acetate.

The oil was examined, in the first place by means of a series of fractional saponifications, and secondly by the fractional distillation of a large quantity of the oil.

The three saponifications were carried out under identical conditions, except that the times allowed for the reaction were 30 minutes, 75 minutes, and 120 minutes respectively. The results obtained were as follows:—

	S.V.
30 minutes' saponification	115·7
75 ,, ,,	122·5
120 ,, ,,	124·8

The increase in values by prolonging the time of saponification is a strong indication of terpinyl acetate, which is, as is well known, an ester much more resistant to the action of alcoholic potash than linalyl acetate.

A much more marked indication is found by taking the differences in the saponification values as suggested by Schimmel & Co. About 2 to 3 grams of the oil are saponified—(1) with 20 c.c. of semi-normal potash for two hours; (2) the same amount of oil is used, but only 10 c.c. of semi-normal potash, diluted with 25 c.c. of alcohol, and the saponification allowed to proceed for one hour. Under these conditions Messrs. Schimmel & Co. have shown that pure or nearly pure terpinyl acetate will show a difference of practically 90 in the saponification value.

On the sample in question the following results were obtained:—

(1) Two hours' saponification	124
(2) One hour's ,, ...	101·5
Difference	23·5

A large quantity of the oil was then repeatedly fractionated under reduced pressure. The difference in boiling temperatures of this group of alcohols and esters is considerably less under very low pressure than at normal pressure. Attempts to fractionate at normal pressure were sufficiently definite to strongly support the whole of the preceding results, although the characters of the fractions in question only showed approximations to those of terpinyl acetate. After a series of fractionations a fraction was finally obtained boiling at 217° to 221°, which had a clear and definite odour of terpinyl acetate, and whose specific gravity was 0·917, and refractive index 1·4773 at 20°. A fractional saponification of this fraction, by the method above described, gave the following results :—

Two hours' saponification	192
One hour's ,, ...	160
Difference	32

By reducing the pressure to about 11 to 14 mm., and collecting the fractions boiling from 100° to 118° and refractionating three times, a fraction was obtained boiling at 110° to 115°, which had the following characters :—

Specific gravity	0·9205	
Refractive index	1·4781	
Saponification value (two hours)		199
,, ,, (one hour)		164
Difference		35

Similar fractionations were made with the saponified oil, and fractions were obtained having a marked odour of terpineol, although the body could not be obtained in a pure condition.

Petit-grain oil contains camphene, *β*-pinene, dipentene, limonene, *l*-linalol, *d*-*a*-terpineol, geraniol, esters of linalol and geraniol, a sesquiterpene, and methyl anthranilate.

Oil of Limes.

There are three distinct oils found in commerce, the one, hand-pressed oil obtained from the *Citrus limetta* (Italian), the others being the distilled and hand-pressed oils respectively from *Citrus medica var. acida* (West Indian). Expressed oil is prepared from the fruit in December and January. In the south-west of Italy the tree is known as *limoncello di Spagna*. The Italian oil somewhat resembles lemon oil, but with a much richer odour, somewhat recalling the odour of bergamot. It contains pinene, limonene, and probably dipentene, linalol, linalyl acetate, and citral. It has the following constants :—

Specific gravity	0·872
Optical rotation	+ 58°
Saponification value	75

Adulterants such as lemon oil and terpenes are frequently used and may be detected by distillation tests as given for lemon and orange oils. Distilled oil of limes, the usual oil of commerce, is obtained from Jamaica, Dominica, and Tahiti, but chiefly from the island of Montserrat, the harvest lasting from September to January. The oil is really a by-product obtained by distillation during the concentration of the lime-juice, and the peculiar odour is due to the partial decomposition of the oxygenated constituents. Citral decomposes into cymene when boiled with dilute acids. It contains pinene, limonene, cymene, dipentene, bisabolene, and *l*-terpineol, melting-point 35°, and another alcohol having a strong lime-like odour forming a urethane, melting-point 131°, but not yet identified. Distilled lime oil has the following constants :—

Specific gravity	0·860 to 0·872
Optical rotation	+ 33° „ + 47°
Refractive index	1·4700 „ 1·4715
Aldehydes	1 to 2 per cent.

Hand-pressed West Indian limes oil is of considerably higher value than the distilled oil. Its characters are as follows :—

Specific gravity	0·878 to 0·902
Optical rotation	+ 30° „ + 38°
Refractive index	1·4820 „ 1·4860
Fixed residue	9 to 18 per cent.
Citral	6 „ 9 „

The fixed residue has a saponification value of 160 to 180. This oil contains citral, bisabolene, methyl anthranilate, and citraptene.

The leaves of the West Indian lime tree yield an oil on distillation, which has the following characters :—

Specific gravity	0·877 to 0·878
Optical rotation	+ 37° „ + 38°
Acid value	3 „ 5
Ester „	22 „ 25
Aldehydes	43 per cent.

The author has examined an oil distilled from the flowers of the Italian lime tree, and found it to have the following characters :—

Specific gravity	0·870
Optical rotation	+ 21° 30′

It contains linalol and methyl anthranilate.

Oil of Bergamot.

This oil is obtained by expression from the fresh peel of the fruit of *Citrus aurantium* subsp. *Bergamia*, the ordinary bergamot. The chief centre of the industry is Southern Calabria, which is practically the only province in which the tree is cultivated, extensive plantations being situated at, and in the neighbourhoods of, Reggio, Melito, Gallico, Arangea, Sancta Catarina, S. Lorenzo, Palizzi, Staiti, etc. Almost all the oil is expressed by machines, but as these are usually adjusted for globular fruits, any that have the oblong lemon shape are treated by hand. Hand-pressed oil is, however, rarely met with in commerce. The trees are best grown on well-irrigated, low-lying lands, and are frequently cultivated amongst the lemon and orange trees. The fruit is gathered in the months of November and December and the first week or two in January, when the manufacturing season is over. The resulting oil is of a green colour, due to the presence of chlorophyll (not to the presence of copper, as sometimes stated).

The bergamot tree is a very delicate one, and any inclemency of the weather will adversely affect the fruit very seriously. Excess of either

FIG. 45.—Manufacture of the oil of Bergamot in Calabria. [*Roure-Bertrand Fils.*

moisture or drought, too much sun, mist, or early autumnal rains will sometimes change the whole aspect of the crop in a few days.

Bergamot oil is one which is adulterated with considerable skill, the use of artificial esters being fairly common for the purpose. A comprehensive examination is therefore necessary, as the physical characters can be, and frequently are, standardised to those of a pure oil.

A pure bergamot oil has the following characters:—

Specific gravity	0·881 to 0·886 (rarely 0·880 to 0·887)
Optical rotation	+ 8° ,, + 22° (in some seasons to + 25°)
Refractive index	1·4640 to 1·4675
Acid value	1 ,, 4
Esters as linalyl acetate	34 per cent. to 42 per cent. (rarely to 45 per cent.)

On evaporation of the oil on a water-bath it leaves a non-volatile residue of 4·5 to 6·5 per cent. (rarely a little outside these limits). This non-volatile residue should have an acid value of 18 to 30, and an ester value of 115 to 180.

The above characters may be somewhat altered if fallen, unripe fruits be used in the preparation of the oil. Oil from such fruits may have a specific gravity as low as 0·879, and an ester content of 26 to 33 per cent. Under the name "Nero de Bergamotto" is understood a bastard bergamot (Bergamottella), which, if mixed with the ordinary fruits when pressed, will raise the specific gravity and lower the ester value. The oil from these bastard fruits has a specific gravity 0·890 to 0·898, and an ester content of 20 to 35 per cent.

Distilled bergamot oil, obtained by distilling the residues from the expression process, is of very low odour value, having a specific gravity as low as 0·865 and an ester content which may fall as low as 5 per cent.

The principal adulterants used to-day consist of artificial esters, with or without a neutral oil such as lemon terpenes. Esters of the non-volatile acids, or which are themselves of low volatility will be found in the evaporation residue of the oil, which if over 6 per cent. will at once arouse suspicion. The saponification value of this residue will therefore be increased if such esters are present. If the saponification value is over 185, it is almost certain, and if it is over 195, quite certain, that such an adulterant is present. Ethyl citrate is the principal ester of this type used as an adulterant. One per cent. of triethyl citrate added to the oil will raise the saponification value enormously—for example, from 175 to 220.

In determining the ester value of the oil, it should be carefully watched for the first five or ten minutes. The clear liquid will very soon become turbid, on account of the separation of potassium citrate (which is not very soluble in alcohol), if citric acid esters are present, and eventually become clear again. Denigès[1] has devised a test by which the presence of citric acid can be positively decided.

Denigès' test consists in vigorously shaking 10 c.c. of the solution with 1 to 1·5 grams peroxide of lead, adding 2 c.c. solution of mercuric sulphate, and filtering. Of the filtrate, 5 c.c. is heated to boiling-point, after which a 2 per cent. solution of permanganate of potassium is added by drops, with stirring, until the latter ceases to become decolorised immediately. If citric acid is present a flocculent, white or pale yellow precipitate shows itself after the first drop.

[1] *Bull. Soc. Ph.*, Bordeaux (**1898**), 33.

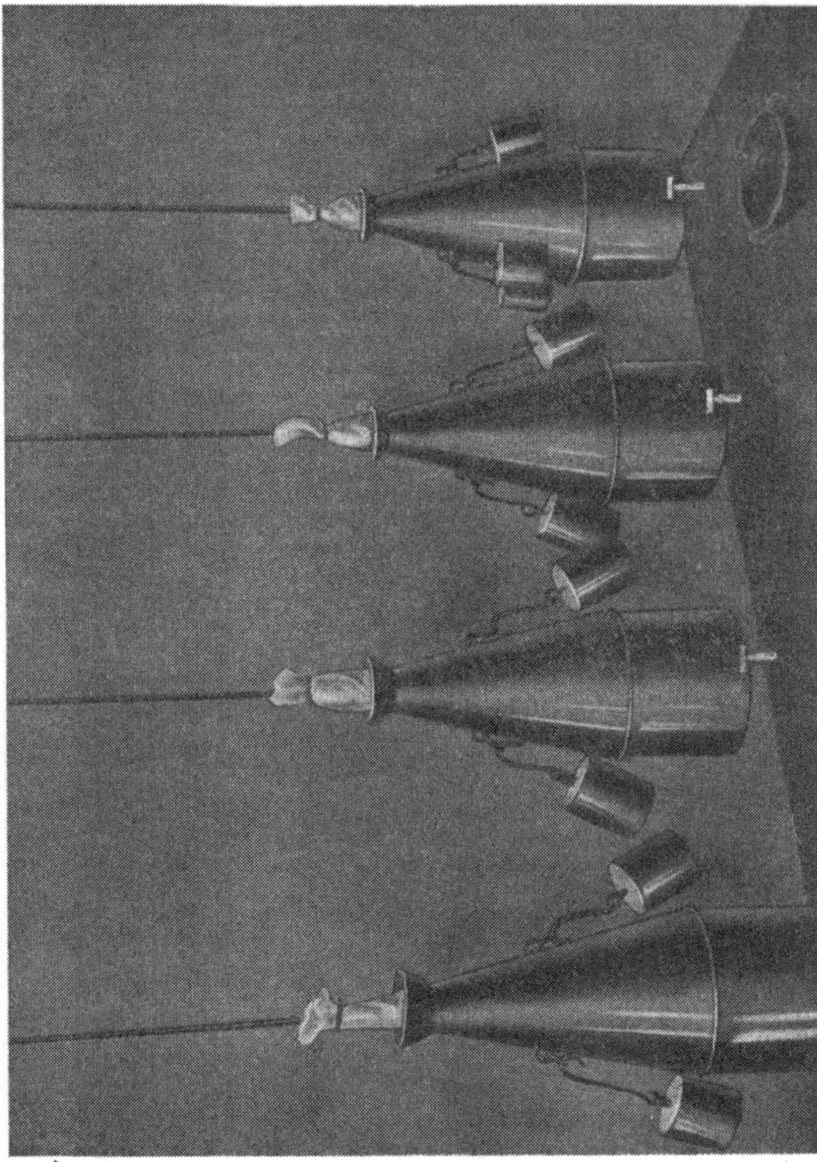

FIG. 46.—Filtering apparatus for the oil of Bergamot.

[*Roure-Bertrand Fils.*

Citric acid can also be detected by precipitating as calcium citrate. This test is carried out by saponifying about 2 grams of the oil, or the residue of evaporation of 5 grams of oil, with alcoholic potash, diluting the solution with water, neutralising with hydrochloric acid, removing the alcohol on the water-bath, and then extracting the solution with ether, and filtering. For the calcium test the solutions are rendered faintly alkaline with a trace of caustic soda, after which a few drops of concentrated chloride of calcium solution are added and the whole boiled. Any citric acid which is present is shown by a precipitate, which is often formed only after some time.

Oxalic acid, resulting from the use of ethyl oxalate as an adulterant, is similarly detected, the calcium precipitate then having the usual characters of calcium oxalate.

With the exception, however, of triethyl citrate, nearly all the non-volatile esters are to a great extent removed mechanically during the evaporation in the oil vapour. To detect, for example, glyceryl acetate, ethyl oxalate, ethyl tartrate, and ethyl succinate, the evaporation should only be proceeded with until about 10 to 15 per cent. of the oil is left in the dish, and this residue tested for the various acids.

Terpinyl acetate is indicated by a difference to be observed in the apparent ester value by different times of saponification. This ester is far more resistant to the action of caustic alkali than is linalyl acetate, and requires two hours at least for complete saponification. Hence, if the oil shows a difference in the saponification value in thirty minutes and in two hours, which amounts to more than from 1 to 2, terpinyl acetate is almost certainly present. The following table[1] shows the effect of this partial-saponification on the two esters and on adulterated oils :—

Time of Saponification.		5 mins.	15 mins.	30 mins.	45 mins.	1 hr.	2 hrs.
Linalyl Acetate	E. No.	191·5	217·5	**223·2**	223·7	**223·1**	224·7
Terpinyl ,,	,,	108·2	166·8	**209·7**	233·4	**245·8**	262·7
Bergamot Oil	,,	80·3	94·5	**97·3**	97·5	**97·8**	98·5
,, ,, + 5 °/。 Terpinyl Acetate	,,	82·5	94·8	**101·2**	102·1	**104·7**	107·2
,, ,, + 10 °/。 ,, ,,	,,	79·9	96·4	**102·8**	105·2	**108·3**	112·5
,, ,, + 25 °/。 ,, ,,	,,	78·8	100·6	**108·1**	116·4	**119·0**	126·8

Fractional saponification, with the use of varying amounts of caustic alkali, will also reveal the presence of terpinyl acetate.

The following table will indicate the differences observed when about 2·5 grams of the oil are saponified (1) with 20 c.c. of $\frac{N}{2}$ alkali for two hours, and (2) with 10 c.c. of $\frac{N}{2}$ alkali, diluted with 25 c.c. of alcohol for one hour :—

[1] Schimmel's *Report*, October, 1910, 60.

Oil.	20 c.c. × 2 hours.	10 c.c. (and 25 c.c. Alcohol) × 1 hour.	Difference.
Bergamot (1)	100·5	98·6	1·9
Bergamot (2)	108	105·5	2·5
,, with 5 per Cent. Terpinyl Acetate	117	102·5	14·5
,, with 10 per Cent. Terpinyl Acetate	121	104·0	17·0

The above table represents the behaviour on fractionation at 3 mm. pressure of two samples of bergamot adulterated with terpinyl acetate and a sample of pure bergamot oil.

The author [1] has recommended the examination of the last 10 per cent. left on evaporation of the oil on a water-bath, since the heavy artificial esters accumulate in this fraction. The refractive index of this 10 per cent. should not be below 1·5090, and the saponification value should not exceed 190. The figures given on p. 446 represent nine samples of adulterated oil, all sold as genuine bergamot oil.

Glyceryl acetate, which is an artificial ester commonly used in the adulteration of bergamot oil, is detected fairly easily on account of its high solubility in dilute alcohol. The test is carried out as follows [2]: Ten c.c. of bergamot oil and 20 c.c. of 5 per cent. alcohol are well shaken in a separating funnel, and after the solutions have separated and become clear the watery solution is run off and filtered. Ten c.c. of the filtrate are exactly neutralised with deci-normal alkali, and then 5 c.c. of semi-normal alkali run in, and the whole saponified under a reflux condenser for one hour. In the case of pure bergamot oil 0·1 or at most 0·2 c.c. of semi-normal alkali will have been used up by the saponification, whilst each 1 per cent. of glyceryl tri-acetate present in the oil will be represented by practically 0·5 c.c. of semi-normal alkali.

Schimmel & Co. have proposed to detect esters of fixed acids by an estimation of the amount of volatile acids obtained by distilling the acidified saponification residues, and comparing this figure with the amount of acid indicated by the saponification value.

In this determination about 2 grams are saponified in the usual manner, and the saponification residue rendered slightly alkaline, and evaporated to dryness on a water-bath. The residue is dissolved in 5 c.c. of water and acidified with 2 c.c. of dilute sulphuric acid. This liquid is now distilled by passing a current of steam through it, and when no further acid comes over, the distillate (about 300 c.c.) is titrated with deci-normal alkali, using phenolphthalein as indicator. The alkali consumed in this neutralisation is nearly identical with that used in the direct saponification, if all the esters present are those of volatile acids, as is the case with pure bergamot oil. The distillation value should not be more than 5 to 10 below the direct saponification value (i.e. milligrams of KOH per 1 gram of oil). When esters of non-volatile acids have been used as adulterants the difference is enormous. For example, an oil containing 2 per cent. of ethyl citrate yielded a direct saponification value of 109·1

[1] *P. and E.O.R.* (1911), 14. [2] Schimmel's *Bericht*, April, 1911, 151.

Fraction.	I.			II.			Pure Bergamot Oil.		
	Per Cent.	α_D.	$n_{D20°}$.	Per Cent.	α_D.	$n_{D20°}$.	Per Cent.	α_D.	$n_{D20°}$.
1. to 40°	2	+ 36° 35′	1·47225	10·6	+ 52° 34′	1·47235	18·6	+ 58° 16′	1·47303
2. 40 ,, 50°	38·8	+ 60° 5′	1·47274	28·5	+ 64° 47′	1·47264	18·9	+ 68° 51′	1·47245
3. 50 ,, 68°				3·9	+ 8° 35′	1·46664	5·6	+ 17° 15′	1·46545
4. 68 ,, 72°	9·1	– 11° 32′	1·46030				21·4	– 15° 20′	1·45781
5. 72 ,, 78°				8·7	– 11° 20′	1 46001	20·2	– 11° 16′	1·45331
6. 78 ,, 82°	12·1	– 8° 56′	1·45871	14·8	– 6° 12′	1·45991	—	—	—
7. 82 ,, 88°	20·1	– 3° 42′	1·46011	19·5	– 2° 16′	1·46229	—	—	—
8. 88 ,, 91°	5·4	– 1° 30′	1·46387	—	—	—	—	—	—
9. Residue	12·5	—	—	14·0	—	—	15·3	—	—

	1.	2.	3.	4.	5.	6.	7.	8.	9.
Specific Gravity	0·885	0·884	0·884	0·8855	0·886	0·886	0·886	0·885	0·878
Optical Rotation	+ 16°	+ 16° 30′	+ 16° 30′	+ 18°	+ 23°	+ 17°	+ 20°	+ 26°	+ 27°
Refractive Index at 20°	1·4660	1·4660	1·4662	1·4658	1·4660	1·4660	1·4681	1·4675	1·4691
Apparent Esters as Linalyl Acetate	39 %	39·5 %	41 %	39 %	38·8 %	39·5 %	40 %	41 %	34 %
Fixed Residue	6·5 %	6·3 %	4·2 %	6·4 %	6·1 %	7·2 %	4·5 %	6·9 %	4 %
,, ,, Saponification Value of	257	225	160	239	236	252	172	242	162
,, ,, of Saponified Oil	5·3 %	5 %	4·1 %	4·9 %	4·7 %	5·4 %	4·2 %	5·4 %	4 %
Refractive Index of last 10 per cent.	1·5040	1·5042	1·5085	1·5042	1·5040	1·5050	1·5070	1·5050	1·5091
Increase in Ester Value in one hour	—	—	3·8 %	—	—	—	4·1 %	—	—

All these were adulterated with ethyl citrate except Nos. 3 and 7, which contained terpinyl acetate, and No. 9, which was adulterated with lemon terpenes.

and a distillation value of 92·8, and one containing 5 per cent. of ethyl succinate gave a direct value of 127·6 and a distillation value of 91·5.

Umney[1] has made a critical study of this method, and recommends the following apparatus to be used in the process (see Fig. 47):—

(*a*) A 3 litre Jena glass flask.

(*b*) A rubber connection, the removal of which, of course, immediately cuts off the steam supply.

(*c*) A long-necked CO_2 flask of Jena glass and 150 c.c. capacity.

(*d*) The most suitable splash head for the operation.

(*e*) A Davies' condenser.

(*f*) A 500 c.c. Erlenmeyer flask.

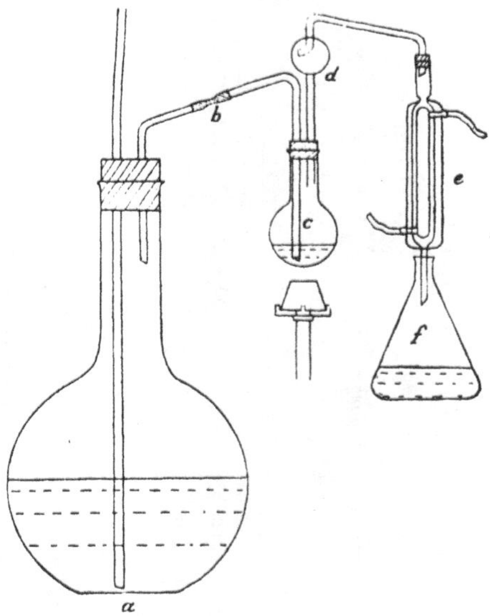

FIG. 47.

The results obtained, unless the special precautions described be adopted when calculated as percentages of ester in the oil, are considerably too high. Whilst some of the causes may be apparent to many, nevertheless the following is a list constructed to include the more important of these causes, and will serve to indicate in what manner the necessary amendments should be made:—

1. The use of methylated spirit (unpurified by further distillation) in the preparation of the standard potash solution employed by some experimenters in the saponification of the oil.

2. The use of hydrochloric acid in neutralising the excess of alkali after saponification.

3. The employment of water in the steam generating flask which has been insufficiently boiled to free it from carbon dioxide and other impurities.

4. The sulphuric acid, used to acidulate before distillation, may be advantageously replaced by phosphoric acid. This modification, whilst

[1] *P. and E.O.R.* **(1914)**, 116.

in many cases not absolutely essential, is desirable on account of the fact that sulphuric acid is liable to become reduced by certain constituents of oils, particularly of old oils, which frequently contain substances of a resinous nature. In such cases the volatile acid products of the reduction pass over along with the true acids of the oil undergoing examination.

The relations which the abnormal results obtained bear to the above outlined conditions are clearly shown by the appended figures :—

Method.	Percentage of Ester found.
1. Oil saponified by solution of potash in unpurified methylated spirit. Excess of alkali neutralised by hydrochloric acid and the acids liberated, previous to distillation by sulphuric acid	47·28
2. As 1, but the excess of alkali after saponification neutralised by sulphuric acid	43·51
3. As 2, but the methylated alcoholic potash replaced by a solution of potash in 96 per cent. (60 o.p.) alcohol .	41·38
4. As 2, methylated alcoholic potash (the spirit being previously purified by distillation over potash) being used instead of the solution of potash in unpurified spirit	41·45
5. As 2, the methylated alcoholic potash being replaced by a solution of potash in absolute alcohol purified by distillation over potash	41·00
6. A "blank" experiment, employing for distillation the residue resulting from the evaporation of 25 c.c. of the alcoholic potash used in 5, previously neutralised by means of sulphuric acid	1·5 c.c. deci-normal sodium hydroxide was required for the neutralisation of the distillate
7. Ester found in 5 less the amount of ester equivalent to the volume of deci-normal sodium hydroxide used up in the blank experiment	39·69

It is evident that, in order to obtain accurate results, the method of working must be clearly and minutely adhered to, especially in view of the fact that the determination of ester by the method of steam distillation is a very valuable indication as to the purity of an oil, serving to detect the fraudulent addition to oils of such esters as diethyl succinate, triethyl citrate, and diethyl oxalate, the free acids of which are non-volatile in steam. It will not detect glyceryl acetate or terpinyl acetate. The method, yielding reliable results and including modifications, devised to remove the sources of error above-mentioned, is as follows :—

About 2 grams of the oil (bergamot or lavender) is accurately weighed into a carbon dioxide flask, and 15 c.c. neutralised alcohol added along with a few drops of phenolphthalein solution, and the whole is just boiled on the steam-bath. The acid number is ascertained by titration with deci-normal alcoholic potassium hydroxide, 25 c.c. semi-normal alcoholic potash (made with 90 to 96 per cent. spirit, preferably distilled over potash) is now added, and the whole boiled under a reflux condenser for one hour, the excess of potash, after saponification and addition of 40 c.c.

of carbon dioxide-free water, being neutralised by means of semi-normal sulphuric acid. This titration gives the figure from which the ester percentage is calculated.

A few drops of semi-normal alcoholic potash are added, and the liquid allowed to evaporate on the steam-bath.

To the residue is added 10 c.c. of dilute phosphoric acid, prepared by mixing about 3·5 c.c. of 88 per cent. acid with 100 c.c. of carbon dioxide-free distilled water.

The carbon dioxide flask is now immediately attached to the apparatus, and the distillation is commenced.

It should here be noted that the distilled water in the steam generating flask must have been allowed to become entirely free from carbon dioxide by at least half an hour's preliminary boiling.

The whole apparatus must be thoroughly cleansed and freed from air by allowing steam from the generator to blow through for a few minutes before attaching the carbon dioxide flask.

Distillation is allowed to proceed, the water in the generator being kept boiling as quickly as possible, and the volume of liquid in the smaller flask being kept at about 10 c.c. by means of a small flame.

The time taken for the collection of the required 250 c.c. of distillate is usually about thirty minutes.

The distillate is collected in a 500 c.c. Erlenmeyer flask having a mark upon it to indicate the level of 250 c.c. Phenolphthalein solution and a sufficient excess of deci-normal sodium hydroxide solution are added to the distillate and the excess of alkali determined by back-titration.

Bergamot oil contains as its principal constituent linalyl acetate, to the extent of up to over 40 per cent. It contains limonene, pinene, camphene, bornylene, bisabolene, *l*-linalol in the free state, dihydrocumic alcohol, nerol, terpineol, and about 4 to 5 per cent. of the solid body bergaptene, $C_{12}H_8O_4$.

OIL OF SHADDOCK.

The shaddock is the fruit of *Citrus decumana*, usually known in this country as the grape fruit. It is a large fruit, native to China and Japan, and introduced into the West Indies by Captain Shaddock, whence its name. The oil, which is not known to any extent in commerce, has a specific gravity of about ·860 and an optical rotation of + 90° to + 95°. It is a typical *Citrus* oil, containing much limonene, but its constituents have not yet been carefully examined.

According to Brooks, the leaves of the tree *Citrus decumana* growing in the Philippines, which is probably a variety of the European tree, yields 1·7 per cent. of an oil having the following characters:—

Specific gravity at $\frac{30°}{30°}$	0·870
Optical rotation at 30°	+ 22·9°
Refractive index at 30°	1·4644
Ester number	10

It contains dipentene, 15 per cent. of linalol, and 1 per cent. of citral.

H. F. Zoller[1] has examined the oil obtained by steam distillation from the peel of *Citrus decumana* grown in the United States. He found it to have the following characters:—

[1] *Jour. Ind. Eng. Chem.* (1918), 10, 5, 364.

	1.	2.
Specific gravity	0·845	0·860
Refractive index	1·4750	1·4785
Optical rotation	+ 72·5°	+ 78·5°

The constituents present, with their approximate quantities, are as follows:—

α-pinene	0·5 to 1·5 per cent.
d-limonene	90 ,, 92 ,,
Linalol	1 ,, 2 ,,
Citral	3 ,, 5 ,,
Geraniol	1 ,, 2 ,,

Geranyl acetate, linalyl acetate, and citronellal are probably present in traces.

Oil of Citron.

This oil is expressed from the fresh peel of the fruit of *Citrus medica* var. *vulgaris*, the ordinary citron, a native of the Himalayas, which was cultivated by the Aryans from very early times. In France the fruit is known as a "cédrat"—the French "citron" being our lemon. Consequently "essence de citrons" must not be confused with citron oil. The tree flourishes in the south of Europe, particularly in Italy, the fruit sometimes weighing from 10 to 14 lb. Several other varieties of this plant exist and appear to be used for expression, when the figures given below will be exceeded; for example, some varieties yield an oil with a specific gravity of about 0·870 and optical rotation about + 67°. This oil, which is seldom met with pure, should have the following constants: specific gravity ·851 to ·854, rotation + 77° to 82°, refractive index about 1·4750. It contains chiefly limonene, dipentene, citral, and a crystalline body; melting-point 145°. The usual adulterants are lemon and orange oils and their terpenes. The odour of this oil is very little superior to that of good lemon oil, whereas the price demanded does not justify its extensive use in commerce.

Oils of Xanthoxylum.

Thoms, about two years ago, reported on the essential oil of the fruit of Japanese pepper, *Xanthoxylum piperitum*, which yielded 4·33 per cent. of the oil. Schimmel & Co. have stated that the chief constituent of the oil is citral. The oil has now been more exhaustively examined by Duruttis.[1] He finds the oil to have the following characters:—

Specific gravity at 20°	0·890
Optical rotation	− 26·5°
Refractive index	1·4732

From the fraction 175° to 177° (specific gravity = 0·8469, optica rotation = + 55·7°) a tetrabromide melting at 120° and a nitrosochloride melting at 93° to 95° were obtained. This fraction appears to be a mixture of dipentene and *dextro*-limonene. The oil consists of terpenes to the extent of about 90 per cent. The fractions boiling over 180° were treated with sodium bisulphite, and the compound formed treated in the usual manner, when a body was liberated which was found to be cumic aldehyde, which was identified by its forming a semi-carbazone melting at 210°, and by oxidation to cuminic acid melting at 115°. In the fraction

[1] *Arbeit. a. d. Pharm. Inst. Univ.*, Berlin, 60.

melting at 180° to 184° free palmitic acid was detected. Geraniol was found in the oil, but in small quantity only.

Asahina and Konno[1] have also examined the oil recently, and state that they find 13 per cent. of citronellal present. The oil which they examined had the following characters:—

Specific gravity	0·8511 at 19°
,, rotation	+ 45·45°

They also identified *d*-β-phellandrene and cumic aldehyde.

There has recently been imported into this country a "Japanese pepper" which Holmes has identified as the fruit of *Xanthoxylum alatum*. Its essential oil has been examined by Umney[2] and found to have a specific gravity 0·889, and an optical rotation − 23°. It contains a crystalline stearoptene melting at 80°.

Shinosaki[3] has examined the essential oils distilled from the leaves of *Xanthoxylum ailanthoides*, and found them to have the following characters:—

	Full-grown Leaves.	Half-grown Leaves.	Young Leaves.
Specific gravity	0·8437	0·8385	0·8390
Optical rotation	− 6° 35′	− 5° 75′	− 6° 63′
Refractive index	1·4474	1·4408	1·4440
Acid value	0·94	—	0·04
Saponification value	2·81	4·21	1·83

The principal constituent of the oil is methyl-nonyl-ketone, terpenes, and about 1·5 per cent. of phenols.

The essential oil of the Formosan plant known as Yamakosho may be mentioned here, its botanical origin being unknown, but the oil having been reported on also by Shinosaki.[4] From the leaves he obtained 1·2 of a brownish-yellow essential oil having an odour of camphor. It had the following characters:—

Specific gravity	0·9049
Optical rotation	− 18·8°
Refractive index	1·4635
Ester value after acetylation	48 66

It contains eucalyptol and hydrocarbons.

The oil from the fruits of *Xanthoxylum acanthopodium* known as "Wartara seeds" has been examined by Schimmel & Co.[5] The oil had the following characters:—

Specific gravity	0·8714
Optical rotation	+ 6° 31′
Saponification value	27·1

On distillation *in vacuo* (14 mm.) the oil gave the following fractions:—

Fraction.	Boiling-point.	Specific Gravity at 22°.	α_D at 18·5°.	Quantity of Distillate.
1	65° to 70°	0·841	+ 0° 26′	45 c.c.
2	70° ,, 80°	0·846	+ 4° 17′	9 ,,
3	80° ,, 90°	0·856	+ 11° 38′	34 ,,
4	90°	0·865	+ 13° 43′	19 ,,
5	90° to 100°	0·865	+ 12° 39′	27 ,,
6	100° ,, 130°	0·939	− 6° 2′	7 ,,

[1] *Apotheker Zeit.*, 32 (**1917**), 556.
[2] *P. and E.O.R.* (**1912**), 37.
[3] *Jour. Soc. Chem. Ind.*, 34 (**1915**), 1270.
[4] *Jour. Chem. Ind. Tokyo*, 18 (**1915**), 1081.
[5] *Report*, April, **1900**, 49.

Fraction 1, at atmospheric pressure, boiled at 175° to 176° and yielded with bromine a tetrabromide melting at 125°. It consisted of almost pure dipentene.

Fractions 3 to 5 were united and again fractionated, yielding a liquid having the properties of *d*-linalol. Its specific gravity was 0·868, optical rotation + 14° 20′, and boiling-point 78° at 14 mm. pressure. The identity of this body with linalol was proved by its conversion into citral and *l*-terpineol.

The citral obtained in the oxidation with a chromic acid mixture yielded the characteristic naphtocinchoninic acid with a melting-point between 197° and 200°. Treatment with formic acid produced *l*-terpineol with a melting-point of 35°. An alcoholic solution containing 3·6 per cent. of this terpineol had the specific rotatory power $[\alpha_D] = -10° 14'$ at 18°. Treatment with carbanil yielded terpinylphenylurethane melting at 112° to 113°.

It is thus proved that *d*-linalol, previously found only in coriander oil, is also a constituent of wartara oil.

The oil from *Xanthoxylum senegalense* (*Fagara Xanthoxyloides*) has been examined by Priess, and found to have the following characters :—

Specific gravity	0·9229
,, rotation	− 1° 20′
Acid value	2·19
Ester ,,	58·5

The oil contains methyl-nonyl-ketone, and, probably, decyl aldehyde· Dipentene, linalol, a sesquiterpene, esters, and the lactone xanthotoxin were also isolated from the oil.

Thoms, who isolated this body, considered it to be a coumarin-like compound, related to citraptene. His latest investigations[1] have confirmed the correctness of this supposition. He obtained a solid substance (melting-point 128°) which was separated by repeated recrystallisation into two bodies, xanthotoxin (melting-point 145° to 146°) and bergaptene (melting-point 190° to 191°), the latter a constituent of bergamot oil. The two bodies are isomeric, but bergaptene is a phloro-glucinol derivative, whereas xanthotoxin is derived from pyrogallol.

From a botanical point of view the occurrence of xanthotoxin and bergaptene in this oil, and of bergaptene in that of bergamot oil, is interesting, for the two plants are closely related, belonging to subdivisions of the same botanical family,

The following formulæ indicate the relationships of the two bodies :—

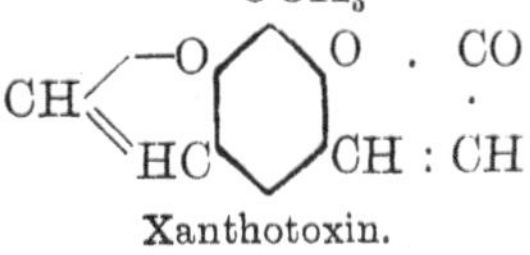

Xanthotoxin.

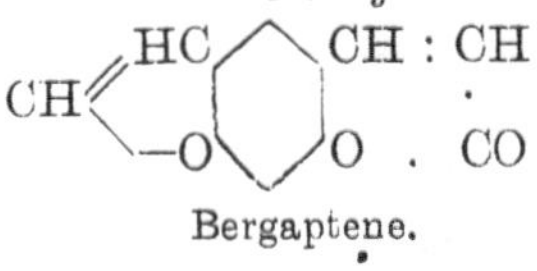

Bergaptene.

Oil of Evodia Simplex.

An oil distilled in Réunion from *Evodia simplex* has been examined by Schimmel & Co.,[2] who found it to have the following characters :—

Specific gravity	0·9737
Optical rotation	− 13° 4′
Acid value	2·1
Ester ,,	16·4
,, ,, (after acetylation)	63·3

[1] *Berichte*, 44 (**1911**), 3325. [2] *Bericht*, October, **1906**, 83.

The oil contains methyl-eugenol and a crystalline paraffin melting at 80° to 81°.

Oil of Boronia.

The herb *Boronia polygalifolia*, a plant found in most of the Australian colonies, yields an essential oil which has been examined by J. C. Umney.[1] It is described as having a specific gravity 0·839 (is this a misprint?), and an optical rotation + 10°. On fractional distillation the following fractions were obtained :—

150° to 170°	31 per cent.
170° ,, 180°	38 ,,
180° ,, 190°	15 ,,
Above 190°	16 ,,

Oil of Rue.

The genus Ruta comprises about fifty species inhabiting the Mediterranean region and Eastern and Central Asia. The oil is distilled from the herb *Ruta graveolens*, yielding the French oil, whilst Algerian oil is distilled from *Ruta montana*, the summer rue, and *Ruta bracteosa*, the winter rue.

The various oils of rue have the following characters :—

French Oil (Ruta Graveolens).

Specific gravity	0·830 to 0·845
Optical rotation	− 0° 50′ to + 3°
Refractive index	1·4300 to 1·4350
Solidifying-point	5° ,, 11°

It is soluble in 3 volumes of 70 per cent. alcohol.

Algerian Oil (Ruta Montana).

Specific gravity	0·830 to 0·838
Optical rotation	0° to + 1°
Refractive index	1·4300 to 1·4325
Solidifying-point	+ 7° ,, + 10°

This oil, which is soluble in 2·5 to 3 volumes of 70 per cent. alcohol, is known as "essence figeable" (congealing oil) in Algeria.

Algerian Oil (Ruta Bracteosa).

Specific gravity	0·837 to 0·845
Optical rotation	− 1° ,, − 5°
Refractive index	1·4300
Solidifying-point	below − 5°

This oil is known as "non-congealing oil" in Algeria. It is soluble, often with turbidity, in 3 volumes of 70 per cent. alcohol.

Spanish Oil.

Specific gravity	0·833 to 0·845
Optical rotation	− 1° ,, + 1°
Solidifying-point	− 3° ,, + 8

[1] *Imperial Inst. Jour.*, 2 (1896), 302.

Syrian Oil.

Specific gravity	0·841
Optical rotation	− 0° 28′
Refractive index	1·4329
Solidifying-point	− 1·9°

A sample of the oil from *Ruta montana*, which had the following characteristics—

Specific gravity	0·8344
Optical rotation	+ 0° 18′
Refractive index	1·4313
Melting-point	10·5°

was fractionated at ordinary pressure,[1] with the following results :—

	Boiling-point.	Specific Gravity.	Refractive Index 20°.	Melting-point.
20 per cent.	219·5° to 224°	0·834	1·4307	11°
30 ,,	,, 226°	0·832	1·43	14°
25 ,,	,, 227°	0·832	1·4297	14·5°
15 ,,	,, 228·5°	0·836	1·4305	14°
10 ,, Res.	above 228·5°	—	1·4471	—

The characteristic constituents of the oils of rue are the ketones methyl-nonyl-ketone, $CH_3 . CO . C_9H_{19}$, and methyl-heptyl-ketone. Of these methyl-nonyl-ketone predominates in the French oil and in the oil of *Ruta montana*, whilst methyl-heptyl-ketone is present in the oil of *Ruta bracteosa*.

Schimmel found traces of a basic substance present in the oil, which is probably the methyl ether of methyl-anthranilic acid, and would account for the fluorescence of the oil. Van Soden has examined all Algerian oil of rue, and found it to have a specific gravity 0·842 and optical rotation − 5°. It did not solidify even at − 15°. Its principal constituent was found to be methyl-heptyl-ketone, whilst only a small quantity of methyl-nonyl-ketone was present. This is evidently the oil of *Ruta bracteosa*. A German oil investigated by Houben was found to contain both ketones, the methyl-nonyl-ketone predominating. Traces of caprylic acid and of a phenyl were detected. Power and Lees have examined an oil, probably of Algerian origin, and found present a basic substance having an odour of quinoline, traces of methyl salicylate and of salicylic acid. They also found valeric acid (as esters?), laevo-pinene, limonene, and cineol. The alcohols corresponding to the ketones were also found present.

Bergaptene has also been found in traces in the oil.

The characters of the ketones above mentioned are as follows :—

Methyl-Heptyl Ketone.

Boiling-point	194·5° to 195·5° (763 mm.)
Specific gravity	d $\frac{14°}{16°}$ 0·8296
Melting-point of the semi-carbazone	119° to 120°

[1] Evans' *Analytical Notes* (**1912**), 65

METHYL-NONYL KETONE.

Boiling-point	229° to 233° (759 mm.)
Boiling of the ketone regenerated from the semi-carbazone	231·5° to 232·5° (761 mm.)
Specific gravity	$d\frac{20·5°}{16°}$ 0·8263
Melting-point of the semi-carbazone	122°

Methyl-nonyl ketone is capable of being estimated quantitatively by the phenyl-hydrazine method (see under lemon oil), and the same applies to methyl-heptyl ketone. It should be remembered that in rue oil both the ketones are always present in varying proportions. Hence the quantity of phenyl-hydrazine which has been used only affords an approximate estimation of the quantity of the ketones, and it must at the same time be ascertained, by determining the solidifying-point, which of the two ketones predominates in the oil, and whether it will be more practicable in estimating to use as a basis the molecular weight of methyl-nonyl ketone (170) or that of methyl-heptyl ketone (142).

OIL OF BUCHU.

Several species of *Barosma* are known by the name *Buchu*, and from the leaves of these the oil is distilled. The chief of these are, *Barosma serratifolia*, *Barosma betulina*, and *Barosma crenulata*. The yield of oil is from 1 to 2 per cent. The recent researches of Kondakoff[1] and Bialobrzeski[2] have proved that the oil contains several constituents. These are diosphenol $C_{10}H_{16}O_2$, the chief optically inactive compound in the oil; the terpenes limonene and dipentene, and a ketone $C_{10}H_{18}O$ which is probably laevo-menthone. Diosphenol is a crystalline solid, melting at 83° and boiling at 232°. It occurs only to a very small extent in the oil distilled from the leaves of *Barosma serratifolia*, whereas that from *Barosma betulina* contains so much that crystals separate at ordinary temperature.

The ketone $C_{10}H_{18}O$ is a slightly laevo-rotary liquid boiling at 206° to 209°, of specific gravity ·902, of agreeable peppermint like odour, and is probably identical with menthone. Pure diosphenol can be obtained from the oil by extraction with soda solution and decomposition of the alkaline liquid with an acid, and recrystallising the crude diosphenol from a mixture of alcohol and ether. Buchu oil is often adulterated with turpentine and possibly other oils. It is used to a small extent in medicine.

The oils have the following characters:—

Barosma Betulina.

Specific gravity	0·935 to 0·970
Optical rotation	– 15° „ – 48°
Refractive index	1·4740 „ 1·4865
Diosphenol (by NaOH absorption)	17 to 30 per cent.

Barosma Crenulata.

Specific gravity	0·936
Optical rotation	– 15° 20′
Refractive index	1·4800

[1] *Jour. fr. Chem.*, ii. 54 (1896), 433.
[2] *Pharm. Zeit. Russ.*, 35 (1896), 417, 433, 449.

Barosma Serratifolia.

Specific gravity	0·918 to 0·960
Optical rotation	− 12° ,, − 36°

In addition to the varieties of Barosma which are used in the preparation of buchu leaf oil the leaves of other species have lately been met with on the London market on several occasions. Some years ago, Sage[1] reported on leaves of *Diosma succulentum* var. *Bergianum*, known as Karoo Buchu, which are reputed to yield an oil equalling that of *Barosma betulinum* in value. In other quarters, however, leaves have also been noticed of which the oil, by its odour, betrays a composition differing materially from that of the ordinary buchu leaf oil. The leaves of *Barosma pulchella* belong to the latter category. The leaves of *Barosma pulchella* have an odour resembling that of citronella oil. The shrub is a native of Cape Colony.

The leaves, on distillation, yield 3 per cent. of essential oil, having the following characters :—

Specific gravity	0·883
Optical rotation	+ 8° 36′
Refractive index	1·4577
Acid number	18·5
Ester ,,	27·2
,, ,, (after acetylation)	237

The oil contains citronellal, methyl heptenone, *d*-menthone, *d*-citronellol, and a phenol in traces.

The leaves of *Barosma venusta* yield, according to Jensen,[2] 1·1 per cent. of an oil, having the following characters :—

Specific gravity	0·8839
Optical rotation	+ 0° 30′
Refractive index	1·4967
Acid value	2·4
Ester ,,	13·4
,, ,, (after acetylation)	52·8
Phenols	16 per cent.

The oil contains about 35 per cent. of a terpene, boiling at about 66·5° at 15 to 18 mm., and having a specific gravity 0·790 and refractive index 1·4778.

On fractional distillation using a four pear-head, at a pressure of 15 mm., the results were as follows :—

No.	Per Cent.	Boiling-point.	Optical Rotation.	Refractive Index.	Specific Gravity.
1	10	to 66·5°	− 0° 25′	1·4785	0·796
2	20	,, 73°	− 0° 45′	1·4815	0·817
3	20	,, 90°	+ 0° 30′	1·489	0·848
4	20	,, 96°	+ 2° 20′	1·508	0·945
5	30	over 96°	+ 0° 30′	1·5135	0·964

Separation was here not very successful, but the extremely low density of the first fractions (portions being as low as 0·79), together with similar boiling-point and optical inactivity, its instability, and the

[1] *Chemist and Druggist*, 65 (**1904**), 506, 717. [2] *Phar. Jour.*, 90 (**1913**), 60.

fact that its composition by combustion is equivalent to $C_{10}H_{16}$, all form a very strong presumption that the olefine-terpene myrcene has here been isolated, and that it in fact almost entirely constitutes the first 50 per cent. of the oil; although it is much more prone to polymerization and oxidation than when similarly obtained from bay oil. In the higher fractions evidence was obtained of the presence of the allied alcohol myrcenol and its acetate; also of methyl chavicol, anethol, and chavicol or similar phenol, with sesquiterpenes.

No indications whatever of diosphenol were observed.

The oil of *Barosma venusta* has recently been examined by Goulding and Roberts.[1]

The oil examined had the following characters:—

Specific gravity (15°)	0·865
Optical rotation	+ 0° 47′
Acid value	5·6
Ester ,,	6·2

Corresponding with 2·2 per cent. of esters.

Ester value after acetylation	55

Corresponding with 15·7 per cent. of total alcohols, or 14·3 per cent. of free alcohols and 2·2 per cent. of esters.

On distillation, under atmospheric pressure, some decomposition occurred, but when distilled under 60 mm. pressure a hydrocarbon having the properties of myrcene was obtained, showing a great tendency to resinify and combining with four atoms of bromine.

A small proportion of aldehyde was separated by shaking the oil with solution of sodium hydrogen sulphite, separating and washing with ether and extracting the aqueous solution with ether, after making alkaline with sodium hydroxide. The aldehyde was identified as anisic aldehyde, being readily converted into anisic acid melting at 183° to 184°. The presence of linalol was confirmed in the fraction boiling at 190° to 205° by oxidation to citral. No diosphenol was obtained, and only 0·2 per cent. of phenol was absorbed by solution of sodium hydroxide after washing with sodium carbonate solution and sodium bisulphite solution to remove acids and aldehydes. The small quantity of phenol extracted gave no coloration with ferric chloride. Methyl chavicol was identified in the fraction boiling at 213° to 218° by conversion into anethol, melting at 22°, and confirmed by determination of the methoxyl group.

The residual fraction contained sesquiterpenes together with polymerides and decomposition products of myrcene.

The approximate composition of the oil is as follows:—

	Per cent.
Hydrocarbons, chiefly or entirely myrcene	43·0
Aldehydes, chiefly or entirely anisaldehyde	0·5
Phenols	0·2
Phenol ethers (methyl chavicol)	21·4
Alcohols partly linalol	14·3
Esters calculated as $C_{10}H_{17}O$. Ac	2·2
Sesquiterpenes loss, etc. (by difference)	18·4

The only important difference from the results of Jensen's investigation is the almost complete absence of phenols, which occurred in Jensen's oil to the extent of 16 per cent., and gave a dark green coloration with ferric chloride. This was presumed to be impure chavicol, although not definitely proved. It should be pointed out, however, that Jensen's oil

[1] *Jour. Chem. Soc. Trans.* (**1914**), 2613.

had a higher specific gravity (0·8839), and might, therefore, be expected to contain more phenols.

Jensen's results were as follows:—

	Per Cent.
Hydrocarbons (myrcene)	35
Phenols	16
Phenol ethers	15
Alcohols	15
Esters calculated as linalyl acetate	4
Aldehydes and ketones	4
Sesquiterpenes, etc.	11

J. C. Umney[1] has recently received from Grahamstown a sample of small leaves stated to be those of *Barosma scoparia.*

An examination of the leaves showed that they correspond very closely with the leaves of *Barosma venusta*, which has already appeared in commerce. The leaves are very small, about $\frac{1}{8}$ in. broad, and a little longer, elliptical, obtuse in a few cases, but mostly obovate. They certainly do not belong to the typical form of *Barosma scoparia.*

The essential oil possesses a characteristic, decidedly pleasant odour, somewhat similar to *Barosma serratifolia*, and the oil shows no separation of diosphenol. It has this feature in common with the oil of *Barosma venusta* and also *Barosma serratifolia*, and for that reason may have less medicinal value than the official buchu, *Barosma betulina.*

West Indian Sandalwood Oil.

This oil is not a true sandalwood oil, the misnomer having largely contributed to the substitution of this oil for the genuine sandalwood oil, and to its use as an adulterant of the latter.

It is distilled from the wood of *Amyris balsamifera*, which yields from 1·5 to 4 per cent. of oil. It is a thick, viscid yellow oil of pleasant odour recalling that of cedarwood oil, with a secondary odour of true sandalwood oil, and has the following characters:—

Specific gravity	0·948 to 0·972
Optical rotation	+ 19° ,, + 30°
Alcohols as $C_{15}H_{26}O$	30 to 55 per cent.

It is soluble in an equal volume of 90 per cent. alcohol.

It contains an alcoholic body which has been named amyrol, of the formula $C_{15}H_{26}O$, or, according to Van Soden and Rojahn,[2] a mixture of two alcohols, having the following characters:—

	1.	2.
Formula	$C_{15}H_{26}O$	$C_{15}H_{24}O$
Specific gravity	about 0·987	—
Optical rotation	,, + 36°	0°
Boiling-point	299°	below 299°

The oil also contains a lactone, amyrolin, of the formula $C_{14}H_{12}O_3$, melting at 117°, *d*-cadinene, and *β*-caryophyllene.

Angostura Bark Oil.

The bark of *Galipea cusparia* (*Cusparia trifoliata*) yields between 1 and 2 per cent. of essential oil. The tree is found in abundance on the mountains in Venezuela, and it is especially plentiful in the country round

[1] *P. and E.O.R.* (**1914**), 373. [2] *Pharm. Zeit.*, 45 (**1900**), 229.

about the Gulf of Santa Fé. The name *angostura* appears to have been given to it on account of its having been found first in the district bearing that name, situated on the Orinoco. The oil has the following characters :—

Specific gravity	0·930 to 0·960
Optical rotation	− 7° ,, − 50°
Refractive index	1·5074
Acid value	1·8
Ester ,,	5·5
,, ,, (after acetylation)	35·7

It has been investigated by Beckurts and his pupils.[1] In their earlier investigations they obtained 1·5 per cent. of oil from the bark, of specific gravity ·956. More recently they obtained the same yield of oil of specific gravity ·941 at 20°, of optical rotation − 50°. The oil decomposes partially, when distilled under ordinary pressure, with the formation of water. A small portion was obtained boiling at 160°, and having an odour resembling pinene. A small quantity of a terpene is thus probably present. The main portion of the oil, however, distils between 260° and 280°. and consists of a sesquiterpene and a sesquiterpene alcohol. The sesquiterpene, which is laevo-rotary, was originally described under the name *galipene*, but has now been shown to be identical with cadinene. The alcohol $C_{15}H_{16}O$ is termed galipol; it is a liquid of specific gravity ·927 at 20°, and is optically inactive. By the action of dehydrating agents this body gives up water and yields cadinene. In addition to these bodies there is also present a small quantity of an inactive sesquiterpene, and traces of a crystalline stearoptene. The oil is used to a small extent, chiefly for the preparation of liqueurs, etc.

Jaborandi Leaf Oil.

The leaves of *Pilocarpus Jaborandi* yield about 5 per cent. of an essential oil having the following characters :—

Specific gravity	0·865 to 0·895
Optical rotation	+ 1° ,, + 4°

It is soluble in 2 volumes of 80 per cent. alcohol. Its composition is not well known, but the terpene dipentene has been identified, and also a crystalline hydrocarbon, belonging to one of the unsaturated series. The latter body sometimes occurs to so great an extent that the oil solidifies on cooling. It also contains methyl-nonyl ketone.

Oil of Evodia Rutæcarpa.

The essential oil distilled from the fruit has been examined by Asahina and Kashiwuki.[2] The authors have isolated from it a terpene which they believe to be different from any of the known terpenes, and which they have named evodene. It is a colourless liquid boiling at 67° (20 mm.), with a sharp odour, and has the characters of an aliphatic terpene of the formula $C_{10}H_{16}$. It has a specific gravity of 0·799 and refractive index 1·4843. On treatment with plantinum black it is converted into dimethyl-octane. It resembles the olefinic terpenes myrcene and ocimene, but differs from them in forming a liquid tetrabromide after reduction to dihydro-evodene. A crystalline substance melting at 278° has also been isolated from the fruit, which the

[1] *Arch. der Pharm.*, 237 (**1897**), 518 and 634; 236 (**1898**), 392; 229 (**1891**), 612.
[2] *Jour. Pharm. Soc. Japan*, November, 1915.

authors have termed evodiamine. On heating with alcoholic potash it yields a base $C_{11}H_{10}N_2$ and methyl-anthranilic acid. Its formula is stated to be $C_{19}H_{17}N_3O$.

In reference to the name "evodene," it should be mentioned that Semmler and Schorsberger had previously used the same name for a terpene which they isolated from the oil of *Xanthoxylum aubertia*.

Oil of Toddalia.

Toddalia aculeata (*Toddalia asiatica*) is a tree growing wild in the Nilgiri mountains, and is also found in the Philippine Islands. It is known as the wild orange tree. The leaves yield 0·08 per cent. of essential oil, which, according to Brooks,[1] has an odour of camphor and lemon-grass. Its characters are as follows:—

Specific gravity	0·9059 at $\frac{30°}{30°}$
Refractive index	1·4620 ,, 30°

It contains linalol to a considerable extent, and about 18 per cent. of a substance resembling camphor and melting at 97°.

Oil of Pelea Madagascarica.

The leaves and fruits of *Pelea Madagascarica*, a Malagasy shrub, yield about 4 to 5 per cent. of essential oil having a powerful aniseed odour. It is yellow in colour and has the following characters:—

Specific gravity	0·953
Optical rotation	+ 32° 22′
Refractive index	1·51469

It contains anisic aldehyde, but not much anethol, since it does not congeal even at − 18°.

Oil of Clausena Anisum-olens.

The leaves of *Clausena Anisum-olens*, a plant growing in the Philippines, yield about 1·2 per cent. of an essential oil having a powerful odour of Russian aniseed oil. The oil has been examined by Brooks,[2] who found it to have the following characters:—

Specific gravity at $\frac{30°}{30°}$	0·963
Optical rotation	0°
Refractive index at 30°	1·5235
Saponification value	3·6

The oil contains 90 to 95 per cent. of methyl-chavicol.

ZYGOPHYLLACEÆ.

Guaiacum Wood Oil.

This oil is identical with that which used to be offered in commerce under the name of champaca wood oil. As a matter of fact neither name can be regarded as correct. The wood from which the oil is distilled was formerly believed to be that known as *Palo balsamo* or guaiacum wood. It was originally believed to be the product of *Guaiacum officinale*, but it is now clearly settled that it is the product of *Bulnesia sarmienti*, and is

[1] *Philipp. Jour. Sc.* (**6**), 333. [2] *Ibid.*, 6 (**1911**), 344.

known as *Palo santo* (holy wood) to the natives. It appears to be indigenous to Argentine and Paraguay, and is especially plentiful along the course of the river Rio Bermejo, a tributary of the Rio Paraguay. The wood yields from 4 to 8 per cent. of essential oil, which is practically solid at ordinary temperatures. This oil has an exceedingly delicate tea odour, and is well suited both for soap perfumery and for blending with stronger perfumes for ordinary scents. The oil, which is of a hard semi-solid consistency, has the following characters:—

Specific gravity	0·965 to 0·975 at 30°
Optical rotation	− 3° to − 8°
Refractive index	1·5030 to 1·5050 at 30°
Acid value	0 to 2
Ester ,,	0 ,, 5
,, ,, (after acetylation)	100 ,, 150

The oil is soluble in 3 to 5 volumes of 70 per cent. alcohol. It melts at from 42° to 50°.

The only constituent yet ascertained has been examined by Wallach and Tuttle.[1] It is a crystalline alcohol of the formula $C_{15}H_{26}O$, which is termed guaiol. This body is identical with that found by Merck and named by him champacol, believing that the wood was really champaca wood. Guaiol melts at 91°, boils at 288° under ordinary pressure, and at 155° under a pressure of 13 mm., and is laevo-rotatory. By heating it with zinc chloride it yields a blue oil, of specific gravity ·910 at 20°, and which appears to be a sesquiterpene. An acetate of the alcohol has been prepared, which boils at 155° under a pressure of 10 mm. In addition to its legitimate use as a perfume, it is sometimes used as an adulterant of otto of roses. It is usually mixed with geranium oil, so as to have the same congealing-point as otto of roses, and is then used under the name "essence de la gaïac à la géranium".

ANACARDIACEÆ.

Oil of Mastic.

The resin from *Pistacia lentiscus*, the mastic tree, an inhabitant of Southern Europe, North Africa, and the Levant, which is known as mastic, consists of a mixture of resins and resinoids, with about 1 to 3 per cent. of volatile oil. This oil, which has a characteristic, pleasant odour, has the following characters:—

Specific gravity	0·850 to 0·903
Optical rotation	+ 22° ,, + 35°
Refractive index	1·4680 ,, 1·4760
Acid value	0 ,, 5
Ester ,,	3 ,, 19

The oil contains *d*-pinene as its principal constituent, with a small amount of *i*-α-pinene.

Oil of Chian Turpentine.

This oil is the product of distillation with water, of Chian turpentine, the resin of *Pistacia terebinthus*, an inhabitant of Syria and the Greek archipelago. The yield is about 15 per cent. of an oil of the following characters:—

[1] *Annalen*, 279 (**1894**), 395.

Specific gravity	0·868 to 0·870
Optical rotation	+ 12° ,, + 20°

It consists almost entirely of α-pinene.

Oil of Schinus Molle.

Schinus molle is a native of America and is widely distributed in Bolivia, Chili, Peru, New Granada, and Mexico. It is very commonly known as the American pepper plant.

It is also met with in Uruguay and the southern regions of Brazil, the provinces of San Paulo and Rio Grande do Sul.

It is extensively cultivated in Senegal, in Teneriffe, and in the whole of the Mediterranean regions. It ornaments numerous gardens in Provence in the neighbourhood of Nice, and is also much cultivated in Algeria where it is very widely distributed.

Roure-Bertrand Fils[1] have examined two essential oils of *Schinus*, obtained from Algeria (Saint-Denis du Sig) in order to compare their constants with those of an essential oil prepared from the plant cultivated at Grasse. One of the Algerian oils was obtained by the distillation of entire branches, leaves, wood, and fruits, and the other by the distillation of the leaves only.

Schinus molle is a tree of small dimensions with small diœcious flowers, carried on articulated peduncles disposed in bunches. The calix is small, with five lobes more or less cut off, with quincuncial praefloration. The corolla has five petals, much longer than the sepals, inserted on a disc; they are erect at first and spread out later on.

Samples of the oil distilled in Mexico have been examined by Schimmel & Co.[2]

These oils had the following characters:—

	Algerian Oils.		French Oils from Grasse.	Mexican Oils.		
	Entire Branches.	Leaves.	Entire Branches.	Leaves.	Fruit.	Unknown.
$d_{15°}$	0·8634	0·8658	0·8696	0·8583	0·8600	0·8492
α_D	+50° 54′	+65° 20′	+46° 13′	+44° 50′	+42° 30′	+56° 27′
$n_{D20°}$	—	—	—	1·47665	1·47909	1·47616
Acid Number . .	—	0·7	2·1	—	—	—
Ester Number .	5·5	3·4	8·2	7·2	25·2	—
Ester Number after Acetylation . .	29·4	40·4	43·4	—	56·5	—
Solubility in 90 per cent. Alcohol	5 vols.	5 vols.	10 vols. with turbidity	Even with absolute alcohol the solution, at first clear, becomes turbid when more alcohol is added		

The oil contains α-pinene, *l*-phellandrene, caroacrol, a crystalline body melting at 160°, and possibly a sesquiterpene.

Oil of Rhus Cotinus.

This oil was first obtained from the young twigs of the tree by Perrier and Fouchet,[3] who obtained 0·1 per cent. of the oil, which was known as "oil of fustic". The oil had the following characters:—

[1] *Bulletin*, April (**1909**), 29. [2] *Report*, April (**1908**), 121.
[3] *Bull. Soc. Chem.*, iv. 5 (**1909**), 1074.

Specific gravity	0·875
Optical rotation	+ 13°
Refractive index	1·4693
Acid number	6·1
Ester value	28·2

A sample examined by Schimmel & Co.[1] had the following characters :—

Specific gravity	0·871
Optical rotation	+ 32° 54′
Refractive index	1·4887
Acid value	0·3
Ester „	20·4

The oil consisted almost entirely of terpenes, of which β-pinene, limonene, and dipentene were detected with certainty. No other constituent has been identified.

Oil of Heeria Paniculosa.

The berries of *Heeria Paniculosa*, a plant indigenous to Zuzuland, known to the natives as Isifeku, yield from 4·5 to 6 per cent. of essential oil, which has been examined by Juritz.[2] A sample examined in the Imperial Institute Laboratories had the following characters :—

Specific gravity	0·852
Optical rotation	+ 3·75°
Ester value	4·2
Acid value	1·6
„ „ (after acetylation)	14·8

The oil consists mainly of terpenes.

BURSERACEÆ.

Oil of Myrrh.

Myrrh is an oleo-resinous substance obtained as a secretion in the parenchyma of the bark of various species of Commiphora. The so-called true myrrh, or Herabol myrrh, is collected principally in Arabia and Somaliland, whilst Bisabol myrrh, or bdellium, occurs as an admixture of true myrrh, especially when collected in Somaliland, and is identical or closely related to Indian bdellium.

Herabol Myrrh is chiefly derived from *Commiphora myrrha*, and yields from 3 to 10 per cent. of a thick, yellow essential oil, having the following characters :—

Specific gravity	0·985 to 1·045
Optical rotation	– 30° „ – 90°
Refractive index	1·5180 „ 1·5280
Acid value	1 „ 8
Ester „	15 „ 45
„ „ (after acetylation)	30 „ 70

It is soluble with, at most, slight turbidity, in 10 volumes of 90 per cent. alcohol.

Lewinsohn[3] emphasises that the composition of myrrh oil differs according to the origin of the resin, the method of production, and the age. He believes that distillers add small quantities of alkali to the

[1] *Report*, April (**1913**), 86. [2] *Chemical News*, 1920, **120**, 277.
[3] *Arch. der Pharm.*, 244 (**1906**), 412.

resin after no more oil passes over for the purpose of obtaining a larger yield, and the distillation is then continued. The constituents of myrrh oil identified by him are: cuminic aldehyde, $C_{10}H_{12}O$ (up to 1 per cent.), boiling-point 116° (12 mm. pressure); oxime, melting-point 56°; semicarbazone, melting-point 201°.—Acids: Of these, only acetic acid and palmitic acid (melting-point 62°) were detected with certainty. Both are present in old oils in the free state, whilst freshly distilled myrrh oil contains the acids as esters.—Phenols (about 1 per cent.): eugenol (about 0·23 per cent.; benzoyl compound, melting-point 69°), and a small quantity *m*-cresol were detected.—Hydrocarbons: Fractional distillation over sodium yielded four hydrocarbons (terpenes) of the formula $C_{10}H_{16}$ (all of which, however, were not found in every oil examined), of which three were identified; pinene (nitrosochloride, melting-point 103°; nitrolbenzylamine and nitrolpiperidine); dipentene (tetrabromide, melting-point 124°); limonene (tetrabromide, melting-point 104°); the fourth hydrocarbon probably belongs also to the limonene group (a_D + 80°, tetrabromide, melting-point 115°, and monohydrochloride, melting-point 6°) but could not be identified. Sesquiterpenes: probably cadinene is present, and a second sesquiterpene of specific gravity 0·911 at 21°, and optical rotation + 30·4°.

Friedrichs[1] distilled an oil from picked myrrh, which had the following characters:—

Specific gravity	1·011
Optical rotation	– 73·8°
Refractive index	1·5359
Acid number	6·15
Ester ,,	47·6

No eugenol could be detected, but *meta*-cresol was present. Cumic and cinnamic aldehydes were also found to be present. A sesquiterpene having the following characters was also isolated:—

Boiling-point	130° to 136° at 16 mm.
Specific gravity	0·943 at 20°
Optical rotation	– 14° 12′
Refractive index	1·5125
Molecular refraction	64·98

To this body he assigns the name herabolene.

Traces of formic and acetic acids were also found, and a non-volatile acid melting at 159°, and an acid of the formula $C_{16}H_{21}O_3 . COOH$, melting at 236°, which he names myrrholic acid.

Bisabol Myrrh.—The principal plant yielding this resin is probably Commiphora erythræa.

Tucholka[2] has extracted the essential oil from a Bisabol myrrh coming from the interior of the Somali country, by removing it from the alcoholic solution of the resin by means of petroleum ether. The oil was purified by steam distillation, and then had a light yellow colour, specific gravity ·8836, and optical rotation – 14° 20′. It boiled at 220° to 270°. The yield was 7·8 per cent. By means of dry hydrochloric acid gas, a crystalline hydrochloride was separated, which, on treatment with boiling acetic solution of sodium acetate, set free a hydrocarbon which the above-named investigator terms bisabolene. This hydrocarbon had a specific gravity ·8914 and boiled at 260°. The same chemist suggests the unlikely formula $C_{56}H_{96}O$ for the residual portion of the oil.

[1] *Arch. der Pharm.*, 245 (**1907**), 427. [2] *Arch. Pharm.*, **235**, 289.

Oil of Opopanax.

The oil from the true opopanax has not been investigated, and the botanical source of the gum-resin is unknown. It is almost certainly derived from an Umbelliferous plant—but not *Opopanax chironium* as usually stated. The commercial product known as oil of opopanax is derived from the exudation of *Commiphora erythræa, var. glabrescens*, which is possibly identical with bisabol myrrh. Samples of the oil have the following characters:—

Specific gravity	0·870 to 0·905
Optical rotation	1·4890 ,, 1·4940
Acid value	0 ,, 3
Ester ,,	7 ,, 20
,, ,, (after acetylation)	40 ,, 55

The oil is soluble to an almost clear solution in 1 to 10 volumes of 90 per cent. alcohol.

The only constituents so far identified are the sesquiterpene bisabolene, and a sesquiterpene alcohol boiling at 135° to 137° at 2 mm. pressure.

Oil of Frankincense.

This oil is distilled from the resinous exudation of various species of *Boswellia*, principally *Boswellia Carterii*. The yield of essential oil varies from 5 to 10 per cent.

It is a colourless or pale yellow liquid of balsamic odour, having the following characters:—

Specific gravity	0·875 to 0·894
Optical rotation	+ 15° ,, + 30°
Refractive index	1·4725 ,, 1·4825
Acid value	0 ,, 4
Ester ,,	5 ,, 20
,, ,, (after acetylation)	40 ,, 50

It is soluble in 4 to 8 volumes of 90 per cent. alcohol, sometimes with turbidity.

Up till about ten years ago, this oil was found to be laevo-rotatory, but apparently the species of Boswellia from which it is obtained have recently been varied, as the oil is almost always dextro-rotatory now. The laevo-rotatory oil contains α-pinene-dipentene, phellandrene, and cadinene.

The dextro-rotatory oil has recently been examined by Schimmel & Co.[1] The oil examined by them had the following constants: d_{15° 0·8775, α_D + 19° 18′, n_{D20° 1·47245, acid value 1·8, ester value 7·5, ester value after acetylation 106·0, soluble in up to 2 volumes of 90 per cent. alcohol, the solution being clear at first, and turning turbid after two volumes. Of the terpenes, 92 per cent. boiled between 156° and 161°, about 5 per cent. between 161° and 163°, and about 4 per cent. between 163° and 181°. In the fractions boiling between 156° and 157° (d_{15° 0·862 to 0·863; α_D + 30° 58′) they established the presence of *i*- and *d*-α-pinene from the nitrosochloride and the nitrol piperidide (melting-point 118°), as well as by oxidation into pinonic acid (boiling-point 170° to 175° at 7 mm.; melting-point of the active acid 69° to 70°; of the inactive acid 104°). The presence of camphene in all the fractions boiling between 158° and 164° was established by conversion into *iso*borneol, melting-point 210° to 211°. Dipentene

[1] *Report*, April, **1914**, 62.

was identified by its tetrabromide, and *para*-cymene by oxidising with potassium permanganate to *p*-hydroxy-isopropyl benzoic acid.

Fromin and Antin [1] isolated from this oil a (ketonic ?) alcohol, of the formula $C_{10}H_{16}O$, boiling at 210° to 211°. This body has been named olibanol. Esters of borneol were also detected.

Boswellia serrata, a closely-related plant, is a native of North-west India, and yields a gum resin which is collected twice a year in the Punjab. It is employed by the natives in the treatment of rheumatism and nervous disease, and is an ingredient in certain ointments. In Gujerat it is burnt as incense in religious ceremonies. Experiments have been conducted at the Forest Research Institute, Dehra Dun, with a view to determining the best method of treating the crude gum resin in order to obtain the essential oil, resin, and other products. The gum and impurities were first separated from the resin and oil by treating a quantity of the raw product in a false-bottomed copper still and passing steam through the mass. The resin and most of the oil passed through the false bottom, leaving the gum and impurities above. The oil was then separated from the resin by steam distillation, and the resulting oil fractionally distilled up to 155° C. yielded an oil consisting almost wholly of dextro-pinene, and a portion consisting mostly of limonene or dipentene. Samples of these were examined at the Imperial Institute.[2] The oil distilled principally between 153° and 160°, had specific gravity 0·8446, optical rotation + 31° 24′, ester value 2·6, ester value after acetylation 36·4. It readily dissolves resins such as colophony, dammar, sandarac, and soft copal, but varnishes thus prepared dried more rapidly than those made with commercial turpentine, and gave a rather less lustrous surface.

Oil of Elemi,

And other Canarium Species.

The elemi of commerce is the balsamic exudation of *Canarium luzonicum*, the Manila elemi, which is produced entirely in the Philippine Islands. Many other species of *Canarium* yield similar products, which, however, are scarcely commercial articles.

Manila elemi yields from 15 to 30 per cent. of essential oil, whose odour is largely that of phellandrene. The oil has the following characters :—

Specific gravity	0·870 to 0·915
Optical rotation	+ 30° to + 55° (rarely + 4° to + 134°)
Refractive index	1·4775 to 1·4900
Acid value	0 ,, 2
Ester ,,	4 ,, 10

The oleoresin is produced in a manner similar to that used for the collection of turpentine, and is then steam distilled.

According to Clover,[3] the oil is very variable in character and composition. He has examined twenty-one different samples of authentic origin, and divided them into the following groups :—

1. Ten samples consisted of practically pure *d*-limonene.
2. Nine samples contained phellandrene, associated with either pinene or limonene.

[1] *Annalen*, 401 (**1913**), 253. [2] *Bull. Imp. Inst.*, xiii., **3**, 351.
[3] *Amer. Chem. Jour.*, 39 (**1908**), 613.

3. Two samples consisted of (*a*) practically pure terpinene ; (*b*) practically pure terpinolene.

The presence of terpinolene had hitherto not been observed in nature, and is remarkable as the oil containing terpinolene showed a very peculiar behaviour. The crude oil boiled within one degree (Clover does not mention the boiling-point), and had a rotation of + 4°. The reactions for terpinene and phellandrene were negative, but a bromide of the melting-point 116° to 117° was obtained in the form of brilliant scales. From this the presence of terpinolene might be assumed. The oil containing terpinolene was purified by treatment with potassium hydroxide, distilled *in vacuo*, and had then the constants : specific gravity at $\frac{30°}{4°}$ 0·8361 ; rotation + 1·7° ; $n_{D_{30°}}$ 1·4701 ; boiling-point 173·5° to 175° (at ordinary pressure) ; 80° to 81° (37 mm. pressure). The rotation decreased considerably between the second and third distillation, although the oil had been prepared only on the previous day from elemi. In order to accelerate a possible change in the principal constituent of the oil, a portion of it was heated for three hours in a sealed tube on an oil-bath to 200°. The distillate now showed a rotation of − 7·5°, and gave a phellandrene reaction (nitrite, laevo-rotatory, melting-point 119° to 120°). Five days later an attempt was made to produce terpinolene tetrabromide from the purified oil, but instead of this there were obtained crystals of the melting-point 125°, which were dipentene tetrabromide. In the meantime the rotation of the oil had also changed from + 1·7° to − 1·6° ; after a further six days it had increased to − 9·8° ; after another month to − 34·5°, a proof that a considerable alteration was proceeding spontaneously in the oil. After three more weeks the oil was distilled with steam ; the distillate now had an optical rotation − 38°, and the conversion had consequently probably come to an end. The tetrabromide produced in various solvents melted at 125°. Phellandrene could no longer be detected by the nitrite-reaction, probably owing to the presence of a large quantity of dipentene. In Clover's opinion the almost pure terpinolene present in the original oil (which had a particularly low boiling-point), had consequently been converted, by standing for several weeks, into dipentene, a small quantity of *d*-phellandrene, and an unknown laevo-rotatory terpene.

The high-boiling oil portions had mostly a feeble laevo-rotation, and their specific gravities fluctuated between 0·900 and 1·030 ; they were chiefly mixtures of sesquiterpenes and oxygenated bodies. From a fraction with the rotation + 71·6° a solid substance of the formula $C_{15}H_{26}O$ was isolated.

In the case of phellandrene, Clover observed that the α-modification of this terpene occurs in elemi oil. The rotation of the phellandrene diminished slowly if left standing at ordinary temperature. When the phellandrene was left standing, crystals separated on the glass walls of the bottle, the formation of which must be attributed to autoxidation of the phellandrene ; after purification from hot acetic ether they melted at 164·5° to 165·5°, and had a composition corresponding to the formula $C_{10}H_{18}O_2$.

Semmler[1] has detected a small quantity of a sesquiterpene alcohol, and a phenol ether to which he assigned the name elemicin. This body has a specific gravity 1·063 at 20°, refractive index 1·5285, and boils at 144° to 147° at 10 mm.

[1] *Berichte*, 41 (1908), 1768, 1918, 2183, 2556.

Elemicin is allyl-trimethoxy-3·4·5-benzene. Bacon[1] has examined numerous samples of the oil, and his results differ somewhat from those of Clover, thus emphasising the variable character of the oil. He only found two samples to contain *d*-limonene out of sixty-two obtained from the district of Calaoag, whilst six out of nine from Gumaca contained this terpene. Bacon considers that β-phellandrene is sometimes present in the oil.

Semmler and Liao[2] have isolated the sesquiterpene alcohol of this oil and find it to boil at 152° to 156° (at 17 mm.) and to have a specific gravity 0·9411 at 20°, optical rotation − 5°, and refractive index 1·5030. They have termed this body elemol. It yields a sesquiterpene, elemene, on dehydration.

An Elemi oil from Southern Nigeria has been found to contain much phellandrene and to have a specific gravity 0·8686 and rotation + 50° 30′.

Uganda Elemi, from *Canarium Schweinfurthii*, yields about 11 per cent. of essential oil of specific gravity 0·8451 and optical rotation + 79° 20′. It contains much phellandrene.

Mauritius Elemi, from *Canarium Colophania*, yields 16 per cent. of essential oil of specific gravity 0·859, optical rotation + 3° 13′, and refractive index 1·4725.

Canarium villorum,[3] another Philippine variety of elemi, yields an oil of specific gravity 0·8627, optical rotation + 11° 3′, and refractive index 1·47245. It contains much *p*-cymene, and the terpenes α-pinene and dipentene.

Canarium microcarpum, from the Moluccas, is a very soft oleo-resin, yielding about 80 per cent. of oil of specific gravity 0·920 to 0·930, and optical rotation + 2° to − 52°.

Canarium euptoron is the principal species yielding "Lagam" balsam in the Dutch Indies. The oil[4] has a specific gravity 0·905 to 0·912, optical rotation about − 8°, and refractive index 1·4972. It consists chiefly of caryophyllene, with a little caryophyllene alcohol. Samples[5] distilled from a balsam of the same name, but obviously of different origin, had specific gravities of 0·882 to 0·883 at 26°.

It is possible that Lagam balsam is not really the product of *Canarium euptoron*, but of a *Dipterocarpus* species.

An oil from Cameroon Elemi[6] has been found to have a specific gravity 0·953, and is materially different from all other elemi oils examined.[7]

West Indian Elemi oil is distilled from *Dacroydes hexandra*, one of the Burseraceæ. The yield of oil is about 16 per cent., the oil having a specific gravity 0·8875 and optical rotation − 13° 20′. More found α-pinene and *l*-sylvestrene in the oil. The presence of the latter terpene has been confirmed by Schimmel & Co.[8]

MEXICAN LINALOE OIL.

The linaloe oil of commerce is distilled in Mexico, and also in Europe, from several species of Bursera, mainly from the wood, but also to some

[1] *Philipp. Jour. Sc.*, 4 (**1909**), 93.
[2] *Berichte*, 49 (**1916**), 794.
[3] Schimmel's *Bericht*, October (**1907**), 24.
[4] *Arch. der Pharm.*, 250 (**1912**), 199.
[5] *Jaarb. dep. Land. Ned. Indii.* (**1909**), 66, 81.
[6] *Pharm. Zeit.*, 58 (**1913**), 774.
[7] *Jour. Chem. Soc.*, 75 (**1899**), 718.
[8] *Report*, April (**1914**), 51.

extent from the fruit. The principal species used for distillation are *Bursera Delpechiana* and *Bursera aloexylon*. Possibly *Bursera glabrifolia* and *Bursera fagaroides* are used to a small extent. The wood is known locally as "Bois de citron du Mexique". The oil is not to be confused with the so-called Cayenne linaloe oil, which is derived from an entirely different species.

According to Dr. Altamirano,[1] Mexican linaloe oil is obtained from the districts Oaxaca, Puebla, Guerrero, Morelos, and Michoacan down to Colima. The two trees which are or have been the principal source of the oil are known as "Spanish linaloe" and "copal limón". The former is now nearly eradicated, but the latter still flourishes abundantly.

The natives distinguish three kinds according to the external appearance and odour: fine, common, and caraway linaloe. Fine linaloe has a very delicate and pleasant odour, the common kind less so, and the third resembles caraway. The wood of the *linaloe* tree is said to be more compact than that of the *copal limón*, which is more spongy. The oil which is met with in commerce is chiefly obtained from the wood of *copal limón*. This tree is very hardy and grows even in very stony soil; although it can most easily be propagated from cuttings, nobody thinks of planting it. The young trees, however, can only be made use of for oil production after about twenty years. It is said that during the last few years oil has frequently been placed on the market which consisted of a mixture of oils obtained from the wood and from the fruit. The fruit is ripe in September, and forms fleshy, greenish to reddish berries of the size of a Spanish pea (*garbanzo*). The yield of oil from the fruit (which is distilled from July to September) amounts to 3 per cent. The fruit is gathered by stripping it off with a very sharp crescent-shaped instrument, care being taken to damage the fleshy part of the berries as little as possible; the oil obtained in the distillation possesses a herbaceous odour, and resinifies very readily. For this reason it is not sold as such, but is mixed with the oil obtained from the wood, which keeps better. In order to increase the quantity and improve the quality of the oil, the Indians used to submit the fruit before the distillation to a kind of fermentation during which the herbaceous odour disappears, by placing the undamaged fruit in cases well covered with sacks, and keeping it thus for three to four days at a constant temperature so as to let the fruit mature.

The oil from the wood is best obtained from trees which are quite old—forty to sixty years; in the case of young trees it is endeavoured to increase the oil-content by cutting notches in the tree, upon which an increased formation of oil takes place, which may be regarded as a pathological product. The tree is treated so that strips 16 ins. long and 8 ins. wide are cut out of the bark and the wood, down to a depth of 2 ins., and running parallel, or preferably slanting, to the direction of the trunk. If it is now found that the tree has a "heart," *i.e.* yellowish, highly aromatic wood, it can be used at once for distillation. If the tree has no "heart," it is left standing with open wounds so that it may develop "heart". With a view to this, the upper extremity of the notch is cut to a point, in order that the rain water may run alongside, but the lower edge is bevelled off in such manner that no foreign material can collect there which would set up decay. If, in spite of this, the wood should commence to rot, the part is at once planed off, so as to arrest the injury which otherwise would cause a cessation of the oil formation. If no

[1] *Report of the Inst. Med. Nac. Mexico* (**1904**), Jan.-Mar., and Schimmel's *Report*.

interruption takes place, the change in the coloration of the wood becomes apparent already after a few days; after the lapse of a month a large portion of the previously white and soft wood has been transformed into yellowish, hard, and much more aromatic wood. The Indians were led to this operation by the observation that cut-off linaloe roots frequently contained so much oil that it could be pressed out with the fingers, and that trees from which branches had accidentally broken off through gales, formed much oil, and had changed after many years. The operation, which has the effect of converting trees poor in oil into trees rich in oil, is carried out in April and September. The April cut is not so effective as the September one, as the individual trees in both cases only "mature" (*i.e.* run to sap) in the month of January. In October to November the trees drop their leaves. The distillation of the wood, for which only the trunk is used, is carried on in the most primitive manner, chiefly from December to June. The smaller distillers are partly still absolutely wild Indians who do not speak Spanish; actual factories are not yet in existence, which is probably due to the difficult conditions of life in the linaloe districts, where vermin abounds, and where among the Indians contagious diseases such as leprosy, etc., prevail. The yield of oil obtained in Mexico from the wood amounts at most to 2·5 per cent. The production of Puebla is said to yield annually about 4000 to 5000 kilos. The oil is put up in kerosene tins containing 35 to 37 lb., of which two are packed in a case.

The still, arranged for heating by direct fire, is (according to Schimmel & Co.), like the other parts of the apparatus, with the exception of the copper conical head, made from galvanised sheet-iron; it is about 5 ft. high, and has a diameter of 45 ins. At the lower part of the still an iron tube has been arranged, which is closed with a wooden plug wrapped up in banana leaves, and by means of it the water can be drawn off when the distillation is completed. The removable head, which is provided with means for cooling, is connected by two tubes with the condenser proper; pails or kerosene tins serve as receivers.

The still is usually built near a mountain stream. The cut wood is so filled into the still (about 5 cwt. of chips) that a space of about 10 ins. from the top of the still is left empty. Water is then added so that the wood is well covered, and a space of about 7 ins. from the top is left empty.

The wood when distilled in Europe yields from 7 to 9 per cent. of oil, but whether this is due to better methods of distillation or to an after development of oil in the wood is a matter of uncertainty.

The oil was investigated by Verneuil and Poisson, who stated that it distilled almost entirely between 189° and 192°, and that it consisted mainly of an oxygenated body of the formula $2(C_{10}H_8)5H_2O$. Semmler,[1] however, was the first to discover the true odorous principle of the oil, which he termed linalol. In addition to this alcohol, geraniol and *d*-terpineol and the ketone, methyl-heptenone, have been found in the oil, together with traces of a sesquiterpene and two terpenes. Barbier and Bouveault give us the composition of an oil examined by them :—

Linalol	90 per cent.[2]
Geraniol	2 ,,
Sesquiterpene	3 ,,
Methyl-heptenone	0·1 ,,
Terpene (1)	0·1 ,,
,, (2)	0·1 ,,

[1] *Berichte* (1891), 207.

[2] This figure is too high.

Schimmel & Co. have isolated *dextro-α*-terpineol from the oil, and also an alcohol not previously found naturally, which they identified as methyl-heptenol.[1] This body has the following characters:—

Specific gravity	0·8579
Optical rotation	− 1° 34′
Refractive index	1·44951
Boiling-point	178° to 180°

Schimmel & Co. in 1908 isolated a body possessing the formula $C_{10}H_{18}O_2$ from this oil. They have since identified it as linalol monoxide, identical with the body prepared by Prileshaeff[2] by the oxidation of linalol by means of benzoyl hydrogen peroxide.

Linalol monoxide is somewhat viscous and is also clearly differentiated from linalol by its mouldy odour, which recalls that of fenchyl alcohol and of camphor. As its boiling-point is only 3° to 4° lower than that of linalol, it is impossible, in view of the very small extent to which it occurs in the oil, to prepare it in the pure form by fractionation. It is, however, possible to obtain it pure, although at a considerable loss of material, by repeatedly oxidising the fractions containing linalol of high specific gravity with 0·5 per cent. permanganate solution at ice temperature, each time driving off the unattacked oil with steam. It also appears to be possible to accumulate the compound by treating the fractions in question with glacial acetic-sulphuric acid at 40° to 50° for several hours. From the formation of esters, it is clear that the body $C_{10}H_{18}O_2$ is an alcohol. Its acetic ester, of which the odour differs altogether from that of linalyl acetate, boils between 60° and 65° (4 mm.), and its benzoic ester between 157° and 160° (3 to 4 mm.). These chemists have also isolated the hydrocarbons octylene[3] (previously found in bergamot oil by Burgess and Page[4]) and nonylene from the oil. An oleginic terpene is also present, possibly identical with myrcene.

Linaloe oil is a colourless or pale yellow oil of sweet, soft odour, that of linalol predominating. It has the following characters:—

Specific gravity	0·875 to 0·898
Optical rotation	− 3° to − 15° (sometimes dextro-rotatory to + 8°)
Refractive index	1·4590 to 1·4655
Acid value	0 ,, 6
Ester ,,	5 ,, 75

The oil is soluble in 3 volumes of 70 per cent. alcohol, or in 4 to 5 volumes of 60 per cent. alcohol.

Occasionally pure oils will be found which vary outside these limits. Thus, oils with a high content of linalol oxide may have a specific gravity of 0·900 or even a little over. Oils with a high ester value naturally contain less free linalol.

The amount of free linalol is the most important feature of the oil. Its actual determination is difficult, if not impossible, on account of the ease with which it decomposes under the action of acetic anhydride. Boulez's modification of the usual acetylation process, however, yields approximately correct results.

A fractional distillation is most useful, and will give a good indication of the amount of linalol present. The following results were obtained on fractionating four pure samples, by the author and C. T. Bennett:—

[1] *Berichte*, October, **1908**, 78.
[2] *Ibid.*, 42 (**1912**), 4811.
[3] Schimmel's *Report*, April, **1909**, 67.
[4] *Jour. Chem. Soc.*, 85 (**1904**), 1327.

No. 1.

No.		Specific Gravity.	Optical Rotation.	Refractive Index.
1	10 %	0·868	− 11°	1·4580
2	20 %	0·870	− 11° 30′	1·4590
3	20 %	0·871	− 13°	1·4605
4	20 %	0·872	− 13°	1·4612
5	20 %	0·876	− 12°	1·4620
6	10 % (residue)	0·913	—	1·4750

No.		No. 2.		No. 3.		No. 4.	
		Specific Gravity.	Optical Rotation.	Specific Gravity.	Optical Rotation.	Specific Gravity.	Optical Rotation.
1	10 %	0·870	− 10° 30′	0·872	− 11° 40′	0·871	− 12°
2	20 %	0·870	− 10° 20′	0·871	− 11° 30′	0·871	− 13°
3	20 %	0·871	− 10° 40′	0·873	− 12°	0·873	− 13° 20′
4	20 %	0·873	− 10° 10′	0·874	− 11° 50′	0·874	− 11° 50′
5	20 %	0·877	− 9° 40′	0·876	− 11° 10′	0·876	− 10° 30′
6	(residue)	0·909	—	0·914	—	0·912	—

The characters of the original oils were as follows:—

No.	Specific Gravity.	Optical Rotation.
1	0·882	− 11°
2	0·882	− 10°
3	0·877	− 11° 30′
4	0·879	− 12° 10′

Schimmel gives the following figures for two dextro-rotatory oils:—

Oil I.: d_{15° 0·8816; α_d + 6° 3′; acid number 1·7; ester number 20·0; n_{d20° 1·46209; soluble in 1·6 and more volumes 70 per cent. alcohol. On distillation *in vacuo* there passed over:—

No.	Pressure.	Temperature.	Per Cent.	Optical Rotation.	Remarks.
1	4 mm.	up to 66°	8·5	α_d + 6° 30′	
2	4 ,,	66 ,, 67·5°	23·5	α_d + 8° 18′	linalol fraction
3	4 ,,	67·5 ,, 68°	32·1	α_d + 9°	linalol fraction
4	5 ,,	73 ,, 76°	17·0	α_d + 7° 30′	linalol fraction
5	5 ,,	76 ,, 82°	2·1	α_d + 3° 20′	
6	5 ,,	82 ,, 96°	8·7	α_d − 10° 5′	

Oil II.: d_{15° 0·8783; α_d + 8°; acid number 1·4; ester number 3·5; n_{d20° 1·46149; soluble in 2·0 and more volumes 70 per cent. alcohol. On distillation *in vacuo* the following passed over:—

No.	Pressure.	Temperature.	Per Cent.	Optical Rotation.	Remarks.
1	10 mm.	60 to 86°	4·5	α_d + 5° 24′	
2	10 „	86 „ 90°	9·0	α_d + 8° 50′	linalol fraction
3	10 „	90 „ 91°	15·0	α_d + 11° 21′	linalol fraction
4	14 „	95 „ 97°	51·0	α_d + 11° 15′	linalol fraction
5	14 „	97 „ 100°	5·0	α_d + 8°	
6	14 „	100 „ 110°	3·0	α_d − 11°	
7	14 „	110 „ 120°	3·0	α_d − 18° 20′	

The oil is sometimes adulterated with fatty oils, which raise the specific gravity, yield a high ester value, and are not volatilized on a water-bath.

Shiu oil, an oil distilled in Central Formosa, has been found as an adulterant of this oil, which it somewhat resembles in odour. As this oil always contains camphor, which is absent from pure linaloe oil, the adulterant can be detected as follows:—[1]

1. By preparation of the semi-carbazone. The sample is steam distilled to remove any fixed residue, and 25 c.c. are treated with a mixture of 1 gram of semi-carbazide hydrochloride, 1 gram of sodium acetate and 25 c.c. of 90 per cent. alcohol. After standing for twenty-four hours, 1 gram of sodium bicarbonate is added, and the whole is steam distilled. The distillation residue is heated on a water-bath and then cooled. The semi-carbazone of camphor separates, if camphor be present, possibly contaminated with hydrazodicarbonamide, which is removed by dissolving the crystals in a little alcohol, which does not dissolve this impurity. On dilution of the alcohol, the semi-carbazone separates, and on treatment with dilute sulphuric acid a marked odour of camphor is perceived.

2. By oxidation of the other constituents by means of permanganate of potassium. Into a 2 litre flask, 300 c.c. of water and 50 grams of permanganate of potassium are placed, and after standing in ice for 30 minutes 10 grams of the oil are added drop by drop. Heat is developed, and the oxidation products are entirely soluble. When the reaction is over (in about two hours) 200 c.c. of water are added, and the unaltered camphor can be steam-distilled off.

As little as 1 per cent. of camphor in the oil can be detected by these methods.

J. C. Umney has described two interesting samples of Mexican linaloe oil,[2] which are unquestionably natural oils, but which may be derived from a different species of *Bursera* from those which are usually distilled.

The oils are entirely different in odour from the normal oil, and one would describe them as a blend of lavender and petitgrain rather than the normal linaloe (lily of the valley) odour.

On examination it was found that these oils have a very high ester value, and by this alone are distinguished from the normal oil. The following figures were obtained:—

Specific gravity	0·897	0·893
Optical rotation	− 2° 30′	− 2° 30′
Refractive index (20°)	1·4600	1·4610
Esters	52·4 per cent.	48·7 per cent.
Total linalol	54·8 „	57·6 „

[1] Schimmel's *Bericht*, October, **1913**, 69.

[2] *P. and E.O.R.* (**1916**), 62.

Particular attention is directed to this oil, because it may be that it would be well to develop the distillation of this particular wood for its content of linalyl acetate and for the purposes for which the oils containing that ester are used, rather than for linalol and the purposes for which oils containing that alcohol are employed.

Roure-Bertrand Fils[1] have described the oil distilled from linaloe seeds. It had the following characters:—

Specific gravity	0·8858 to 0·8883
Optical rotation	+ 1° 30′
Linalyl acetate (esters)	10·2 per cent.
Refractive index	1·4655 at 18°

They have detected the following bodies in the oil: *l*-linalol, *l*-terpineol, nerol, and geraniol. They consider that this dextro-rotatory seed oil is present in the dextro-rotatory linaloe oils of commerce above referred to.

LEGUMINOSÆ.

Oil of Copaiba.

This oil is obtained by the distillation of the oleo-resin (balsam) of copaiba, which is the product of several species of Copaifera. The trees yielding this valuable commercial product are principally found round about the Amazon and its tributary streams, as far north as Guaiana, Venezuela, and Columbia. The principal varieties of the balsam are the following:—

Bahia, Cartagena, Maracaibo, Maranham, Para, Cayenne, and Angostura. The following are probably the sources of several of these (Umney):—

Bahia, *C. coriacea.*
Cartagena, *C. officinalis.*
Maracaibo, *C. officinalis.*
Maranham, *C. landsdorffii.*
Para, *C. multijuga.*
Cayenne, *C. guyanensis.*

But there is no doubt that there is very little balsam in commerce which has been obtained entirely from one species, and it must be emphasised that genuine commercial balsam or oils may well vary somewhat outside the limits quoted for a comparatively small number of specimens examined. The following figures are, therefore, to be accepted subject to the above limitation.

The average yields of oil from the various types of balsam are as follows:—

Maracaibo balsam	38 to 55 per cent.
Maranham	35 ,, 55 ,,
Para	58 ,, 78 ,,
Cartagena	40 ,, 60 ,,
Bahia	40 ,, 60 ,,
Angostura	45 ,, 55 ,,
Maturin	40 ,, 55 ,,
Surinam	40 ,, 72 ,,

[1] *Report*, October, **1907**, 16.

The essential oils have the following values :—

MARACAIBO OIL.

Specific gravity	0·898 to 0·906
Optical rotation	− 4° ,, − 13°
Refractive index	1·4955 ,, 1·4980

MARANHAM OIL.

Specific gravity	0·896 to 0·905
Optical rotation	− 4° ,, − 22°
Refractive index	1·4958 ,, 1·4980

PARA OIL.

Specific gravity	0·886 to 0·910
Optical rotation	− 4° ,, − 33°
Refractive index	1·4930 ,, 1·5020

BAHIA OIL.

Specific gravity	0·885 to 0·910
Optical rotation	− 8° ,, − 28°
Refractive index	1·4940 ,, 1·4980

CARTAGENA OIL.

Specific gravity	0·894 to 0·910
Optical rotation	− 4° ,, − 23°
Refractive index	1·4938 ,, 1·4980

MATURIN OIL.

Specific gravity	0·899 to 0·904
Optical rotation	− 0° 48′ ,, − 10° 10′
Refractive index	1·4970 ,, 1·5008

SURINAM OIL.

Specific gravity	0·903 to 0·910
Optical rotation	− 7° ,, − 12°
Refractive index	1·4978 ,, 1·5015

ANGOSTURA OIL.

Specific gravity	0·905 to 0·916
Optical rotation	− 2° ,, − 10°
Refractive index	1·4990 ,, 1·5020.

The following figures are of value, since they are the results of the examination of forty-one samples, many of which were imported direct by the firm examining them,[1] and of whose authenticity there is no doubt :—

Essential Oil of Bahia Copaiba—

	1	2	3	4	5	6
p. Gr.	0·898	0·899	0·900	0·900	0·902	0·902
a_D	− 11°	− 11° 26′	− 10° 32′	− 10° 30′	− 8° 30′	− 8° 0′
Yield	44 p. cent.	60 p. cent.	46 p. cent.	54 p. cent.	56 p. cent.	54 p. cent.

[1] Evans, *Analytical Notes* (**1910**), 26.

Sample (1) was a consignment *via* New York. Sample (2) was a direct import of undoubted purity.

Essential Oil of Carthagena Copaiba—

Sp. Gr.	α_D.	Yield p. Cent.	Sp. Gr.	α_D.	Yield p. Cent.
0·894	− 19° 30′	40	0·903	− 6° 0′	52
0·8972	− 19° 30′	50	0·904	− 2° 30′	45
0·899	− 22° 30′	40	0·905	− 7° 30′	45
0·901	− 7° 30′	45	0·905	− 10° 30′	50

The abnormally low optical rotation of several of these oils indicates probably some change in the botanical source, with possibly a wider collection than in former years, no adulteration being detected. Some samples contained from 1·2 to 10 per cent. of admixed water.

Essential Oil of Maracaibo Copaiba—

Specific gravity 0·900 ; α_D − 6° ; yield, 48 per cent.

Essential Oil of Maranham Copaiba—

Sp. Gr.	α_D	Yield p. Cent.	Sp. Gr.	D.	Yield p. Cent
0·896	− 18° 30′	36	0·900	− 14° 30′	40
0·899	− 18° 30′	45	0·9005	− 16°	40
0·899	− 19° 0′	43	0·901	− 16°	45
0·8995	− 15° 30′	45	0·901	− 13° 20′	56
0·900	− 13° 30′	48	0·9016	− 12° 18′	50
0·900	− 14° 0′	52	0·902	− 13° 30′	50

Essential Oil of Para Copaiba—

Sp. Gr.	α_D.	Yield p. Cent.	Sp. Gr.	α_D.	Yield p. Cent.
0·886	− 29° 0′	52	0·8916	− 27° 30′	50
0·891	− 25° 12′	5	0·8965	− 17° 30′	60
0·891	− 24° 0′	50	0·897	− 21° 20′	70
0·891	− 23° 0′	45	0··97	− 18° 44′	64
0·8912	− 24° 14′	50	0·906	− 30° 0′	64
0·8915	− 24° 30′	50	0·908	− 29° 0′	72

Cocking[1] considers that the first 10 per cent. distilled *in vacuo* of a copaiba oil should have a higher laevo-rotation than that of the oil itself, and that a lower value indicates the presence of an adulterant. Although the author dissents from Cocking's conclusions, the fact that the British Pharmacopœia has based an official test on it causes it to be of considerable importance.

Cocking's conclusions are as follows :—[2]

To detect adulteration of copaiba with the African oleo-resin, the sample is submitted to distillation either by steam, or *in vacuo*, and the volatile oil so obtained is dried, then fractionally distilled to dryness *in vacuo*. Ten equal fractions are collected, and the α_D of each of these determined in a 100 mm. tube.

If the sample be pure the figures obtained will all be negative, and they will increase arithmetically from the first to the last fraction (that is, each successive fraction is more strongly laevo-rotatory than the preceding one), although not regularly. If, now, the rotation of the first fraction be subtracted from that of the tenth, a figure will be obtained which varies very little for genuine samples, and is always a negative quantity. This figure, the "difference value," will only vary − 3·7° to − 7·6°.

[1] *Chemist and Druggist*, 77 (**1910**), 119. [2] *Year Book of Pharmacy* (**1911**), 128.

When African copaiba is examined in this manner, the rotations of all the fractions are dextro-rotatory, and the rotations of the successive fractions increase, but to a much greater extent than with the South American copaiba, in consequence of which the difference value is much greater than copaiba and is a positive figure. The figures also show a curious feature in that the tenth fraction has a considerably lower rotation than the ninth. As would be expected from the fact that the range of boiling-points of the constituents of the volatile oils from the two varieties are practically identical, a mixture of the two will distil over containing proportional parts in each fraction, and the presence of African balsam will be shown at once by the difference value being positive. In some cases where only a small percentage of the adulterant is present, all the fractions will be laevo-rotatory but the difference value will be positive. When the same process is applied to gurjun oil, like copaiba, it gives laevo-rotatory fractions, but, unlike it, they successively decrease instead of increasing, and thus give a positive difference value, similar to African copaiba.

With the true copaibas the rotation of the first fraction is in every case lower than that of the original oil, but in the adulterated samples it is higher. It is important that the distillation of the oil should be conducted *in vacuo*, since, if carried on under ordinary pressure, the higher temperature necessary causes some decomposition, which entirely alters the optical rotation.

The principal adulterations of copaiba and its oil are the corresponding balsams and oils known as African copaiba (probably obtained from *Oxystigma Mannii* and gurjun balsam).

The author dissents from this view entirely, and considers that there is considerable objection to fixing this arbitrary standard, based on the examination of a few samples, and then jumping to the conclusion that samples of a product of known variability to a very high degree must necessarily be adulterated. The table[1] on the opposite page illustrates the behaviour on fractionation, and the general characters of a number of samples of balsams and oils, of which samples 0 to 5 are stated to be *apparently* genuine, whilst samples 6 to 12 are assumed to be adulterated, and the remaining few being quoted as illustrations of the difficulty of drawing an opinion on fractionating results

The last column in the opposite table, headed "difference value," indicates the optical rotation of the last 10 per cent. of the oil left as a residue on distillation, minus that of the first 10 per cent. distilled, which should show a minus figure if Cocking's deductions are correct. In regard to this point the authors above quoted remark as follows :—

"That generalisation which is conveniently referred to as Cocking's (its most persistent advocate) for the detection in particular of admixed African oil, whereby such dextro-rotatory substances become evident on fractionation through their influence on the optical rotations of the first and last 10 per cent.—the positions of maximum divergence owing to opposite progressions—would appear to be under some suspicion.

"This method of analysis has again been of much service, and in the experience of this laboratory, has been sufficiently reliable for deciding commercially the purity of all balsams except that from Central America (Maracaibo), which variety, as pointed out a year ago, is at times subject

[1] Evans, *Analytical Notes* (**1912**), 26.

Fractionally Distilled Copaibas.

	Balsam Values.			Essential Oil and Vacuum Fractionation Figures.									
								Optical Rotations.					
	Acid Value.	Acid Value of Resin.	Ref. Ind. 15°.	Per Cent. Oil.	S.G. / Ref. Ind	Optical Rotation.	B. Pt. of 90 % / At Pressure	10 %.	40 %.	20 %.	20 %.	10 % Residual.	Difference Value.
Para— 0	72·1	148·1	1·5132	52	0·8893 / 1·4935	− 30°	121°-129° at 15 mm.	− 25° 30′	− 27° 14′	− 29° 20′	− 32° 40	− 34° 50′	− 9° 20′
1	21	70	1·4992	70	0·9052 / 1·50	− 29°	120°-122° at 10-12 mm.	− 23° 20′	− 25° 48′	− 29 20′	− 33° 56′	− 28° 30′	− 5° 10′
2	44·8	135·7	1·5088	67	0·90 / 1·498	− 21° 40′	121°-126° at 15 mm.	− 18° 20′	− 19° 20,	− 20° 52′	− 24°	− 26° 20′	− 8°
Bahia— 3	51	121	1·5095	58	0·9055 / 1·4952	− 14°	133°-135°	− 10° 55′	− 12° 16′	− 14° 4′	− 17° 40′	—	− 6° 45′
4	53·2	106	1·5124	49	0·899 / 1·4949	− 15° 40′	141°-154°	− 13° 45′	− 15° 10′	− 16° 30′	− 13° 30′	—	+ 0° 15′
Maranham— 5	82·6	140·1	1·5168	41	0·900 / 1·4971	− 17° 18′	128°-133·5° (30 %)	− 13° 48′	− 14° 20′ (20 %)	—	—	—	—
Maturin— 6	91	151·7	(S. G. 0·997)	40	0·912 / 1·5008	− 0° 48′	133°-146·5° at 15-20 mm.	− 9° 40′	− 8° 5′	− 3° 36′	+ 3° 30′	+ 17° 50′	+ 27° 30′
Via Hamburg— 7	70	140	1·5135	51	0·912 / 1·4985	+ 0° 50′	118°-132° at 12 mm.	− 2° 58′	− 2° 15′	+ 0° 48′	+ 4° 15′	+ 9° 35′	+ 12° 37′
Central American (Maracaibo)— 8	84	145	1·5076	42	0·90 / 1·4972	− 8° 20′	128°-137° at 20 mm.	− 9°	− 9° 20,	—	− 9° 40′ (40 %)	+ 1°	+ 10°
9	89·6	144·5	1·515	38	0·9005 / 1·4985	− 7° 30′	125°-134° at 15 mm.	− 9° 56′	− 9° 10′	− 9°	− 8° 55′	− 4°	+ 5° 56′
10	81·2	156·2	1·518	48	0·903 / 1·498	− 7° 40°	132°-141° at 20 mm.	− 9° 10′	− 8° 50′	− 8° 10′	− 6° 50′	− 4° 30′	+ 4° 40′
11	84	140	1·519	40	0·904 / 1·498	− 10° 22′	130°-139°	− 11° 23′	− 11° 33′	—	− 12° 30′ (40 %)	− 7° 30′	+ 3° 50′
12	91	151·6	1·518	40	0·90 / 1·4975	− 8°	131°-141° at 25 mm.	− 10°	− 9° 55′	− 9° 30′	− 9° 20′	− 6°	+ 4°
13	86·4	132·5	1·5168 (S. G. 0·985)	35	0·917 / 1·495	− 9° 30′ (Balsam + 6°)	120° at 12 mm.	− 9° 30′	—	—	—	—	—
13A	85·8	141·6	1·5072 (40°) S. G. 0·9838)	39·4	0·8991 / 1·4933	− 8° 36′	—	− 9°	—	—	—	—	—
14	84	161·5	1·518	48	0·899 / 1·4975	− 10° 46′	123°-125° at 15-20 mm.	− 12° 20′	− 9° 56′	− 10° 11′	− 10° 55′	− 12° 5′	—
14A	—	—	—	—	—	—	at 40 mm.	− 13°	− 13° 30′	− 13° 30′	− 14° 55′	− 10° 58′	—
15	84	142·4	1·5176	41	0·8988 / 1·4978	− 10° 14′ (Balsam + 35°)	121°-126° at 12 mm.	− 11° 10′	− 11° 30′	− 11° 35′	− 12° 10′	− 12° 5′	—

to inexplicable changes under such treatment. The decomposition effect of a slightly too high boiling-point is made clear in the comparison 14, 14A, these figures being from the same sample.

"On the contrary, the temperatures at which portions of sample 3 and many further samples of Maranham and other types were distilled rose still higher without any such changes (owing to accidental increase of pressure). Sample 4, also a Bahia variety (as 3), was moreover further anomalous since the rotation of the first 10 per cent. came within Cocking's abbreviated criterion of purity, whereas on complete distillation to obtain the difference value, this became positive owing to the marked drop in rotation of the highest fraction, an unpublished feature for any standard genuine oil—certainly with the correct mean rotation—confirming the absence of decomposition changes."

In examining this essential oil, it appears certain that the optical rotation should always be laevo-rotatory, and very rarely below $-4°$ or $-5°$.

African copaiba oil is always dextro-rotatory, and gurjun balsam oil is always very highly laevo-rotatory, usually about $-80°$ or more. Gurjun oil may be detected with certainty by adding five or six drops of the oil to 10 c.c. of glacial acetic acid containing five drops of nitric acid. In the presence of gurjun oil a purple violet colour is developed within a minute or so. Fatty oils are indicated by a high ester value, the adulterant being recognisable by separation of the products of saponification.

The oil consists principally of sesquiterpenes. The early investigations are quite unreliable, the first authentic investigation being that of Wallach,[1] who stated that the principal constituent of the oil was the sesquiterpene caryophyllene. Deussen[2] and his colleagues have, since then, assisted in the elucidation of the chemistry of this oil, and found that the sesquiterpenes present consisted principally of the inactive α-caryophyllene, but *l*-cadinene and a little β-caryophyllene are also present. Deussen[3] has described a sensitive test for caryophyllene. When β-caryophyllene nitrosite is heated with petroleum ether there results, in addition to dehydrocaryophyllene nitrosate, a body which is distinguished by its insolubility and which can be purified by dissolving it in acetone and precipitating with light petroleum. It crystallises in the pure state in white needles, melting-point 159°. The same body is generated when nitrous acid gas is passed into an ethereal solution of caryophyllene. A blue coloration quickly results, and when the gas is continued for some time a voluminous yellowish-white body is precipitated, while the blue colour disappears. Careful recrystallisation from acetone yields white, silky needles, melting at 160°. This body is optically active: $[\alpha_D] - 133° 50'$. Combustion established the formula $C_{12}H_{19}N_3O_6$. Owing to its insolubility in most organic solvents, this compound is excellently adapted for the detection and determination of β-caryophyllene. For this purpose from 2 to 3 grams of the oil under examination are treated in a 10 per cent. ethereal solution with nitric oxides, the reaction-mixture (which should be protected from light) not being cooled, and the passage of the gas being stopped as soon as the separation of the body in question ceases. The precipitate is washed with ether and weighed. Deussen claims that by this method Para copaiba oil shows the presence of 5·15

[1] *Annalen*, 271 **(1892)**, 294. [2] *Chem. Zeit.*, 34 **(1910)**, 873; 36 **(1912)**, 561.
[3] *Annalen*, 388 **(1912)**, 136.

per cent., and Maracaibo oil 2 per cent. of β-caryophyllene. African copaiba oil contains only 0·87 per cent. of this hydrocarbon.

Several investigations have established the presence in African copaiba balsam oil of considerable proportions of a dextro-rotatory sesquiterpene, possessing the properties of cadinene, but affording a lævo-rotatory hydrochloride. Schimmel & Co.[1] have examined the first runnings of the sample, and submitted 600 grams thereof to careful fractionation. They obtained the following principal fractions with more or less constant boiling-points :—

1. 119° to 120° (10 mm.) ;	246° to 248° (750 mm.)	10 per cent. ;		α_D − 13° 7′.
	248° ,, 250° (,, ,,)	70 ,, ,,		
	250° ,, 251° (,, ,,)	10 ,, ,,		
2. 123° ,, 125° (,, ,,) ;	258° ,, 263° (,, ,,)	—	;	α_D − 6° 52′.
3. 130° ,, 133° (,, ,,) ;	267° ,, 270° (,, ,,)	20 ,, ,,	;	α_D + 50°.
	270° ,, 273° (,, ,,)	70 ,, ,,		

The principal fraction (No. 3) had the same boiling-point as cadinene. The other characters of the "cadinene-fraction" isolated from the oil were: d_{15° 0·9261, $[\alpha]_D$ + 60° 40′, n_{D20° 1·51112.

Fraction 2 (20 gr., α_D − 6° 52′) was tested for caryophyllene. With glacial acetic and sulphuric acids it afforded no caryophyllene alcohol, but when nitric oxide was passed into the ethereal solution according to Deussen's method described above they obtained 2·33 per cent. of a nitrogenous body, derived from β-caryophyllene (melting-point 155° to 156° from acetone).

Fraction 1 (100 gr., d_{15° 0·9077, $[\alpha]_D$ − 13° 21′, n_{D20° 1·48943) only afforded the smallest traces of the nitrogenous body from β-caryophyllene, and can therefore have contained next to nothing of this sesquiterpene. It yielded neither a solid nitrosochloride nor a nitrosite, but it afforded 30 per cent. of a solid hydrochloride, melting-point 117° to 118° and $[\alpha]_D$ − 36·38° in a 5·09 per cent. chloroform solution. These characters, as well as the fact that no depression in the melting-point was obtained by admixture with *l*-cadinene hydrochloride, showed it to be identical with this body.

When treated with sodium acetate in a glacial acetic acid solution the hydrochloride yielded a cadinene possessing the following properties: boiling-point 273° to 275°, d_{15° 0·9229, $[\alpha]_D$ − 116·73°, n_{D20° 1·50651; when treated with hydrochloric acid gas cadinene hydrochloride, melting-point 118, was again formed.

The liquid chlorides freed by suction from the cadinene hydrochloride of fraction I were examined for the presence of other chlorides. As they could not be distilled *in vacuo* they were boiled with their own weight of sodium acetate in glacial acetic acid solution, for the purpose of eliminating the hydrochloric acid, and the resulting oil was fractionated *in vacuo*. Of three equal fractions thus obtained—

(*a*) 111° to 121° (7 mm.), (*b*) 121° to 130°, (*c*) 130° to 135°,

(*b*) and (*c*) afforded copious proportions of cadinene hydrochloride, melting-point 117·5° and 118°, only fraction (*a*) failed to yield a solid product. The liquid chlorides of fraction 1 were practically identical in character with the solid chloride, and must therefore have consisted in the main of cadinene hydrochloride.

[1] *Report*, April (**1914**), 48.

The investigation showed that the first runnings of African copaiba balsam oil contain a lævo-rotatory sesquiterpene, which, although differing materially from cadinene in regard to its physical properties,

	Boiling-point.	d_{15°.	$[\alpha]_D$.	n_{D20°.
Sesquiterpene . .	246° to 251°	0·9077	− 13° 21′	1·48943
Cadinene . . .	271° ,, 273°	0·9215	− 105° 30′	1·50647

yet affords cadinene hydrochloride. It is probably closely related to cadinene.

So far, three sesquiterpenes are known which, when treated with hydrochloric acid gas, afford lævo-rotatory cadinene hydrochloride; namely:—

1. *l*-cadinene proper.
2. The *d*-sesquiterpene of West Indian sandalwood oil and of African copaiba balsam oil.
3. The *l*-sesquiterpene, boiling-point 246° to 251°, of African copaiba balsam oil.

For the following reasons this result is considered by the investigators to be of considerable interest: in by far the great majority of instances the identification of cadinene in essential oils has not been achieved by isolating the body as such, but by preparing its hydrochloride from fractions of which the boiling-point usually lay between 250° and 270°, or frequently "from the sesquiterpene fraction". It is therefore not impossible that in many cases cadinene has been assumed to be a constituent of essential oils which possibly contained another sesquiterpene which yield cadinene hydrochloride. Hence, the isolation of cadinene hydrochloride by itself can no longer be regarded as proof of the presence of cadinene.

The presence of gurjun balsam oil in copaiba oil may be confirmed by the preparation of gurjene ketone semi-carbazone. The oil is fractionated under 10 to 12 mm. pressure. The portion boiling above 145° is rejected. The lower fractions are oxidised with potassium permanganate in acetone solution. The semi-carbazone melts at 234° C. and has a specific rotation + 317° in chloral hydrate solution.

Van Itallie and Nieuwland have identified a sesquiterpene alcohol, $C_{15}H_{20}O$, melting at 113° to 115° in Surinam oil.

Semmler and Stenzel[1] have examined the laevo-rotatory sesquiterpene of African copaiba oil, which yields cadinene hydrochloride, and have named it copæne. By oxidation with potassium permanganate it yields a ketonic acid, which forms a semi-carbazone melting at 221°.

Oil of African Copaiba.

The so-called African copaiba oil has been referred to somewhat fully under oil of copaiba, so that only a few further details are necessary. African balsam of copaiba is the product of a plant whose botanical origin is not quite certain. It is probably *Oxystigma Mannii*, or *Hardwickia Mannii*.

The balsam yields from 35 to 47 per cent. of essential oil having the following characters:—

[1] *Berichte*, 47 (**1914**), 2555.

Specific gravity	0·915 to 0·930
Optical rotation	+ 5° „ + 30°
Acid value	·5 „ 10
Ester „	0 „ 5·6
„ „ (after acetylation)	10

It contains *l*-cadinene and *β*-caryophyllene.

For further details of this oil (which is not employed commercially except in so far as the balsam is used as an adulterant of ordinary balsam of copaiba, so that oil of copaiba may be found adulterated with the oil of the African balsam) see under Oil of Copaiba.

Oil of Hardwickia Balsam.

Hardwickia balsam, the oleo-resinous exudation of *Hardwickia pinnata*, a tree found in further India, yields about 28 to 50 per cent. of essential oil, which has the following characters:—

Specific gravity	0·904 to 0·906
Optical rotation	− 7° 42′ „ − 8° 24′
Acid value	0·85
Ester „	2·88

The oil consists principally of sesquiterpenes.

Oil of Cassie.

The perfume of cassie flowers, *Acacia Farnesiana*, has for many years been well known in the form of a pomade. The essential oil is scarcely a commercial oil, but has been prepared by the steam distillation of the petroleum ether extract.

Acacia Farnesiana is found in the West Indies, tropical Africa, the South of France, Egypt, South America, and the Philippines. It is a native of the West Indies, and was first naturalised in Europe in the Farnesian gardens in Rome. In Cuba, where it is known as "Aroma francesca," it has become a troublesome weed. In the Sandwich Islands also it grows wild, and has proved an obstacle to cultivation owing to its tough, firm roots and bristling spines. It is there known as "klu," and suggestions have been made for its employment for perfume extraction.[1]

In India large quantities of cassia pomade used to be prepared and exported to all parts of the world, including America, the South of France, and Germany. The pomade was exceedingly fine and strong in odour, improving with age. The production, however, gradually got less and less with competition from other parts of the world.

In Syria a factory has been established for the manufacture of pomades near Beyrouth, at the foot of Mount Lebanon, where there is quite a forest of cassie bushes. The climate is temperate and admirably suited for the development of the perfume of the flower, and Syrian cassie is competing favourably with the product of the South of France.

In the South of France the tree flourishes best in the neighbourhood of Grasse and Cannes, where it is known as "cassie ancienne," "cassier de Levant," and "casillier de Farnèse". Another species, *Acacia cavenia*, is known as "cassie romane".

The flowering season is from October to January or February, and the blooms open successively, some being ready for picking whilst others are in course of development. The flowers are usually gathered twice a

[1] *P. and E.O.R.* **(1914)**, 53; **(1916)**, 71.

week during the day, and brought to the factories in the evening. Each full-grown tree will produce about 2 lb. of flowers.

The essential oil has the following characters:—

Specific gravity	1·040 to 1·0580
Refractive index	1·5133 „ 1·5150
Solidifying-point	About 18°
Optical rotation	0° to − 1°
Acid value	25 „ 42·5
Ester „	114 „ 229

The following substances have been detected in cassie oil: farnesol, geraniol, linalol, benzyl alcohol, methyl salicylate, decylic, cuminic and anisic aldehydes, and a ketone having a violet odour.

Oil of Balsam of Peru.

Balsam of Peru is the balsamic, oleo-resinous exudation of the Leguminous tree *Myroxylon Balsamum* (var. *β-Pereiræ*).

The essential oil obtained from the balsam is a highly aromatic, viscous liquid, having the following characters:—

Specific gravity	1·100 to 1·125
Optical rotation	+ 0° 30′ to + 3°
Refractive index	1·5710 to 1·5800
Acid value	25 „ 48
Ester „	200 „ 254

This oil consists principally of benzyl cinnamate and benzyl benzoate, with a small amount of free benzoic and cinnamic acids. Dihydrobenzoic acid is also probably present in the form of esters. Cinnamic alcohol, vanillin, and farnesol are also present in very small amount.

Thoms has isolated an alcohol[1] from Peru balsam oil, which he termed peruviol. This body was stated to have powerful antiseptic properties, but has not been further investigated until Schimmel & Co. took up the subject. The sample of oil upon which they have worked had the following characters: Specific gravity, 1·1200; optical rotation, + 0° 55′; refractive index, 1·57177; acid value, 36·4; and ester value, 228·2. The oil after saponification was fractionated, and after benzyl alcohol had distilled over, a light oil with characteristic balsamic odour passed over. It boiled at 125° to 127° at 4 mm., and had a specific gravity, 0·8987; optical rotation, + 12° 22′; and refractive index, 1·48982. This body appeared to be identical with Hesse's nerolidol, whilst in physical and chemical properties it closely resembles Thoms' peruviol. The characters of the various preparations were as follows:—

	Boiling-point.	Specific Gravity.	Rotation.	Refractive Index.
Peruviol (Thoms)	140° (7 mm.)	0·886	+ 13°	—
Schimmel's body, 1	133° (7 „)	0·882	+ 14°	—
„ „ 2	127° (4 „)	0·899	+ 12°	1·4898
Nerolidol (Hesse)	129° (6 „)	0·880	+ 14°	—
„ (Schimmel)	127° (5 „)	0·880	+ 13°	1·4802

It proved that the impure alcohol isolated from balsam of Peru was, in fact, identical with nerolidol. When allowed to stand for three to

[1] *Arch. der Pharm.*, 237 (1899), 271.

four weeks with phenylisocyanate both alcohols yielded a phenylurethane melting at 37° to 38°. A mixture of the two bodies suffered no depression in melting-point. The alcohols have the formula $C_{15}H_{26}O$. The alcohol from balsam of Peru is therefore mixed with a small quantity of an alcohol of higher specific gravity, the nature of which is still undetermined. Traces of benzyl alcohol were found in it, but not in sufficient quantity to account for the differences observed. Oxidation experiments did not throw any light on the question. It may therefore be safely assumed that the peruviol of Thoms consisted in the main of nerolidol, but contaminated with a substance of the same boiling-point to such an extent that its combustion figures pointed to the formula $C_{13}H_{22}O$ instead of $C_{15}H_{26}O$.

Oil of Balsam of Tolu.

Balsam of Tolu is obtained from *Myroxylon Balsamum*, var. *genuinum* It yields from 2 to 7 per cent. of essential oil having an odour recalling that of hyacinths, and having the following characters :—

Specific gravity	0·949 to 1·080
Optical rotation	+ 1° „ – 2°
Refractive index	1·5440 „ 1·5600
Acid value	5 to 30
Ester „	175 to 210

The oil contains a terpene (phellandrene ?), benzyl benzoate, benzyl cinnamate, and farnesol.

Myroxylon Balsamum, var. *punctatum*, yields a similar essential oil.

Oils of Trifolium.

The flowers of *Trifolium incarnatum* yield 0·029 per cent. of essential oil, which, according to Rogerson,[1] has the following characters :—

Specific gravity $\frac{20°}{20°}$	0·9597
Optical rotation	– 1° 48′

Power and Salway [2] obtained 0·028 per cent. of oil from the flowers of *Trifolium pratense*, having the following characters :—

Specific gravity $\frac{20°}{20°}$	0·9476
Optical rotation	+ 4° 10′

Oil of Amorpha Fruticosa.

The leaves and fruits of *Amorpha fruticosa* yield essential oils which have been examined by Pavesi.[3]

The fruits yield 0·15 to 0·35 per cent. and the leaves 0·05 to 0·08 per cent. of oils. The leaf oil has a refractive index, 1·5003 to 1·5008 at 17·5°, and the fruit oil has the following characters :—

Specific gravity	0·9019 (unripe fruit) 0·9055 (ripe fruit)
Refractive index	1·4995 (unripe fruit) 1·5004 (ripe fruit)
Optical rotation	slightly laevo-rotatory

[1] *Jour. Chem. Soc.* 97 **(1910)**, 1004. [2] *Ibid.*, 232.
[3] *Ann. Soc. Chim.*, Milano, 11 **(1904)**, 1 and 2.

The leaf oil contains a terpene not yet characterised. Cadinene is present, and possibly a second sesquiterpene not hitherto isolated, which Pavesi has named amorphene.

Shinosaki and Hoshino[1] have more recently examined the oil obtained by distillation of the fruits, from which they obtained a yield of 1·1 per cent. The oil had a pepper-like odour; its characters were as follows:—

Specific gravity	0·9125
Optical rotation	± 0°
Refractive index	1·5032
Acid value	0
Saponification value after acetylation	40·5

The higher boiling fractions contain a sesquiterpene alcohol melting at 118°.

The oil, distilled, apparently, from the fruits only, has recently been examined by Shinosaki and Hoshino,[2] who obtained a yield of 1·1 per cent. The oil, which has an odour of pepper, has the following characters:—

Specific gravity	0·9125
Optical rotation	± 0°
Refractive index	1·5032
Acid value	0
Saponification value after acetylation	40·51

Cadinene and another sesquiterpene were detected, and a sesquiterpene alcohol melting at 118°.

Oil of Robinia Pseudacacia.

The flowers of this tree yield an essential oil by extraction with a volatile solvent. The oil has a specific gravity 1·050, and is fluorescent in alcoholic solution. It contains methyl anthranilate, indol, heliotropine, benzyl alcohol, linalol, terpineol, and traces of nitrogenous bodies of the pyridine series.

Oil of Daniella Thurifera.

Lenz[3] has obtained from the oleo-resin of *Daniella thurifera* 23 per cent. of essential oil, having a specific gravity 0·9845 at 20° and optical rotation + 6°. Cadinene was detected in the oil. An oil distilled by Schimmel & Co. had the following characters:—

Specific gravity	0·9216
Optical rotation	− 5° 43′
Refractive index	1·5067
Acid value	0
Ester "	1·9

GERANIACEÆ.

Oil of Geranium.

True geranium oil is distilled from the fresh herb of several species of pelargonium, chiefly *Pelargonium odoratissimum*, *Pelargonium capitatum*, *Pelargonium graveolens*, and *Pelargonium Radula*, and its variety, *roseum*.

[1] *J. Chem. Ind. Tokyo*, 21 (**1918**), 774. [2] *Jour. Soc. Chem. Ind.*, **1919**, 38, 27.
[3] *Berichte*, 47 (**1914**), 1989.

As a matter of fact, however, the plants cultivated are varieties of these descriptions, not corresponding exactly with the plant when cultivated true to name in England. The plants are cultivated in open fields in many parts of Algeria, notably at La Trappe de Staoüeli, Castiglione, Sahel, and near Algiers and Constantine. It is also largely cultivated in the South of France, in Spain (chiefly near Valencia and in the province of Almeria), in Italy, Corsica, Réunion (Bourbon), and in Provence.

The plants are gathered a short time before the opening of the flowers, when the rose-like odour commences to develop—the leaves at the same time commencing to turn yellow. The odorous oil is contained entirely in the green parts of the plant, the coloured petals being practically odourless. The whole plant is, however, put into the still, and sometimes rose petals are added, and the resulting fine oil is sold as oil of rose-geranium.

According to Holmes [1] the plants actually used for distillation are not only varieties of the above-mentioned species, but possibly hybrids also, the particular varieties being kept as secret as possible by the growers.

Holmes states that most of the species of *Pelargonium* are natives of dry rocky slopes in South Africa, and under these conditions the plants are stunted in growth, but in order to obtain a more luxuriant growth and a larger yield of oil, plantations in Southern Europe and North Africa are mostly established on low-lying ground, kept moist by a system of irrigation, which causes a luxuriant growth, the plants attaining a height of about 2½ ft., and the stems nearly an inch in thickness, so that three crops of leaves can be distilled annually instead of only one. But although the oil is thus obtained in larger quantity, it is inferior in quality to that derived from plants grown on dry rocky hill-sides. The plant is usually gathered a little before the opening of its flowers, when the citral-like odour it possesses gives place to a geraniol or rose odour. The best time for distillation is recognised by the leaves beginning to turn yellow. About 300 kilos of the plant yield a kilogram of the oil.

The plant used at Grasse, so far as can be judged from the leaf of the plant, appears to be *Pelargonium capitatum*. *Pelargonium Radula*, var. *roseum*, is apparently also employed there. That cultivated in Bourbon is stated to be the same species, but the oil distilled there corresponds in odour with the leaf of the nearly allied species, *Pelargonium graveolens*. At Grasse the plants are propagated by cuttings, set in October in well-sheltered beds, which in time of frost are protected by straw matting placed over them. The cuttings are taken up in April and planted out in rows in fields, or on terraces for facility of irrigation, and thus treated, grow three or four feet high. At Nice they flower in August, but at Grasse, and in cooler places, in October. For distillation, stem, flowers, and foliage are cut down and put into the still.

Ducellier [2] gives the following interesting information in regard to the culture of the geranium plant in Algeria. There are about 850 hectares (1 hectare = about 2·5 acres) devoted to the cultivation of the plant, Rovigo having about 250, Chelbi about 210, Boufarik about 200, Bouinan about 160, and Mouzaiaville about 45 hectares. In addition to the above there are plantations in Staouéli, Rivet, Bourkika, Chéragas, Oran, Morocco and elsewhere. For the purpose of preparing oil only one species, viz.

[1] *P. and E.O.R.* (**1913**), 239.
[2] *Le Géranium rosat, sa culture en Algeria*, Algeria (**1913**).

Pelargonium graveolens, is grown. *Pelargonium capitatum* is usually named in literature as the parent plant of "géranium rosat," but according to R. Knuth this is an error, *Pelargonium capitatum* being a hybrid of *Pelargonium graveolens*.

The geranium is propagated in Algeria by cuttings, but it is not necessary to renew the plantations every year, as is done in the South of France, for the climate of Algeria is rather warmer, and the plants are therefore not killed by frost in the winter. It is said that in Mitidja the plant sometimes attains an age of twelve years. It often happens, as for instance in 1911-12, that the plant is in leaf throughout the year, but there are occasions in Algeria when the geranium plantations suffer severely from frost. In the South of France the geranium is treated as an annual; that is to say, every spring the fields are planted with fresh cuttings, which have been taken during the autumn and left to winter in protected spots. In the South of France there is only one crop yearly (in the autumn), but in Algeria three crops are gathered every year. In the South of France it is necessary to have recourse to intensive culture by irrigation and rich manuring, in order to make the single crop pay. In Algeria the flowers are cut for the first time in early spring; the second cut is in June, and the third, if any, in October and November.

In addition to suffering from the ravages of insects, the pelargonium plants are frequently attacked by *Cuscuta*, a parasitic plant belonging to the family of the *Convolvulaceæ*. It is also said that species of *Orobanche* (N.O. *Scrophulariaceæ*) thrive upon the geranium plants. On the other hand, according to Ducellier, the *Oidium*-species, which afflicts garden geraniums, has neither been observed in Algeria nor in other districts where the geranium is grown for distilling.

The oil yield varies according to the conditions of the soil. The average estimate in Algeria is from 20 to 30 kilos per hectare, of which the first cut yields 20 kilos and the second (and eventually also the third) from 10 to 12 kilos. Yields of 40, and even of 50 kilos oil per hectare, have been recorded.

Holmes has examined the following well-marked species. The odours have been classified by Umney and Holmes as under:—

Species Examined.	Description of Odour.
Pelargonium capitatum	Rose and diphenyl ether
,, Abbotsbury Seedling	Similar odour
,, Rose Unique	Fine rose odour
,, Rollison Unique	Rose
,, Attar of Rose	Strong rose
,, Shrubland Pet	Rose and faint strawberry
,, *Radula* "Syon"	Strong diphenyl ether and rose
,, ,, var. *roseum*	,, rose and faint ladanum
,, ,, var. *majus*	,, geranium and faint peppermint
,, *Blandfordianum*	Faint rose
,, *odoratissimum*	Tansy and faint rose
,, *fulgidum*	Butyric
,, *graveolens*	Rose and rue, slight peppermint
,, Little Gem	Peppermint, no rose
,, Lady Plymouth	,, ,,
,, *denticulatum*	Rose and ladanum, slight butyric
,, ,, var. *Sandbach*	Citronellal, ,, ,, ,,
,, *quercifolium*	Ladanum
,, ,, Clorinde	Pepper and southernwood
,, ,, (unnamed var.)	Parsley

Species Examined.	Description of Odour.
Pelargonium quercifolium minus	Southernwood and Tarragon
,, *glutinosum*	Ladanum
,, ,, Lady Lumsden	,, (slight odour)
,, *pumilum*	,, and citral
,, *cucullatum*	,,
,, *ovale*, var. *Blattaria*	Slight ladanum
,, *albescens*	Ladanum and weak citronellal
,, *vitifolium*	Citronellal and caproic ester
,, *abrotanifolium*	Diphenyl ether and butyric ester, with trace of southernwood
,, *Zonale*	Diphenyl ether
,, *crispum*, var. *maximum*	Citral and melissa
,, *crispum*, var. *minimum*	Strawberry
,, Lady Scarborough	Citral
,, *citriodorum*, var., *dumosum*	Weak citronellal and rose
,, *citriodorum*, var. *maximum*	Strong ,, ,,
,, *citriodorum*, Prince of Orange	Lemon
,, *tomentosum*	Peppermint.
,, *exstipulatum*	Pennyroyal
,, *tetragonum*	Indefinite, fruity

Charabot and Laloue[1] have published an important contribution to that portion of the chemistry of essential oils which Dr. Charabot has so especially made his own. To study the evolution of the various organic bodies found in the essential oils, a study which may eventually lead to the possibility of favourably modifying the odour of oils by artificial means, it is necessary to have an accurate conception of the actual distribution of these bodies. The paper in question deals with the distribution of the organic constituents in the geranium.

In determining the free volatile acids present, they find that the amount diminishes as one proceeds from the leaves down to the stem, the leaves themselves containing 50 per cent. more than the stems. A careful examination of the different portions of the flowering plant shows that the terpenic compounds (by which the authors evidently mean the geraniol compounds as well) are found to be entirely absent in the petioles and in the stems, being found only in the leaves. It is therefore clear that the odorous substances are elaborated in the leaves themselves. This at once explains the unusual circumstance of the complete absence of essential oil in the flower; for the oil does not at any time traverse the stems, and therefore has no opportunity of reaching the flowers.

The essential oil obtained by separation from the distillation waters was examined in comparison with that extracted from the water itself, and with the whole of the oil yielded by the plant. The following results were obtained:—

	Separated Oil.	Oil from Water.	Total Oil.
Specific gravity	0·8979	0·8930	0·8971
Rotation	− 11° 2′	− 5° 13′	− 10° 6′
Acidity	47	16·5	24
Saponification No.	63·6	23·2	57
Ester No.	16·6	6·7	15
Alcohols as geraniol	70·7 per cent.	77·5 per cent.	71·8 per cent.
Esters (*geranyl tiglate*)	7·0 ,,	2·8 ,,	6·3 ,,

[1] *Bull. Soc. Chim. Paris* (**1903**), 838.

Geranium oil differs to a certain extent in character according to the locality in which it has been produced, and the difference in odour value is even more marked than the differences in physical characters. The principal types of oil have the following characters :—

Bourbon Geranium Oil.

Specific gravity	0·888 to 0·8965
Optical rotation	− 8° ,, − 14°
Refractive index	1·4620 ,, 1·4677
Acid value	1·5 ,, 10
Ester ,,	50 to 78 (rarely below 55)
,, ,, (after acetylation)	205 to 230

Algerian Geranium Oil.

Specific gravity	0·892 to 0·904
Optical rotation	− 7° to − 11° 30′
Refractive index	1·4650 to 1·4720
Acid value	1·5 ,, 10
Ester ,,	40 to 70 (rarely as low as 36)
,, ,, (after acetylation)	200 to 230

French Geranium Oil.

Specific gravity	0·896 to 0·905
Optical rotation	− 7° 30′ to − 11°
Acid value	5 to 10
Ester ,,	50 to 68 (rarely as low as 46)
,, ,, (after acetylation)	210 to 230

Spanish Geranium Oil.

Specific gravity	0·896 to 0·906
Optical rotation	− 7° 30′ to − 11°
Acid value	2 to 10
Ester ,,	60 ,, 100
,, ,, (after acetylation)	205 ,, 235

Corsican Geranium Oil.

Specific gravity	0·896 to 0·901
Optical rotation	− 8° ,, − 11°
Acid value	3·5 ,, 5
Ester ,,	56 ,, 63

Sicilian Geranium Oil.

Specific gravity	0·894
Optical rotation	− 11°
Ester value	84
,, ,, (after acetylation)	218

According to Puran Singh[1] the geranium plant growing wild in the Nilghiris is *Pelargonium graveolens*. An experimental distillation of 500

[1] *Indian Forest Records*, vol. v., pt. viii.

lbs. of the plant, consisting of the entire stems with leaves and flowers, yielded 0·044 per cent. of oil having the following characters :—

Specific gravity at 16°	0·888
Optical rotation	− 5·39°
Refractive index at 18°	1·4600
Saponification value	96·25
" " (after acetylation)	226·08
Total geraniol	74·79 per cent.
Combined geraniol	28·19 "
Free "	46·60 "
Esters as geranyl tiglate	40·12 "

According to Furukawa,[1] three varieties of geranium, *Pelargonium graveolens*, *P. radula*, and *P. denticulatum*, have been cultivated on the Oi experimental farm near Tokyo. Of these only *P. denticulatum* resembled closely the geranium oil of commerce. This had a sp. gr. 0·886 at 20° C.; acid value, 7·17; ester value, 12·8; free alcohol as geraniol 60 per cent.; total geraniol, 63·5 per cent. The yield of oil was 5 per cent. The oil of *P. graveolens* had a sp. gr. 0·9178 at 20° C.; α_D + 4° 6′; acid value, 183·6; ester value, 4·8; total geraniol, 22·1 per cent.; free alcohol (as geraniol), 22 per cent. The yield of oil was from 1·75 to 2 per cent. The oil was found to contain a comparatively large amount of an organic acid, identified as dextro-citronellic acid. Laevo-menthone and laevo-citronellol were also isolated, by fractional distillation. *P. radula* yielded 1·5 per cent. of oil, having a sp. gr. 0·9234; α_D + 3°; acid value, 5·6; ester value, 31·7; total geraniol, 26·3 per cent.; free alcohol (as geraniol), 17·26 per cent.

Geranium oil is soluble in 2·5 to 3 volumes of 70 per cent. alcohol.

The relative proportions of the two alcohols, geraniol and citronellol, in oil of geranium oil is a matter of considerable interest, and the determination is frequently of importance. The only method available in practice for this purpose is the combination of the acetylation process, which gives the measure of the total alcohols and the formylation process, by which the geraniol is decomposed, and the citronellol estimated.

This method has been referred to under Otto of Rose, but has been more critically studied by Simmons[2] in reference to geranium oil. Simmons has shown that the determination of citronellol in the presence of geraniol by the formylation process is not strictly accurate, but the results are exceedingly useful for comparative purposes.

The formylation process, as recommended by MM. Jeancard and Satie, consists in heating 10 c.c. of oil with 20 c.c. of 98 to 100 per cent. formic acid, in a flask attached to a reflux condenser for one hour on a sand-bath. It may be mentioned that the addition of 2 grams anhydrous sodium formate per 10 c.c. oil, while it enables the mixture to be boiled steadily on a sand-bath, does not affect the result.

To prove whether the results obtained represent the true citronellol-content Simmons determined the "total alcohols" by acetylation, and the "citronellol" by formylation of (1) pure geraniol, (2) pure citronellol, and (3) a mixture of (1) and (2) in equal proportions, with the following results :—

[1] *Jour. Chem. Ind. Tokyo* (**1919**), 22, 83.
[2] *Year Book of Pharmacy* (**1913**), 565.

	Total Alcohols as Geraniol.	Citronellol.
1. Geraniol	99·6 per cent.	13·7 per cent.
2. Citronellol	[1] 100·4 „	83·4 „
3. Mixture of 1 and 2 in equal proportions	—	47·3 „

Formylation of a sample of Palmarosa oil also gave an apparent citronellol-content of 14 per cent. From these results it is evident that, assuming the above samples to represent 100 per cent. geraniol and 100 per cent. citronellol, formylation does not do either of the things it is assumed to do, *i.e.* it does not completely convert geraniol into terpene, and it does not completely esterify citronellol.

To ascertain the value of the process as a comparative test a number of African and Bourbon geranium oils have been examined by the two processes, with the following results:—

	Total Alcohols as Geraniol.	Citronellol.	Percentage Composition of Alcohols.	
	Per Cent.	Per Cent.	Geraniol.	Citronellol.
African	72·8	40·0	45	55
	70·0	42·8	39	61
	79·5	33·4	58	42
	69·3	32·0	54	46
	69·6	34·8	50	50
	76·8	33·6	56	44
Bourbon	73·0	50·2	31	69
	71·7	44·0	39	61
	70·4	46·8	34	66
	69·7	51·0	27	73

These results may be tabulated as follows:—

	Citronellol, Per Cent. on Oil.	Percentage Composition of Alcohols.	
		Geraniol.	Citronellol.
African	32 to 48	39 to 58	42 to 61
Bourbon	44 „ 51	27 „ 39	61 „ 73

compared with the following figures given by MM. Jeancard and Satie:—

[1] Or calculated as citronellol, 101·6 per cent.

	Citronellol.	Percentage Composition of Alcohols.	
		Geraniol.	Citronellol.
African	37 to 43	40 to 50	50 to 60
Bourbon	50 ,, 65	20 ,, 40	60 ,, 80

The following results are obtained from other varieties of geranium oil :—

	Total Alcohols as Geraniol.	Citronellol.	Percentage Composition of Alcohols.	
	Per Cent.	Per Cent.	Geraniol.	Citronellol.
Asian . . .	68·4	63·9	6·6	93·4
Corsican . .	69·8	30·3	57	43
Trappe de Staoüeli	71·5	27·9	61	39

Umney[1] gives the following values for a number of geranium oils :—

French	72·7	39·8
Algerian	74·1	32·9
Bourbon	73·0	44·3
Corsican	73·7	45·9
Asian (1)	72·1	51·0
,, (2)	69·1	62·3

It would appear from these results that the citronellol-content of the Asia Minor oil is undoubtedly the highest, whilst that of Algerian oil is the lowest. These observations are confirmed by the odour, indicating that the determination of citronellol-content is not only a useful factor, but closely follows the nose in judging the value of a particular oil.

The constituents of the various geranium oils are practically identical, the differences observable being almost entirely of a quantitative nature.

The principal compounds present are the alcohols geraniol and citronellol, together with esters of these two bodies. The term reuniol was at one time applied to alcohols present in this oil, but it has long been shown to be merely an impure form of one or other of the two alcohols mentioned. Linalol, borneol, terpineol, phenylethyl alcohol, and probably amyl alcohol are also present in small amounts. The esters present are principally those of tiglic acid, but small quantities of esters of acetic, butyric, valerianic, and caproic acids are also present. It is usual to express the esters of geranium oil in terms of geranyl tiglate.

L-α-pinene and β-phellandrene have been isolated from the oil, as well as menthone and traces of citral.

The oil is often adulterated with artificial esters which can be searched for as described under Oil of Bergamot.

[1] *P. and E.O.R.* (1913), 328.

TROPÆOLACEÆ.

OIL OF TROPÆOLUM.

The Nasturtium (*Tropæolum majus*) yields about ·04 per cent. of a strongly smelling essential oil, when the green plant is distilled with steam. The vessel should be well tinned inside, on account of the sulphur contained in the oil. Hofmann investigated this oil in 1874, probably distilling it without previous crushing. In the oil he found benzyl cyanide. Gadamer[1] distilled the plant, first well minced, with steam, and extracted the small quantity of oil from the distillate with ether. He found that it contained 86 per cent. of benzyl-thiocarbimide. He considers this is due to the fact that there is in the plant an enzyme which decomposes the glucoside present, with the formation of this body. He suggests that by not crushing the plant the enzyme is destroyed before it can decompose the glucoside which is distributed all through the cells, and that the benzyl cyanide obtained by Hofmann was a product of decomposition of the glucoside through the distillation.

MELIACEÆ.

CEDRELA WOOD OILS.

One of the most useful woods in Jamaica is that obtained from *Cedrela odorata*, an immense forest tree. It is a native of the Caribbee Islands and the Barbadoes. Its timber has a pleasant odour of cedar, hence it is frequently spoken of as Jamaica or Honduras cedar, and from the wood most of the ordinary cigar boxes are made. The wood yields from ·5 to 1 per cent. of oil of a light blue colour, and having the following characters :—

Specific gravity	0·923 to 0·950
Optical rotation	very variable
Refractive index	1·5040
Acid value	2 to 4
Ester ,,	40 ,, 45

The cedar-wood oils from La Plata and Punta Arenas (Costa Rica), which have been referred to under true cedar-wood oil, are probably obtained from closely related trees of this order. The oil from La Plata is of a light blue colour, has a specific gravity ·928, and is optically inactive. That from Punta Arenas is yellow, of specific gravity ·915 and optical rotation − 6°. It consists chiefly of cadinene.

Corinto cedrela wood yields an oil of specific gravity 0·906 and optical activity − 17. A Cuban wood gave an oil of specific gravity 0·923 and optical activity + 18°, consisting largely of cadinene.

An oil has also been distilled from the Brazilian cedrela wood (*Cedrela brasiliensis*), which yields about 0·5 per cent. of oil of a pale bluish colour, of specific gravity 0·9348 and optical rotation − 0° 22′. There is also a cedrela tree in Brazil, *Cedrela Velloziana*, which yields a small amount of essential oil having an odour recalling that of asafœtida, but whose characters have not been investigated.

[1] *Arch. Pharm.* (**1899**), 237.

CRUCIFERÆ.

OIL OF MUSTARD.

The essential oil of mustard is obtained by distilling the seeds of several species of *Brassica* with water. *Brassica nigra* is the principal species employed, but there is no doubt that the mustard seeds of commerce contain a substantial amount of the seeds of *Brassica juncea* and *Brassica rapa*. The seeds contain the glucoside potassium myronate, which, under the influence of a ferment, myrosin, also present in the seeds, in the presence of water, is decomposed according to the following equation :—

$$\underset{\text{Potassium myronate.}}{C_{10}H_{16}NS_2KO_9} + H_2O = \underset{\text{Glucose.}}{C_6H_{12}O_6} + SO_4KH + \underset{\text{Mustard oil.}}{CS . N . C_3H_5}.$$

The chief product of the reaction is thus allyl iso-thiocyanate (allyl thiocarbimide), a pungent and disagreeable liquid. A small quantity of the normal allyl thiocyanate is also formed, together with traces of allyl cyanide and carbon disulphide. As mustard oil consists almost entirely of allyl iso-thiocyanate, and the latter body is easily prepared artificially, there is an artificial oil on the market. This is made by distilling allyl iodide or bromide with alcoholic solution of potassium thiocyanate—a molecular rearrangement to the iso-thiocyanic radicle taking place. Thus—

$$CN . SK + C_3H_5I = CS . N . C_3H_5 + KI.$$

Pure allyl iso-thiocyanate is a liquid of specific gravity 1·017 at 10°, boiling at 151°.

Essential oil of mustard is an unpleasant smelling liquid of great pungency, having the following characters :—

Specific gravity	1·015 to 1·025 (rarely 1·030)
Refractive index	1·5267 to 1·5281
Optical activity	0°

It boils almost entirely between 148° and 155°.

A sample believed to be distilled from the seeds of *Brassica juncea* only has been examined by Schimmel & Co.,[1] and found to have the following characters :—

Specific gravity	0·995
Optical rotation	+ 0° 12′
Refractive index	1·51849

The greater part of the oil at above 160°.

Seven hundred and fifty grams oil, being fractionated three times, gave :—

From 40° to 150°	53	grams or about	7	per cent.
,, 150° ,, 160°	200	,, ,,	30	,,
,, 160° ,, 174°	160	,, ,,	20	,,
,, 174° ,, 178°	290	,, ,,	40	,,

The above figures by themselves are sufficient indication that the oil is not a uniform body. The fact that the thio-urea obtained from it, which solidified very slowly, had no clearly defined melting-point (67° to 70°) pointed to a mixture of several substances.

[1] *Report*, October, **1910**, 81.

Dimethyl sulphide, $CH_3 . S . CH_3$. The first drops of the initial fraction (boiling-point 40° to 119°) contained small proportions of a highly volatile sulphur compound of a disagreeable sweetish odour, which is probably dimethyl sulphide, from the fact that it gave precipitates with solutions of mercuric chloride and of platinum chloride, as well as because of its very low boiling-point.

Allyl cyanide, $CH_3 . CH : CH . CN$. The analytical values of a fraction amounting to a few grams, which passed over between 120° and 123°, and which also contained sulphur ($d_{15°}$ 0·8570; saponification number 47·7) pointed to the probable presence of allyl cyanide.

This suspicion was confirmed by the formation of an acid boiling at about 184°, with an odour recalling that of fatty acids. This acid was generated during the process of saponification with alcoholic potash. It was oily, and when left to stand in the cold for several days it solidified in crystals; melting-point 70°. Unfortunately, the quantity available was too small for repeated purification and for the positive identification of the substance as crotonic acid, of which the melting-point is 72°.

Allyl iso-thiocyanate, $CH_2 : CH . CH_2 . NCS$, was found to form the principal portion of the fraction, boiling between 150° and 163°. The proportion present did not exceed 40 per cent.

About 50 per cent. of the oil was found to consist of a crotonyl iso-thiocyanate, C_4H_7NCS. This substance has the following characters:—

Boiling-point	175° – 176°
Specific gravity	0·9941
Optical rotation	+ 0° 3′
Refractive index	1·52398

The thio-urea prepared from this substance melts at 69° to 70° and has the formula $C_4H_7 . NH . CS . NH_2$.

Grimme[1] records the following percentages of essential oil in the mustard and allied seeds set out below:—

Brassica oleracea acephala vulgaris	0·079
,, ,, ,, *quercifolia*	0·051
,, ,, ,, *crispa*	0·103
,, ,, *gemmifera*	0·144
,, ,, *sabanda*	0·129
,, ,, *capitala alba*	0·258
,, ,, ,, *rubra*	0·259
,, ,, *gongylides*	0·142
,, ,, *botrytis*	0·104
,, ,, *asparagoides*	0·119
,, *Rapa oleifera annua*	0·085
,, ,, ,, *hiemalis*	0·093
,, ,, *rapifera*	0·166
,, ,, *teltoviensis*	0·132
,, *Napus oleifera annua.*	0·097
,, ,, ,, *hiemalis*	0·071
,, ,, ,, *flora alba*	0·056
,, ,, *rapifera* ("Erdkohlrabi")	0·062
,, ,, ,, ("Wruke")	0·012
Raphanus sativus albus	0·164
,, ,, *niger*	0·108
,, ,, *Radiola*	0·106
,, ,, *oleiferus*	0·133

Numerous methods for the determination of "mustard oil," that is, of the thiocyanate compound, in the essential oil, or in preparations thereof,

[1] *Pharm. Zentral.*, 53 **(1912)**, 733.

or in the seeds, have been proposed. The following are amongst those of the greatest accuracy. Dieterich has proposed the following method :—

Three grams of the oil and 3 grams of alcohol are shaken in a flask with 6 grams of a 10 per cent. solution of ammonia. It should become clear after standing for a few hours, or rapidly if warmed to 50° C., and deposit crystals of allyl thio-urea (thiosinamine)

$$CS\begin{matrix}\diagup NH \,.\, C_3H_5 \\ \diagdown NH_2\end{matrix}$$

To determine the quantity, decant the mother liquor and evaporate it slowly on the water-bath in a tared capsule, adding fresh portions slowly as the smell of ammonia disappears. Then add the crystals from the flask to those in the capsule, rinsing the flask with a little alcohol, and heat the capsule on the water-bath to a constant weight. Three grams of oil should yield between 3·25 and 3·5 grams of thiosinamine, which should melt at 70° to 74°. One hundred and sixteen parts of thiosinamine correspond to 99 parts of allyl iso-thiocyanite.

Gadamer [1] recommends the following process. The mustard oil is dissolved in alcohol to form an exactly 2 per cent. solution. Five c.c. (4·2 grams) of this solution are allowed to remain with 25 c.c. of deci-normal solution of silver nitrate and 5 c.c. of ammonia for twenty-four hours in a well-stoppered 50 c.c. flask. It is then made up to 50 c.c. with water and filtered from the precipitated silver sulphide; 25 c.c. of the filtrate are mixed with 4 c.c. of nitric acid and a few drops of ferric sulphate solution, and titrated with deci-normal ammonium thiocyanate solution, until the characteristic red colour of the ferric thiocyanate appears.

From 4·1 to 4·5 c.c. of the solution (corresponding to 1·85 to 2·0 per cent. of allyl thiocyanate in the alcoholic solution) should be required.

Grützner [2] converts the thiocyanate into thiosinamine, which he oxidises with peroxide of sodium, and weighs the resultant sulphuric acid as barium-sulphate. From the figures obtained in his analysis, Grützner concludes that a mustard oil containing 28·60 per cent. of sulphur (equivalent to 88·48 per cent. of iso-thiocyanate) may be regarded as pure. P. Roeser [3] proposes to modify Gadamer's method for the determination of the sulphur content of oil of mustard so as to determine the excess of silver nitrate, in an ammoniacal solution, instead of an acid solution according to Volhard's method, as is usually done. According to this, the process should be as follows: When the conversion of the thiosinamine with silver nitrate, after twenty-four hours' standing, has taken place, an excess of one-tenth normal solution of potassium cyanide is added to 50 c.c. of the clear filtrate, and the excess of potassium cyanide titrated back with one-tenth normal solution of silver nitrate, in the presence of a few drops of a weak ammoniacal solution (5 per cent.) of potassium iodide.

Schimmel & Co. recommend the following method: About 5 grams of a solution of 1 gram mustard oil in 49 grams alcohol are mixed in a measuring flask of 100 c.c. capacity with 50 c.c. deci-normal solution of silver nitrate and 10 c.c. of solution of ammonia ($d_{15°}$ 0·960); the flask is then closed, and with frequent agitation left standing for twenty-four hours with the light excluded. The flask is then placed for half an hour in water at 80°, during which time it is again repeatedly shaken, then

[1] *Arch. Pharm.* (**1899**), 110, 237. [2] *Ibid.*, 195.
[3] *Jour. Pharm. Chim.*, vi. 15 (**1902**), 361.

cooled down to the temperature of the room, filled up with water to the mark, shaken up, and filtered. Fifty c.c. of the filtrate are titrated with one-tenth normal solution of ammonium sulphocyanide, after adding 6 c.c. nitric acid (d_{15° 1·153) and a small quantity of solution of iron alum, until a change of colour from white to red takes place. In order to ascertain the whole quantity of silver solution which has entered into reaction, the number of c.c. of ammonium sulpho-cyanide solution used up is doubled, and the product subtracted from 50. The percentage of allyl iso-thiocyanate in the mustard oil is obtained by means of the following formula:—

$$\text{Per cent. } CSNC_3H_5 = \frac{a24{\cdot}7875}{b}$$

a = number of c.c. of deci-normal solution of silver nitrate used up, b = spirit of mustard used, in grams. Mustard oil determinations carried out by them in the manner described showed in the case of natural oil a content of about 94 per cent. allyl iso-thiocyanate, whilst in artificial oil about 98 per cent. was determined.

Kuntze[1] has examined the differences to be observed in the results obtained by the gravimetric and the volumetric processes above described, those of the gravimetric process being higher than those of the volumetric processes.

Kuntze finds that the precipitate of silver sulphide, which is formed by both methods under the same conditions, is always contaminated with more or less silver salt of allyl-thiocarbaminic acid ester.

$$C{=}S \begin{cases} OC_2H_5 \\ NHC_3H_5 \end{cases}$$

so that the ester which is always present even in freshly-prepared alcoholic solutions of mustard oil does not decompose quantitatively with ammoniacal silver solution, with formation of sulphide of silver, but is partly precipitated as a silver salt of the above-mentioned ester. Whilst this admixture, owing to its molecular weight 252, which differs but little from that of silver sulphide 248, cannot cause an appreciable error in the gravimetric determination, it caus s a considerable difference in the volumetric estimation, because in the formation of this ester salt only 1 atom of silver corresponds to 1 molecule of mustard oil, whereas for the conversion into silver sulphide, 2 atoms of silver are required, so that the titration results are obviously too low.

Kuntze has found that the separation of the ester salt can be avoided by immediately heating the alcoholic mixture for one hour, because the salt is unstable at a high temperature, and proposes the following volumetric method to deal with this difficulty:—

Five c.c. of the alcoholic solution of mustard oil are mixed in a graduated tube of 100 c.c. capacity with 10 c.c. of solution of ammonia and 50 c.c. deci-normal silver nitrate solution, and the whole is immediately heated for one hour on a briskly boiling water-bath under a reflux condenser. After cooling to 15°, and filling up to the mark with distilled water, adding nitric acid until a feeble acid reaction takes place, and 1 c.c. ferri-ammonium sulphate solution, deci-normal ammonium thiocyanate solution should be added to the clear filtrate, until a red coloration is produced.

[1] *Arch. der Pharm.*, 246 (**1908**), 58.

Brioux[1] describes a method of estimating allyl and crotonyl iso-sulphocyanate in oil-cakes and mustard-flours. He places 25 grams of the finely crushed oil-cake in a flask of one litre capacity, adds 500 c.c. of distilled water, 2 grams of sodium fluoride dissolved in a little water, then heats in the water-bath at 37° to 40° C. after having closed the flask with a cork. If the oil-cakes or flours are derived from foreign varieties, it is advisable to prolong the duration of heating to three or four hours. To the same flask are then added 20 to 25 c.c. of 95 per cent. alcohol, and the liquid is distilled on the sand-bath; 150 c.c. of the distillate are collected in a 250 c.c. marked flask, previously containing 10 c.c. of pure ammonia; then 25 to 50 c.c. of deci-normal silver nitrate solution are added and the flask is heated for an hour under a reflux condenser in a water-bath maintained at about 85° C. After cooling, the volume is made up to 250 c.c. the liquid is shaken and filtered, and finally 125 c.c. are taken for the estimation of the excess of silver by means of deci-normal ammonium sulpho-cyanate. Before making this latter titration care must be taken to add 10 c.c. of pure nitric acid and 10 c.c. of a 10 per cent. solution of iron alum. From the quantity of silver necessary for the formation of the precipitate of silver sulphide, the percentage of crotonyl or allyl mustard oil is readily calculated.

Huber and Van der Wielen[2] determine the volatile oil (thiocyanate) in mustard seeds as follows. Their experiments were directed towards determining to what extent the time of maceration of the crushed seeds in water influences the result. Five gram samples of the mustard were macerated for 1, 2, 4, 18, and 20 hours respectively, with 100 c.c. of water, after which were added 20 c.c. of alcohol and 2 c.c. of olive oil. Of the mixture about 50 c.c. were distilled into a 100 c.c. measure containing 10 c.c. of ammonia, taking care that the delivery tube was immersed in the ammonia. After adding 20 c.c. of deci-normal silver nitrate solution the whole was heated over a water-bath until the silver sulphide had aggregated and the liquid was clear. The liquid was then cooled and made up with water to 100 c.c., the excess of silver nitrate was determined by titration with deci-normal ammonium thiocyanate, using iron alum as an indicator.

The following results were obtained:—

After 1 hour of maceration	0·9 per cent.	and	0·9 per cent.
,, 2 hours ,,	1·15 ,,	,,	1·15 ,,
,, 4 ,, ,,	1·03 ,,	,,	1·03 ,,
,, 18 ,, ,,	0·7 ,,	,,	0·7 ,,
,, 24 ,, ,,	0·77 ,,	,,	0·77 ,,

From the above table the ideal period of maceration is two hours. Having once added the alcohol to arrest fermentation, the time the liquid is allowed to stand before distillation seems to have no influence upon the results; after the two hours' maceration and addition of alcohol, it was allowed to stand overnight.

White mustard seeds, from *Brassica alba*, contain the glucoside sinalbin, $C_{30}H_{42}N_2S_2O_{15}$, which on decomposition in the same manner as the glucoside of black mustard yields glucose, sinapine sulphate, and the evil-smelling oil, acrimyl-thiocarbimide (*p*-hydroxy-benzyl-iso-thiocyanate). The reaction takes place as follows:—

$$C_{30}H_{42}N_2S_2O_{15} + H_2O = C_6H_{12}O_6 + C_{16}H_{23}NO_5 . H_2SO_4 + C_7H_7O . NCS.$$

[1] *Ann. Chim. Anal. Appliquée*, **17**, 3. [2] *Pharm. Weekblad* (**1915**), 39.

Acrimyl iso-thiocyanate, or "white mustard oil," is a yellowish oily liquid of pungent odour and unpleasant hot taste. It is prepared synthetically by treating *p*-hydroxy-benzylamine with carbon disulphide, and the resulting compound with mercuric chloride.

Oil of Spoonwort.

Cochlearia or spoonwort oil is distilled from the fresh-flowering plant *Cochlearia officinalis* (scurvy grass) which yields a very small quantity of the essential oil. The specific gravity of the oil is about 0·940 to 0·960, and it is dextro-rotatory to the extent of + 70° or thereabouts. Its refractive index is about 1·4900. Hofmann showed many years ago that the principal constituent was secondary butyl-iso-thiocyanate, of the formula.

$$CS \,.\, NCH \begin{matrix} \diagup C_2H_5 \\ \diagdown CH_3 \end{matrix}$$

Gadamer[1] has shown that if the dried plant without flowers is cut up, and some white mustard flower is added, the yield of oil is much increased. From 28 kilos of the dried plant, which is known as scurvy grass, 66 grams of essential oil were obtained. The oil boiled at 150° to 162°, and the bulk of the oil was found to be secondary butyl-iso-thiocyanate. A little benzyl-iso-thiocyanate is also present. The amount of iso-thiocyanate was estimated by dissolving a weighed quantity of the oil in alcohol, adding excess of deci-normal solution of silver nitrate, allowing the whole to remain for twenty-four hours in a well-stoppered bottle, filtering from the precipitated silver sulphide, and titrating the excess of silver with ammonium thiocyanate solution. The fractions of the oil of higher boiling-point (156° to 162°) had a lower specific gravity and higher optical rotation than the earlier fractions. They possibly contained a little limonene.

In order to determine the purity of the oil a quantity may be heated on the water-bath with 10 per cent. of its weight of ammonia for a few hours under a condenser, and the liquid evaporated to dryness. The mass is then extracted with alcohol and the liquid filtered and evaporated on a watch glass; the resulting crystals of crude secondary butyl-thiourea melt at 125° to 135°. Pure secondary butyl-thiourea melts at 136° to 137°. An artificial cochlearia oil exists, but appears to consist of iso-butyl iso-thiocyanate. The crystalline thiocarbimide yielded by this melts at 93·5°. The usual adulterants can be thus detected, as the thiocar-bimides (thioureas) formed from them melt at the following temperatures:—

Allyl (mustard oil)	72° to 74°
Normal butyl	79°
Iso-butyl	93·5°
Tertiary butyl	165°

Oil of Wallflower.

The flowers of the wallflower, *Cheiranthus Cheiri*, yield about 0·06 per cent. of essential oil by extraction with a volatile solvent and distilling the extract with steam. This oil has been examined by Kumnest.[2] It is of a yellow colour, and in alcoholic solution has a blue fluorescence, and has the following characters :—

[1] *Arch. Pharm.* **(1899)**, 237. [2] *Chem. Zeit.* **(1911)**, 667.

Specific gravity at 15°	1·001
Distils at 3 mm. pressure	from 40° to 150° C.
Acid value	0·35
Ester ,,	20·0
Saponification value	20·35

A very small amount distilled at 40°, and on analysis was found to contain both sulphur and nitrogen. It is an evil-smelling compound, and is probably related to the alkaloid cheiranthin.

Crystalline semi-carbazones were prepared, which, when decomposed by oxalic acid, yielded aldehydes and ketones, having odours recalling those of the ketone, irone, and of anisaldehyde. The oil, freed from these bodies and saponified, was then treated by the phthalic acid method, and a mixture of alcohols separated. Crystalline diphenyl-urethanes were obtained, which melted sharply at 50° and at 82°, thus proving the presence of nerol and geraniol. The presence of benzyl alcohol was proved by the preparation of a phthalic acid ester melting at 106°. Linalol is also probably present in traces. From the saponification residues traces of a phenol (probably para-cresol) lactone (of coumaric acid ?), acetic and salicylic acids were obtained. Indol and methyl-anthranilate were also isolated in a state of purity.

MAGNOLIACEÆ.

Star-Aniseed Oil.

The aniseed oil of commerce, imported from the Far East, is distilled from the fruit of *Illicium verum*, and is known as star aniseed oil. A certain amount of aniseed oil is also distilled from *Pimpinella Anisum*, one of the Umbelliferæ, and is described here as a matter of convenience.

Pimpinella anisum is indigenous to Asia Minor and Egypt, and is also cultivated in Russia, Spain, Malta, Greece, Bulgaria, Chili, India, and several other parts of Europe. Russia produces the greater part of the anise fruit used for distillation, after which Asia Minor comes chiefly into consideration.

In Russia, from which country most of the fruit used for distillation is obtained, farmers in the districts of Walcysk, Birjutschensk and Ostrog, and to a less extent Podolia, Kursk, Charkow, Tauria and Cherson, cultivate the plant, a portion of the fruit being distilled on the spot, or used for domestic purposes, the remainder being exported. The commercial centres of the anise trade are Krassnaja and Alexejewskaja. The yield of oil varies from about 2 to 3 per cent., rarely, a little higher.

The star anise (*Badiane*) is chiefly found in Southern China and Tonkin. The earliest reliable information as to the actual habitat of this plant was furnished by Mr. Piry in his report on the trade at Pakhoo for the years 1878-79, in which he states that the fruit is brought for export to Kin Chow and Pakhoi from Kwang-Si, chiefly from Lung-chow on the borders of Annam, and Po-Sé on the West River close to Yunnan. The oil is chiefly distilled by the natives, and sold to merchants who transfer it to Hong-Kong, where it is bought by the exporters and sent to Europe, sometimes in a grossly adulterated condition. According to Simon, the adulteration is practised not by the distillers, but by the "Bande Noir," as he terms the Chinese merchants who combine to buy the oil from the native distillers, keeping them more or less under their dominion, by means of money advances in the early part of the season.

European merchants, however, now have their agents both in the Kwang-Si and Tonkin districts, who collect the oil immediately it is distilled and transport it either to Hong-Kong, Pakhoi, or Haiphong, from where it is shipped to Europe. The greatest care is necessary in purchasing this oil on account of the heavy adulteration, especially with petroleum, practised by the Orientals. In Annam the process of distillation is as follows. About 20 lb. of the fruit are placed in an iron pan and covered with water. Upon this is placed, bottom upwards, so as to form a cover, a second iron pan pierced with a circular opening, over which is placed an earthen vessel with three small orifices in the lower part, which allow the access of vapour into this vessel. These orifices are covered on the inside by small ear-shaped hoods which cause the vapour to be spread over to the sides of the vessel. This earthen vessel is covered with an iron pan, which performs the part of a refrigerator, and into which a continuous current of cold water is admitted. The joints of the first two iron pans are luted and the earthen vessel and the refrigerator are jointed with rag bandages. The vapour reaches the earthen vessel and is condensed when it strikes the bottom of the refrigerator, and falls into a circular trough at the bottom of the vessel, whence it escapes through a small pipe. The average annual production of oil is as follows:—

District of Po-Sé	90,000 kilos.
,, Lung-Chow	30,000 ,,
,, Lang So	30,000 ,,

The Chevalier d'Hanoi gives the following account of the tree as grown in the Tonquin districts:—

"The particular species producing the oil belongs to the N.O. *Magnoliaceæ*, the *Illicium verum* (Hooker), also named by Loureiro, *Illicium anisatum*, but this must not be confused with the *Illicium anisatum* of Linnæus and Gaertner. This latter, which grows wild in Japan and China, and has been met with also in Tonquin by Eberhardt, yields an essence, not only useless, but, according to certain writers, actually poisonous. There are also other species of *Illicium* equally valueless, notably *Illicium Griffithii*, of India, and its Tonquin variety, *Illicium Cambodianum*.

"*Illicium verum* is a little tree from 8 to 15 metres high. The trunk, frequently bifurcate at the base, is about 25 to 30 centimetres thick, and rises to a height of 1½ or 2 metres without branches. The branches are much subdivided, and have very heavy foliage, giving to the tree a pyramidal appearance like the cypress.

"The leaves are evergreen, the flowers appear twice a year, and the young fruits are star-shaped, giving to the tree the name of 'star anise'. The various species may be distinguished by the appearance of the fruits. Those of *Illicium anisatum* are covered with hairs, those of the other species above named, and notably *Illicium verum*, are smooth, but while the fruits of *Illicium Griffithii* and its varieties have stars of ten to thirteen points, *Illicium verum* has stars of eight points (rarely seven or nine). The last named—the only one of interest to us—has certain variations which it might be desirable to study in view of a possible selection. For instance, we have observed in the neighbourhood of Langson certain rare specimens producing exclusively fruits of nine carpels, and another form which has unusually fat carpels, presumably richer in essence.

"In Tonquin the natives cultivate it in the north-east on the Chinese frontier, chiefly within a radius of 20 to 30 kilometres of Langson."

The yield obtained is from 2·5 to 5 per cent. In Europe the oil distilled from the following growths of *true* anise yielded the following results (Schimmel):—

Chilian . .	1·9 to 2·6	Italian . .	2·7 to 3·5
Macedonian .	2·2	Moravian . .	2·4 „ 3·2
Mexican . .	1·9 „ 2·1	East Prussian .	2·4
Russian . .	2·4 „ 3·2	Spanish . .	3·0
Syrian . .	1·5 „ 6·0	Thuringian . .	2·4

The following interesting account of the Russian anise seed trade is due to the *Handelsgesellschaft Anis* :—[1]

"In Russia anise is grown in the county of Voronetz, in the districts of Biriutch, Ostrogojsk, and Valuiki. The seed was originally introduced from Spain by a Prince Tcherbatoff, and experimentally cultivated by him on his estate at Krasnoie (district of Valuiki), whence its cultivation gradually extended until it covered a radius of about 200 versts, within which it has remained restricted for many years. Experiments in anise growing further north and south have been unsuccessful owing to climatic conditions; in the north because of the long time which anise requires to reach maturity (about four and a half months in the climate there prevailing), and in the south because there the anise will not develop fully owing to the great heat and drought. It is therefore only possible to grow the plant within certain zones where the climatic conditions approximate to those of the county of Voronetz (average temperature about 15° C. in April, 18° in May, 20° in June and July, and 18° in August).

"The seed begins to sprout about three or four weeks after sowing, the plants developing very slowly; when they have gained some little strength the field is carefully weeded, a second weeding taking place a fortnight later, when the plants must be thinned out where necessary. The central umbel is the first to make growth, the branch umbels only developing later; hence the former flowers first. The all-round flowering begins about two months after the germination of the seed and lasts two or three weeks. During the flowering time the anise plants are exceedingly sensitive, and most of the crop failures are traceable to unfavourable weather conditions during this period. Powerful sunshine following quickly after rain or dew causes the flowers to dry up, and the same result is brought about by the cold east and hot south winds (the latter being known as 'viuga,') which are particularly frequent in the first half of July, just when the plants are flowering. These winds only prevail locally, but scarcely a season passes in which they do not cause some damage. In the year 1903, for instance, a 'viuga' at the beginning of July destroyed almost all the anise fields, and the estimate of the crop fell within a few days from 250,000 to 87,000 poods. The seed harvested in that year was mostly that of the central umbels, very dark, light in weight, and poor in oil. After the flowering is over about one month passes before the seed is ripe; the seed sets first, then it develops, and finally it matures. As soon as the seed is ripe its apex turns a greyish-green colour.

"Within a week after the gathering the ripening of the fruit is completed, and the seed is then threshed upon a flat floor, which has previously been prepared from clay mixed with water. The seed is mostly threshed with flails, less frequently by running a cart over it. The threshed anise is then mechanically freed from stalk, chaff, and the

[1] Schimmels' *Report*, October, 1913, 22.

larger particles of earth in a winnowing machine. The average yield of seed is from 40 to 50 poods per dessiatine.

"The seed is of a greyish-green colour. Quite green seed is obtained when the anise is harvested in an unripe state. It is extremely rare to find anise of a quite uniform colour, some proportion of dark seed, more or less, being generally present; this seed is usually that of the central flowering-head which has ripened first, or of plants which have been affected by the conditions of the weather. For even after the fruit has set the anise plant is still greatly affected by the weather, being liable to damage by rain or dew immediately followed by hot sun; where such damage occurs the colour of the seed changes quickly from green to yellow, brown, and deep blackish-brown.

"The seed is rarely cleaned effectively, owing to the great cost of cleaning (1·50 roubles per pood on an average). It is invariably admixed with some immature seed (from the branch-umbels), chaff, particles of earth (either the remains of earth adhering to the roots or collected on the clay threshing floor) and the seed of various weeds, especially of coriander, a plant which cannot be entirely eradicated from an anise field.

"Apart from these natural admixtures (about 7 to 8 per cent. of foreign substances), deliberate adulteration of anise is much practised by the farmers. Not only do they add earth-particles of a size and colour corresponding with those of anise, but also the seeds of different weeds which are indistinguishable from anise at the first sight. Recently the seed of another umbelliferous field-plant has also been added, and this form of sophistication is no doubt the most dangerous, because this seed closely resembles anise in its colour, and is extremely difficult to recognise even by the most careful examination. This seed has a faintly bitter taste.

"In order to form an estimate of the coming crop it is necessary to make two journeys through the anise country, the first at the beginning of July, with the object of collecting particulars of the area sown from the 'volostines' (district administrators). These figures are generally readily supplied by the officials for a small consideration. When all the figures have been collected it is most important to form a correct estimate of the fields in the various districts, and to do this, practical experience is necessary. The particulars relating to the sowing cannot be gathered before July, as they are not prepared until then, and the second journey must be made when the harvesting is in full swing, because then the anise crop is no longer subject to any further dangers. Moreover, the evaluation of the seed while the crop is standing is a safe guide, in particular because it is also possible then—that is to say at gathering time—to find out the number of bundles to a given superficial area, and to see the size, weight, and colour of the seed. When the latter has been good, and the prices also have been satisfactory, the farmers usually increase their sowings in the following season, and *vice versa*."

The two oils are practically identical, except that the true aniseed oil has a rather more delicate odour and flavour than the star aniseed oil. The oils have the following characters:—

Star Aniseed Oil.

Specific gravity at 20° .	0·980 to 0·990
Optical rotation . . .	0° to − 2° (rarely slightly dextro-rotatory)
Refractive index . .	1·5530 to 1·5565
Congealing-point . .	15° ,, 17°
Melting-point . . .	16·5° ,, 19°

The oil is soluble in 3 volumes of 90 per cent. alcohol.

True Aniseed Oil.

Specific gravity at 20°	0·980 to 0·990
Optical rotation	0° to − 2°
Refractive index	1·5560 to 1·5590
Congealing-point	16° ,, 18°
Melting-point	17° ,, 20°

The oil is soluble in 3 volumes of 90 per cent. alcohol.

The usual adulterants are petroleum, fennel oil, or its stearoptene, and the waste liquid portion of the oil obtained in the manufacture of anethol (the solid odorous constituent of the oil). In addition, spermaceti has been said to be added, but the author has never met with a sample adulterated with this body. The value of the oil depends on the quantity of anethol it contains, and as this melts at 21° to 22° and boils at 232°, the melting-point and the behaviour on fractionation of the oil yield most valuable information as to its purity. With regard to the latter, the frac ion obtained between 225° and 235° should not be less than 80 per cent.—usually varying from 80 to 85 per cent.—corresponding to practically that quantity of anethol. Anethol has a great tendency to exist in a state of superfusion, so that the oil can be easily cooled down, if not agitated, below its normal solidifying-point, consequently it is necessary either to solidify the oil in a very thin test-tube, and taking its melting-point, or to determine its solidifying-point in Beckmann's apparatus.

Schimmel & Co. describe the method they adopt in the following terms, and the author considers it yields very concordant results:—

"For the determination of the point of solidification of essential oils, the well-known Beckmann's apparatus for the determination of the molecular weight from the decrease of the freezing-point may well be used. A few small modifications make it particularly suited for the examination of anise and star-anise oils. These slight alterations consist mainly in the removal of the cork-connections, by which the full view of the mercurial column of the thermometer is impeded. The lateral branch of the freezing tube of the Beckmann's apparatus, serving for the introduction of the substance to be examined, can also be omitted. The battery jar serves for receiving the refrigerating fluid. The test tube suspended in the metal cover serves as aircover around the freezing tube, preventing the premature solidification of the oil to be examined. This tube is wider at its top, narrowing down at the junction where it rests on the rim of the lower tube. In order to support the freezing tube, three glass protuberances are affixed on its inside about 5 c.c. below its upper rim. The thermometer, divided into ½ degrees, is supported in a metallic cover by three small springs, permitting a ready displacement both upwards and downwards.

"For making the determination for anise and star-anise oils, the jar is filled with iced water and broken ice, for fennel oil with a freezing mixture consisting of broken ice and common salt. Then the sample of oil to be tested is introduced into the freezing tube, reaching to a height of about 5 cm., and the thermometer is inserted into the oil with the precaution not to touch the wall of the tube in any place. During the cooling process any concussions have to be avoided, as they tend to cause a premature solidification of the oil. When the thermometer has reached about 10° below the point of solidification, this being at 6° to 8° for anise and star-anise oils, the solidification is induced by gentle friction with the thermometer upon the wall of the tube holding the oil. Should this simple procedure fail in its object, a minute crystal of solidified oil or of anethol may be introduced into the fluid, whereupon the crystallisation sets in at once, disengaging considerable heat. It may be accelerated by constantly stirring with the thermometer, whose mercurial column rapidly rises, soon reaching the maximum height, called the solidification-point of the oil."

The solidification point taken in this way, or the melting-point, should not be below 15°. It usually varies in pure samples between 15° and 18°, according to the percentage of anethol. A quite satisfactory method of determining the solidifying-point is to carefully supercool a small amount in a test tube—say to 10°, add a crystal of anethol and stir with the thermometer, which will now rapidly rise to the correct solidifying-point. Small quantities of petroleum do not bring this point below the lowest limit for a pure oil, but even 2 or 3 per cent. greatly interferes with the solubility in alcohol. The pure oil is soluble in 3 volumes at most of 90 per cent. alcohol, whereas an oil adulterated with 5 per cent. of petroleum will not dissolve in 10 volumes. On keeping for any considerable time, the oil becomes partially oxidised, with the formation of anisic aldehyde and anisic acid with a lowering of the melting-point and raising of the specific gravity. A few years ago some samples of aniseed oil appeared on the German and London markets, which possessed an abnormally low melting-point. No adulterant could be detected in these, although the melting-point varied from 8° to 11°. Two views have been taken on this point. Umney has examined a sample of star-anise leaf oil, which, according to Simon, is largely distilled in the Po-Sé district. He finds its specific gravity to be ·9878, its rotation + 1°, and its solidification-point below that of the ordinary fruit oil. A larger percentage of anisic aldehyde was also indicated by its behaviour on fractionation, and towards the usual aldehydic reagent—fuchsine decolorised by sulphurous acid. Possibly the oil in question was distilled from leaves and fruits. The other view is that the flowers, or very early fruits, are distilled and yield an oil with a much lower anethol content. It is believed that some of the flowers of very young fruits are removed in order to allow the remainder to develop to the best advantage, and instead of being thrown away, these are distilled. Some oil from Tonkin, said to be oil of this description, absolutely free from petroleum and apparently quite authentic, gave the following results:—

Specific gravity	·9916
Rotation	+ 0° 23′
Solidifying-point	12°.

Gross adulteration with up to 40 per cent. of petroleum oil was found in a number of samples a few years ago, but the great alteration in the physical characters of the oil render this easy of detection.

The following figures represent samples of this type:—

Specific gravity	0·950
Refractive index	1·5280
Congealing-point	8°
Melting-point	10°

Petroleum oil separates in a nearly pure condition on shaking the oil with 10 volumes of 90 per cent. alcohol.

As indicating the effect of anise-leaf oil, Umney [1] gives the following comparative figures for the leaf and fruit oils :—

	Leaf Oil.	Fruit Oil.
Specific gravity at 15·5°	0·9878	·975 to ·990
Optical rotation	+ 1°	0 „ − 2°
Fractionation :—		
Below 225° C.	10 per cent.	20 per cent.
From 225° to 230° C.	60 „	65 „
Above 230° C.	30 „	15 „

The author [2] and J. C. Umney [3] have examined a number of commercial samples, which had abnormal characters, and which had either been adulterated with fractions of camphor oil, or had had part of the anethol abstracted.

The samples examined by the author had the following characters :—

Specific gravity at 20°	0·9575 to 0·9740
Refractive index at 20°	1·5460 „ 1·5475
Melting-point	12° „ 15°
Congealing-point	10·5° „ 12°
Optical rotation	+ 0° 15′ „ 0° 35′

It will be noticed that the specific gravity is abnormally low, and does not correspond with pharmacopœial requirements, while the melting and congealing-points are also very low. The refractive index is lower than has ever been found for pure aniseed oil, which has a value between 1·5530 and 1·5565, or thereabouts. The slight dextro-rotation indicates nothing, as some samples of pure star-anise oil are faintly dextro-rotatory.

A large sample from one shipment was examined more fully, and finally a bulked sample made from a number of the above-described samples, and fractionally distilled, side by side with a pure anise oil. The results of this distillation are embodied in the following table :—

	Adulterated.	Pure.
Specific gravity at 20°	0·9720	0·982
Optical rotation	0°	− 1° 10′
Refractive index at 20°	1·5469	1·5547
Melting-point	13°	18°
Congealing-point	11°	15°
Refractive index of 1st 10 per cent.	1·5160	1·5346
Melting-point „ 10 „	− 3° (partial solidification)	8°
Refractive index of 2nd 10 „	1·5208	1·5496
„ „ last 20 „	1·5403	1·5560

All the immediate fractions had lower refractive indices and lower melting-points than the corresponding fractions of the pure oil, and the amount distilling between 225° and 235° was only 69 per cent., against 87 per cent. for the pure oil. The residue of 20 per cent. was less soluble in alcohol than the corresponding fraction of pure aniseed oil. The odour of the adulterated oils was characteristic, and quite different from that of pure oil. These figures proved conclusively that these samples were not genuine anise oil, and, taken in conjunction with the peculiar odour and taste, indicate strongly that the adulteration took the form of the addition of some foreign oil. The most probable adulterant is some fraction of camphor oil.

[1] *P. and E.O.R.* **(1914)**, 73. [2] *Chemist and Druggist* **(1910)**, 687.
[3] *P. and E.O.R.* **(1910)**, 236.

The following bodies have been detected in star aniseed oil: *Dextro*-α-pinene, phellandrene (in several modifications), cymene, cineol, dipentene, *l*-limonene, α-terpineol, methyl-chavicol, hydrokinone ethyl ether, safrol, a sesquiterpene, anise ketone $C_6H_4(OCH_3)(CH_2 . CO . CH_3)$, and anethol.

True aniseed oil has not been so fully investigated, and the only bodies so far detected in this oil are anethol, methyl-chavicol, anise ketone, and traces of acetic aldehyde.

The fruit of a plant, *Seseli Harveyanum*, known in Victoria (Australia) as aniseed, but which more closely resembles fennel, yields an oil which has been examined by Umney, who found its specific gravity to be ·914 and its rotation + 14°. It was still liquid at 4°. The oil thus resembles fennel more than anise oil.

A false aniseed, also belonging to the N.O. *Magnoliaceæ*, is found in Japan under the name "Shikimi no Ki". It is the *Illicium religiosum*, and its oil is known as Shikimi or Japanese star-anise oil. This oil has the following characters:—

Specific gravity	0·980 to 20
Optical rotation	0° to − 4° 30′

It contains eugenol, cineol, and safrol, and a terpene, probably limonene, with some sesquiterpene. The oil from the leaves has a specific gravity 1·006, and optical rotation − 8·6°. It contains eugenol, safrol, and a terpene. The so-called anise bark oil, from the bark of *Illicium parviflorum*, is an oil of spicy taste, and odour recalling sassafras and tarragon. It has a specific gravity ·969 and optical rotation − 0° 46′. It consists chiefly of methyl-chavicol, containing very little anethol.

Oil of Magnolia Kobus.

The oil obtained from the fresh leaves and branches of the Kobushi tree (*Magnolia Kobus* D.C.) to the extent of about 0·45 per cent., has a bright yellow colour, and possesses the following characters:—[1]

Specific gravity	0·964
Optical rotation	− 1° 6′
Acid number	1·5
Ester "	8·87

It is soluble in 1·2 volumes 80 per cent. alcohol; the greatly diluted solution shows opalescence. The odour of the oil reminds of sassafras oil, and gives rise to the presumption that it contains a large proportion of safrol; the oil also contains among others small quantities of citral. The species of magnolia which yields this oil occurs chiefly in the central districts of Japan; the distillation of the oil takes place from July to September.

A sample examined by Charabot and Laloue[2] had a specific gravity 0·943 and optical rotation − 1° 20′. These chemists detected citral (about 15 per cent.) and anethol, with small quantities of alcohols and esters. Methyl-chavicol is also probably present. Cineol is also, according to Schimmel, a constituent of the oil.

Oil of Champaca Flowers.

The so-called champaca wood oil, commonly known as guaiacum wood oil, must not be confused with the true champaca oil. The first-

[1] Schimmel's *Report*, October, **1903**, 78. [2] *Compt. Rend.* 146 (**1908**), 183.

FIG. 48.—Young Champaca tree growing in a garden of Manila.
[*Roure-Bertrand Fils.*

named is the product of an entirely different plant. The true champaca oil is distilled from the flowers of *Michelia Champaca*, a plant cultivated, and also growing wild, in the forests of the Himalayas, from Nepul and Kumaon eastwards; and also in the Nilghiris and Travancore, Java and the Philippines. The flowers, which are of an exquisite odour, are not unlike a double narcissus. The essential oil distilled from the flowers is known in India as *Pand* or *Champa-ka-utter*. The oil is known as *Yellow Champaca*, whilst that from *Michelia longifolia* is known as *White Champaca*.

The following interesting account of the plant is due to Professor Perrot,[1] director of the laboratory of Materia Medica at the Higher School of Pharmacy of Paris:—

"*Michelia Champaca* L. (*Michelia rufinervis* D.C., *Michelia aurantiaca* Wall, *Michelia Rheedi* Wight, Magnoliaceæ) is a native of India. It is found there in all parts of the country more or less plentifully, but it occurs chiefly in tropical India: in Nepaul, Assam, and Bengal. In the South and in the Himalaya district it still exists in the wild state, but is less generally distributed. The tree is applied to numerous uses and is known by different names according to the district: *Champaca*, *Champaka* in Bengal, *Tita-sapa* in Assam, *Oulia Champ.* in Nepaul; it is also known in other parts of the country under the names of *Shampang*, *Shimbu*, *Sempangam*, *Sampige*, *Champakam*, *Saga*, *Sapu*, etc. It is also cultivated in the Philippine Islands. This large tree, which is an evergreen, prefers sandy soils. It is of fairly large size, since a man can scarcely encircle the trunk with his arms.

"The flowers, of a yellow colour, are produced at the tops of the branches and are carried by stout peduncles of a yellowish-green colour, polished and erect. The number of flower heads may vary slightly; in general there are eight heads to the verticillaster.

"There are two flowerings annually, and the tree only yields fruits after some years.

"The numerous applications of the *Michelia Champaca* make it a valuable plant, which is regarded in some districts as sacred.

The flowers adorn the turbans of the men and the hair of the women. The natives of Malaya, Java, Macassar, and the Philippines use them in a like manner. They wear the flowers in their clothing, and employ them for the decoration of their nuptial beds.

"These flowers likewise serve as offerings in the Hindu temples. The trees also are frequently found in the vicinity of the temples.

"The perfume of the flowers of the Philippine *Champaca* is stronger and sweeter than that of the Singapore, Penang, Colombo, and Peradeniya flowers. It would appear that it is the same with this tree as with the ylang, which does not yield the same essential oil in Java, the Straits, or Ceylon, as in the Philippines."

It is doubtful whether the commercial oil is ever distilled from the yellow flowers alone, so that the champaca oil of commerce is probably the product of the flowers of *Michelia Champaca* and *Michelia longifolia*.

Oils produced by maceration of the true champaca flowers from Java plants in petroleum oil and distillation by steam at very low pressure were found by Bacon[2] to have the following characters:—

[1] Roure-Bertrand Fils, *Bulletin*, April, **1909**, 24.
[2] *Philip. Jour. Sc.* (**1909**), 131.

Specific gravity $\frac{30^\circ}{30^\circ}$	0·954 to 1·020
Refractive index at 30°	1·4550 ,, 1·4830
Saponification value	160 ,, 180

Oils distilled in Manila were found by Brooks [1] to have the following characters :—

Specific gravity $\frac{30^\circ}{30^\circ}$	0·904 to 0·9107
Refractive index	1·4640 ,, 1·4688
Ester value	124 ,, 146
,, ,, (after acetylation)	199

The oil distilled from the white flowers of *Michelia longifolia* have the following characters :—

Specific gravity	0·883 to 0·897
Optical rotation	− 12° 50′
Refractive index	1·4470 at 30°
Ester value	180

Commercial oils are therefore to be expected to have characters intermediate between the above, the white flowers probably being employed to a greater extent than the yellow in distillation.

Champaca oil contains benzyl alcohol, linalol, geraniol, esters of methyl-ethyl acetic acid, methyl anthranilate, isoeugenol, benzoic acid, methyl eugenol, and a crystalline body melting at 165° to 166°, of the formula $C_{16}H_{20}O_5$.

Brooks states that this substance reacts quantitatively with bisulphite solution, but that it cannot be recovered from the bisulphite compound. This renders it probable that the group CH : CH . CO is present in this body. The fact that the substance is a ketone is proved by its behaviour towards ammoniacal silver solution and magenta solution. The phenylhydrazone melts at 161°. The ketone does not react with acetanhydride, and with acetyl chloride it resinifies. When heated with an excess of alcoholic potash solution it gives rise to neutral succinate of potassium. It would seem that the succinic acid occurs in the ketone-molecule in the form of a neutral ester. In order to identify the alcohols which were associated with the succinic acid, the ketone was saponified with aqueous solution of caustic soda, but only ethyl alcohol could be detected. It appears that one of the carboxyl groups of the succinic acid is esterified with ethyl alcohol and the other with an alcohol $C_{10}H_{12}O_2$.

An oil has been distilled from champaca leaves in the Malay Peninsula. The yield was 0·04 per cent. of an oil having the following characters :—

Specific gravity	0·922
Optical rotation	+ 12° 30′
Acid value	1·9
Saponification value	15·2
,, ,, (after acetylation)	63·5

Oil of Winter's Bark.

The bark of *Drimys Winteri* yields between ·5 and 1 per cent. of this oil. This plant was originally discovered by Captain Winter in the Straits of Magellan, and is now found in various countries from Mexico to Cape Horn. The oil, which has been examined by Avata and Canzoneri,[2] has a specific gravity of about ·945. It consists of a mixture of

[1] *Philip. Jour. Sc.* 6 (**1911**), 333 [2] *Jahr. f. Pharm.* (**1889**), 70.

several bodies from which these chemists state that they have isolated a sesquiterpene. It distilled between 260° and 265° and had a specific gravity ·9344 and a rotary power $[\alpha]_d = + 11° 2'$. They term this body winterene, but its chemical identity cannot be regarded as settled.

ANONACEÆ.

OILS OF YLANG-YLANG AND CANANGA.

Ylang-Ylang oil ("flower of flowers") or oil of Cananga is the product of distillation of the flowers of *Cananga odorata*, a native of Ava and Tenasserim, and generally distributed and cultivated throughout Southern

FIG. 49 —(Ylang-Ylang in the main street of Malabon).
[*Roure-Bertrand Fils.*

Asia. When wild, the tree grows to its maximum height, but its flowers are said to be almost odourless. The cultivation of the plant reaches its highest degree of perfection in the Philippine Islands, and the best variety of the oil is distilled in Manila, where the industry is carried on in a scientific manner. There is also a considerable industry in this oil in the island of Réunion. A large quantity of oil also comes from Java and the neighbourhood, but the oil is not nearly of so fine an odour as that from the East Indies. How far this is due to climatic conditions, and how far to the less scientific methods used in the distillation of the oil, it is difficult to say, but there appears to be no doubt that, botanically, the plant is the same in both districts. The oil is the most pronounced example of the product of the same plant having entirely different values according to the locality of production. The oil from the Philippine

Islands is known as Ylang-Ylang oil, and is of the finest odour of all the Cananga oils. The Réunion oil is also known as Ylang-Ylang oil, but is of much lower value; and the oil from Java and the neighbourhood is known as Cananga oil commercially, and is of much lower value than Ylang-Ylang oil.

The most recent and authentic account of the Ylang-Ylang industry in the Philippine Islands is contained in a paper by R. F. Bacon,[1] who has made a special study of the plant and its essential oil :—

" In Manila, where the conveniences for careful distillation are greater

FIG. 50.—Cananga (in front of a Pagoda Bangkoh).
[*Roure-Bertrand Fils.*

than in outlying districts, fine oils as well as second-class oils, which are frequently known even there as ' Cananga ' oil, are produced, but in the provinces the distillers, who are compelled to work with less scientific apparatus, generally produce only a second-class oil. As the distillers usually do not possess any ylang-ylang plantations of their own, they are compelled to purchase the flowers. In order to increase the weight, the blossoms, picked by the natives during the night and brought to the distilleries in the morning, are frequently moistened as much as possible with water and mixed with leaves and twigs. Moreover, quite three-fourths of the blossoms delivered at Manila are green and unripe, although the use of ripe yellow blossoms is of the greatest importance for the yield and quality of the oil. In Manila the best blossoms are picked in May and June, although in 1907 the trees blossomed so late that the dis-

[1] *Philipp. Jour. Sc.*, 3 (108), 65, and Schimmel's *Report.*

tillation lasted from August to the following February. Of the provinces, where the ylang-ylang tree is also found, the oil is distilled in the Camarines, Mindoro, and Albay; the tree also grows in Bohol, but the distillation has not yet been introduced there. The oil produced in the provinces is generally considered in Manila as inferior, but this can only be attributed to the primitive method of production. In most cases the flowers are quite as good as those of the district round Manila, and a few Manila houses also distil in the provinces an oil for which they obtain

FIG. 51—Ylang-Ylang (Malabon near Manila).
[*Roure-Bertrand Fils.*

just the same high prices as for their other make. There can be no doubt that the installation of scientific distillation methods would bring about a powerful advancement of the ylang-ylang oil industry in the provinces.

"A good deal of mystery is made of the distillation of ylang-ylang oil in Manila, and it is said that the manufacturers possess valuable trade secrets. Bacon, who has visited several plants, and has himself carried out distillation experiments, does not believe that there are any trade secrets, although it is clear that not every one can distil first-class oil. But with a little experience in the distillation of essential oils it will, in his opinion, not be difficult to find out the special small precautions which have to be observed, provided proper apparatus is employed. The great mistake made by the distillers in the provinces is in the wrong choice of fractions and in burning the blossoms, owing to which the oil shows too large a content of resin. A matter requiring special attention is, that the oil, which, as a rule, only represents a fraction of the total oil contained in the blossoms, is distilled slowly with steam. The provincial distillers

FIG. 52.—Cananga (environs of Batavia).
[*Roure-Bertrand Fils.*

usually make the mistake of distilling 1 kilo oil from only 150 to 200 kilos blossoms, so as to obtain a good yield, whilst the average quantity to be used for such purpose should be about 400 kilos. After distilling out 1 kilo of fine quality oil, it is then still possible to obtain the same quantity inferior oil. The distiller usually judges by the odour of the distillate whether the time has arrived to change the receiver. The oil itself is collected in Florentine flasks, of which two or more are connected in series; the distillation-water is used for future distillations. It is useful to tin the inside of the distilling apparatus, although some distillers prefer to nickelplate it. The oil is very sensitive towards air and light, and must therefore be kept in a dark place and in well-filled vessels. Experiments with steam distillation *in vacuo* should be made with proper apparatus, and the blossoms should also be macerated beforehand; this would no doubt favour a rapid production of oil.

"The valuation of ylang-ylang oil by means of an analytical determination of the individual constituents is a matter of considerable difficulty and doubt, as it owes its perfume to a large number of odoriferous substances, and, as a matter of fact, the oil is judged in the trade chiefly by the odour so long as analysis shows that it is pure."

The ylang-ylang tree was introduced into Réunion about 1770, but it was not exploited as a source of perfume material until about 1880. According to Desruisseaux,[1] the plant succeeds well in the island in all porous and substantial soils which are met with in the tropical and equatorial zones. It dreads above all the presence of stagnant water, as also violent winds, which in some countries blow almost incessantly. Consequently in Réunion, below the 21st degree of South latitude, it has been recognised that at an altitude of 400 metres and upwards, the ylang-ylang did not flourish. But it is certain that in the equatorial zone the tree may accommodate itself to higher altitudes and flourish there.

The ylang-ylang in Réunion begins to produce flowers at the age of eighteen months to two years. At the age of four years, a tree may yield 5 kilos of flowers if it is planted in a rich soil. Trees have been known of ten years of age to produce, under the action of strong manuring, as much as 10 and 15 kilos of flowers annually.

The flowers succeed each other on the tree continuously, so that it is possible, so to speak, to pluck them all the year round. But it is during the winter season that they are particularly plentiful, that is from November to March in the tropical countries of the Southern hemisphere.

The flowers do not all develop at the same time on the bunches; only those which are at the right stage should be plucked, with a portion of their peduncle. For this purpose the gatherer should make use of his nails or of a pair of small scissors. The flowers are placed in small baskets or cloths.

When plucking is in progress, it is necessary to avoid shaking the branches of the tree and the bunches of flowers too roughly, so as not to spoil those which remain on the tree, and when the latter is too high, the gatherers should not climb on the branches which break rather easily, but should use a double ladder. Generally this delicate work is performed by the women and children.

The flower should be delivered to the stills a short time after gathering, otherwise its yield of essential oil will decrease and the fineness of the perfume will be lost. The plucking should be made as far as possible in the early morning, and it is necessary to avoid bruising the flowers. Bruised, faded, and blackened flowers yield an essential oil of inferior quality.

The distillation is carried out in presence of water, and may be performed in a simple still. The type of still which provides for the return of the waste waters of distillation is the most practical.

The flowers of ylang-ylang contain from 1·5 to 2·5 per cent. of their weight of essential oil. They are richer when they are derived from lands well exposed to the sun and not very moist, and when the plucking is

[1] *Bulletin du Jardin Colonial* (1910).

performed in dry weather. In Réunion they have succeeded in obtaining a kilogram of essential oil per 40 kilos of flowers, and by simple but prolonged distillation.

This yield is too high and the oil which is obtained is then not of its best possible value, and its value is much decreased.

In order to have a good essential oil it is necessary only to remove from the flower 50 to 60 per cent. of its perfume, and in the island of Réunion a good product is obtained by only taking 1 kilo of essential oil per 100 kilos of flowers. For this purpose, in practice, a still of one hectolitre capacity is charged with 15 kilos of flowers which are distilled only for nine hours, and yield 150 grams of essential oil.

The analyses which have been made show that if, with a distillation of short duration a comparatively small yield is obtained, it is compensated by the quality of the essential oil, which is the first condition necessary for the commercial success of this product. It is probable that in order to obtain more satisfactory results, the distillation with water should be replaced by distillation with steam, which not only gives a better product but decreases the cost of the operation, and enables several operations to be performed in a day with the same still.

The distillation conducted for a short time removes only the finest and best essential oil. Some remains in the flowers, and the product which is distilled in six or nine hours, once obtained, should be placed on one side. Then distillation may be continued in order to collect an inferior or second quality essential oil, which will be worth rectifying for the production of a medium quality which will still be commercially saleable.

The essential oil of good quality is amber-coloured. It is of lower value if it is brown or whitish.

Immediately after distillation it should be filtered, then preserved in opaque vessels of glass, tin, or tinned copper. The vessels should if possible be fitted with ground-in stoppers.

Attempts have been made to cultivate the ylang-ylang tree in Amani, by the German colonists there, but from a commercial point of view the experiments were unsuccessful.

Madagascar, the Seychelles, and various other places have produced the oil, but so far not in any important quantities.

The characters of ylang-ylang and cananga oils are very variable, not only on account of the variability in cultivation, climatic conditions, etc., but also owing to the fact that the oil is prepared by a fractionation process, the individual distiller stopping the distillation at such point as he conceives to yield the best combination of quality and quantity.

The following values, which must be accepted with the above qualifications, cover most authentic samples of the types mentioned:—

Manila Ylang-Ylang Oil.

Specific gravity	0·908 to 0·967
Optical rotation	− 27° ,, − 87°
Refractive index	1·4800 ,, 1·5082
Ester value	42 ,, 155

It is soluble in 0·5 to 2 volumes of 90 per cent. alcohol.

Bacon has examined twenty-three samples of oils considered to be of the best quality, and thirty-six samples of second grade, which he found had the following characters:—

	First Quality Oils.	Second Quality Oils.
Specific gravity $\frac{30°}{4°}$	0·911 to 0·958	0·896 to 0·942
Refractive index at 30° . .	1·4747 ,, 1·4940	1·4788 ,, 1·5082
Optical rotation	– 27° to – 49·7°	– 27·4° to – 87°
Ester number	90 ,, 138 (in one case 169)	42 to 94

According to Bacon, in first-class oils the ester number is mostly 100 or more, the refractive index rarely above 1·4900, optical rotation rarely above – 45°, but mostly between – 32° and – 45°. In the case of the second grade or cananga oils, however, the ester number is rarely higher than 80, the refractive index approaches 1·5000 owing to the larger content of sesquiterpenes and resins, and the optical rotation amounts to – 60° and more. Low values for optical rotation and ester number usually point to second-class oils, which have mostly also a low specific gravity. Oil with a low refractive index, low rotation, and a high ester number can with a fair amount of certainty be considered as good, whilst a high refractive index, high rotation, and low ester number point to oils of lower quality. The figures for the specific gravity did not indicate any special regularities, but in an oil with a high specific gravity and a high ester-content (ester number above 100), the refractive index also will usually be found high, and the oil can therefore be considered as good.

Attempts made to arrive at oils of better quality by distilling inferior oils *in vacuo* gave no favourable results, as the odour suffered distinctly, and could only be improved by passing air through the oil—a process which is also employed to remove the burnt odour caused when the blossoms are allowed to be slightly burnt in the distillation. It appears that the highest boiling portions of the oil fix the fragrant volatile constituents, for the resinous distillation-residues always showed an odour which lasted for a remarkably long time. It was further found that first-class oils contained over 50 per cent. of portions boiling below 100° at 10 mm. pressure.

Réunion Ylang-Ylang Oil

Specific gravity	0·930 to 0·967
Optical rotation	– 34° ,, – 65°
Ester value	96 ,, 160
Refractive index	1·500 ,, 1·5100

Java Cananga Oil.

Specific gravity	0·906 to 0·956
Optical rotation	– 15° ,, – 52°
Refractive index	about 1·4900 to 1·5050
Ester value	10 to 60 (rarely above 35)
,, ,, (after acetylation) . . .	45 to 100

Four samples of ylang-ylang oil distilled in Amani (East Africa) were examined by Schimmel & Co.[1] with the following results:—

[1] *Report*, October, 1911, 102.

Condition of the Material.	1 Fresh Flowers.	2 Faded Flowers after 24 hours' keeping.	3 Faded Flowers kept for 2 days and partly somewhat shrivelled.	4 Faded Flowers kept for 4 days and dried in the shade.
$d_{15°}$	0·9366	0·9190	0·9170	0·9382
α_D	− 17° 1′	− 20° 4′	− 19° 50′	− 18° 46′
$n_{D20°}$	1·48451	1·48431	1·48267	1·49394
Acid Number .	1·1	1·1	1·9	2·8
Ester ,, .	136·3	103·8	96·9	114·2
Solubility . .	in 8 vols. 70 °/。 alcohol with opalescence	in 1 vol. 80 °/。 alcohol dilute sol. opal.	in 1 vol. 80 °/。 alcohol dilute sol. turbid	in any prop. 90 °/。 alcohol dilute sol. opalescence

Apart from the strikingly low rotation, which would correspond with that of cananga oil, the general constants of the four samples agree with those which have thus far been recorded for ylang-ylang oil, but some of the samples in question are very much more soluble, which shows that they contain very little sesquiterpene. That would be an advantage rather than otherwise, but notwithstanding this, samples 2, 3, and 4 were quite useless for perfumery purposes because their aroma was wholly insufficient, and steadily deteriorated according to the age of the raw material used.

Roure-Bertrand Fils[1] record the following values for ylang-ylang oil obtained from various sources:—

	Oil of Ylang-Ylang obtained from			
	Réunion.	Nossi-Bé.	Mayotte.	Manila.
Density at 15° C.	0·9492	0·9673	0·9324	0·911 to 0·958
Rotatory Power	− 38° 8′	− 42° 12′	− 47° 40′	− 27° to − 49·7°
Solubility in 90 per Cent. Alcohol	½ volume, then cloudy	1 volume, then cloudy	1 volume, then cloudy	,,
Acid Number	,,	1·4	1·0	,,
Coefficient of Saponification .	96·03	129·5	113·4	90 to 138
Esters, as Linalyl Acetate .	33·6 °/。	45·3 °/。	39·7 °/。	,,
Coefficient of Saponification of the Acetylated Oil . . .	132·11	154·7	134·4	,,
Total Alcohols (as Linalool) .	40·3 °/。	42·7 °/。	41·0 °/。	,,
Combined Alcohols . . .	26·4 °/。	35·6 °/。	31·2 °/。	24·7 to 38 °/。
Free Alcohols	13·9 °/。	7·1 °/。	9·8 °/。	,,

Three further samples from Mayotte were examined[2] and had the following characters, the third sample being quite different from the other two, and having a rotation − 4° 4′ :—

[1] *Bulletin*, April, **1910**, 61
[2] *Ibid.*, October, **1911**, 40.

	I.	II.
Density at 15° C.	0·9594	0·9651
Rotatory Power	– 53° 56′	– 45° 16′
Solubility in 95 per Cent. Alcohol	1 volume, then cloudy	
Acid Number	1·3	1·4
Ester ,,	129·7	131·6
,, Contents expressed as $C_{10}H_{17}COOCH_3$	45·4 %	46·0 %
Coefficient of Saponification of the Acetylated Oil	167·0	180·8
Total Alcohols expressed as $C_{10}H_{18}O$	52·5 %	57·5 %

Ylang-ylang oil from the Seychelles has been examined by the chemists at the Imperial Institute. Sample No. 1 was examined in 1913; No. 2 was examined in 1910, and No. 3 is the same as No. 2, but kept for three years and then re-examined.

	1.	2.	3.
Specific Gravity at $\frac{15° C.}{15° C.}$	0·9567	0·9200	0·9250
Optical Rotation α_D	– 28° 5′ at 20° C.	– 30° at 21° C.	– 28° 36′ at 24° C.
Acid Value	3·3	4·2	2·6
Ester ,,	126·0	42·2	53·5
Saponification value	129·3	46·4	56·1
,, ,, after Acetylation	181	—	—
Solubility in 90 per Cent. Alcohol	Gives a clear solution with less than 0·8 vol.; with 0·8 vol. or more becomes turbid	Soluble in 1 vol.; becoming turbid with a larger quantity	Soluble in $\frac{1}{3}$ vol.; becoming turbid with 1 or more vols.

A Madagascar ylang-ylang oil has been examined by Schimmel & Co.[1] and found to have the following characters:—

Specific gravity	0·9577
Optical rotation	– 49° 55′
Refractive index	1·51254
Acid number	1·8
Ester ,,	113·2
,, ,, (after acetylation)	160·2

On fractionation it gave the following fractions:—

	Per Cent.	$d_{15°}$.	α_D.	Saponification No.
45 to 70°	10·7	0·9848	– 5° 20′	207·2
70 ,, 100°	18·0	0·9572	– 49° 16′	166·5
100 ,, 110°	27·0	0·9000	– 93° 6′	37·3
110 ,, 125°	12·5	0·8932	– 72° 40′	28·0
Residue	31·8	1·0725	– 3° 45′	181·0

[1] *Report*, November, **1908**, 135.

Umney[1] gives the following figures for ylang-ylang oils, which are not intended to be exhaustive, but represent some oils examined, showing the percentage of non-volatile matter when the oil (about 2 or 3 grams) is heated on the water-bath for two hours:—

	Manila.	Bourbon.	Madagascar.	Mayotte.
Specific Gravity (15°)	·927 to ·969	·964 to ·967	·961 to ·981	·947
Optical Rotation .	− 38° ,, − 51°	− 39° ,, − 41°	− 30° ,, − 42°	− 48°
Refractive Index (20°)	1·494 ,, 1·505	1·5130	1·5122	1·5070
Ester Number . .	99·6 ,, 153·2	142·6 to 160·1	123·2 to 171·2	114·7
Non-volatile . .	4·3 to 16·4 %	25·1 to 40·3 %	16·9 to 29·3 %	14·0 to 19 %

It is doubtful whether these high non-volatile residues are due to natural constituents of the oil, or to added substances, such as benzyl benzoate. If the oil be heated at 110° to 120° until no loss in weight occurs, any residue above 5 to 7 per cent. should be regarded with suspicion and carefully examined. The saponification value of the non-volatile residues (at this temperature) for normal oils is usually from 65 to 100. If much fatty oil is present as an adulterant, this figure will be greatly increased, and the residue should be examined for fatty acids, which can be separated from the saponification liquid, and the molecular weight determined by titration with alkali, when an indication of the nature of the acids is obtained.

Umney finds that in the case of the finest Manila oils the residue obtained by evaporation does not usually exceed 6 per cent., whilst inferior oils frequently contain much more non-volatile matter. It should be noted that the loss by evaporation on a water-bath is not constant even after six hours, and that if a larger quantity of oil than 1 to 2 grams be taken the rate of volatilisation is much slower.

Benzyl benzoate, one of the natural constituents of this oil, is very slowly volatilised when heated on a water-bath, as shown by the following experiment: 5 grams of pure benzyl benzoate heated in a flat bronze dish over a water-bath showed the following loss:—

After	3 hours the loss was		9·3 per cent.	
,,	6 ,,	,,	30·9	,,
,,	9 ,,	,,	45·4	,,
,,	12 ,,	,,	59·3	,,
,,	15 ,,	,,	72·3	,,
,,	18 ,,	,,	85·7	,,
,,	21 ,,	,,	98·3	,,

It would appear, therefore, that the presence of benzyl benzoate in considerable quantity in ylang-ylang oil would be shown by heating the oil on a water-bath for a limited time. This ester, which has a high saponification value, is comparatively inodorous, and, as it might be employed as an adulterant, the usual tests for ylang-ylang oil should be supplemented by a determination of the non-volatile matter to guard against such sophistication.

A sample of pure ylang-ylang oil distilled in Mauritius had the following characters:—

Specific gravity	0·993
Optical rotation	− 28°
Refractive index (25°)	1·5072
Saponification number	180
Non-volatile residue	16·6 per cent.

[1] *P. and E.O.R.* (**1914**), 37, 85.

Both the specific gravity and saponification number are high, and these factors, together with the figure for non-volatile matter, indicate a high proportion of benzyl benzoate. The odour was somewhat coarse, and the value of the oil was considered inferior to that of the finest Manila oils, thus showing that the odour value is not proportional to the saponification value.

Petroleum is sometimes found as an adulterant, especially of the cheaper cananga oils. This is usually detectable by its odour, but the following are the figures for a number of samples examined during 1915-16 by the author, which all contained petroleum :—

Specific gravity	0·865 to 0·886
Optical rotation	− 9° „ − 15°
Refractive index	1·4708 „ 1·4885

Attempts have been made by Jahrling[1] and by Gibbs to value ylang-ylang oils in a more or less absolute manner by means of their analytical figures. Jahrling's classification is as follows :—

Quality.	Extra.	I. (*a*)	I. (*b*)	II.
α_D . . .	Below − 35°	Below − 48°	Below − 60°	Over − 60°
n_D . . .	„ 1·4900	„ 1·4950	„ 1·4990	„ 1·4990
Saponification No.	Over 145	Over 120	Over 100	100 or less
Solubility . .	In 80 per cent. Alcohol	In 90 per cent. Alcohol	In 90 to 96 per cent. Alcohol	In 96 per cent. Alcohol

This classification will not bear careful examination, and must only be accepted in a very general way, the samples proving exceptions to it being very numerous.

This oil contains a large number of aromatic compounds. The earliest investigation of this oil was that of Gal, who showed the presence of benzoic acid in the form of esters. Reychler[2] has since shown that linalol and possibly geraniol exist, together with the sesquiterpene cadinene. Pinene, *p*-cresol methyl ether, and a crystalline solid melting at 138°, probably a sesquiterpene alcohol, were also shown to be present. Eugenol has been found in cananga oil, and iso-eugenol in ylang-ylang oil. Both these bodies are probably present in the form of their methyl ethers.

Methyl benzoate, methyl salicylate, benzyl acetate, benzyl benzoate, and free benzyl alcohol are present in small amounts, as well as methyl anthranilate and creosol in traces. Bacon[3] has also recently identified formic acid, and safrol *or* iso-safrol.

Elze[4] has also isolated the alcohols nerol and farnesol from Java cananga oil, but only in very small amount.

OIL OF POPOWIA CAPEA.

This plant grows in the Ivory Coast Colony, where it is known as "capé". It has been examined by E. G. and A. Camus, who found

[1] *Comptes Rendus*, 76 (**1873**), 1482.
[2] *Bull. Soc. Chim.*, iii., 11 (**1894**), 407, 576, 1045; and 13 (**1895**), 140.
[3] *Philipp. Jour. Sc.* 3 (**1908**), 65. [4] *Chem. Zeit.*, 34 (**1910**), 857.

that it was a species hitherto unnamed, and they assigned its name to it. Messrs. Roure-Bertrand Fils [1] give the following account of its essential oil. Forty-two kilos of the leaves were distilled with steam, with cohobation of the condensed waters, and yielded 246 grams of volatile oil, of which 160 grams had collected at the bottom of the receiver and 86 grams floated on the surface.

The last water of condensation, when exhausted with petroleum spirit, yielded to this solvent only a very small quantity of essential oil (less than 2 grams).

In all, the yield of essential oil from the dried "capé" leaves was therefore 0·59 per cent.

They determined separately the constants of the *heavy* and *light* portions, then those of a mixture of these two liquids, which they call the *complete* essential oil, and which was prepared by uniting the first two in proportion to their yields.

Below are the analytical characters of the *complete essential oil of Popowia Capea*:—

Specific gravity at 20° C.	1·00416
Optical rotation	+ 76° 56′
Solubility in 80 per cent. alcohol	1 vol., then cloudy
,, 95 ,, ,,	0·5 vol., then a very slight opalescence
Acid value	2·8
Saponification number	166·1
,, ,, of the acetylated oil	239·9

By working separately with the heavy and the light component portions of this oil, the following constants were found :—

	Heavy Portion.	Light Portion.
Specific gravity at 20° C.	1·00808	0·99596
Optical rotation	+ 90° 54′	+ 51° 26′
Solubility in 80 per cent. alcohol	1 vol., then very strong cloudiness	1 vol., then strong cloudiness
,, 95 ,, ,,	1 vol.; after 3 vols. a distinct opalescence	0·5 vol., then very slight opalescence
Acid value	1·5	3·7
Saponification number	192·3	123·2
,, ,, of the acetylated oil	248·3	218·4

The constituents of this oil have not been investigated, but there appear to be present one or more sesquiterpene alcohols or alcohols of the benzene series. Sodium bisulphite solution absorbs 6 per cent. of the oil.

Oils of Monodora.

The seeds of *Monodora myristica,* a tree 50 to 60 ft. high, found in West Africa, but also growing elsewhere, notably in Ashanti and in the West Indies, where it is called "Calabash Nutmeg," yield an essential oil. A small sample of the seeds from the Gold Coast yielded on steam distillation 5·9 per cent. of a colourless volatile oil, possessing an agreeable lemon-like odour and taste, and appearing to consist mainly of terpenes.[2] A larger but poorer sample (also from the Gold Coast, obtained to make a more thorough examination) gave only 2·2 per cent. of essential

[1] Roure-Bertrand Fils, *Bulletin*, October, **1913**, 4, 15.
[2] *Bull. Imp. Instit.*, July, **1915**.

oil, the aroma of which was inferior, due probably to the low condition of the seeds. The laboratory figures of the oils from the first and second samples were:—

	First Sample.	Second Sample.
Specific gravity $\frac{15° C.}{15° C.}$	·849	·8592
Optical rotation	− 44·7° at 20° C.	− 57·8° at 25° C.
Acid value	1·2	1·4
Ester value—		
Before acetylation	1·9	6·4
After "	52·9	33·50
Soluble in 90 per cent. alcohol	In 6 or more vols.	In 4·5 or more vols.

Thoms [1] obtained a yield of 7 per cent. of oil from the seeds, which had a specific gravity of 0·896 and optical rotation − 58·5°. The oil, from which traces of free acids and phenols were first removed, was fractionated *in vacuo*.

The terpene fraction had the specific gravity 0·842 at 20° and consisted chiefly of limonene; boiling-point 74° to 76° at 16 mm.; α_D = − 105·68°, and $[\alpha]_D$ = − 125·5°; melting-point of the nitrosochloride 103° to 105°. In the third fraction, which represents 20 per cent. of the oil, and which boils at 110° to 116° (at 16 mm.), a body $C_{10}H_{16}O$ could be detected, which is probably identical with myristicol.

Schimmel & Co. obtained 5·37 per cent. of essential oil from the seeds, which had the following characters: $d_{15°}$ = 0·859; α_D = − 117° 40′; acid number = 1·36; ester number = 3·4; ester number after acetylation 27·11. Soluble in about 4 volumes and more 90 per cent. alcohol. It consisted chiefly of phellandrene.

Leimbach has examined the corresponding oil from the seeds of *Monodora grandiflora*. This oil has the following characters:—

Specific gravity	0·8574
Optical rotation	− 46° 15′
Acid value	3·9
Saponification value	7 to 12

It contains *l*-phellandrene, camphene (?), *p*-cymene, palmitic acid, carvacrol (?), an uninvestigated compound of the formula $C_{10}H_{16}O$ (specific gravity 0·935; optical rotation − 9° 14′; and boiling-point 130° to 154°; which suggests that the body is not a chemical individual); a sesquiterpene (specific gravity 0·9138; optical rotation + 24°; and refractive index 1·5051), and a crystalline body melting at 160° to 163°.

PITTOSPORACEÆ.

OILS OF PITTOSPORUM.

The fruit of *Pittosporum undulatum*, a plant indigenous to South-east Australia, yields, according to Power and Tutin,[2] 0·44 per cent. of essential oil, whose characters are as follows:—

Specific gravity	0·8615
Optical rotation	+ 74° 4′

The oil contains *d*-α-pinene, *d*-limonene, traces of palmitic acid, salicylic acid, and a phenol, small quantities of esters of formic and valerianic acids, and a sesquiterpene of specific gravity 0·910, and refractive index 1·5030.

[1] *Ber. deutsch. pharm. ges.*, 14 (**1904**), 24. [2] *Jour. Chem. Soc.*, 89 (**1906**), 1083.

The fruit of *Pittosporum resiniferum* yields 0·5 per cent. of essential oil, obtained by expression, having the following characters :—

Specific gravity	0·883
Refractive index	1·4577

According to Bacon,[1] the oil contains heptane, and a dihydroterpene, $C_{10}H_{18}$, having the following characters :—

Boiling-point	168° to 170°
Specific gravity at $\frac{30°}{40°}$	0·8050
Optical rotation	+ 1° 6′
Refractive index at 30°	1·4460

HAMAMELIDACEÆ.

Oil of Storax.

The oleo-resinous substance obtained from the bark of the tree *Liquidambar orientale*, a plant found chiefly in Asia Minor, constitutes the liquid storax of commerce.

The oleo-resinous balsam yields from 0·5 to 1 per cent. of essential oil having the following characters :—

Specific gravity	0·950 to 1·050 (sometimes lower)
Optical rotation	+ 1° to − 35°
Refractive index	1·5395 ,, 1·5653
Acid value	1 ,, 26
Ester ,,	up to 130

The very variable characters of this oil are due to differences in distillation methods, as the further the distillation is carried, the more the amount of slightly volatile, high gravity constituents coming over.

The principal odorous constituent of the oil is styrol, or phenylethylene $C_6H_5 . CH : CH_2$, an optically inactive hydrocarbon boiling at 146°. The cinnamic acid esters of ethyl, benzyl, phenyl-propyl, and cinnamyl alcohols are also present, as well as the free alcohols, and traces of vanillin.

The oil is adulterated with artificial esters, notably benzyl benzoate.

The volatile oil from the leaves and twigs of *Liquidambar Formosana* has been examined by Kafuku.[2] A yield of about 0·05 per cent. was obtained. It was greenish-yellow in colour, having a specific gravity of 0·8655 at 20° C., optical rotation − 3·3°, refractive index 1·4755 at 20° C., acid value nil, ester value 5·9, ester value after acetylation 25·2. It consisted principally of terpenes; camphene, α-pinene, and dipentene were identified with certainty, nopinene and phellandrene being also indicated. Traces of aldehydes and ketones were found to be present.

Oil of Hamamelis Virginiana.

The oil obtained from *Hamamelis Virginiana* has the following characters :—

Specific gravity	0·900 to 0·906
Optical rotation	+4° ,, + 5°
Refractive index	1·4830 ,, 1·4895
Ester value	about 4
,, ,, (after acetylation)	,, 30

[1] *Philipp. Jour. Sc.*, 4 (**1909**), 115. [2] *Jour. Chem. Ind. Tokio* (**1916**), 19, 516.

Jowett and Pyman [1] found the oil to contain a phenol having the odour of eugenol, free fatty acids, and a sesquiterpene having the following values:—

Specific gravity	0·897
Refractive index	1·4916
Optical rotation	+ 14·9°

CISTINEÆ.

Oil of Ladanum.

Ladanum or labdanum resin is a secretion from the leaves of several species of *Cistus*, especially *Cistus creticus* and *Cistus ladaniferus*. *Cistus creticus* is the Cretan rock rose, a native of rocky ground in Macedonia, Thrace, Greece, and the islands of Crete, Rhodes, Sicily, and Cyprus. Its botanical relationships are hardly decided, as it is regarded by some as being merely a variety of *Cistus villosus* Linn. In Cyprus the ladanum (*labdanum*) is chiefly collected by the shepherds from the fleeces of the sheep, who unconsciously collect it from the plants whilst pasturing, the leaves becoming very viscid with the secretion.

There seems to be a considerable amount of confusion as to the exact botanical relationships of *Cistus ladaniferus*, but this has recently been cleared up by E. M. Holmes [2] in dealing with the source of Spanish ladanum. He finds that the plant used for collecting the oleo-resin in Spain is *Cistus ladaniferus*, var. *macellatus*. There are about sixteen species of *Cistus* known in Spain, and the name *Cistus ladaniferus* has been applied by different botanists to at least four different species, so that a few remarks on the distinctive characters of the true *Cistus ladaniferus* may be useful.

The plant figured under this name in Curtis's *Botanical Magazine*, tab. 1112, is *Cistus Cyprius* Lam. and is a native of Cyprus.

Cistus ladaniferus of Gouan is *Cistus Ledon* of Lamarch and the *Cistus glaucus* of Poures, and is a native of the South of France.

Cistus ladaniferus of Stokes's *Botanical Materia Medica*, vol. iii., p. 209, is *Cistus polymorphus* Wilkoum, and the *Cistus creticus* of Limnœus.

From all these the true *Cistus ladaniferus* of Limnœus is distinguished by the ovary having ten cells; in the other five only are present.

The true *Cistus ladaniferus* has been described in Stokes's *Botanical Materia Medica* under the name of *Cistus viscosus*, and by Hoffmansegg under the name of *Cistus ladanosma*.

The oleo-resin yields from 0·7 to 2·0 per cent. of essential oil having the following characters:—

Boiling-point	50° to 185° at 15 mm.
Specific gravity	0·928 to 1·011
Refractive index	1·5100 ,, 1·5140

On keeping the oil becomes oxidised somewhat rapidly and deposits a considerable amount of crystals.

According to Masson, [3] the oil contains acetophenone, a compound not hitherto found naturally occurring in essential oils; a second ketone of the formula $C_9H_{16}O$, and having the following characters:—

[1] *Pharm. Jour.*, 91 **(1913)**, 129.
[2] *P. and E.O.R.* **(1911)**, 155.
[3] *Comptes Rendus*, 154 **(1912)**, 517.

Boiling-point	178° to 179°
Specific gravity at 0°	0·922
Optical rotation	0°
Refractive index at 23°	1·4494
Melting-point of semi-carbazone	220° to 221°

Alcohols, phenols, and esters are also present in the oil. Guaiol is probably present.

The ketone $C_9H_{16}O$ does not react with bisulphite of sodium. It yields a monobromide, melting-point 41°, and a semi-carbazone, melting-point 220° to 221°. With sodium and moist ether it is reduced to the corresponding secondary alcohol, which forms large crystals, melting-point 51° (boiling-point 87° at 28 mm.). With cold 3 per cent. potassium permanganate solution the ketone is oxidised into geronic acid (boiling-point 190° to 191° at 31 mm.; melting-point of the semi-carbazone 164°), the latter, when oxidised with sodium hypobromite, giving rise to bromoform and α, α-dimethyladipic acid. This shows the ketone $C_9H_{16}O$ to be trimethyl-1, 5, 5-hexanone-6, a body which has not been described up to the present.

$$CH_3 . CO . CH_2 . CH_2 . CH_2 . C(CH_3)_2 . CO_2H$$

Geronic acid.

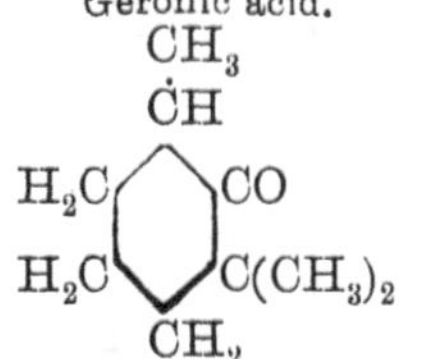

Trimethyl-1, 5, 5-hexanone-6.

Emmanuel[1] isolated a colourless crystalline substance to which he assigned the formula $C_{17}H_{30}O$, and the name ladaniol, but it is possible that it is an impure form of guaiol.

The essential oils from the following two species of *Cistus* have been examined by Schimmel & Co.:—[2]

Oil from *Cistus monspeliensis* (Cistaceæ). Yield 0·015 per cent.; $d_{15°} = 0·9786$; $\alpha_D = + 1° 40'$; acid number = 15·7; ester number = 31·51. The light brown oil separates off between 20° and 25° abundant quantities of paraffin of the melting-point 64°.

Oil from *Cistus salviifolius* (Cistaceæ). Yield 0·024 per cent.; $d_{15°} = 0·9736$; $\alpha_D = + 17° 20'$; acid number = 16·86; ester number = 22·73. This oil is yellowish-green, and behaves like the one mentioned before with regard to the separation of paraffin.

RESEDACEÆ.

Oil of Mignonette.

The fresh flowers of the mignonette, *Reseda odorata*, yield traces, about ·002 per cent., of an essential oil of exceedingly powerful odour. This oil is semi-solid at ordinary temperatures, and its chemistry is not understood, as the oil is in itself so very rare. Indeed, commercially, an ordinary floral extract is generally used in preference to any direct

[1] *Arch. der Pharm.*, 250 **(1912)**, 111.

[2] *Report*, October, **1903**, 77.

preparation of the oil. An oil obtained by extracting the flowers with petroleum ether, and distilling the extract with steam (the water soluble oil[1] being recovered by ether) was examined by von Soden and found to have the following characters:—

Specific gravity	0·961
Optical rotation	+ 31° 20′
Acid value	16
Ester ,,	85

The fresh roots of this plant yield from ·01 to ·04 per cent. of an unpleasantly-smelling oil recalling the odour of radishes. This oil contains phenyl-ethyl-thiocarbimide, $C_2H_4(C_6H_5)NSC$, and is a liquid of specific gravity 1·01 to 1·09 and optical rotation about + 1° to + 2°.

TURNERACEÆ.

Oil of Damiana.

The leaves of various species of the Damiana plant, principally *Turnera diffusa* and *Turnera aphrodisiaca*, yield about 0·5 to 1 per cent. of essential oil. The oil has the following characters:—

Specific gravity	0·943
Optical rotation	− 23° 25′
Saponification value	41·8

Another sample, distilled by Schimmel & Co., was found to have a specific gravity 0·970 and to boil between 250° and 310°. By standing in the cold the oil deposits crystals which probably consist of paraffin hydrocarbons.

CANELLACEÆ.

Oil of Canella Bark.

The bark of *Canella alba* yields from ¾ to 1 per cent. of essential oil. In the drug trade this bark is known frequently as "wild cinnamon," and is found in the south of Florida, the Bahamas, Cuba, Jamaica, Martinique, Barbadoes, etc. The whole tree is very aromatic, especially the flowers, which perfume the neighbourhood for a considerable distance, although they seldom open. The bark is brought to Europe in the form of long quills, rather thicker than those of cinnamon. The taste of the bark somewhat resembles a mixture of cloves and cinnamon, being hot, bitter and aromatic. Meyer and Reiche[1] investigated the oil, of which they obtained ·94 per cent. from the bark, so long ago as 1843, and stated that they isolated four distinct bodies from it. The first was eugenol, the second was described as being closely allied to the principal constituent of oil of cajuput, and the other two were not closely investigated. Flückiger and Hanbury obtained ·74 per cent. of oil from the bark, and stated that it consisted of 2 parts of an oxygenated body mixed with 1 part of a hydrocarbon resembling a mixture of mint and cajuput oils. Eugenol, however, remained as the only well-defined body yet ascertained as a constituent of the oil, till 1890, when Schimmel & Co. showed that the body referred to by Meyer and Reiche as resembling the principal constituent of cajuput oil was actually identical with this, being the now well-known cineol. Williams[2] has later examined the oil, and has

[1] *Annalen*, 47 (**1843**), 234. [2] *Pharm. Rund.* (New York), 12 (**1894**), 183.

identified pinene and caryophyllene in it. The specific gravity of the oil ranges from ·920 to ·935 and the optical rotation from + 1° to + 3°.

Oil of Warburgia.

The wood of *Warburgia stuhlmanni*, a Zanzibar tree, yields about 0·6 per cent. of essential oil, having an odour recalling that of sandalwood oil. Its characters are as follows:—[1]

Specific gravity	0·9864 at 20°
Refractive index	1·5127
Optical rotation	− 41·77°
Saponification value	11·2
,, ,, (after acetylation)	100·3

The oil contains aldehydes and (sesquiterpene ?) alcohols, together with esters and traces of free acids.

DIPTEROCARPACEÆ.

Oil of Borneo Camphor.

The wood of *Dryobalanops aromatica*, or Borneo camphor tree, yields this essential oil. The tree is a native of Sumatra and North-West Borneo, but its essential oil is rarely seen in commerce.

It has the following characters:—

Specific gravity	0·9180
Optical rotation	+ 11° 5
Refractive index	1·48847
Acid value	5·6
Ester ,,	0
,, ,, (after acetylation)	50·5

The oil contains α-pinene, camphene, *d*-borneol, β-pinene, dipentene, terpineol, and sesquiterpenes.

Schimmel & Co.[2] have investigated the oil with the following results.

The terpenes have the following characters:—

Boiling-point	155° to 156°
Specific gravity	0·8631
Optical rotation	+ 7° 32′
Refractive index	1·46668

These figures indicate the presence of α-pinene.

A fraction possessing the boiling-point of camphene gave the following figures: $d_{15°}$ 0·8639; α_D + 5°37′. On hydration with glacial acetic-sulphuric acid it afforded a large proportion of *iso*borneol, but in spite of repeated purifications from light petroleum the melting-point of the last-named body remained as low as about 208°. Small quantities of β-pinene were found in a fraction possessing the following physical properties: boiling-point 163° to 167°; $d_{15°}$ 0·8607; α_D − 0° 20′; $n_{D20°}$ 1·47102. When treated with permanganate and alkali it yielded in the form of an oxidation-product the characteristic leaflets of sodium nopinate, from which nopinic acid, melting-point 124° to 126°, was obtained. Oxidation of the sodium salt with permanganate in acid solution afforded nopinone, of which the semi-carbazone had melting-point 186° to 188°.

d-Borneol was found to occur in the following fraction, which distilled between 55° and 85° (5 mm.). When fractionated once more it congealed in the condenser. Filtered by suction the crude product, when tested

[1] *Ber. Pharm.* (**1910**), 20, 351. [2] *Report*, April, **1913**, 33.

for camphor with semi-carbazide, proved to be camphor-free. The borneol, after being purified by recrystallisation from light petroleum, melted at 204°. Its chloroform solution was dextro-rotatory.

With the object of setting free the borneol from the higher boiling fractions of the oil and from the mother liquors, these latter were benzoylated with benzoyl chloride, with the aid of pyridine. The oil which was driven off in the steam-current by this manipulation, and which was not esterified, proved to consist substantially of *l*-α-terpineol, melting-point 35°. A fraction gave the following constants: d_{15° 0·9379, α_D − 8° 2′. The resulting phenylurethane melted at 112°. After saponification of the residue of steam distillation no alcohols other than borneol were discovered. The highest boiling fractions, boiling-point 102° to 112° (4 to 5 mm.), consisted only of sesquiterpenes. None of them, when warmed, reacted with strong formic acid; hence no sesquiterpene alcohols were present. One fraction was found to possess the following constants: boiling-point 102° (4 to 5 mm.), 256° to 260° (ordinary pressure); d_{15° 0·9200; α_D − 137° 12′; $n_{D_{20^\circ}}$ 1·50111. This is probably cadinene.

Oil of Gurjun Balsam.

The so-called "wood oil" or gurjun balsam oil is the product of the distillation of the oleo-resin from several species of *Dipterocarpus*. Gurjun balsam resembles copaiba in its general characters, and is often used to adulterate the latter drug. On distillation from 45 to 70 per cent. of essential oil is obtained, which has the following characters:—

Specific gravity	0·915 to 0·930
Optical rotation	− 35° ,, − 135°
Refractive index	about 1·5050
Acid value	0 to 1
Ester ,,	4 ,, 8
,, ,, (after acetylation)	6 ,, 10

This oil is not employed commercially to any extent, but is used from time to time for adulteration purposes. For example, it has recently been detected as an adulterant of palmarosa oil, and in otto of rose in traces, due to the adulteration of the last-named oil with geraniol manufactured from adulterated palmarosa oil.

It can be detected by dissolving 5 to 10 drops of the suspected oil in 5 c.c. of glacial acetic acid containing 5 drops of nitric acid. In the presence of gurjun oil a red-violet colour is developed, which appears within a minute if more than mere traces are present.

The principal constituent of the oil is a sesquiterpene or mixture of sesquiterpenes. These have been investigated by Deussen and Philipp,[1] who showed that the sesquiterpene mixture consisted principally of two sesquiterpenes which they named α-gurjunene and β-gurjunene. Semmler, Spornitz, and Jakubowicz[2] have also investigated these sesquiterpenes, and consider that both of them are tricyclic compounds. Deussen and Philipp also showed that, by the oxidation of the oil in acetone solution by means of potassium permanganate, a ketone, $C_{15}H_{24}O$, which they named gurjun-ketone, is formed, and which yields a semi-carbazone melting at 234°, which serves as a useful means of identifying the oil in mixtures.

The following method of identifying this oil (for example in otto of

[1] *Annalen*, 369 (**1909**), 56; 374 (**1910**), 105. [2] *Berichte*, 47 (**1914**), 1029, 1141.

roses) has been described by Schimmel & Co.[1] The fraction boiling at 95° to 96° at 4 mm., or about 255° to 260° at normal pressure, is treated with sodium and again fractionated until the boiling-point is as nearly 258° to 261° as possible, and the laevo-rotation at its highest point.

The fraction is dissolved in about 200 c.c. acetone, diluted with 35 c.c. water, and to the clear solution 15 grams of powdered potassium permanganate are gradually added. The unattacked oil, which boils at 104° to 105° (5 mm.) is removed by fractionation and the distillation-residue treated with semi-carbazide. The reaction-product is left standing overnight, when, after the addition of water, it is deposited as an oily precipitate, solidifying in part after prolonged cooling in ice when stirred with alcohol. The semi-carbazone, repeatedly recrystallised from hot alcohol, melts at 234°, and, with a semi-carbazone of the same melting-point and constants, prepared from gurjun balsam oil for purposes of comparison, suffers no depression of melting-point.

Two species of *Dipterocarpus* are mentioned by Pearson[2] as being used for the preparation of gurjun balsam, namely *Dipterocarpus turbinatus* and *Dipterocarpus tuberculatus*. Both of these have been examined by Schimmel & Co.,[3] who report thereon as follows :—

"*Dipterocarpus turbinatus* is a large tree which is found throughout tropical Burma, in Bengal, and on the Andaman Islands. Its balsam is specially known there by the name of 'gurjan oil'. It is collected in large quantities and is used as a paint for houses and ships, as well as a preservative for articles of bamboo. The balsam is exported from Chittagong (Bengal). The sample received by us consisted of a faintly acid, milky liquid; acid value 10·9; specific gravity 0·9811 (15°). When allowed to stand, especially in the warmth, it separated out into a brown oil which floats on the surface and a viscous, whitish-grey, emulsion-like mass. The constants of the oily layer were as follows: d_{15° 0·9706; α_D − 10° 8′; n_{D20° 1·51200; acid value 7·3; ester value 1·9. By steam distillation we succeeded in separating out from the total balsam 46 per cent. of a pale yellow oil of a balsamic odour, possessing the following constants; d_{15° 0·9271; α_D − 37°; n_{D20° 1·50070; acid value 0; ester value 1·9; soluble in 7 volumes a.m. 95 per cent. alcohol.

"*Dipterocarpus tuberculatus* yields a balsam of very different appearance. It is known in Burma as 'in oil,' and plays a very subordinate part compared with 'gurjan oil'. The balsam is of a pale-brown colour; its consistency is that of turpentine, and its specific gravity at 15° is 1·029; acid value 17·8; ester value 0. The oil which was separated out from it by steam distillation (yield 33 per cent.) was of a yellow-brown colour, and dissolved in 6 volumes a.m. of 95 per cent. alcohol. Its constants were as follow: d_{15° 0·9001; α_D − 99° 40′; n_{D20° 1·50070. It did not contain saponifiable constituents (saponification value 0). With Turner's colour-reaction both the balsams and the oils gave the characteristic violet colour. This test consists in dissolving 3 or 4 drops of the balsam or oil in 3 c.c. of glacial acetic acid, adding to the solution 1 drop of freshly prepared 10 per cent. sodium nitrite solution, and pouring the mixture very carefully on 2 c.c. of concentrated sulphuric acid."

[1] *Report*, April, **1912**, 108.
[2] *Commercial Guide to the Great Economic Products of India* (**1912**), 140.
[3] *Report*, April, **1913**, 68.

THEACEÆ.

Oil of Tea.

The leaves of the tea plant, *Thea chinensis*, yields a minute quantity of essential oil. Van Romburgh[1] obtained 0·006 per cent. of oil which had a specific gravity 0·866, and optical rotation − 0° 11′. The principal constituent was found to be methyl alcohol, with possibly a little acetone (which were recovered from the watery liquid, not forming a portion of the oil proper). In the oil, methyl salicylate and a body of the formula $C_6H_{12}O$, which appeared to be of an alcoholic nature, were detected. The latter body yielded compounds with acetic and benzoic acids.

MALVACEÆ.

Oil of Musk Seed.

The seeds of *Hibiscus abelmoschus* yield from 0·1 to 0·6 per cent. of essential oil of very aromatic odour. It is known as musk seed or ambrette seed oil, and has the following characters:—

Specific gravity	0·900 to 0·905 at 15°
,, ,,	0·880 ,, 0·885 ,, 45°
Acid value	75 to 140
Ester ,,	70 ,, 130

It is semi-solid at ordinary temperature, the melting-point varying according to the amount of free fatty acids present.

The oil contains the alcohol farnesol, together with free palmitic acid and constituents not yet identified. A sample of Java distilled oil, however, was liquid at ordinary temperatures, and a French oil of specific gravity ·908 did not solidify at − 10°.

RANUNCULACEÆ.

Oil of Pæonia Moutan.

The root bark of this plant, which is found in Japan and China, yields a small amount of essential oil, which at ordinary temperatures is of a solid consistency. It has the following characters:—

Specific gravity (superfused) at 15°	1·500
Optical activity	0°
Refractive index (superfused) at 20°	1·5646
Acid value	12·6
Ester ,,	24·5
,, ,, (after acetylation)	220·7

The oil consists chiefly of pæonol, which, according to Nagai,[2] forms colourless glancing prisms, melting at 50°. It has the formula $C_9H_{10}O_3$ and the constitution $C_6H_3(OH)(OCH_3)(COCH_3)$. It forms a phenylhydrazone melting at 170°, and is *p*-methoxy-*o*-oxyphenyl-methyl-ketone.

Oil of Nigella.

The seeds of *Nigella damascena* yield from 0·4 to 0·5 per cent. of a fluorescent essential oil having the following characters:—

Specific gravity	0·895 to 0·915
Optical rotation	+ 1° 4′ ,, − 7·8°
Refractive index	1·5582

[1] Schimmel's *Bericht*, April, **1897**, 42.

[2] *Berichte*, 24 (**1891**), 2847.

The fluorescence of the oil is due to a basic compound which was isolated by Schneider[1] and termed by him damascenine. This body has the formula $C_{10}H_{13}NO_3$, and melts at 26°. It is probably the methyl ester of methylamino-methoxybenzoic acid of the constitution.

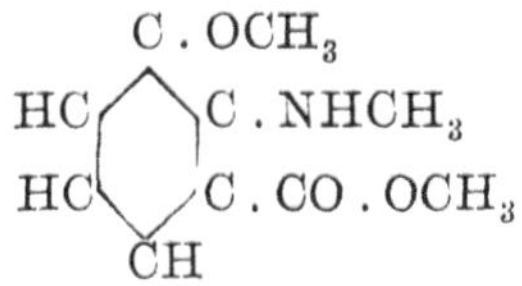

The seeds of *Nigella sativa*[2] yield 1·4 per cent. of oil having the following characters :—

Specific gravity	0·875 to 0·886
Optical rotation	+ 1° 26′ to + 3°
Refractive index	1·4836 to 1·4844
Acid value	0 „ 2
„ „	up „ 32
„ „ (after acetylation	„ „ 73

FLACOURTIACEÆ.

Oil of Phoberos Cochinchinensis.

According to Roure-Bertrand Fils[3] the green leaves and stalks of this tree, found wild in Annam, yield a small quantity of an oil resembling sweet orange oil in odour. It has the following characters :—

Specific gravity	0·9042
Optical rotation	+ 3° 2′
Acid value	2·5
Ester „	51·1
„ „ (after acetylation)	127·9

[1] *Pharm. Central.*, 31 (**1890**), 173, 191.
[2] Schimmel's *Bericht*, April, **1895**, 74.
[3] *Report*, April, **1914**, 7.

APPENDIX.

OIL OF CHENOPODIUM.[1]

THE distillation of chenopodium oil requires special precautions, because the chief ingredient, ascaridole, is decomposed gradually on boiling with water. Consequently the distillation must be carried on rapidly with steam at a good pressure, the condenser kept warm, and the warm distillation water separating from the oil in the receiver, discarded. When chenopodium oil is properly made, the residue on fractionation will be small, but if it has been manufactured with low-pressure steam and without the other precautions mentioned the percentage of ascaridole and the specific gravity will be low, and an excessive distillation residue will be found. The distillation residue appears to consist in part of water-soluble glycol-like substances resulting from the hydration and re-arrangement of the ascaridole molecule. This indicates that during a protracted distillation with low-pressure steam a part of the ascaridole must be decomposed and the soluble products in part pass away in the condensation water. In the producing region of Carroll County, Maryland, there are at least seven distilling plants in operation during the season. From five of these factories E. K. Nelson took authentic samples last October, and on examination the highest distillation residue was only 2 to 3 per cent., whereas in eight samples experimentally distilled with very low pressure steam the distillation residues amounted to from 7 to 25 per cent., and the ascaridole content was very low. The amount of fractional distillation residue, therefore, gives an idea of the care with which the oil has been made. In carrying out a fractional distillation, it should be remembered that ascaridole suffers a molecular re-arrangement when heated to 150°, and therefore it is necessary to maintain the bath at as low a temperature as possible, never allowing it to reach 150°. Most of the author's distillations were at 3 to 6 mm. pressure. During the distillation of the ascaridol fractions a bath temperature of 115°, and during that of the terpene fractions one of 80°, was found sufficient to maintain an even distillation. The first distillation gave a good separation of the terpenes from the ascaridole, and only two fractions were registered to yield ascaridole of constant boiling-point. A very simple test of the purity of the ascaridole fraction was found in the readiness with which a few drops will ignite when poured on an electric hot plate heated to about 250°. It should be noted that the estimation of ascaridole in the way just outlined gives results somewhat low, as it is difficult to recover all of the ascaridole from the lower fraction. The analytical results obtained by Nelson are presented below. They show that his seven authentic samples of oil from Carroll County conformed to the requirements of the U.S.P., and contained 60 to 77 per cent.

[1] *P. and E. O. R.* (**1920**), 247 ; see also page 121.

ascaridole. An oil distilled in Buitenzorg, Java, was found to be very similar to the oils of American origin, and the analysis of five American oils returned from Brazil where they had been used by the International Health Board, and which were at least one year old, exemplified the keeping qualities of the oil under varying conditions: the distillation residues, while higher than those found in fresh oils, were not excessive. Chenopodium oils distilled from wild plants collected around Orlando, Florida, one from dried and the other from green plants (both oils being produced under the supervision of Mr. H. L. Funk, of the Bureau of Plant Industry) were found to contain less ascaridole than the oil distilled from cultivated plants in Maryland, but the same constituents were found in it as in the Maryland oil. Another sample examined was from wild plants in the Eastern United States transmitted by the International Health Board. The results of these analyses are given below. It should be noted that the U.S.P. (IX.) specifies that the oil should be soluble in 8 volumes of 70 per cent. alcohol, should have a specific gravity of 0·955 to 0·980 at 25°, and an optical rotation varying between $-4°$ and $-10°$ in a 100 mm. tube at 25°. An approximate check on the ascaridole estimation was obtained by calculating the ascaridole content of each oil from its density, assuming that the density of the terpenes is 0·8466, and that of the ascaridole 1·005, and that no other constituents are present. The resulting figure is influenced, of course, by any variations in the density of the terpene mixture and by the amount of distillation residue which consists, of course, neither of terpenes nor of ascaridole. The figures thus obtained will be seen to run roughly parallel with those of the direct estimation of ascaridole by distillation.

Sample.	$d\frac{25^\circ}{25^\circ}$.	$[\alpha]_D$ (100 mm. Tube).	Solubility in 8 Volumes of 70 per Cent. Alcohol.	Ascaridole (per Cent. by Vol.). By Distillation. Per Cent.	Ascaridole (per Cent. by Vol.). Calculated from Density. Per Cent.	Optical Rotation of Ascaridole.	Distillation Residue.
Carroll County 1 . .	0·9758	− 4·25°	Complete.	74	81·5	− 2·1°	2·3
2 . .	0·9792	− 3·9°	do.	77	83·7	− 2·15°	1·1
3 . .	0·9713	− 5·5°	do.	75	78·7	− 3·05°	1·1
[1] 4 . .	0·9538	− 6·2°	do.	60	70·0	− 1·65°	1·7
5 . .	0·9592	− 6·2°	do.	65	71·0	− 2·5°	1·1
6 . .	0·9783	− 3·6°	do.	76	83·1	− 2·0°	0·9
7 . .	0·9564	− 6·5°	do.	64	69·3	− 2·45°	1·4
Java . . .	0·9762	+ 0·25°	do.	70	81·8	—	1·3
Florida dry wild plants . .	0·9148	− 7·75°	Not completely soluble.	41	43	—	2·2
Green wild plants	0·8940	− 10·25°	do.	26	30	—	2·4
Eastern (U.S.) wild plants .	0·8922	+ 1·65°	do.	20	28·8	− 4·0° (α_D of terpenes + 4·15°)	—
American origin, returned from Brazil—							
1 . . .	0·9628	− 5·0°	Complete.	68	73·4	—	2·7
2 . . .	0·9645	− 7·5°	do.	61	74·4	—	7·8
3 . . .	0·9470	− 6·45°	Slightly turbid.	56	63·4	—	2·8
4 . . .	0·9589	− 6·6°	do.	62	70·9	—	3·1
5 . . .	0·9730	− 5·33°	Complete.	72	79·8	—	9·1

[1] Sample of first distillation of the season.

INDEX.

ABERDEEN: THE UNIVERSITY PRESS

www.ingramcontent.com/pod-product-compliance
Lightning Source LLC
LaVergne TN
LVHW091515200926
848483LV00005B/419/J
* 9 7 8 1 4 2 7 6 1 9 4 2 6 *